Microearthquake Seismology and Seismotectonics of South Asia

Microearthquake Seismology and Seismotectonics of South Asia

J.R. Kayal
Former Deputy Director General (Geophysics)
Geological Survey of India, Kolkata;
Emeritus Scientist (CSIR)
Jadavpur University, Kolkata
and
Adjunct Professor
Indian School of Mines, Dhanbad

Capital Publishing Company
NEW DELHI KOLKATA BANGALORE

Capital Publishing Company
7/28, Mahaveer Street, Ansari Road
Daryaganj, New Delhi 110 002

Microearthquake Seismology and Seismotectonics of South Asia

J.R. Kayal

ISBN–81-85589-26-7

Typeset by Innovative Processors, New Delhi
Printed in India at Chaman Enterprises, New Delhi

Foreword

Hardly a week passes without our learning of natural geologic disaster somewhere in the world, be it a volcanic eruption, landslide, or destructive earthquake. The prominent public notice given to such events is not only the result of better communications, but also results from the increased impact of these events on a growing human population. In recent years, the population has increased greatly in regions of active tectonics. Northern India and the surrounding areas are prime examples. The consequence is that people and their man-made structures are concentrated close to active faults and steep, landslide-prone terrains. In just the past several years, even moderate earthquakes with seismic magnitudes less than 6.5 have killed as many as 20,000 people precisely because these earthquakes occurred directly beneath population centres in central India.

The greater Himalayan region, including the Ganges Plain, is a prime example of the coexistence of a pronounced geological hazard with a growing human population. Due in part to the spectacular topography, the region has long attracted scientific investigations, and may be considered as the birthplace of modern studies of earthquake hazards. R.D. Oldham (1858-1936) of the Geological Survey of India played a prominent role in the development of modern studies of historical seismicity, active faulting and seismic wave analysis. Oldham published extensively on the earthquakes and the geology of India, including his report entitled "Catalogue of Indian earthquakes from the earliest time to the end of A.D. 1869" (Mem. Geol. Surv. India, 19, 1882, 52 pp.) and his massive "Report on the great earthquake of 12th June, 1897" (Mem. Geol. Surv. India, 29, 1899, 1379 pp.). Later, Middlemiss reported on the great Kangra earthquake of 1905 (Mem. Geol. Surv. India, 32, 1905, 36 pp.). D.N. Wadia (1883-1969) was a pioneer in geologic and tectonic studies, and the Wadia Institute of Himalayan Geology in Dehradun bears his name. A.N. Tandon and his colleagues at the India Meteorological Department have carried out detailed studies of major earthquakes that have occurred since the 1950's.

The present comprehensive monograph by J.R. Kayal is, therefore, a continuation of the proud tradition established by several outstanding geologists and seismologists who have devoted their scientific lives to studying the underlying causes of earthquake hazards in the greater Himalayan region. Undoubtedly, investigations such as the present one by J.R. Kayal will provide the necessary basis for a safer coexistence of mankind and geological hazard in this region. It is also my wish that this monograph, like that of Oldham (1899), will inspire young readers to join this exciting path of scientific discovery.

Walter D. Mooney, PhD
Senior Research Seismologist
United States Geological Survey (USGS)
Menlo Park, California USA
and Seismology Section President
American Geophysical Union (AGU)

Preface

This book is a humble effort to share my experiences in the field of seismology and seismotectonics. During my PhD study in early 1980s at the Victoria University of Wellington (VUW), New Zealand, I came across only one book on *Microearthquake Seismology* by Lee and Stewart (1981). We get hundreds of books on theoretical seismology and thousands of research papers on theoretical as well as on observational seismology, but not many books on observational seismology or seismotectonics. I felt this demand in my student life as well as in my professional life. I think it is appropriate to have such a reference book on South Asia region, that may help to foster more communication between seismologists and geologists at national and international levels.

Instrument seismology in microearthquake networks has developed very rapidly, from analog system to 24-bit digital to broadband system within a few years, say within the time span of 1970-2000. Understanding and interpretation of observational seismology, however, need to keep an equal pace with the higher quality of seismograms and earthquake data. The author has worked with both systems, analog and digital. Now various software systems are also available for seismic data analysis. This requires upgraded training and understanding. For example, fault-plane solutions of earthquakes may now be quickly obtained using software, but its judgement and interpretation need proper understanding, training and experience. Particularly with respect to newcomers in the field, I noticed this gap.

In this humble effort, I tried to bridge the gap. I tried to explain the basics and then the interpretation with many case histories of earthquake study in the Indian subcontinent. This work took much longer than I expected. My official commitments and field as well as research and teaching activities kept me too busy to complete this book earlier, but made the book richer with time as more and more data were available. During the last decade there were seven devastating earthquakes in the Indian subcontinent (1991 Uttarkashi, 1993 Killari, 1997 Jabalpur, 1999 Chamoli, 2001 Bhuj, 2004 Andaman-Sumatra and 2005 Kashmir earthquakes). The author happened to lead all the investigations, for the Geological Survey of India, except the 2005 event. Investigations of these events made the book more interesting with more case histories and more understanding of the earthquake processes in different tectonic regions of South Asia.

I share my lifetime experiences in this book, and have tried to review seismotectonic researches, as much as possible, in this region of South Asia. It is a huge job to review all research publications, and I confess that a complete review was not possible. Much remains to be done in the next volume of the book. Any comments and feedback from readers are welcomed very much and may help me in my future endeavour.

I thank my professor, guide and philosopher, Late Prof. Frank F. Evison who introduced me to the field of Seismology as his PhD student under a Commonwealth Scholarship scheme at the Victoria University of Wellington, New Zealand and Prof. R.K. Verma (Retd. Professor, Indian School of Mines, Dhanbad), who first encouraged me to take up this work about 10 years ago. I have made an attempt to honour their esteemed expectation and love to me. My family support provided me to carry out this huge task, without which it would not have been possible. Thanks to my colleagues and coauthors of my publications who worked with me in the field and in the laboratory.

This book reviews some principles and applications of microearthquake networks and an overview of the published results of the field surveys carried out by the author in the Geological Survey of India and by others in different organizations/national and international institutes in various parts of the Indian sub-continent. In this volume, seismotectonics of the Himalayas, peninsular India, northeast India and adjoining areas including Bangladesh, Sri Lanka, Indo-Burma ranges, and Andaman-Sumatra regions are discussed.

Different types of earthquakes in the light of *plate tectonics* and various seismic waves generated by earthquakes are described in Chapter 2. In addition, magnitude and intensity scales, earthquake hazard and the great earthquakes of India are briefly discussed in this chapter. Chapter 3 contains microearthquake field surveys and data analysis. It describes microearthquake recording system, data acquisition, data processing, microearthquake location, seismic tomography etc. Since excellent books are available on theoretical seismology, theoretical part is dealt in minimum. A special emphasis on fault-plane solution is given in Chapter 4. Readable accounts on fault-plane solution, particularly for the microearthquake network data, seems to be lacking in the present literature. Although software is now available to obtain fault-plane solution, an attempt is made to present this topic in an introductory manner for clear understanding of this very important technique. Chapter 5 sheds some light on seismotectonics of the northwestern, western, central (Nepal) and eastern Himalayas (Sikkim and Bhutan), the 2500 km long seismic belt, from Hindu Kush in the northwest to Arunachal Himalaya in the northeast. The recent aftershock investigations of the Uttarkashi earthquake of October 20, 1991, the Chamoli earthquake of March 28, 1999 and results of several permanent and temporary microearthquake network data shed new light on the seismotectonic models of the Himalayas.

Chapter 6 deals with seismotectonics of northeast India and adjoining south Asia region, which includes northeastern Himalaya, Burmese Arc, Shillong Plateau, Assam Valley, Tripura fold belt and adjoining areas like Tibet, southeast China, Indo-Burma ranges, Bengal basin (Bangladesh) etc. Seismotectonics of the Andaman-Sumatra region is briefly discussed in this chapter in view of the 2004 mega thrust (M_w 9.3) event; a detailed discussion of the Andaman-Sunda arc needs to be dealt separately. The seismological data and the results of several temporary microearthquake surveys and permanent digital networks in various parts of northeast India are reviewed. A detailed study of 3-D velocity structure, fractal dimension, b-value mapping in northeast India region is also given in this chapter. Pop-up as well as transverse tectonics in the Shillong Plateau is critically reviewed. Chapter 7 describes seismotectonics of the *intraplate* earthquakes of peninsular India with particular references to the Killari (Latur) earthquake (M 6.3) of September 30, 1993, Jabalpur earthquake (M 6.0) of May 22, 1997 and the devastating Bhuj earthquake (M 7.7) of January 26, 2001. Results of a few temporary microearthquake surveys and permanent network data in the shield areas are also discussed. The seismic tomography results of the 1993 Killari earthquake and the 2001 Bhuj earthquake source areas shed new light on the generating process of the intraplate earthquakes.

To make this book more useful a glossary and a subject index are given. I hope that this work will provide some guidance and useful data to geoscientists and teachers who are involved in this field, and to the beginners who are aiming to start their career as a field seismologist or earthquake geologist.

J.R. Kayal

About the Author

J.R. Kayal, former Deputy Director General (Geophysics), Geological Survey of India, Kolkata, is Emeritus Scientist (CSIR), Jadavpur University, Kolkata and Adjunct Professor, Indian School of Mines, Dhanbad, and guest faculty at the Tezpur University, Assam, and UNESCO training courses in South Asia. He did his M.Sc. Applied Geophysics in 1969 from the Indian School of Mines, Dhanbad, and PhD from the Victoria University of Wellington, New Zealand on a Commonwealth Scholarship (1979-83). He has been awarded National Mineral Award (1994) by Govt of India and awarded Fellow by several scientific societies in India and abroad. He is a visiting scientist/professor to several national and international universities/institutes, and leads several national and international projects in earthquake seismology.

Acknowledgements

Since this book focuses on developments in our understanding of earthquake science, with a particular reference to Indian subcontinent, most of the earth scientists, whose work is discussed in this book, are my colleagues and friends, and I am proud to be a part of this respectable geoscience community in India and abroad. I owe my sincere gratitude to my friends and earth science colleagues all over the world whose names are sprinkled through the pages of this book. My attempt to name them all would exceed the space limitations of this short section.

The list of my immediate colleagues who worked with me all along may not be too long and more amenable to enumeration. G. Karunakar, Deven Chowdhury and Moly Mitra gave me working hand to carry a big part of the load in preparing the manuscript. Without Karunakar's help in graphics, so many figures would not have been possible to illustrate. I owe my sincere gratitude to my dear colleagues late Reena De, Sagina Ram, Pankaj Mala Bhattacharya, S.G. Gaonkar, O.P. Mishra, O.P. Singh, Pranab Chakraborty, G.K. Chakraborty, S.N. Chowdhury, B. Srirama, P.R. Rao, Balaram Das, Debangshu Banerjee, P.C. Das, M.K. Rai, R.P. Singh, A.K. Pawa, C.S. Verma, C.B.K. Sastry, L.K. Das, B. Banerjee, C.S. Venkiteswaran, T.R. Goswami, T.K. Sinha, A. Roy, S.K. Roy, D.R. Nandy, R.K. Chaturbedi, Prabhas Pande, Sujit Dasgupta and many others who worked with me in field investigations. A major part of this book is based on the field investigations that the author led during the last three decades.

The UNESCO and USGS nominated me as a member of the South Asia Seismic Analysis team since 2001 and as a visiting faculty for its training courses since 2005 that provided many information and direct interactions with many friends of South Asian countries. Leadership opportunities in several international projects like IGCP 411, IGCP 474 and Indo-Russian ILTP project helped me in making a comprehensive review work to the extent possible.

I must acknowledge the direct encouragement, help and inspiration that I received from many respectable leaders and earth scientists in India and abroad, like R.N. Bose, S.K. Acharyya, S.K. Mazumdar, Late S.N. Saha, R.K. Verma, V.K. Gaur, T.M. Mahadevan, H.K. Gupta, B.K. Rastogi,

V.P. Dimri, S.K. Guha, K.N. Khattri, Y. Sreedahar Murthy, Sagarika Mukhopadhyay, Saurabh Baruah, S.N. Bhattacharya, Manoj Mukhopadhyay, C.P. Rajendran, Kusala Rajendran, N. Purna Chandra Rao, V. K. Gahalaut, S.S. Rai, Ravi Kumar, P. Mandal, Paramesh Banerjee, S.K. Nath, Pradeep Talwani, Shamita Das, Walter Mooney, Peter Suhadolc, Dapeng Zhao, Euan Smith, Martin Reyners, Russel Robinson, S.K. Singh, David Gubbins, James Mori, Roger Bilham, E.R. Engdahl, Susan Haugh – to name a few.

I was fortunate to have all supports from my family, brothers, sisters, and relatives. I dedicate this book to my wife (Sunanda), son (Saurav) and daughter (Dahlia), for simple reasons that any husband or parent who has written a book would understand.

Contents

Abbreviations

ABF	Allah Bund Fault
ADC	Analog to Digital Convertor
ADMB	Aravalli Delhi Mobile Belt
AIN	Angle of Incidence
AK	Achankovil
AMD	Atomic Mineral Division
ANF	Alokananda Fault
ARL	Anjar Rapar Lineament
ASR	Andaman Spreading Ridge
BARC	Bhaba Atomic Research Centre
BARS	Belt of Active Rift Systems
BA	Burmese Arc
BB	Broad Band
BC	Bhandra Craton
BIS	Bureau Indian Standard
BN	Bundelkhund
BL	Bhachau Lineament
BS	Bastar Craton
BSF	Barwani Sukta Fault
BT	Basement Thrust
BTF	Basement Thrust Front
CBR	Carlsberg ridge
CCR	Central Core Region
CESS	Centre for Earth Science Studies
CGS	Coast and Geodetic Survey
CIS	Central India Suture
CMB	Core Mantle Boundary
CMP	Common Midpoint
CMT	Centroid Moment Tensor
CRB	Cambay Rift Basin
CRS	Central Recording Station
CSIR	Council of Scientific and Industrial Research
CWPRS	Central Water and Power Research Station
DAFB	Delhi Aravalli Fold Belt

Db.F	Dhubri Fault
DD	Double Difference
DF	Dauki Fault
DSS	Deep Seismic Soundings
DST	Department of Science and Technology
DVP	Deccan Volcanic Province
EDC	East Dharwar Craton
EGF	Empirical Green's Function
EGGB	Eastern Ghat Granulite Belt
EGMB	Eastern Ghat Mobile Belt
ER	East Ridge
ERI	Earthquake Research Institute (Japan)
ESC	European Seismological Commission
GBF	Great Boundary Fault
GDSN	Global Digital Seismic Network
GERI	Gujarat Engineering Research Institute
GL	Goalpara lineament
GMT	Greenwich Mean Time
GPS	Global Positioning System
GR	Godavari Rift
G-R	Gutenberg and Richter
GSHAP	Global Seismic Hazard Assessment Programme
GSI	Geological Survey of India
GSN	Global Seismographic Network
GTF	German Task Force
HFF	Himalayan Frontal Fault
HFT	Himalayan Frontal Thrust
HFU	Heat Flow Unit
HRV	Harvard University
HSB	Himalayan Seismic Belt
HVZ	High Velocity Zone
HYB	Hyderabad
IASPEI	International Association of Seismology and Physics of the Earth's Interior
IBF	Island Belt Fault
IGAP	Indo-Gangetic Alluvial Plains
IGP	Indo-Gangetic Plains
IIT (K)	Indian Institute of Technology (Kharagpur)
IIT(R)	Indian Institute of Technology (Roorkee)
ILTP	Integrated Long Term Program (Indo-Russian)
IMD	India Meteorological Department
INDEPH	International Deep Profiling in Tibet and Himalaya
IRIS	Incorporated Research Institute for Seismology
ISC	International Seismological Centre
ISPP	Indo Stanvac Petroleum Project
ISR	Institute of Seismological Research

IST	Indus Suture Thrust
J-B	Jeffreys-Bullen
J&K	Jammu & Kashmir
JHD	Joint Hypocentre Determination
KaF	Karakuram Fault
KC	Karnataka Craton
KLP	Kalpa
KMF	Kutch Mainland Fault
KRB	Kutch Rift Basin
KSEB	Kerala State Electricity Board
KSZ	Koyna Seismic Zone
LET	Local Earthquake Tomography
LP	Long Period
LTA	Long Term Average
LVZ	Low Velocity Zone
MBT	Main Boundary Thrust
MCS	Mercalli-Cancani-Sieberg
MCT	Main Central Thrust
MERI	Maharashtra Engineering Research Institute
MEQ	Micro Earthquake
MG	Mahanadi Graben
MHSB	Main Himalayan Seismic Belt
MIL	Milne
M-K	Malda Kishenganj
MKN	Makni
MKT	Main Karakuram Thrust
MM	Modified Mercalli
MMT	Main Mantle Thrust
Moho	Mohorovičić
M-S	Munger-Saharsa
MSK	Medvedev-Sponheuer-Karnik
MSR	Munger Saharsa Ridge
MT	Magnetotelluric
NCSN	Northern California Seismic Network
NEIC	National Earthquake Information Center
NEZ	North Escarpment Zone
NGRI	National Geophysical Research Institute
NIED	National Research Institute for Earth Science and Disaster Prevention
NKF	North Kathiwar Fault
NNF	Narmada North Fault
NPF	Nagar Parkar Fault
NRB	Narmada Rift Basin
NSF	Narmada South Fault
NSL	Narmada-Son Lineament
OBS	Ocean Bottom Seismograph

OIL	Oil India Limited
ONGC	Oil & Natural Gas Commission (now Corporation)
PGA	Peak Ground Acceleration
PF	Patna Fault
RF	Rossi Forel
RIS	Reservoir Induced Seismicity
RMS	Root Mean Square
RRL - J	Regional Research Laboratory - Jorhat
RSVP	Rajmahal-Sylhet Volcanic Province
RTS	Reservoir Triggered Seismicity
SC	Singhbhum Craton
SCB	Saurashtra Continental Block
SCR	Stable Continental Region
SCSI	Small Computer System Interface
SCSN	Southern California Seismic Network
SEPB	South East Platformal Block
SEZ	South Escarpment Zone
SGT	Southern Granulite Terrain
SIRT	Simultaneous Iterative Reconstruction Technique
SMA	Strong Motion Array
SOI	Survey of India
SONATA	Son-Narmada-Tapti Lineament
SP	Short Period
SPS	Samples Per Second
SRN	Srinagar
SRO	Seismic Research Observatories
STA	Short Term Average
STF	Source Time Function
SVD	Singular Value Decomposition
TAN	Tandon
THDC	Tehri Hydro Development Corporation
THF	Trans-Himadri Fault
TL	Tapti Lineament
UNESCO	United Nations Educational, Scientific and Cultural Organization
UPS	Uninterrupted Power Supply
USGS	United States Geological Survey
VBB	Very Broadband
VUW	Victoria University of Wellington
WAF	West Andaman Fault
WCF	West Coast Fault
WDC	West Dharwar Craton
WIHG	Wadia Institute of Himalayan Geology
WK	Wajrakarur
WSZ	Warna Seismic Zone
WWSSN	World Wide Seismograph Station Network

1

Introduction

Seismology is the study of generation, propagation and recording of *elastic waves* or *seismic waves* in the Earth (and other celestial bodies) and of the source, which produces them. The sources can be natural earthquakes or man-made sources of deformational energy that generate the seismic waves. Study of the seismic waves of the Earth provides the highest resolution of the internal Earth structure that no other geophysical method can provide. It is a branch of geophysics with the highest societal impact, both in assessing and reducing earthquake hazards, and in understanding *seismotectonics,* the structure and dynamic processes of the Earth.

1.1 CLASSIFICATION OF EARTHQUAKES AND FREQUENCY-MAGNITUDE RELATION

Seismologists had used the words like *small, medium* or *large* to describe an earthquake size. After introduction of the Richter magnitude-scale (Richter, 1935), the classification has been more definite. Hagiwara (1964) classified the earthquakes, according to magnitude, as follows:

Magnitude (M)	**Classification**
$M \geq 8$	Great earthquake
$7 \leq M < 8$	Major or Large earthquake
$5 \leq M < 7$	Moderate earthquake
$3 < M < 5$	Small earthquake
$1 \leq M < 3$	Microearthquake
$M < 1$	Ultra-microearthquake

All earthquakes of magnitude less than 3 are generally called *micro-earthquakes.* Gutenberg and Richter (1941) found that the frequency of earthquake occurrence is related to their magnitudes. The frequency-magnitude relation is given by: $\log_{10} N = a - b\mathrm{M}$, where N = number of earthquakes of magnitude M or greater, a and b are constants. The parameter b is often referred to *b-slope* or *b-value.*

Many studies of earthquake statistics show that in the normal cases the value of b is approximately 1.0, which means that the number of earthquakes increases tenfold for each decrease of one magnitude unit. Thus, if an area produces one earthquake of magnitude 4 in a month, it would produce 10 earthquakes of magnitude 3, 100 earthquakes of magnitude 2 and 1000 earthquakes of magnitude 1 in one month. Therefore, smaller the magnitude of the earthquakes, greater the amount of seismic data one can record. Microearthquake networks are designed to take an advantage of the above frequency-magnitude relation. Table 1.1 illustrates the global observations of frequency of occurrence of earthquakes since 1900, and Table 1.2 illustrates the number of earthquakes located by the National Earthquake Information Center (NEIC), US Geological Survey (USGS) for the period 1990-2002 and the estimated loss of lives.

Table 1.1: Frequency of occurrence of earthquakes based on observations since 1990, NEIC, USGS

Description	*Magnitude*	*Average Annually*
Great	8 and higher	1
Major	7 - 7.9	18
Strong	6 - 6.9	120
Moderate	5 - 5.9	800
Small	4 - 4.9	6,200 (estimated)
Minor	3 - 3.9	49,000 (estimated)
Microearthquakes	<3.0	Magnitude 2 – 3: about 1000 per day Magnitude 1 – 2: about 8000 per day

(Note: Description or classification as per USGS)

Table 1.2: Number of earthquakes worldwide, 1990-2000, NEIC, USGS

Magnitude	*1990*	*1991*	*1992*	*1993*	*1994*	*1995*	*1996*	*1997*	*1998*	*1999*	*2000*	*2001*	*2002*
8.0 to 9.9	0	0	0	1	2	3	1	0	2	4	1	1	0
7.0 to 7.9	12	11	23	15	13	22	21	20	14	23	14	15	13
6.0 to 6.9	115	105	104	141	161	185	160	125	113	113	158	126	133
5.0 to 5.9	1635	1469	1541	1449	1542	1327	1223	1118	979	982	1345	1243	1037
4.0 to 4.9	4493	4372	5196	5034	4544	8140	8794	7938	7303	6998	8045	8084	9034
Estimated Deaths	51916	2326	3814	10036	1038	7949	419	2097	8928	22711	231	21357	1711

1.2 HISTORICAL DEVELOPMENT OF EARTHQUAKE SEISMOLOGY AND GLOBAL NETWORK

Seismology is relatively a young science, which is basically developed with the evolution of theory of elasticity and development of instrumental and observational data base. Although the Chinese had a first operational seismic-wave detector (seismoscope) around 132 AD, the theoretical seismology was considerably ahead of observational seismology until late 1800s. The evolution came with the Hook's law indicating proportionality between stress and strain, which was followed by equations of elasticity by Navier and Canchy in 1921-22. In 1830, Poisson used the equations of motion and laws of elasticity to show two fundamental types of waves propagate through the interior of homogeneous solids. These two waves are *P wave* (Primary or Compressional wave) and *S wave* (Secondary or Shear wave). In 1887, Lord Rayleigh discovered existence of surface wave called *Rayleigh wave*, and in 1911 a second type of surface wave was discovered by A.E. Love, called *Love wave*. Both these waves are the results from the interaction of P and S waves with boundary conditions on the body with no shear stress of the surface.

International efforts led to the invention of the first *seismometer* by Filippo Cecchi in Italy in 1875 (Lay and Wallace, 1995). The sensitivity of early seismometer improved rapidly, and in 1889 accurate recording of seismic waves from a distant earthquake was possible in Potsdam, 15 minutes after the earthquake in Japan, demonstrating that the Earth behaves as a nearly elastic body in the frequency band of most seismic observations.

In early 1960s, a World Wide Seismograph Station Network (WWSSN) was established with identical sets of short-period and long-period three-component seismographs. This global network was equipped with very accurate timing by crystal clocks and standardized instrumentation; about 120 stations were in operation. The impact of the WWSSN was tremendous, coming at the time of *plate tectonics* revolution, when accurate seismic recordings and earthquake location were critical. In around 1975, the WWSSN was superseded by the Seismic Research Observatories (SRO), and eventually the network was digitally upgraded to the Global Digital Seismic Network (GDSN) in 1980s jointly by the Incorporated Research Institute for Seismology (IRIS) and the US Geological Survey (USGS). Due to large ocean surface as well as political situations, the network was not well distributed. The network is, however, growing continuously in collaboration with the French project GEOSCOPE together with instruments deployed by Australia, Canada, Europe and Japan. The ocean basins, of course, cause substantial gaps in the coverage, and Ocean Bottom Seismographs (OBS) are planned to minimize these gaps. In mid 2004, the Global Seismographic Network (GSN) surpasses its design goal (Park et al., 2005); a total of 136 GSN stations are now in operation from south pole to Siberia, and from Andaman basin to the sea floor of the northeast Pacific ocean. All GSN broadband data are freely available through the internet (www.iris.edu).

1.3 HISTORICAL DEVELOPMENT OF MICROEARTHQUAKE NETWORK

In terms of historical development, microearthquake networks evolved from temporary expeditions for aftershock study of large earthquakes to reconnaissance surveys, and finally mobile arrays to permanent telemetered networks. By the discovery of the earthquake frequency-magnitude relation (Gutenberg and Richter, 1941), methods and techniques to study microearthquakes have been developed by various groups of seismologist in several countries, like Japan, South Africa, USA and USSR. Although implementation of technological advances in instrumentation, data transmission and data processing in microearthquake networks was made in 1970s, many pioneering investigations were carried out in many countries much early. A detailed account of historical development is given by Lee and Stewart (1981). A brief outline is given here.

Japanese seismologists first developed portable and sensitive seismographs to study aftershocks, and Asada (1957) first reported a detailed study of microearthquakes. In the 1960s many microearthquake networks were established in Japan. These networks consisted 5-15 stations each and were reported by many groups (e.g. Suzuki et al., 1979).

In USA, the first microearthquake study was reported by Sanford and Holmes (1962) in a manner similar to Asada (1957). J.P. Eaton first developed a telemetered network for precise location of microearthquakes in Hawaii (Eaton, 1962). Since then several groups in the United States began microearthquake surveys, notable among them were J. Oliver and his group (e.g. Oliver et al., 1966), J.N. Brune and his group (e.g. Brune and Allen, 1967). A detailed study of the aftershocks of the 1966 Parkfield earthquake by Eaton et al. (1970a) demonstrated the usefulness of microearthquake networks in understanding the earthquake generating process, and made a great impact to develop the Central California Telemetric Network on large scale (Eaton et al., 1970b).

In India, microearthquake investigation started in 1978 for geothermal exploration in Puga valley and also to cater the need of river valley projects in 1980s in northwestern Himalaya (IMD, 1980 and 1988). K.N. Khattri and his group in the University of Roorkee (now Indian Institute of Technology, Roorkee), and S.N. Saha and his group in the Geological Survey of India (GSI) made a collaborative microearthquake investigation in northeast India region in 1980s (Khattri et al., 1983). The investigation in northeast India was extended by the author and his group in the GSI (e.g. Kayal, 1987; Kayal and De, 1987). Two permanent telemetric networks were established in late 1980s on experimental basis, one by the then University of Roorkee in Garhwal Himalaya, and the other by the NGRI in northeast India. The aftershock study of the 1991 Uttarkashi earthquake in the Garhwal Himalaya, and the aftershock study of the 1993 Killari (Latur) earthquake in Maharashtra, in particular, made tremendous impact on modernisation of the permanent seismological observatories as well as microearthquake networks in India. A great deal of effort has been made by the Department of Science and Technology (DST) to upgrade the permanent seismological observatories of the India Meteorological Department (IMD) to GSN standard, and to establish more permanent observatories and telemetric networks in different earthquake prone zones of the country. Many institutes and universities in the country are also funded by the DST for running microearthquake networks and permanent observatories.

1.4 APPLICATIONS OF MICROEARTHQUAKE NETWORK

Microearthquake networks are powerful tool for detailed seismological investigations. These networks generate large quantity of seismic data because of high occurrence-rate of microearthquakes. Applications of microearthquake networks are numerous; to name a few:

(i) mapping seismicity and active faults/lineaments,
(ii) studying focal mechanism of earthquakes, tectonic stress etc.,
(iii) investigating velocity structure of the crust and upper mantle,
(iv) exploring geothermal resources,
(v) investigating aftershocks, swarm activity etc.,
(vi) developing mining industry, hydel-dam projects, nuclear power plants etc.,
(vii) investigating Reservoir Induced Seismicity (RIS) and other man-made induced seismicity,
(viii) monitoring volcanic seismicity,
(ix) monitoring earthquake precursors, etc.

Results from these studies are integrated with other geophysical and geological field investigations for better understanding of earthquake processes, seismicity, seismotectonics, crustal structure, precursors etc. in South Asia region. A large quantity of local microearthquake and regional earthquake data are analysed to understand seismotectonics of the Himalaya from the westerm syntaxis, Pamir-Hindu Kush to eastern syntaxis, Arunachal Himalaya, and a 2500 km long Himalayan seismic belt. The seismic activity in the northeast India region as well as the intense activity along the Indo-Burma and Andaman-Sumatra arcs are well studied using the recent digital seismic network data at the atypical continental subduction zone beneath the Indo-Burma ranges and at the typical oceanic plate (Indian Ocean) subduction beneath the Andaman-Sumatra arc. The complex intraplate tectonics of peninsular India, Bangladesh and Sri Lanka are also well studied; the typical intraplate seismicity, RIS, aftershocks, swarms and active faults are investigated using the microearthquake network data.

2

Earthquakes and Seismic Waves

EARTHQUAKES

Earthquakes are exciting and sometimes devastating. A large earthquake attracts general attention because it causes damage to houses and other works of man, or breaks up and changes the ground surfaces. To the engineers, the earthquakes of great interest are those which affect the stability and safety of the buildings and other constructions. To the geologists, the earthquake phenomena of interest are the evidence of faulting or other tectonic process and the relation of earthquakes to the geological structures. And for the geophysicists or seismologists, every earthquake, large or small, presents a problem of explanation and interpretation of every wriggle of the seismic waves produced by the earthquakes. Earthquake seismology is thus defined to be a dual science, a border line field between geology and geophysics. It calls for sound physical and mathematical thinking, and a strong background of geological concepts.

2.1 TYPES OR CAUSES OF EARTHQUAKES

Earthquakes can be divided into four categories based on their mode of generation. These are discussed below:

2.1.1 Tectonic Earthquakes

Tectonic earthquakes are by far the most common earthquakes. These earthquakes are produced when the rocks fracture by various geological forces because of constant geological reshaping of the Earth. Tectonic earthquakes are most important to study the structure and interior of the Earth, and they are of much social significance as they pose greatest hazard.

Sensitive instruments register major tectonic earthquakes at large distances demonstrating that such events disturb the whole Earth.

Plate tectonics concept is the most recent and quite satisfying explanation for majority of the earthquakes. The plate tectonic hypothesis is defined in a series of papers between 1967 and 1971 (e.g. Elsasser, 1967; McKenzie, 1967 and 1969; Oliver and Isacks, 1967; Isacks and Molnar, 1971 etc.). Readers are referred to read a few excellent books on Plate Tectonics, e.g. Le Pichon et al. (1976), Condie (1982) etc. The basic idea in plate tectonics is that the Earth's outermost part, called *lithosphere*, consists of several large and fairly stable slabs; these are called plates. Boundaries of these plates are the seismic belts of the world. The major plates and seismic belts are shown in Fig. 2.1. The plates move horizontally relative to adjoining plates on a layer of softer rock, called the *asthenosphere. Lava* is continually upwelling at the *mid-oceanic ridges*. This rock slowly moves across the Earth's surface as a new sea-floor on either side of the ridge. In this way plates move like a conveyer belt over the asthenosphere, and get cooled as they get further away from the ridges. The mid-oceanic ridges are called *spreading zones* or *accreting plate boundaries* or *divergent plate boundaries* (Fig. 2.2).

Fig. 2.1 World map showing relation between tectonic plates, seismicity (red dots) and volcanoes (triangle). Red boundary indicates spreading zone, yellow subduction zone, blue transcurrent and white diffuse zone (website: http://www.geol.binghamton.edu/faculty/jones).

Now the question arises, if a new plate is constantly being created at mid-oceanic ridges what happens to older plate? Since the Earth probably remains the same size over quite a long period of geological time, the moving plates must be absorbed at some places; these are *convergent plate boundaries*. The burial ground of plates is believed to be the *ocean trenches,* where the surface layers of rock plunge into Earth's interior. This process is known as *subduction*. The other type of convergent plate boundary forms the continent-continent collision zone. The third type of plate boundary is the *transcurrent boundary*, where the plates move past one another, without either convergence or divergence. *Transform faults* or *strike-slip* faults are the examples of transcurrent boundaries.

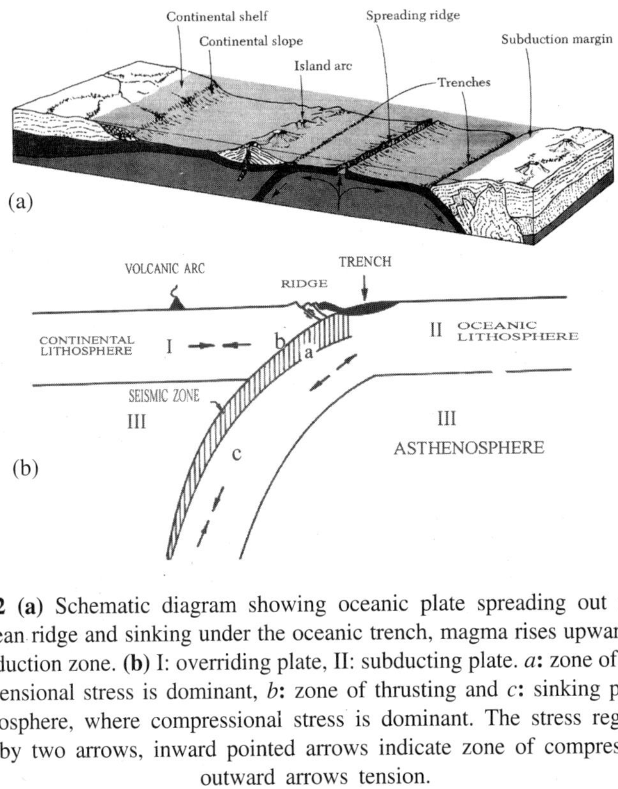

Fig. 2.2 (a) Schematic diagram showing oceanic plate spreading out from the mid-ocean ridge and sinking under the oceanic trench, magma rises upwards above the subduction zone. **(b)** I: overriding plate, II: subducting plate. *a***:** zone of bending, where tensional stress is dominant, *b***:** zone of thrusting and *c***:** sinking portion of the lithosphere, where compressional stress is dominant. The stress regimes are shown by two arrows, inward pointed arrows indicate zone of compression and outward arrows tension.

Subduction Zone Earthquakes

As the oceanic plate bends downwards at the ocean trenches, fractures cause shallow earthquakes within the *subducting plate* as well as in the *overriding plate* (Fig. 2.2); these are called *interplate* earthquakes. In the process of its downward movement, additional forces are generated causing further deformation and fracturing, thus giving rise to deep focus earthquakes. The deeper earthquakes, which occur in the subducting plate, define a clear dipping seismic zone, called *Benioff zone*. On the average, frequency of occurrence of earthquakes in this zone declines rapidly below 200 km depth.

Some earthquakes are, however, as deep as 700 km in some of the subduction zones, e.g. the Tonga Islands, Java Islands, the New Hebrides etc. Earthquakes, generally, with foci between 70-300 km deep are called *intermediate focus* and those below this depth are called *deep focus*. Along with the deep focus earthquakes, the chemical composition of rocks in the subducting plate changes. The molten fraction of rocks may be stored for a time in *magma chambers,* which are huge reservoirs underneath volcanic vents. From these pressure chambers magma moves upwards from time to time as lava. Thus, a volcanic arc is a typical characteristic in a subduction zone (Fig. 2.2). Further, interaction between the two plates, overriding continental plate and the subducting oceanic plate, near the trench causes folding, faulting and thrusting in the continental crust, and makes an exposure of ophiolite (oceanic crust) at the surface.

Collision Zone Earthquakes

Tectonic earthquakes are also produced at the edges of the interacting plates where the plates collide head on, and rise as mountain chains, e.g. the Himalayan belt in Indo-Eurasian collision zone, the Alpine belt in the eastern Europe, the Zagros thrust system in Iran (Fig. 2.1). The interplate earthquakes in collision zones are usually shallow (<100 km), and dominated by low angle thrusting, which is a manifestation of lithospheric shortening. These low angle faults/thrusts can generate large earthquakes as we see in the Himalayan belt.

Divergent and Transcurrent Plate Boundary Earthquakes

In general, divergent and transcurrent plate boundaries are characterized by much shallower seismicity (focal depth less than 30 km). The mid-oceanic ridges, the divergent zones, are the manifestation of shallow extensional faulting (Fig. 2.3). One of the interesting features of mid oceanic ridges is that they are offset by lineament known as fracture zones. Wilson (1965) proposed the concept of *transform faults* or *strike-slip faults* that accommodate ridge spreading. *Continental strike-slip* or *transcurrent plate boundary* may also be observed in continent-continent or oceanic-continental situation (Fig. 2.3). The most famous example in oceanic-continental situation is the San Andreas fault in California, which separates the North America and Pacific plates. Transcurrent plate boundary in continental settings is extremely important from seismic hazard point of view. A few examples of such continental transcurrent faults are the Anatolian fault in Turkey, Alpine fault in New Zealand and the Sagaing fault in Burma (Fig. 2.3). These transcurrent-fault earthquakes are shallow and sometimes produce large fatal earthquakes, e.g. the 1989 Loma Prieta earthquake on the San Andreas fault (Catchings and Kohler, 1996).

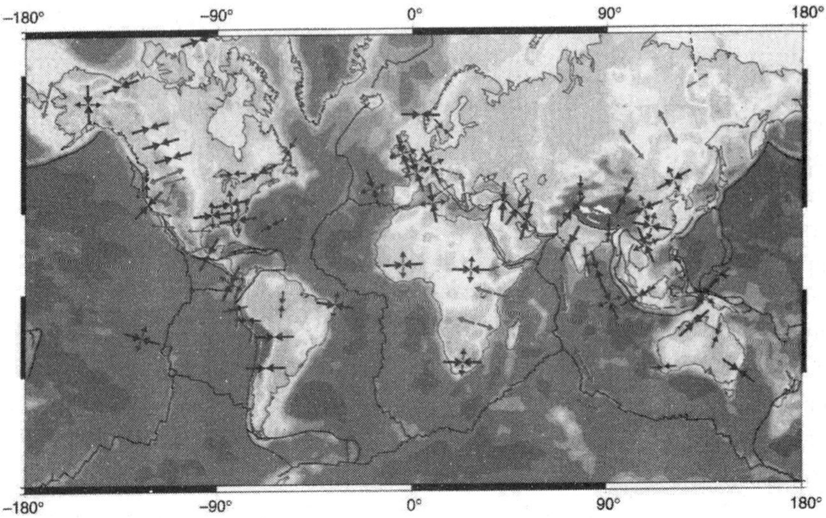

Fig. 2.3 World stress map (Zoback, 1992); inward pointed arrows indicate compression (thrust faulting), outward arrows tension (normal and strike slip faulting).

Intraplate Earthquakes

About 60 percent of the world's seismicity is mapped by the *interplate earthquakes* at the plate boundaries as stated above, and about 10 percent is mapped at the *spreading zones.* The world's seismicity map, however, shows that earthquakes do take place within the plate, far from the plate boundaries as well (Fig. 2.1). These earthquakes are called *intraplate earthquakes,* and they arise from the localised system of forces. The intraplate and interplate earthquakes differ in two important ways. First, the recurrence interval of intraplate earthquakes is generally much longer than those of the interplate earthquakes, and second, the intraplate earthquakes typically have much higher stress drops. Three large/fatal intraplate earthquakes (M 6.0~7.7) that have occurred in peninsular India during the last decade are discussed in Chapter 7.

Hypocentres of the earthquakes in the intraplate zones rarely occur deeper than 15-20 km. Deepest earthquakes indicate the base of the seismogenic zone, or bottom of the fault. Within the seismogenic zone the crust deforms either by stable or unstable sliding on the existing fault(s), or as a brittle material that fails generating a new fault, when subjected to stresses greater than the strength of the material. The strength depends on both temperature and pressure. The strength increases to a depth of about 15-20 km, and then decreases rapidly. Below the seismogenic zone stable sliding or ductile deformation takes place.

2.1.2 Volcanic Earthquakes

Volcanic earthquakes are defined as the earthquakes which occur or are linked up with the volcanic activity. These earthquakes occur in two ways.

First, often before an eruption minor seismic activity increases in the vicinity of the volcano. Some kilometres below the volcanic vent, very hot viscous magma moves sluggishly under high steam pressure through a network of veins and arteries from one storage chamber to another. Due to this motion, various parts of the surrounding rock become hotter and more strained as the magma pushes them. These forces fracture the neighbouring rocks, and the strain is relieved by small or moderate earthquakes. Second, sometimes fault rupture precedes the motion of magma and eruption of lava. The earthquake waves from the rupturing fault may shake up the molten material in the storage reservoir beneath the volcano. In a way, similar to the violent shaking of a bottle of soda pop, steam and gas that have previously dissolved in the magma begin to boil off and accelerate the escape of lava and gaseous material. This, in turn, would disturb the unstable equilibrium of the magma below the vent, and stimulate local volcanic earthquakes.

2.1.3 Induced Earthquakes

Some earthquakes are induced; this means that human activity triggers the earthquakes. These are: (i) *Reservoir Induced Seismicity* (RIS), (ii) *Nuclear explosion,* (iii) *Rock burst, fluid injection,* etc.

Reservoir Induced Seismicity

Reservoir induced earthquakes have been reported in conjunction with the impoundment of water or rapid water level changes behind large dams (Gupta and Rastogi, 1976). Since these earthquakes are triggered along pre-existing and pre-stressed tectonic faults, they show typical characteristics of *double couple sources* (see Section 4.6.1).

The December 10, 1967 Koyna earthquake (M 6.7) in India that caused a loss of about 200 human lives, is attributed to RIS for the Koyna dam (Gupta and Rastogi, 1976). It is unlikely that simply weight of the water, which would only add a tiny fraction of the total stress at a depth 3 or 5 km below the surface, causes the RIS. A more likely explanation is that the *pore pressure* increases because of the hydrostatic head of the reservoir. The RIS was first observed with the filling of Lake Mead behind Hoover Dam beginning in 1935 (Simpson, 1976). Before the reservoir was filled, the background seismicity was low; the seismicity began to rise in 1936, with occurrence of an earthquake M 5.0 in 1940. RIS is mostly shallow focus, less than 6 km. Damaging earthquakes M>6.0, that are referred to the RIS, occurred at Hsinfeugkiang, China in 1962, at Kariba, Zambia-Zimbabwe border in 1963, at Kremasta, Greece in 1966 and at Koyna, India in 1967 (Gupta, 1992). In a recent book on Reservoir Induced Earthquakes by Gupta (1992) all the necessary information on RIS has been given. In his recent review Gupta (2002), however, preferred the term RTS (Reservoir Triggered Seismicity) rather than the RIS.

Nuclear Explosion

The other man-made earthquake is the *nuclear explosion*. Theoretically the seismic waves generated by an underground nuclear explosion should be very different from shear motion that characterizes double-couple sources for natural earthquake. The nuclear explosion produces outward directed compressional motions in all directions while tectonic earthquake produces motions of different amplitude and polarity in different directions (see Section 4.6.2). So, the seismograms from an explosion should not have S_H or *Love waves*, but many explosions do have S_H type energy. This energy is thought to be the release of pre-existing strain by the explosion. A very good case study of the Indian nuclear explosion at the Pokhran-test site, Rajasthan on May 11, 1998, was made by Sikka et al. (1998). The estimated magnitude of this explosion was $m_b = 5.4$. They did not report S_H wave.

Rock Burst, Fluid Injection etc.

The third type of induced earthquakes are generated by rock burst, fluid or stream injection etc. Collapsing of the roof of mines and caverns, the so-called phenomena of *rock-burst*, occurs due to induced stress around the mine workings, and it produces seismic waves. Readers are referred to a review book on Mining Seismology by Gibowicz and Kijko (1994) for detailed information. Similarly, fluid or steam injection process in the oil fields for secondary production also causes minor tremors (e.g. Talebi and Cornet, 1987; Talebi et al., 1998). In a recent book, Guha (2000) has given a detailed review of induced seismicity.

2.2 TELESEISMIC, REGIONAL AND LOCAL EARTHQUAKES

Earthquakes may be further classified into three types depending on the distance from source to the *seismograph* station. Since the character of *seismograms* depends on *epicentral* distance, the nomenclature for seismic phases is also distant dependent.

2.2.1 Teleseismic Earthquakes

The earthquakes, which are recorded by a seismograph station at a greater distance, are called *teleseismic earthquakes*. These are very often called *teleseisms*. By international convention the epicentral distance is required to be more than 1000 km for a teleseism. Lay and Wallace (1995), however, define teleseismic distance as $\Delta \geq 30°$. An example of a tele-seismogram is shown in Fig. 2.4. Depending on the magnitude of the earthquake, teleseismic amplitudes can range from barely perceptible to those that saturate the instrument. These earthquakes provide very useful seismic phases for the crust as well as for the interior of the Earth. The direct P and S-wave arrivals recorded at teleseismic distances between 30° and 95° are relatively simple

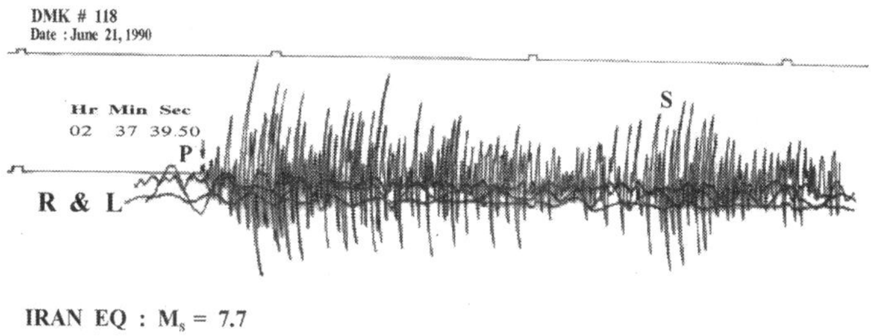

Fig. 2.4 Part of a tele-seismogram showing seismic phases of the Iran earthquake of June 21, 1990, Ms 7.7, recorded by a microearthquake recorder (PS-2) at a temporary seismograph station at DMK (Doimukh), Arunachal Pradesh, northeast India during a microearthquake survey. Minute marks are indicated by square pulses above. The seismic phases continued for more than 40 minutes. Long period surface waves (R & L) are also recorded (Kayal et al., 1993).

indicating a smooth velocity distribution. Beyond 95°, the direct phases become complicated due to interaction with the Earth's core (see Section 2.17).

2.2.2 Regional Earthquakes

The earthquakes, which occur beyond say 500 km but within 1000 km of a seismograph station, are called *regional earthquakes*. An example of a regional earthquake, recorded by a microearthquake recorder at a distance of about 700 km, is shown in Fig. 2.5. Like teleseismic events, amplitudes of regional

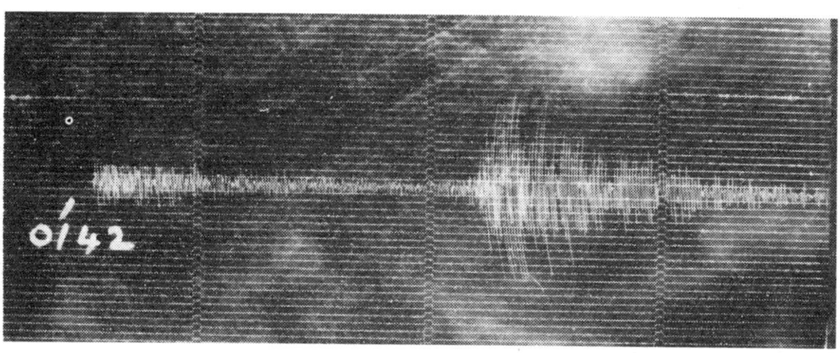

Fig. 2.5 Part of an original seismogram showing a regional earthquake recorded by a microearthquake recorder (PS-2) in northeast India region. The event is annotated by hr. min. (0142). The square pulses indicate the minute marks. The S-P interval is more than one minute (Kayal et al., 1993).

earthquakes can range from barely perceptible to large, but their periods are less than those of the teleseism. These earthquakes also provide seismic-wave data for the Earth's crust and mantle.

2.2.3 Local Earthquakes

Earthquakes occurring within a distance of few hundred km, say 500 km, from a seismic station are called *local earthquakes*. Local earthquakes are often characterised by impulsive onsets and high frequencies. A seismogram showing local earthquakes recorded by a microearthquake station at a distance ranging from 50 to 250 km is illustrated in Fig. 2.6. The local earthquake signal has typically an exponential decreasing tail. The seismic-wave data are very useful to study the local geological structure/velocity structure of the Earth's crust and upper mantle.

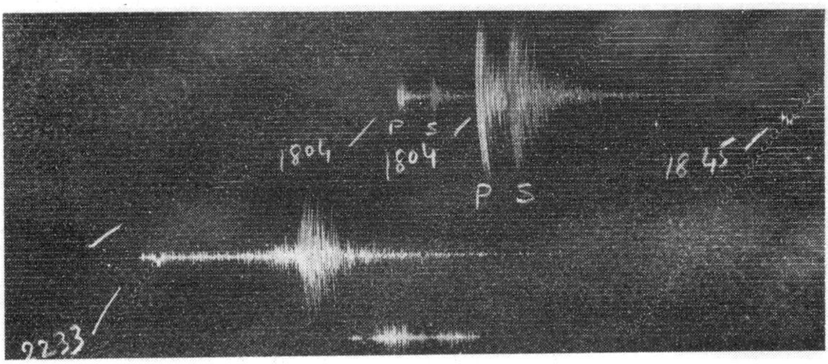

Fig. 2.6 Part of an original seismogram showing local earthquakes including microearthquakes; P and S phases are indicated. The S-P intervals are few seconds to less than a minute (Kayal, 1983). Other explanations are same as Fig. 2.5.

2.3 FORESHOCKS, AFTERSHOCKS AND EARTHQUAKE SWARMS

An earthquake of consequence is never an isolated event. A larger shock is likely to be preceded by a few smaller shocks; these are called *foreshocks*. There are, almost certain, to be many shocks after a main shock, which are called *aftershocks*. The aftershocks gradually decrease in frequency and magnitude. The foreshocks and aftershocks are expected to occur in the same epicentre area, and are, generally, associated with the same fault system of the main shock. On the other hand, a long series of small shocks with no main event is frequently recorded in certain localities; these sequences are called *earthquake swarms*.

2.3.1 Foreshocks

Evidence of *foreshocks* is hard to judge before the occurrence of a main event since there is nothing to distinguish foreshocks from the ordinary small shocks in an active region. Suspicious series of small earthquakes may be followed by no main shock or by a sizable earthquake. Of earthquakes with M>7.0, about 70% are preceded by foreshocks. The nature of foreshock activity varies greatly, but it typically begins 5-10 days before the main shock.

Oldham (1899) reported no foreshock for the 1897 great Shillong earthquake (M 8.7); neither were any reported for the 1934 great Bihar-Nepal earthquake (M 8.4), nor for the 1950 great Assam-Tibet earthquake (M 8.7). By contrast, 1915 Nevada (USA) earthquake was preceded by many small foreshocks. Foreshocks subsequently are reported for many large earthquakes at various parts of the world and are used for earthquake precursor study. Jones and Molnar (1979) summarized the characteristics of foreshocks associated with major earthquakes of the world.

2.3.2 Aftershocks

Aftershocks are almost always reported after a sizable earthquake. The number, magnitude and the duration of aftershock activity depend on the size of main shock, depth and tectonic setting of the area. The aftershocks are apparently related to the fault plane that slipped during the main event. Båth (1979) noted that in many instances the largest aftershock is about 1.2 less in magnitude than that of the main shock. Aftershocks are more often journalistically called settling shocks, that is readjustment to equilibrium. Internal readjustment may be expected to go on for some time with decreasing frequency and magnitude of the aftershocks.

Aftershock Decay: Omori law

The decay of aftershock event rate $N(t)$ with time t was initially found to obey a simple hyperbolic law known as Omori law (Omori, 1894). More generally, Utsu (1957) showed that the rate of aftershock decay could be described by power law as:

$$N(t) \propto t^{-p} \tag{2.1}$$

where p is the attenuation factor or rate, constant of aftershock decay. The eq (2.1) is called modified Omori formula (Utsu, 1957). Thus, the Omori law is equivalent to modified Omori formula with exponent $p = 1$. Several researchers have empirically shown that the p-values of large earthquakes are close to 1.0, but it ranges from 0.6 to 2.5 (e.g. Mogi, 1962; Utsu, 1969; Kisslinger and Jones, 1991; Guo and Ogata, 1995, 1997).

Aftershocks of second order may be closely followed by a trail of small shocks which fall off more rapidly than the general aftershock activity. Occurrence of second order aftershocks is explained by a larger aftershock event, which dynamically may be an independent earthquake.

Fault Area

The distribution of aftershocks is often used to infer the fault area of the main shock. For most earthquakes, the fault area or aftershock area scales with magnitude. Utsu and Seki (1954) developed an empirical formula:

$$\log A = 1.02\ M_S + 6.0 \tag{2.2}$$

where A is measured in cm^2, and M_S is surface wave magnitude of the earthquake (see Section 2.4). In general, the fault area is estimated from the extent of aftershock zone, one or two days after the main shock. This limit is imposed because sometimes aftershock zones grow continually for a month or more, presumably involving outward expansion of the main shock rupture zone (Lay and Wallace, 1995). The total seismic moment release by an aftershock sequence rarely exceeds 10% of the moment of the main shock.

As mentioned in Chapter 1, aftershock studies have greatly stimulated the development of microearthquake network. Vast quantities of data are recorded by deploying temporary seismograph stations for detailed studies of damaging earthquakes. Investigations of aftershocks for the recent damaging earthquakes during the last decade in India are discussed in Chapters 5–7.

2.3.3 Earthquake Swarms

Swarms are sequence of earthquakes that are clustered in space and time, and are not associated with identifiable main shock. The *b*-value for a swarm can be as high as 2.5, which implies that no large earthquake accompanies the occurrence of huge small magnitude earthquakes. The high *b*-value may be explained by weak crust, incapable of sustaining high strain levels and heterogeneous stress system in the shallow crust.

Earthquake swarms are common in volcanic regions, they occur before and during the eruptions. Swarms are also observed in non-volcanic areas, as in Comrie in Scotland and in the Vogtland on the southern border of Germany (Lee and Stewart, 1981) and in New Zealand (Kayal, 1983). Swarm activities in the Himalaya and in central India are also reported (see Chapters 5 and 7). Swarms of special pattern in time and space may indicate a precursory signal to an impending large earthquake (Evison, 1982). Earthquake swarm used as a precursor tool in forecasting a large Indo-Burma border earthquake is discussed in Chapter 6.

2.4 EARTHQUAKE MAGNITUDE, INTENSITY AND ENERGY

Magnitude is one of the basic and important parameters of an earthquake. It defines the size of an earthquake. The common people are, in general, confused about different scales of magnitude, and sometimes they mix-up earthquake intensity with its magnitude. Journalists often report the magnitude value of an earthquake as its intensity; this is wrong.

Intensity of an earthquake is a measure of its effect, i.e. degree of damage; for example broken windows, collapsed houses etc. produced by an earthquake at a particular place. The effect of the earthquake may cause collapsed houses at city A, broken windows at city B and no damage at city C. Intensity observations are, thus, subject to personal estimates and are limited by the circumstances of reported effects (Bolt, 1999). Intensity varies from place to place for the same earthquake. Therefore, it is desirable to have a scale for rating earthquakes in terms of energy, independent of the effects produced at a particular area. In response to this practical need, Richter (1935) first proposed a magnitude scale based solely on amplitudes of ground motion recorded by a seismograph.

2.4.1 Earthquake Magnitude

There are now different magnitude scales to define the size of an earthquake. After Richter (1935), various magnitude scales are proposed. The Richter scale and the recent theoretical based scales are discussed below.

Richter Magnitude (or Local Magnitude) M_L

Richter (1935) defined the local magnitude M_L of an earthquake observed at a station to be

$$M_L = \log A - \log Ao\ (\Delta) \tag{2.3}$$

where A is the maximum amplitude in millimetres recorded on the *Wood-Anderson seismograph* for an earthquake at epicentral distance of Δ km, and Ao (Δ) is the maximum amplitude at Δ km for a standard earthquake. The local magnitude is thus a number characteristic of the earthquake, and independent of the location of the recording station.

Three arbitrary choices are made in the above definition: (i) the use of standard Wood-Anderson seismograph, (ii) the use of common logarithms to the base 10, and (iii) selection of the standard earthquake whose amplitudes as a function of distance are represented by Ao (Δ). The zero level of Ao (Δ) can be fixed by choosing its value at a particular distance. Richter chose the zero level of Ao (Δ) to be 1 μm (or 0.001 mm) at a distance of 100 km from the earthquake epicentre. Thus, an earthquake with trace amplitude A = 1 mm recorded on a standard Wood-Anderson seismograph at a distance of 100 km is assigned magnitude 3. Richter arbitrarily chose –log Ao = 3 at Δ = 100 km so that the earthquakes do not have negative magnitudes. In other words, to compute M_L a table of –log Ao as a function of epicentral distance in kilometres is needed. Based on observed amplitudes of a series of well located earthquakes the table of –log Ao as a function of epicentral distance is given by Richter (1958).

In practice, we need to know the approximate epicentral distance of an earthquake, which can be estimated from S-P time. The maximum trace amplitude on a standard Wood-Anderson seismogram is then measured in millimetres, and its logarithm to base 10 is taken. This number is then added to the quantity tabulated as –log Ao for the corresponding station-distance from the epicentre. The sum is a value of local magnitude for that seismogram. Since there are two components (EW and NS) of Wood Anderson seismograph, average of the two magnitude values may be taken as the station magnitude. Then average of all the station magnitudes is an estimate of the local magnitude M_L for the earthquake.

A graphical procedure for estimating the Richter magnitude (M_L) is then developed; it is exemplified in Fig. 2.7. The S-P time and the maximum trace amplitude on the seismogram are used to obtain M_L = 5.0 in this example. In Richter's procedure, the largest amplitude recorded on the seismogram is taken.

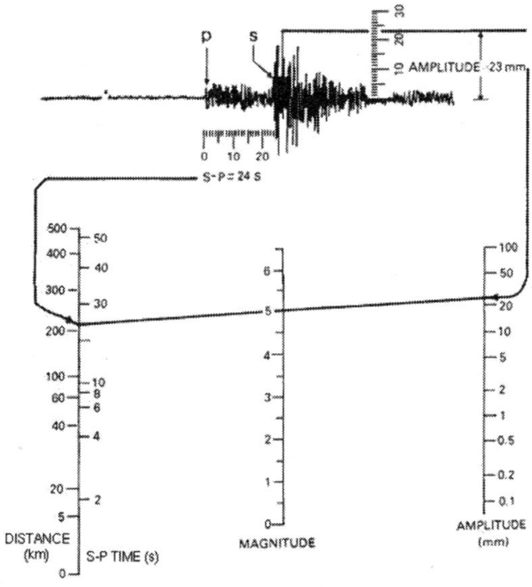

Fig. 2.7 Example of estimating Richter magnitude (M_L) of a local earthquake. Procedure: measure the S-P interval (=24 seconds), maximum amplitude (= 23 mm), and draw a straight line between the appropriate points on the distance (left) and amplitude (right) scales to obtain M_L = 5.0 (Bolt, 1999).

It follows from its definition that there is no upper or lower limit in the magnitude scale. The size of an earthquake is, however, limited by the strength of rocks in the Earth's crust. The maximum size of an earthquake that has been recorded in south Asia is 8.9 in the Richter magnitude scale (December 26, 2004 Sumatra earthquake, IMD report, 2005). The Richter magnitude, however, gets saturated at this level. In such cases moment magnitude may be more stable, and the moment magnitude of the Sumatra earthquake was estimated to 9.3. The moment magnitude scale is discussed below.

Body Wave Magnitude (m_b)

It is now a routine practice in seismology to measure the amplitude of the P-wave which is not affected by the focal depth, and thereby determine P-wave or body-wave magnitude (m_b). Gutenberg (1945a) defined body-wave magnitude m_b for teleseismic body waves P and PP in the period range 0.5 to 12 s:

$$m_b = \log (A/T) - f(\Delta, h) \tag{2.4}$$

where A/T is amplitude-to-period ratio in micrometres per second, and $f(\Delta, h)$ is a calibration function of epicentral distance Δ in degree and focal depth h in kilometre. Gutenberg and Richter (1956) published a table for the calibration function.

It is recommended that the largest amplitude be taken within the first few cycles instead of considering the whole P-wave train (Willmore, 1979). Both the ISC and NEIC, however, determine body wave magnitude only from vertical component short period P-wave readings of $T \leq 3$ s.

Surface Wave Magnitude (M_s)

For shallow and distant earthquakes, a surface wave train is present that is used for estimation of surface wave magnitude M_s. Gutenberg (1945b) defined the surface wave magnitude M_s as:

$$M_s = \log A_{Hmax} - \log Ao\ (\Delta^o) \tag{2.5}$$

where A_{Hmax} is the maximum combined horizontal ground amplitude in micrometres for surface waves with a period (T) of 20 $\pm$ 2 second, and (–log Ao) is a calibration function that is tabulated as a function of epicentral distance Δ in degrees in a similar manner to that for local magnitude (Richter, 1958).

In a collaborative research Karnik et al. (1962) proposed a new M_s scale as below:

$$M_s = \log (A/T)_{max} + 1.66 \log \Delta + 3.3 \tag{2.6}$$

for epicentral distances $2° < \Delta < 160°$ and source depth $h < 50$ km. The IASPEI committee on magnitudes recommended at its Zurich meeting in 1967 the use of this formula as standard for M_s determination for shallow seismic events ($h \leq 50$ km). Today, both ISC and NEIC use this eq. (2.6) for determination of M_s. The ISC accepts surface waves with period 10-60 s from stations at a distance range 20°-160°.

A difficulty in using the surface-wave magnitude-scale is that it can be applied only for the shallow earthquakes that generate observable surface-waves. For shallow focus earthquakes, an approximate relation between m_b for P-waves and M_s is given by:

$$m_b = 2.5 + 0.63\ M_s \tag{2.7}$$

Moment Magnitude (M_w)

There are some problems that have been encountered with the magnitude scales. For large earthquakes the Richter as well as body wave magnitude scales saturate. No matter how large the earthquake is, the magnitude computed from body waves tend not to get much above 6.0 to 6.5. The surface-wave scale is less affected by this problem, but for very large earthquakes M>8 the surface-wave scale also gets saturated. It turns out that the limitation is in the instrument recording the earthquake. As new long period instruments as well as digital seismographs have come into use, it has been possible to assign a better measure of the size of these very large earthquakes using the moment magnitude scale. Hanks and Kanamori (1979) proposed the moment magnitude scale by:

$$M_w = 2/3 \log M_o - 10.7 \tag{2.8}$$

where M_o is *seismic moment* of the earthquake in dyne cm. The seismic moment is defined as

$$M_o = \mu A\ \Delta u \tag{2.9}$$

where μ = shear modulus, A = fault area and Δu = average slip over the fault area (Aki, 1966).

Hence the seismic moment of an earthquake is a direct measure of the strength of an earthquake caused by fault slip. If an earthquake occurs with surface faulting, we may estimate its rupture length L and average slip Δu. The source area A may be approximated by Lh where h is the focal depth. A reasonable estimate for μ is 3×10^{11} dynes/cm^2. With these quantities we can estimate the seismic moment from eq. 2.9. For more discussion on seismic moment readers may refer to Brune (1976) and Aki and Richards (1980).

The moment magnitude scale is consistent with M_L: 3-6, M_s: 5-8. The moment magnitude M_w has the advantages that it does not saturate at the top of the scale, and it has a sound theoretical basis than M_L or M_s. However, for moderate magnitude shallow focus damaging earthquakes, it is sufficient for engineering purposes to take M_L, M_s and M_w to be roughly the same.

Duration Magnitude (M_d)

Analog paper or film recordings have limited dynamic range. These records are often clipped for strong or even medium magnitude local seismic events. This makes magnitude determination from A_{max} impossible. Therefore, alternative magnitude scale such as M_d was developed. This scale is based on signal duration. It is almost routinely used in microearthquake surveys. Details of this scale is given in Section 3.15.

Macroseismic Magnitude (M_{ms})

Macroseismic magnitudes (M_{ms}) are particularly important for analysis and statistical treatment of historical earthquakes, and were initially proposed by Kawasumi (1951). There are three main ways to compute M_{ms}:

(i) M_{ms} is derived from maximum reported intensity as:

$$M_{ms} = a\ Io + b \tag{2.10}$$

or when focal depth (h) is known

$$M_{ms} = c\ Io + \log h + d \tag{2.11}$$

(ii) M_{ms} is derived from the total area (A) of perceptibility as:

$$M_{ms} = e \log A_{Ii} + f \tag{2.12}$$

where A_{Ii} in km^2 shaken by intensities I_i with i ≥ III, and a, b, c and d are constants.

Examples of regionally best fitting relationships are published for California (Toppozada, 1975), Italy (Tinti et al., 1987), Australia (Greenhalgh et al., 1989). For Europe Karnik (1969) reported the best results using

$$M_{ms} = 0.5\ I_o + \log h + 0.35 \tag{2.13}$$

(iii) Another M_{ms} is related to the product P = Io × A (km^2), which is independent of the focal depth (Galanopolous, 1961):

$$M_{ms} = \log P + 0.2\ (\log P - 6) \tag{2.14}$$

Magnitude and Frequency of Earthquake Occurrence

As discussed in Chapter 1, the frequency-magnitude relation is a widely accepted concept in seismology. The frequency of earthquake occurrence as a function of magnitude is represented by Gutenberg and Richter (1941) relation:

$$\log_{10} N = a - bM \tag{2.15}$$

where N is the cumulative number of earthquakes of magnitude M or greater, and *a* and *b* are numerical constants. The constant *a* represents the seismic level of activity; it is not an independent quantity. It depends upon the largest magnitude in the given group of earthquakes and on the constant *b*. The constant *b* describes the rate of increase of earthquakes as magnitude decreases. Studies in many areas of the world have shown that the *b-values* are usually within the range of 0.5-1.5 (e.g. Pacheco et al., 1992; Wiemer et al., 1998). This range corresponds to an increase in number of earthquakes 4 to 16 times for each decrease of magnitude by one unit. For a given sample of earthquake magnitude data, if we plot $\log_{10}$ N versus M, the *b-value* can be estimated from the slope of the log-linear relation, the least-square fit line. The relation equally holds good for a aftershock sequence (Utsu, 1961 and 1969).

Readers interested in more detailed discussion on b-value estimation and its significance in earthquake statistics are referred to Utsu (1965), Aki (1965), Shi and Bolt (1982), Frohlich and Davis (1993) and Wiemer and Wyss (1997). A detailed study of b-value in northeast India region is given in Chapter 6.

2.4.2 Earthquake Intensity

Rossi-Forel Intensity Scale

The first intensity scale of modern times was developed by De Rossi of Italy and Forel of Switzerland in 1880s. This scale, which is still sometimes used in describing damage effect of an earthquake, has values I to X. The 1906 San Francisco earthquake was rated with the Rossi-Forel intensity scale. For description of this scale readers are referred to Richter (1958).

Modified Mercalli (MM) Intensity Scale (1956 version)

The Italian seismologist and volcanologist Mercalli made certain changes in the Rossi-Forel scale in 1902. Cancani and Sieberg made further changes to develop Mercalli-Cancani-Sieberg (MCS) scale in 1923, and the scale was expanded to 12 degrees i.e. I to XII. Wood and Neumann gave a new version of the MCS scale, which came in use in USA as Modified Mercalli (MM) Scale. Richter (1958) gave a rewritten version of the MM scale, which is referred to MM scale (1956 version). Like the Richter scale for estimating M_L, the Modified Mercalli (MM) scale is popularly used for estimating the earthquake shaking intensity. The 1956 version of this scale is given in Appendix 2.1.

Medvedev-Sponheuer-Karnik (MSK) Intensity Scale (1992 Version)

The MCS and MM scales were thoroughly revised and the MSK scale was approved at the UNESCO meeting on Seismology and Engineering in 1964 in Paris. Later it was, however, realised that introduction of the sophisticated MSK scale would be of less practical use. A working group, European Seismological Commission (ESC), was established in 1988 for logical version of the MSK scale. A modified version of the scale was finalised and adopted as MSK scale at the XXIII ESC General Assembly in 1992 in Prague.

The MSK and MM scales are almost equivalent, only difference is in the sophistication employed in the formulation. It may, however, be noted that although these scales have 12 degrees, in practice only 8 degree scales are used. Intensity I means not felt and intensity II is too weak to be reported; so, these two ratings are rarely used. At the other end of the scale, intensity XII is defined in a manner which cannot necessarily be reached in an earthquake. Again intensities X and XI are hard to differentiate in practice; so, intensity XI is rarely used. Thus the working range of these scales is usually from intensity III to intensity X.

In the macroseismic study of an earthquake, the following simple steps are followed:

(i) *Data acquisition*: This may be done by questionnaire survey, field visit, appeals for information, literature search or by other means.
(ii) *Data sorting*: The data may be organised into a form in which it can be interpreted. This may be done by arranging the questionnaire indicating the place of origin.
(iii) *Intensity assignment*: Data are interpreted using the intensity scale, and a table indicating places with intensities may be prepared.

Isoseismals

Isoseismals are the curved lines joining the localities of same intensity. Isoseismals often show elliptical elongation in the direction of major structural trends/damage. Generally the areas, the *isoseists,* between the isoseismals are marked with intensity numbers (say IV or V), and the curved lines are drawn to separate out the isoseists. An old practice was to assume epicentre, in absence of seismographs, in the centre of *meizoseismal* area i.e. in the maximum intensity zone. Since the inception of the instrumental seismology, such practice is, however, discontinued. It is rather observed that instrumentally located epicentre is mostly outside the meizoseismal area. Isoseismal maps for the great Indian earthquakes are exemplified in Sec. 2.7. The geologists look at an isoseismal map with a view point of nature and extent of faulting. The engineer's interest in such map is to judge the performance of various types of construction under such conditions. An isoseismal map becomes an index of weak or danger spots to be avoided for future construction or warrant safety measures.

Intensity and Acceleration

Richter (1958) has given an empirical relation between intensity and acceleration of an earthquke as follows:

$$\log a = \frac{I}{3} - \frac{1}{2} \tag{2.16}$$

where a is the acceleration in cm/sec^2 and I is the MM intensity. If one assumes $I = 1\frac{1}{2}$ which represents the limit of perceptibility between I and II, then $\log a = 0$, or $a = 1$ cm/sec^2. Such an acceleration may reach the level of shaking ordinarily perceptible to persons. Similarly, if one lets $I = 7\frac{1}{2}$, then log $a = 2$ or $a = 100$ cm/sec^2 which is equal to 0.1g approximately. This value is appreciable as it damages ordinary structures not designed to be resistant. One gets $a = 1$ g, for $I = 10\frac{1}{2}$, which is rather rare.

2.4.3 Earthquake Energy

It is tempting to correlate the *energy release* of an earthquake with its *magnitude* or *intensity*. Although the correspondence is very approximate, it is nevertheless very useful for estimating the amount of energy released by an earthquake.

Gutenberg and Richter's (1954) elaborate calculations produced the formula which relates energy release with magnitude as follows:

$$\log E = 12 + 1.8\ M \tag{2.17}$$

This relation is fair enough for the earthquakes of magnitude range $4 < M < 7$, but for the large earthquakes, energy given by this formula is too high. Gutenberg and Richter (1956) revised the formula after extensive study of strong motion seismograms, and preferred unified magnitude m derived from body waves recorded at teleseismic distances, and it took the form:

$$\log E = 5.8 + 2.4\ m \tag{2.18}$$

where $m = 2.5 + 0.63\ M_s$, thus it is equivalent to

$$\log E = 11.4 + 1.5\ M_s \tag{2.19}$$

Putting M = 8, eqs 2.17 and 2.19 give log E = 26.4 and 23.4 respectively. Thus the revised relation (eq. 2.19) greatly reduced the values of energy for the larger shocks ($M > 7$).

Shebalin (1955) derived formulas to relate earthquake energy, intensity and depth as follows:

$$0.9 \log E - I = 3.8 \log h - 3.3 \tag{2.20}$$

and

$$0.9 \log E - I = 3.1 \log h - 4.4 \tag{2.21}$$

where h is the hypocentral depth in kilometres and I is the maximum intensity (MM scale) at the surface. Equation 2.20 applies to hypocentres from the surface down to 70 km, and eq. 2.21 to depth of 80 km or more.

It is worth noting, mainly of journalistic interest, that an official figure for the energy release by a nominal atom bomb of the Hiroshima type is 8×10^{20} ergs, and a great earthquake ($M \geq 8$) might have an energy of 8×10^{26} ergs, which is comparable with million atom bombs. Since the energy of the annual *heat flow* from the interior of the Earth is roughly 8×10^{27} ergs, the two numbers are close enough to suggest various geophysical speculations. Revision for seismic energy release now gives a figure 9×10^{24} ergs per year, which is hardly more than a thousandth of the heat energy. Table 2.1

illustrates changes in ground motion and energy with magnitude change. It shows, for example, that a magnitude 7.0 earthquake produces 10 times more ground motion than a magnitude 6.0 earthquake, but it releases 32 times more energy. The energy release best indicates the destructive power of an earthquake.

Table 2.1: Magnitude versus ground motion and energy

Magnitude Change	*Ground Motion Change (Displacement)*	*Energy Change*
1.0	10.0 times	about 32 times
0.5	3.2 times	about 5.5 times
0.3	2.0 times	about 3 times
0.1	1.3 times	about 1.4 times

It can be shown that a magnitude 9.7 earthquake is 794 times bigger on a seismogram than a magnitude 6.8 earthquake. The magnitude scale is logarithmic, so

$$10^{9.7}/10^{6.8} = 10^{9.7-6.8} = 10^{2.9} = 794.328 \tag{2.22}$$

The magnitude scale is really comparing amplitudes of waves on a seismogram, not the energy of the earthquakes. So, a magnitude 9.7 is 794 times bigger than a M 6.8 earthquake as measured on seismograms, but the M 9.7 earthquake is about 23,500 times STRONGER than that of the M 6.8! This means that it would take about 23,500 earthquakes of magnitude 6.8 to equal the energy released by one magnitude 9.7 earthquake. Since it is really the energy or strength that knocks down buildings, this is really the more important comparison. This explains why big earthquakes are so much devastating than small ones. The amplitude ("size") differences are big enough, but the energy ("strength") differences are huge.

2.5 SOME USEFUL DEFINITIONS IN EARTHQUAKE SEISMOLOGY

As stated in the introduction of this chapter, earthquake seismology being a dual science some useful terms are defined here from the view point of geologist investigating the earthquake effect in the field and the geophysicist/ seismologist interpreting the instrumental records of their observatory/field survey in the laboratory.

Accelerometer is a seismograph that records acceleration as a function of time. The c.g.s. unit of acceleration is the Gal (= 1 centimetre per second per second), one thousandth of a Gal is a *milliGal.* Acceleration of earthquake motion is expressed in terms of g (g = acceleration due to gravity = 980 cm/ sec^2). A *strong-motion seismograph,* which does not record continuously, is triggered into motion by the first wave of an earthquake above threshold intensity. These instruments are designed to trigger during strong acceleration of the ground, and continues for some minutes or until the ground shaking falls again to the imperceptible levels. The acceleration of earthquake motion or g value is obtained from this record, which is a very useful parameter for earthquake engineers for assigning design factor of engineering construction.

Asperity is defined as a strong section of a fault system, that is locked during the decades of stress build up in the fault system, and it releases most of its energy during an eventful earthquake. The surrounding areas, *barriers,* are weakly coupled, possibly drag along during an earthquake. The effective stress normal to the fault surface in asperity is believed to be relatively higher than in the barrier. Alternatively, barriers may be regions of ductile deformation in which rupture dies out. This type of barrier is known as *relaxation barrier. Aseismic creep,* i.e. continuous slip on a fault, may also produce relaxation barrier surrounding an asperity, that limit the rupture dimension when the asperity fails. The best known *creeping fault* is the 100 km long segment of the San Andreas between San Juan and Parkfiled, California. Faults that creep require special frictional behaviour; either the fault has unusually low normal stresses or the fault is lined with very ductile or weak materials, the *fault gouge.*

Asthenosphere is the layer below the lithosphere (Fig. 2.2). The asthenosphere is marked by low seismic velocity and high seismic attenuation. It is a soft layer, probably partially molten. *Lithosphere* is the outer rigid shell, and consists of crust and the upper mantle.

Compression and dilatation, corresponding to the nature of Primary (P) seismic-waves, refer to the direction of the first arrival of P-wave on a seismogram. Conventionally, for vertical component seismometer, upward motion indicates a compression of the ground, and downward motion a dilatation. When the source of P-wave is a small explosion at a point in the Earth the first P-wave generated would push outward on a spherical surface, and the seismographs would detect this P-wave as upward movement, referred to as compression. If the P-wave arises from rupture on a fault, as in the case of an earthquake, the first motions will not be all compressions. Depending on the directions in which P-wave left the fault and depending on the positions of the seismographs, both up and down motions of the P-wave would be recorded.

Epicentre and hypocentre indicate the earthquake location. Epicentre is the point on the Earth's surface directly above the focus or hypocentre of an earthquake. Focus or hypocentre is the origin point at which the rupture initiates, and *focal depth* is the depth of the focus below the surface of the Earth (Fig. 2.8). *Origin time* is the instant at which an earthquake commences at the hypocentre, and *transit time* is the elapsed time between the origin time and the arrival of a given seismic wave at a seismograph station. Most researchers term this as *travel time* (more details in Chapter 3).

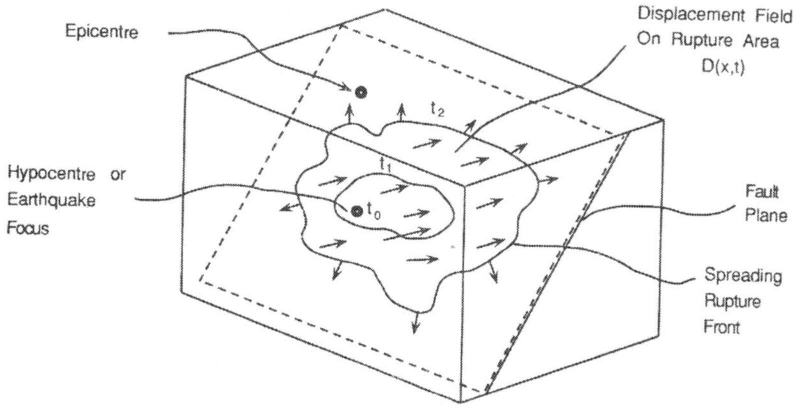

Fig. 2.8 Diagram showing epicentre, focus (hypocentre), rupture area, fault-plane etc. (after Bolt, 1999).

Gravity anomaly is the deviation of the observed acceleration of gravity (g) from an expected value calculated from the general gravitational field of the Earth. Gravity anomaly is expressed in terms of *Free-air anomaly, Bouguer anomaly* and *Isostatic anomaly*. The free-air anomaly takes into consideration of the free-air correction, which compensates for the fact that the attraction of gravity above sea level decreases with height because of the increased distance from the Earth's centre. The residual between the gravity so corrected and the value calculated from the International Formula (which considers latitude) is referred to as free-air anomaly. The *Bouguer anomaly* considers the attraction of the material between the elevation of the observing station and the sea-level, which is called Bouguer correction. The International Formula, however, makes no allowance for lateral variation of density of the Earth's crust. An *Isostatic correction* is made on the basis of surface elevation data. The form of the correction depends on the assumed isostatic model of Airy (1855) or Pratt (1859). Once the data are corrected for predicted isostatic effects along with the free-air and Bouguer corrections, the residual between the observation so corrected and the calculated value at the latitude of the observing station is referred to as the isostatic anomaly. If perfect isostatic compensation exists, the anomaly is zero. If the anomaly is positive over a land, it infers to a deficit of compensation. If it is negative, the area is considered to be over compensated.

Heat flow is the outward flow of Earth's internal heat through the crust. Determination of heat flow involves measurement of temperature gradient in the crust and measurement of thermal conductivities of the rocks in which the gradient is measured. Heat flow unit (HFU) is 10^{-2}cal/m^2 sec. The mean heat flow of the Earth is 1.2-1.5 HFU. Heat flow ranges from about 0.9 in the shield areas to over 2 HFU in the Cenozoic volcanic areas. In mid-ocean ridges the values reach upto 8 HFU. A sinking lithospheric plate, on the

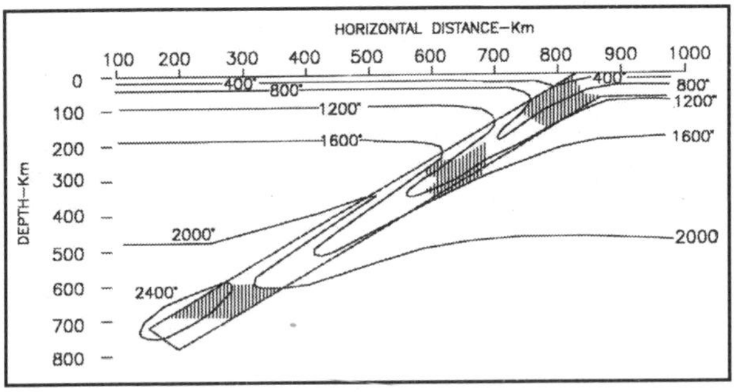

Fig. 2.9 Temperature distribution in a down going plate, shaded areas indicate zone of phase changes (after Toksöz et al., 1971).

other hand, is a very efficient heat sink; the mean heat flow in a sinking plate is of the order of – 10 HFU. Consequently, the continuous sinking of the plate should induce considerable cooling of the surrounding mantle. An example of temperature distribution in a down going plate is shown in Fig. 2.9.

Macroseismic effects of earthquakes are those which are observed on the large scale in the field without any instrumental aid. Earthquake damages, crustal deformations etc. are referred to as macroseismic effects.

Microseismic effects, on the other hand, are those that can be observed in small scale only with instruments. Quite illogically, the term *microseism* is used in a different way. These are not small earthquakes. More or less continuous disturbances or seismic noises in the ground recorded by seismographs are called *microseisms*. Most of these are associated with weather, wind action on topography and in trees, other natural movements such as breaking of surf on the beach etc.

Q factor or quality factor describes the attenuation of seismic waves. It is inverse of attenuation coefficient, and an important parameter in earthquake engineering. *Q* is a dimensionless quantity that expresses the wave amplitude decay, which occurs when a wave propagates through a medium, and is not attributable to geometrical spreading. It is a combination of intrinsic attenuation, the loss of elastic energy to heat or other forms of energy, scattering and deflection and/or mode conversion of seismic energy due to randomly distributed inhomogeneities in the transmitting medium. *Q* values are generally determined from strong-motion accelerograms and broadband seismograms. High seismic velocity and high *Q* values are observed along the sinking lithospheric plate. *Q* is about 500 for the crust, 150 for the upper mantle and 1000 along the sinking plate. A schematic distribution of *Q* in a subduction zone is shown in Fig. 2.10.

Seismograph is an earthquake recording instrument which measures, in details, the ground shaking. The first earthquake recorder was an artistic device invented by a Chinese scholar, Chang Heng, about 132 A.D. (Fig. 2.11). The instrument was a *seismoscope*, unlike seismograph it did not record the complete time history of the earthquake waves but did record the

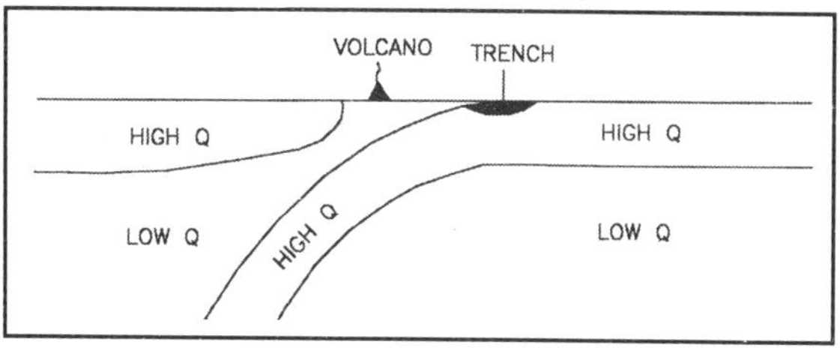

Fig. 2.10 Schematic diagram showing distribution of Q (quality factor) in a subduction zone.

Fig. 2.11 Left: photograph showing Chinese scholar Chang Heng with the seismocope he designed (about 132 A.D.). Right: Maori god of earthquake (carving and photo, courtesy: Terry Ball, New Zealand).

direction of principal impulse due to an earthquake. The first effective seismographs were constructed just before the beginning of the twentieth century.

Seismometer is a sensor part of a seismograph. A variety of suitable components (vertical and/or horizontal) high-gain seismometers are now commercially available to meet various network requirements. A brief theory and practice of the seismometers are discussed in Chapter 3. *Seismograms* are the detailed graphical records of the earthquake waves. Photographic, magnetic, pen-ink, micro-film and smoked-paper recordings are common till the development of digital instruments in the recent times.

2.6 EARTHQUAKE EFFECTS AND HAZARDS

2.6.1 Damage to Man-made Structures

Effects of earthquakes on artificial structures are of prime importance to the public, engineers and architects. Description of earthquake damages includes the degree of damage made on small weak mud-hut to modern structures. Masonry structures, however, have been classified into four categories to devise the intensity scale (see Appendix 2.1). Some damages are caused by the effect of shaking not directly on the masonry structure but on the foundation. Warping or fissuring can make a considerable damage even to exceptionally strong building. Relative motions of the two sides of an active fault may destroy a structure of any kind by bending or crushing. Other engineering structures like railways, bridges, dams etc. are also subject to severe damage by shaking, slumping or faulting. Proper earthquake-resistant design for large or tall structures is still vigorously debated in the engineering profession.

Causes of most earthquake-damages are due to the horizontal motion rather than the vertical part of the motion. Vertical motion operates against gravity, whereas horizontal motion meets no such resistance. Earthquake effects are often evaluated exclusively in terms of acceleration. With an amplitude of one inch and a frequency of one per second, we have an acceleration of 0.1g, which is considered as damaging. If such motion continues only few seconds, it will not damage ordinary structure, but if the motion continues for 15 or 20 seconds, as it may in a great earthquake, damage may be much greater. Long duration, reasonably high acceleration and considerable amplitudes are the combination which cause maximum damage in buildings and loss of lives. Some killing earthquakes of the world are tabulated in Table 2.2.

Table 2.2: Some killing earthquakes of the world: 1500-2005 (compiled from Utsu, 2002)

Year, Month & Date	*Region*	*Magnitude/Intensity*	*Loss of Lives*
1	2	3	4
1500, January 04	China (Yunnan)	7.0, X	10,000
1509, September 14	Turkey	7.4, XI	13,000
1531, January 26	Lisbon Portugal		30,000
1549, February 16	Qayin, Iran	6.7, IX	3,000
1556, January 23	Shanxi, China	8.3	830,000
1570, February 08	Chile	8.3, XI	2,000
1575, December 16	Chile	8.5, X	1,500
1605, February 03	Japan	7.9	1,000
1605, July 13	China	7.5, XI	1,000
1611, September 27	Japan	6.9, IX	3,700
1611, December 02	Japan	8.1	5,000
1622, October 25	China	7.0, X	12,000
1626, June 28	China	7.0, IX	5,200
1627, July 30	Italy	6.8, X^{+}	5,000
1639, March 27	Italy	6.9, XI	30,000

(Contd.)

Table 2.2 (*Contd.*)

1641, February 05	Iran	6.8, IX	12,600
1647, May 14	Chile	8.5, XI	2,000
1648, April 01	Turkey	6.7, X	2,000
1652, July 13	Yunnan, China	7.0, IX+	3,000
1653, February 23	Izmir, Turkey	X	15,000
1654, July 21	Gansu, China	8.0, XI	31,000
1667, April 06	Croatia	7.2, X	5,000
1667, November 18	Iran	6.9, X	12,000
1668, January 04	Shemakha	7.0	80,000
1668, July 25	Shandong, China	8.5, XI	47,600
1668, August 17	Turkey	8.0	8,000
1673, July 30	Iran	7.1	5,600
1679, September 02	Hebei, China	8.0, XI	45,500
1687, October 20	Lima, Peru	8.2, X	5,000
1688, June 05	Italy	6.6, XI	10,500
1688, July 10	Izmir, Turkey	7.0, X	17,500
1693, January 11	Sicily, Italy	7.4, XI	54,000
1694, September 08	Basilicala, Italy	6.8, X	4,800
1695, May 18	Shaxi, China	7.8, X	52,600
1703, January 14	Norcia, Italy	6.7, XI	9,800
1707, October 28	Japan	8.6	5,000
1709, October 14	China	7.5, X	2,000
1718, June 19	Gansu, China	7.5, X	75,000
1721, April 26	Tabriz, Iran	7.4, XI	40,000
1727, November 18	Tabriz, Iran	7.2, X	77,000
1733, August 02	Yunnan, China	7.8, X	1,200
1739, January 03	Ningxia, China	8.0	50,000
1746, October 29	Lima, Peru	8.4, X	18,000
1752, July 21	Syria	7.0	20,000
1755, November 01	Lisbon, Portugal	8.5	62,000
1759, November 25	Syria	7.4	30,000
1771, April 24	Japan	7.4	12,000
1773, June 03	Guatemala		20,000
1778, December 15	Iran	6.2, IX	8,000
1780, January 08	Tabriz, Iran	7.4, XI	50,000
1783, February 04	Calabria, Italy	6.9, XI	35,000
1784, July 18	Turkey	7.6	5,000
1797, February 04	Quito, Ecuador	8.3	40,000
1805, July 26	Italy	6.6, X+	5,570
1810, February 16	Greece	7.8, X	2,000
1812, March 26	Venezuela	6.3, IX	20,000
1815, October 23	Shanxi, China	6.8, IX	13,000
1816, December 08	Sichuan, China	7.5, X	3,000
1819, June 16	Kutch, India	8.3	1,543
1822, August 13	Syria	7.4	20,000
1822, September 05	Syria		22,000

(*Contd.*)

Table 2.2 (*Contd.*)

1	2	3	4
1828, June 06	Srinagar, India	X	1,000
1828, December 18	Japan	6.9, IX	1,700
1833, September 06	Yunnan, China	8.0	6,700
1837, January 01	Israel	7.0	5,700
1943, February 08	Guadaloupe	7.8, IX	5,000
1847, May 08	Japan	7.4	8,170
1948, December 03	Taiwan	6.8, IX	2,000
1850, September 12	Sichuan, China	7.5, X	24,000
1853, May 05	Iran	6.2, X	9,000
1854, December 23	Japan	8.4	2,000
1955, February 28	Turkey	7.3, X	1,900
1955, November 11	Japan	6.9, X	7,500
1857, December 16	Italy	7.0, XI	10,900
1861, March 20	Argentina	7.0	18,000
1868, August 13	Peru and Bolivia	8.5, XI	25,000
1868, August 16	Ecuador	7.7	40,000
1870, April 11	Sichuan, China	7.3, X	2,300
1871, Dec. 23	Iran	7.2, IX	2,000
1872, April 03	Turkey	7.2	1,800
1875, May 18	Colombia	7.3	16,000
1879, July 01	Gansu, China	8.0, XI	29,500
1881, April 03	Greece	6.5, IX	8,000
1885, May 30	Srinagar, India	IX	3,000
1887, December 16	Yunnan, China	7.0, IX+	2,250
1891, October 28	Japan	8.0	7,300
1893, March 31	Turkey	7.0	1,500
1893, November 17	Iran	7.1, X	10,000
1896, June 15	Riku-Ugo, Japan	8.2	22,000
1897, June 12	Shillong, India	8.7, XII	1,500
1899, September 29	Indonesia	7.4	3,900
1902, April 19	Guatemala	7.5	2,000
1902, August 22	Xinjiang, China	8.3	5,600
1903, April 28	Turkey	7.0, IX	3,600
1905, April 04	Kangra, Himalaya	8.6, IX	19,000
1906, January 31	Colombia	8.6	1,000
1906, April 18	San Francisco, California	8.3	700
1906, August 17	Chile	8.4	3,800
1907, October 21	Uzbekistan	7.4	15,000
1908, December 28	Messina, Italy	7.5, XI	1,20,000
1909, January 23	Iran	7.3	5,500
1912, August 09	Turkey	7.4, X	2,800
1913, December 21	Yunnan, China	7.0	1,300
1914, October 03	Turkey	7.0, X	4,000
1915, January 13	Avazzano, Italy	7.0, XI	32,600
1917, July 31	Yunnan, China	6.8, IX	1,900

(*Contd.*)

Table 2.2 (*Contd.*)

1920, December 16	Haiyuan, China	8.5	2,35,500
1922, November 11	Chile	8.3	1,000
1923, September 01	Kanto, Japan	8.2	1,43,000
1925, March 16	Yunnan, China	7.0	5,800
1927, March 7	Kyoto, Japan	7.3	3,000
1927, May 23	Gansu, China	8.0	41,400
1929, May 01	Iran	7.2	3,300
1930, May 06	Iran	7.3, X	2,500
1930, July 03	Italy	6.7, X	1,400
1931, February 02	Hawke's Bay, New Zealand	7.9	256
1931, August 11	Xinjiang, China	8.0	10,000
1932, December 26	Kansu, China	7.6	70,000
1933, March 03	Japan	8.1	3,100
1933, August 25	Sichuan, China	7.5, X	7,000
1934, January 15	Bihar, India	8.4	10,700
1935, April 21	Taiwan	7.1	3,300
1935, May 31	Quetta, Pakistan	7.5	60,000
1937, August 01	Shandong, China	7.0, IX	3,900
1939, January 25	Chile	7.8, X	30,000
1939, December 26	Erzincan, Turkey	8.0, XII	32,700
1942, December 30	Turkey	7.3	3,000
1943, July 23	Indonesia	8.1	213
1943, November 26	Turkey	7.6	4,000
1944, January 15	Argentina	7.4	8,000
1944, February 01	Turkey	7.6	4,000
1944, December 7	Japan	7.9	1,300
1946, November 10	Peru	7.3	1,400
1946, December 21	Japan	8.0	1,300
1948, June 28	Fukui, Japan	7.1	3,800
1948, October 05	Turkmenistan	7.3	19,800
1949, July 10	Tajikistan	7.4	12,000
1949, August 05	Ecuador	6.8, XI	6,000
1950, August 15	Assam, India	8.7	1,500
1953, March 18	Turkey	7.4	1,100
1955, March 31	Philippines	7.6	465
1957, December 13	Iran	7.2	2,000
1960, February 29	Agadir, Morocco	5.9, X	13,000
1960, May 22	Chile	8.5	5,700
1962, September 01	Iran	7.2	12,200
1963, July 26	Skopje Yugosalvia	6.1, X	1,200
1964, March 27	Alaska	8.6, X	130
1966, March 22	Hebei, China	7.2	8,000
1966, August 19	Turkey	6.8, IX	2,500
1967, December 10	Kyona, India	6.7, IX	180
1968, August 31	Iran	7.3	15,000
1970, January 05	Yunnan, China	7.8	15,600
1970, March 28	Turkey	7.1, IX	1,100

(*Contd.*)

Table 2.2 (*Contd.*)

1	2	3	4
1970, May 31	Peru	7.8	66,800
1972, April 10	Iran	6.8, IX	5,000
1972, December 23	Nicaragua	6.2, IX	6,000
1973, February 06	Sichuan, China	7.6, X	2,200
1974, May 11	Yunnan, China	7.1	1,500
1974, December 28	Pakistan	6.2, IX	5,300
1975, February 04	Haiching, China	7.3, IX^{+}	1,300
1975, September 06	Turkey	6.7, IX	2,400
1976, February 04	Guatemala	7.5, IX	22,900
1976, June 25	Indonesia	7.1	6,000
1976, July 28	Tangshan, China	7.8, XI	2,43,000
1976, August 16	Philippines	7.9, X	8,000
1976, October 29	Indonesia	7.2	6,000
1976, November 24	Turkey	7.3	4,000
1978, September 16	Iran	7.4	18,200
1980, October 10	Algeria	7.3	3,500
1980, November 23	Italy	6.7, IX	2,500
1981, June 11	Iran	6.7, IX	3,000
1981, July 28	Iran	7.1	1,500
1982, December 13	Yemen	6.0, IX	2,800
1983, October 30	Turkey	6.9, IX	1,400
1985, September 19	Mexico	8.1	9,500
1986, October 10	San Salvador	7.0	1,400
1987, March 06	Colombia	6.9, IX	5,000
1988, August 20	Bihar-Nepal	6.6, X	1,400
1988, December 07	Armenia	7.0	25,000
1989, October 17	Loma Prieta, California	7.1	100
1990, June 20	Iran	7.7	35,000
1990, July 16	Philippines	7.8	2,500
1991, October 20	Uttarkashi, India	6.6, IX	1,500
1992, December 12	Indonesia	7.5	1,700
1993, September 30	Killari, India	6.4, IX	8,000
1995, January 17	Kobe-Osaka, Japan	7.2, IX	6,400
1995, May 27	Sakhalin Island, Russia	7.5	2,000
1997, February 28	Iran	6.1	1,100
1997, May 10	Iran	7.3	1,600
1998, February 04	Afganistan	6.1	2,300
1998, May 30	Afghanistan	6.9	4,000
1998, July 17	Papua New Guinea	7.1	2,700
1999, March 28	Chamoli, India	6.8, VIII	105
1999, August 17	Turkey	7.8	17,200
1999, September 20	Taiwan	7.7	2,400
2001, January 26	Bhuj, India	7.7, X	20,000
2003, December 26	Bam, Iran	6.5, IX	50,000
2004, December 26	Sumatra	9.3 X	3,00,000
2005, October 8	Kashmir	7.8 X	80,000

2.6.2 Effects on Ground

The primary effects on ground by an earthquake are regional *warping, faulting, development of scarps, offsets, fissures, cracks*, changes in *elevation* or *depression*, changes in *coast line* etc. The geographical extent of these macroseismic effects can be vast, may be upto 1000 km for a great earthquake. Among the larger secondary effects, *landslides* in the form of *earth slumps, earth flows, earth avalanches* and *earth lurches* are most notable.

Liquefaction of ground is often seen in alluvial deposits. Solid sand formations, if saturated with water, are transformed into a state of suspension during the vibration of seismic waves, and behave as a viscous liquid. Such consequences can cause total collapse of structures, buildings etc. All these ground effects can be preserved in the geologic record, and can be recognized hundreds and sometimes thousands of years later as *paleoseismic* indicators, called *seismites*.

2.6.3 Effects on Water (Ground and Surface)

Oscillation in wells is produced during the passage of large long period surface waves of a teleseism. The general accepted theory of oscillation is that when a passing seismic wave compresses an *aquifer* it squeezes additional water into the well, raising its level, while the opposite half of the wave dilates the aquifer and draws out water from the well. Where plenty of groundwater is present, *earthquake fountains, spouts, geysers* are produced during the strong shaking of an earthquake. Ejection of sands are caused by the spouts.

Seiches are produced by the earthquake waves. This is a standing wave set up on the surface of an enclosed body of water such as pond, lake etc. The word *seiche* comes from Switzerland, used by Forel, which means bound water is made to slosh from side to side. If the motion is maintained it is a standing wave with an antinode at one side and a node in the middle; this is called a *uninodal seiche*.

Tsunamis, the long waves on the ocean, are produced by large earthquakes. The height of tsunamis, sometimes, increases greatly and they crash down upon the shore with disastrous effects. A comparison of historical earthquakes and tsunamis shows that a large tsunami washing onto a stretch of populated coastline is likely to be much more destructive than the shaking from all but exceptionally large earthquakes. Most tsunamis are caused by fault rupture along a submerged fault. The killing tsunami of the December 26, 2004 Sumatra earthquake is the most recent example, that was generated due to the mega thrust event (M 9.3) in the Indian ocean subduction zone. It caused a ~1300 km long rupture from Sumatra to north Andaman along the submerged fault and killed about 3,00,000 people in the coastal areas of Indonesia, Sri Lanka, Thailand, Andaman and Nicobar Islands, Tamil Nadu, Andhra Pradesh,

Pondicherry, Kerala, etc. There are other causes like submarine, landslide, avalanche of soil and rock on mountain sides into a bay, volcanic eruptions etc. that generate tsunami.

2.6.4 MISCELLANEOUS EFFECTS

Earthquake sound is directly produced by the transfer of elastic wave energy from the ground to the air (Thomson, 1929). Small amplitude of vibration in the air is heard as a large sound. It is certain that the sound is produced by the P-waves, while shaking begins to be felt by the later arrival of S-waves. More reports of earthquake sound are obtained from outdoors rather than indoors where miscellaneous noises obscure the earthquake sound.

Visible waves moving over the ground during an earthquake are reported, particularly from the meizoseismal areas of great earthquakes. Richter (1958) has quoted some of the observations reported by Oldham (1899) and Dutton (1888). It is suggested that elastic waves, emerging into air, may locally change the refractivity index of the air sufficiently to detect a light-ray reaching an observer at some distance; rapid changes in ray path would give the ground an appearance of motion.

Earthquake lights, a still more elusive earthquake effect, are reported, which are usually compared to summer heat-lightning. Some authors have suggested that relative displacement of large crustal blocks may perhaps manifest itself to some extent in the form of static electricity, which causes earthquake lights.

Stopping of pendulum clocks is an earthquake effect, which may be even used to fix the time of occurrence of an earthquake. Thus it acts as an excellent *seismoscope.* The pendulum clocks are stopped mainly because of the motion at right angles to the face causing friction between the pendulum and the escapement. An excellent example is obtained in the Bhuj earthquake of January 26, 2001 (see Chapter 7).

Fishes are killed in many numbers by tsunamis or seaquakes created by earthquake in the sea. There are many reports of unusual *animal behaviour* before and during earthquakes, e.g. horses snort, cattle stampede, dogs bark and whine, and cats spit.

Human beings can feel an earthquake when the acceleration approaches 1 Gal (1 cm/sec^2). Fright and panic are the most common effects.

2.7 GREAT EARTHQUAKES OF INDIA

The Indian subcontinent can be broadly divided into three sub-regions, viz., the Himalaya, the Gangetic Plain and the Peninsular Shield (Fig. 2.12). The Himalayan Arc is covexed southward and fronting the alluviated depression of the Gangetic Plain. In front of the Himalaya are the foothills, the Siwaliks

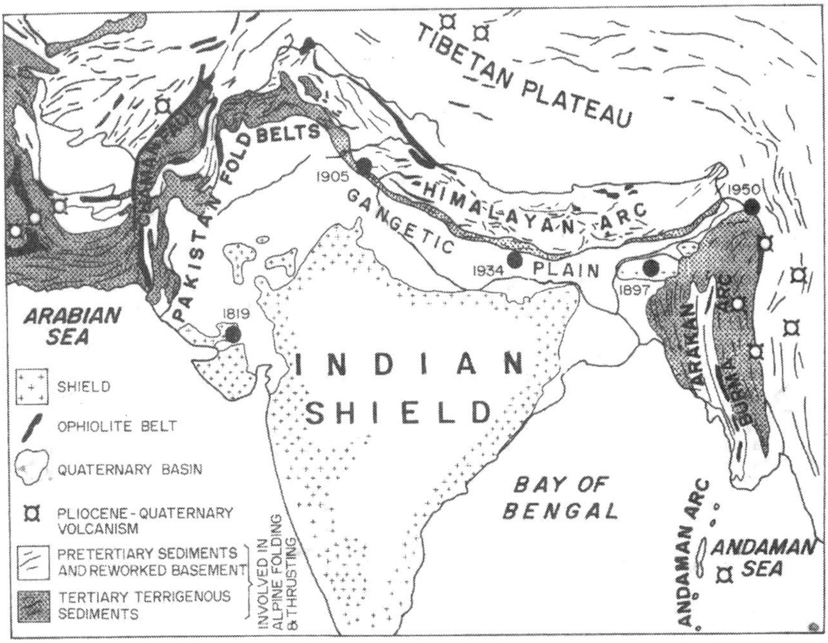

Fig. 2.12 General geology of the Indian subcontinent (Valdiya, 1976), and its great earthquakes. Epicentres of the great earthquakes M ~8.0 and above are shown by solid circles.

and the Tertiary metasediments. The Gangetic Plain separates the Himalaya from the Peninsula. Archaean rocks are over more than half of the Peninsula, and a large part of the remainder is covered by basaltic flows of Deccan Traps.

Five great earthquakes (M > 8.0) have occurred in India during the last 100+ years since 1897; three in the Himalayan Arc (1905 Kangra, 1934 Bihar and 1950 Assam), one in the northwestern margin of peninsular shield (1819 Kutch) and one in the Shillong Plateau shield (1897) in northeast India. The epicentres of these great earthquakes are shown in Fig. 2.12, and brief descriptions of these earthquakes are given below.

2.7.1 The 1819 Kutch Earthquake

The Kutch (also spelled Kachchh and Cutch) earthquake that occurred on June 16, 1819 in the northwestern corner of Peninsular India is considered to be the largest event of Stable Continental Region (SCR), and the first for which crustal deformation was quantified (Wynne, 1872). The maximum intensity was reported to be XI on MM scale (Oldham, 1928). This earthquake occurred before the invention of the seismographs. Based on the reported intensity, Gutenberg and Richter (1954) assigned M_L=8.4, while Johnston and Kanter (1990) reviewed the stable continental earthquakes and assigned

M_W=7.8 for this event. A remarkable feature of this earthquake was the creation of an 80-90 km long and 2-3 m elevated tract of land, known as 'Allah Bund', dam of God. The fault scarp appeared in the Rann of Kutch, close to the international border between India and Pakistan. ('Rann' means uninhabited salt flats that are neither sea nor land and are flooded periodically). The epicentre of the earthquake was given at 23.6°N and 69.6°E (Chandra, 1977). The loss of life was over 1500. The earthquake was not felt all over the country as those of 1897 or 1934; it was, of course, violent at the Kutch area. Due to lack of instrumental data its exact source parameters are not known. There are some damage reports of this earthquake (e.g. Grant, 1819); but no source mechanism was reported. Rajendran et al. (1998), however, suggested a fault dislocation model for the 1819 event and inferred fault length 95 km, width 16 km, strike 290° and dip 30°, and M_W = 7.6. They also suggested that the epicentre should lie to the north of Allah Bund. Based on geodetic measurements Bilham et al. (1998), on the other hand, inferred fault dip 67 ± 5°, reverse slip = 11.5 ± 1 m and down dip width 10-20 km. Ground excavations in the area revealed large multiple liquefaction features with cross-cutting relations at different stratigraphic levels, and recurrence of earthquakes ($M \geq 7.5$) on multiple segments in the region has been suggested (Rajendran and Rajendran, 2002 and 2003).

2.7.2 The 1897 Shillong Earthquake

Among the great earthquakes of the world, the June 12, 1897 earthquake of the Shillong Plateau holds a very prominent place in seismology. Dr. R.D. Oldham, the then head of the Geological Survey of India, directed and personally carried out most of the investigations of this great earthquake. Loss of life was only 1,542 compared to the magnitude of the earthquake. The loss of life was fortunately less because the earthquake occurred at 5.15 p.m. local time, when most of the people were outdoor. The epicentre was reported to be at 25.90°N and 91.80°E (Milne, 1912), and the maximum intensity reached XII on MM scale as rated by Richter (1958) from Oldham's report (Oldham, 1899).

Damage to property had been very great. Within an area of 30,000 square miles, all brick and stone buildings were practically destroyed. The ground rocked so violently that it was impossible to stand. Rumbling sound, visible waves and seiches were observed. Fissures were abundant over the whole area. Secondary effects like ejection of water and sand, rotation of pillars, rising of river height, crushing of soils into which houses sank were reported in the epicentral area.

It was the first strong earthquake in the country, which was instrumentally recorded outside the country. The seismographs which recorded the 1897 earthquake, however, were not of modern type. So, it was difficult to use the records to determine its magnitude. In a detailed study of the large global earthquakes, Gutenberg and Richter (1954) assigned a magnitude M_L = 8.7 for this earthquake. A comprehensive damage report of this earthquake is given by Oldham (1899). Based on damage survey and other evidences like upthrown boulder etc., Oldham (1899) reported that acceleration resulting from the 1897 great earthquake exceeded that of gravity, which is widely accepted by the modern seismologists. There was evidence of a surface fault, called Chedrang fault, in the epicentre area, which extended over 18-20 km with throws upto 35 feet in crystalline rocks (see Chapter 6). Largest dimension of the meizoseismal area was reported to be 160 miles (~230 km) with highest intensity XII in the MM scale (Fig. 2.13). Based on Oldham's detailed report, Molnar (1987) suggested that the length of the rupture was 200 ± 40 km. The southern edge of the rupture was more likely at the southern edge of the Plateau than south of it. The possibility that the rupture extended beneath the Himalaya was not ruled out.

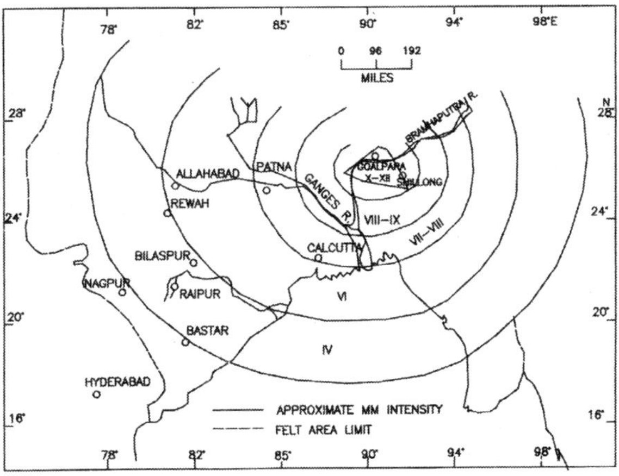

Fig. 2.13 Isoseismal map of the great Shillong earthquake 1897; maximum intensity XII (MM scale), (modified from Richter, 1958).

2.7.3 The 1905 Kangra Earthquake

This is a well documented earthquake which occurred in the Himachal Pradesh on April 4, 1905; the epicentre was reported to be at 32°N and 76°E. The initial magnitude was estimated to be M_L = 8.6 (Gutenberg and Richter, 1954), and the revised magnitude M_s = 7.8 ± 0.05 (Ambraseys and Douglas, 2004). The loss of life was maximum, about 19,000. The maximum intensity was reported to be X on MM scale (Fig. 2.14). The isoseismals were oriented along the Main Boundary Thrust in the Himalaya (see Chapter 5). Two zones of high intensity, with an intervening region of low intensity, were identified; one in Kangra area and the other in Dehradun (Fig. 2.14). The Kangra area had larger and higher intensity. A detailed study of the earthquake

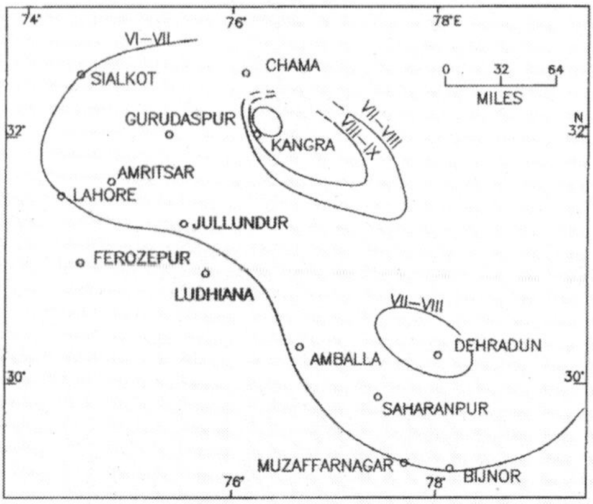

Fig. 2.14 Isoseismal map of the great Kangra earthquake 1905; maximum intensity X (MM scale), (Middlemiss, 1910).

was done by Middlemiss (1910). He estimated the depth of focus at 21-40 km. Post earthquake elevation changes were reported by Chander (1988).

2.7.4 The 1934 Bihar Earthquake

The January 12, 1934 Bihar earthquake is well documented (e.g. Dunn et al., 1939). The seismograms of the National Seismograph Stations (India Meteorological Department) and Global Stations made it possible to locate the epicentre at 26.5°N and 86.50°E, and to assign the magnitude M_L 8.4, and focal depth 20-30 km (Richter, 1958). The extent of the meizoseismal area was about 120 km long and 35 km wide. Two meizoseismal spots, separated by almost 150 km, one at Munger, east of Patna (Bihar) to the south and the other at Kathmandu, Nepal to the north, were identified (Fig. 2.15). From the distribution of damages reported by Rana (1935) for Nepal and by Dunn et al. (1939) for India, Pandey and Molnar (1988) showed that the damage was greatest in the Lesser Himalaya of eastern Nepal. The maximum intensity was reported to be X on MM scale. The loss of life in India is given about 7,250 and in Nepal 3,400.

Seeber and Armbruster (1981) inferred that slip occurred on a nearly flat plane lying beneath the Indo-Gangetic Plains. Pandey and Molnar (1988) assumed that the epicentre lay beneath the Himalaya than south of the ranges as commonly believed (Dunn et al., 1939; Richter, 1958; Seeber and Armbruster, 1981; Singh and Gupta, 1980). Chen and Molnar (1977) relocated the epicentre which lies within the area of high intensity zone (Fig. 2.15).

2.7.5 The 1950 Assam Earthquake

This is one of the few great earthquakes to which instrumentally a well determined magnitude M_S = 8.7 is assigned. On August 15, 1950 at 19.00 hr. 39 min. (local time) the earthquake occurred off the northeastern boundary

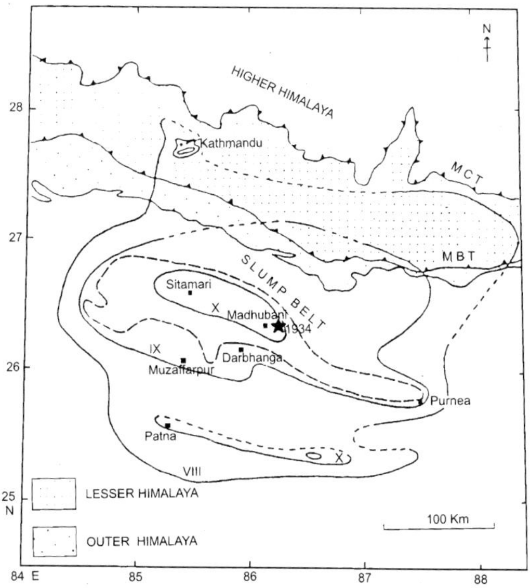

Fig. 2.15 Isoseismal map of the great Bihar earthquake 1934; maximum intensity X (MM scale), (GSI, 1939).

of Assam. The epicentre was reported at 28.5°N and 96.70°E, and maximum intensity to XII on MM scale (Tandon, 1954), (Fig. 2.16). This earthquake caused about 1520 casualties, and was more damaging in terms of property losses than the great earthquake of the 1897. The USGS determined the epicentre at 28.5°N and 97.0°E, and focal depth at 20 km. There was no strong-motion seismograph station in the affected area to calculate the acceleration. An acceleration of the order of 0.5g was, however, estimated from the damage survey in the epicentral region. Foreshocks and aftershocks were reported by many researchers (e.g. Tandon, 1955; Chen and Molnar, 1977; Molnar and Pandey, 1989). Fissures and sand vents occurred in many localities in the alluvial plain. Railway lines and road suffered a considerable

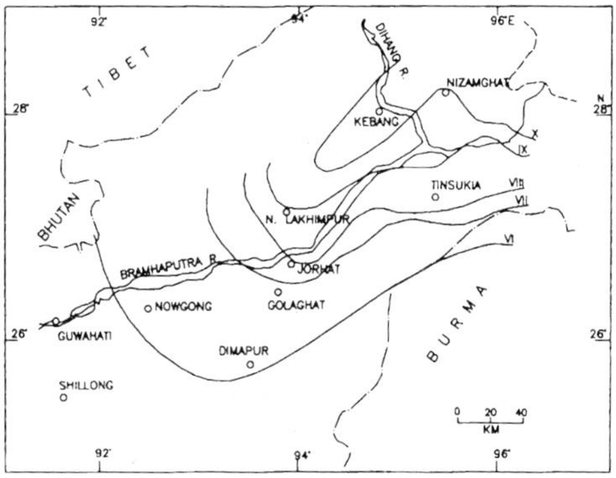

Fig. 2.16 Isoseismal map of the great Assam earthquake 1950; maximum intensity XII (MM scale), (Poddar, 1950).

damage. Landslides were observed at various places. A tragic death was reported in North Lakhimpur where a nine-year old girl was caught in a fissure which had immediately closed up burying her alive.

2.8 LARGE AND DAMAGING EARTHQUAKES OF INDIA

Figure 2.17 shows the epicentres of the large earthquakes of magnitude ≥ 7.0 and or intensity VIII and above, and also the great earthquakes of M ≥ 8.0. Details are given in Table 2.3. One can discuss an arcuate belt of large earthquakes following the Himalayan mountain belt and the Indo-Burma ranges. Maximum number of large earthquakes are, however, in the Indo-Burma ranges and many of these are deep focus earthquakes; so, loss in terms of casualties were less for these earthquakes. Peninsular India presents a quiet picture except the 1819 great earthquake and some recent damaging earthquakes (6.0 ≥ M < 8.0), which have caused considerable damages and loss of lives. Epicentres of these events are also shown in Fig. 2.17. There were a number of moderate magnitute earthquakes M ≥ 5.0 but less than 6.0, which caused considerable damages and loss of lives. These earthquakes are discussed in the following chapters.

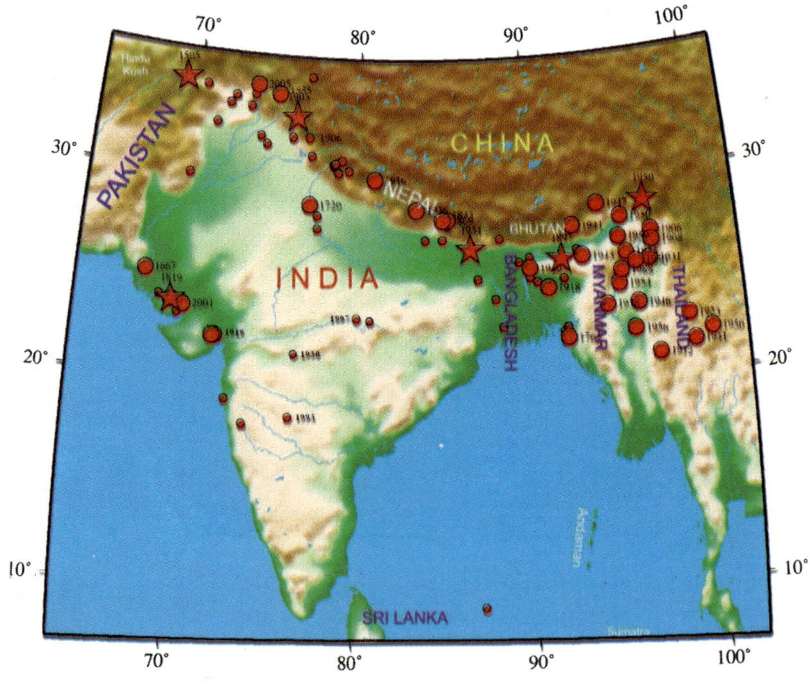

Fig. 2.17 Map showing topography of the Indian subcontinent and epicentres of the great (stars), large M ≥ 7.0 (larger circles) and earthquakes M>6.0 and/or intensity VIII or greater (smaller circles) in India since 1505 (Table 2.3).

Table 2.3 Large Earthquakes (Mag > 7.0, Int > VIII) in India and its neighbourhood (1505-2005) (25°-36°N, 69°-83°E; 20°-30°N, 83°-100°E)

Date	*HMS*	*Lat (North)*	*Long (East)*	*Depth (km)*	*Int*	*Mag*	*Location*	*Ref*
1	2	3	4	5	6	7	8	9
1505 07 06		34.50	69.20		XII		Near Kabul	OLD
1618 05 26		18.9	72.90		IX ?		Bombay	OLD
1667 05		25.0	68.0		X		Delta of Indus	OLD
1669 06 04		33.40	73.30		XI		Mandra, Upper India	OLD
1669 06 22		35.00	77.00		IX		Kashmir	OLD
1669 06 23		33.90	72.30		IX		Attock	OLD
1720 07 15	12	28.70	77.20		X		Delhi	OLD
1762 04 02	17	22.00	92.00		XI		Bengal, Arakan and Burma	OLD
1764 06 04		24.00	88.00		VII		Banks of Ganga	OLD
1803 05 22	15	30.60	78.60				Upper Ganga	OLD
1803 09 01		30.30	78.80		IX		Upper Ganga, Sirmoor and Garhwal	OLD
1803 09 01	00 30	27.50	77.70		IX		Mathura	OLD
1808 04 13	02	22.60	88.40				Calcutta, Chandernagore	
1809		30.70	78.50		VIII		Garhwal	OLD
1816 05 26	22	30.90	79.00		VIII		Gangotri, U. Valley of Ganga	OLD
1819 06 16		23.60	69.60		XI		Kutch	OLD
1827 09		31.60	74.40		IX		Lahore	OLD
1831		33.50	72.00		VIII		Peshawar and Valley of Indus	OLD
1832 02 21		32.00	76.00		VIII		Lahore, N.W. India	OLD
1833 08 26	05 30	28.00	85.00		X		Nepal, North and East India	OLD
1833 10 04		27.00	85.00		IX		Kathmandu, Monghyr, Allahabad	OLD
1833 10 18	04 40	27.00	84.00		VIII		Kathmandu, Gorakhpur, Allahabad	OLD
1834 07 08		25.80	89.40		VIII		Rangpur	OLD
1834 07 21		25.80	89.40		VIII		Rangpur	OLD
1839 03 23		21.90			XI		Amarapoora, Burma	OLD
1840 01 26		34.50	69.20		VIII		Kabul	OLD
1842 02 19		34.30	70.50		IX		Kabul, Peshawar, N.W. India	OLD
1842 11 11	21 38	25.00	90.00		IX		Calcutta, Darjeeling, Gauhati	OLD
1845 04 19		23.80	68.90		VIII		Lakhpat	OLD
1846 10 17		24.80	90.40		VIII		Mymensingh	OLD

(*Contd.*)

Table 2.3 (*Contd.*)

1	2	3	4	5	6	7	8	9
1846 12 10		26.00	93.00		VIII		Assam	OLD
1847		26.30	92.70		VIII		Assam, Nowgong	OLD
1849 02 27	21	27.00	88.30		VIII		Darjeeling	OLD
1851 01 08		22.30	91.80		VIII		Chirragong	OLD
1851 01 21		32.00	74.00		VIII		Lahore and all Punjab	OLD
1852 01 24	03 45	34.00	73.50		VIII		Upper Sind, Murree Hills	OLD
1852 05		27.00	88.30		IX		Darjeeling	OLD
1858 08 11		31.10	77.20		VIII		Simla	OLD
1865 12 19	19 30	22.50	92.00		IX		Chittagong	OLD
1866 05 23		25.00	87.00		VIII		Bengal	OLD
1866 05 23		27.70	85.30		VIII		Kathmandu	OLD
1868 06 30		24.90	91.90		VIII		Sylhet	OLD
1868 11 10		32.50	71.30		VIII		Marwat, Bannu	OLD
1869 01 10		26.00	90.00		VIII		Cachar	OLD
1869 01 10		28.00	85.00		X		Nepal	OLD
1871 11		34.50	69.20		VIII		Kabul	MIL
1874 10 18		34.50	69.20		IX		Kabul	MIL
1875 12 12		31.60	74.40		VIII		Lahore, Peshawar	MIL
1882 01		08.60	87.20		VIII		Srilanka	MIL
1885 05 30		34.10	75.00		X		Sopor, Srinagar	MIL
1885 06 06		34.00	75.00		IX		Dubgaon, Jamalpur, Ovan, Kashmir	MIL
1897 06 12		25.90	91.80		XII	8.7	Shillong, Gauhati	MIL
1899 09 25		27.00	88.30		VIII		Darjeeling	MIL
1905 04 04	00 55	33.00	76.00	25	X	8.6	Kangra Valley	G-R
1906 02 26		32.00	77.00		VIII		Kangra, Simla	SRS
1906 08 31	14 57 30.00	27.00	97.00	100		7.0	N. Burma	G-E
1908 12 12		26.50	97.00			7.5	N. Burma	TAN
1912 05 23	02 24 06.00	21.00	97.00	25		7.9	Burma	G-R
1916 08 28		30.00	81.00			7.5	N.W. Nepal	TAN
1918 07 08	10 22 07.00	24.50	91.00			7.6	Srimangal, Assam	G-R
1919 04 21		22.00	72.00		VIII		Gujarat	TAN
1923 06 22	06 44 43.00	22.75	98.75			7.3	Shan Plateau	G-R
1930 07 02	21 03 42.00	25.50	90.00			7.1	Dhubri, Assam	G-R
1931 01 27	20 09	25.60	96.80			7.6	Cochin	G-R
1932 08 14	04 39 32.00	26.00	95.50	120		7.0	East of Nagaland	G-R
1934 01 12	08 43 18.00	26.50	86.60	25		8.4	Bihar	G-R
1936 05 27	06 19 19.00	28.50	83.50			7.0	Dhaulagiri	G-R

(*Contd.*)

Table 2.3 (*Contd.*)

1938 08 16	04 27 50.00	23.50	94.25	100		7.2	Sagaing	G-R
1941 01 21		27.50	92.50			7.1		
1941 12 26	14.48 04.00	21.50	99.00			7.0	Shan Plateau	G-R
1943 10 23	17 23 16.00	26.00	93.00			7.2	N.E. Assam	G-R
1946 09 12	15 17 15.00	23.50	96.00			7.5	Sagaing	G-R
1946 09 12	15 20	23.50	96.00			7.7	Sagaing	G-R
1947 07 29	13 43 22.00	28.50	94.00			7.9	N.E. Assam	G-R
1950 02 02	23 33 39.00	22.00	100.00			7.0	Burma-China border region	G-R
1950 08 15		28.50	96.70			8.7	Assam	TAN
1950 08 16	01 07 34.00	25.50	96.00			7.0	N. Burma	CGS
1950 08 26		26.80	95.10			7.0	Patkai hills	TAN
1950 09 13		27.80	95.30			7.0	Lakhimpur	TAN
1954 03 21	23 42	24.50	95.00	180		7.4	Mainpur-Burma	CGS
1956 05 13	07 50 33.00	29.90	70.00		VIII	6.5	Pakistan	ISC
1956 07 16	15 07 11.00	22.24	95.73	39		7.0	Sagaing	ISC
1956 07 21		23.00	70.00		VIII		Anjar	CGS
1956 10 10	15 31 36.00	28.15	77.67		VIII	6.0	Near Khurja	ISC
1967 12 10		17.70	73.90		VIII	6.0	Koyna	CGS
1988 08 06	00 36 26.90	25.12	95.17	115		7.5	Burma-India	USGS
1988 08 20	23 09 10.10	26.72	86.63	65	VIII	6.6	Bihar-Nepal	ISC
1991 10 20	21 23 17.37	30.75	78.60	12	VIII	6.6	Uttarkashi	NGRI
1993 09 29	22 25 48.60	18.07	76.45	6.8	VIII	6.3	Latur	USGS
1997 05 21	22 51 30.08	23.08	80.06	35	VIII	6.0	Jabalpur	IMD
1999 03 28	19 05 13.4	30.41	79.42	21	VIII	6.8	Chamoli	IMD
2001 01 26	08 46 41.00	23.40	70.28	25	X	7.7	Bhuj	IMD
2004 12 26	00 58	03.29	95.77	30	X	9.3	Sumatra	USGS
2005 10 08	03 50 40.00	34.49	73.63	26	X	7.8	Kashmir	USGS

The following abbreviations are used in the Ref. Column:

OLD : Oldham, 1883
MIL : Milne, 1912
G-R : Gutenberg and Richter, 1954
SRS : Srivastava and Ramachandran, 1985
TAN : Tandon, 1956
CGS : Coast and Geodetic Survey
ISC : International Seismological Catalogue

2.9 PERMANENT SEISMOLOGICAL OBSERVATORIES IN INDIA

Prior to development of instrumental seismology in India, the GSI has been engaged to carry out field investigation of every large earthquake, and published the report in the form of GSI Memoirs or Special Volumes. The first scientific study of an Indian strong earthquake was carried out for the Cachar earthquake of 1869 by Dr. T. Oldham, the first Director General of the GSI. His illustrious son Dr. R.D. Oldham made a very thorough study of the great Shillong earthquake of 1897. The report of the 1897 earthquake (Oldham, 1899), is a classical work, and a great foundation for the present day modern seismology. After the occurrence of the great Shillong earthquake in 1897, the necessity of installing seismographs in the country was very much felt. The seismological Committee of the British Association recommended the installation of a few seismographs in India in 1898.

2.9.1 National Network

The first seismological station in India was established in Calcutta (Alipore) on December 1, 1898 under the auspices of the India Meteorological Department (IMD). During 1898-99 two more observatories were started, one at Bombay (Colaba) and the other at Kodaikanal. These observatories used Milne seismograph. After the great Kangra earthquake in 1905, a seismological observatory was started in Simla with an Omori Ewing seismograph. In 1929 an observatory was started in Agra with a Milne-Shaw seismograph. During 1930s two more observatories were started, one at Dehradun and the other at Nizamiah (Hyderabad). In 1941, the Agra observatory was shifted to Delhi. The number of observatories increased to eight in 1950, and later rose to 15 in 1960 when more sensitive instruments like Benioff, Sprengnether and Wood-Anderson seismographs were added. At present, the number of IMD observatories under the national network in the country is about 60 (Fig. 2.18). Out of these four stations, one each at Delhi, Poona (Pune), Kodaikanal and Shillong, were equipped with sensitive seismographs under the World Wide Standard Seismograph Network (WWSSN) programme in 1964. The national network is modernised with 10 GSN (Global Standard Network) digital instruments with broadband seismometers in 1996, and 10 more are upgraded with broadband instruments in 1998. In these upgraded observatories high quality broadband digital data are being obtained.

In addition to this, IMD maintains a few permanent small networks like a six-station network in Delhi, six observatories in Punjab and Himachal Pradesh since 1965-66, and three in Jammu and Kashmir since 1980s. The Delhi network is presently upgraded to a digital telemetric network.

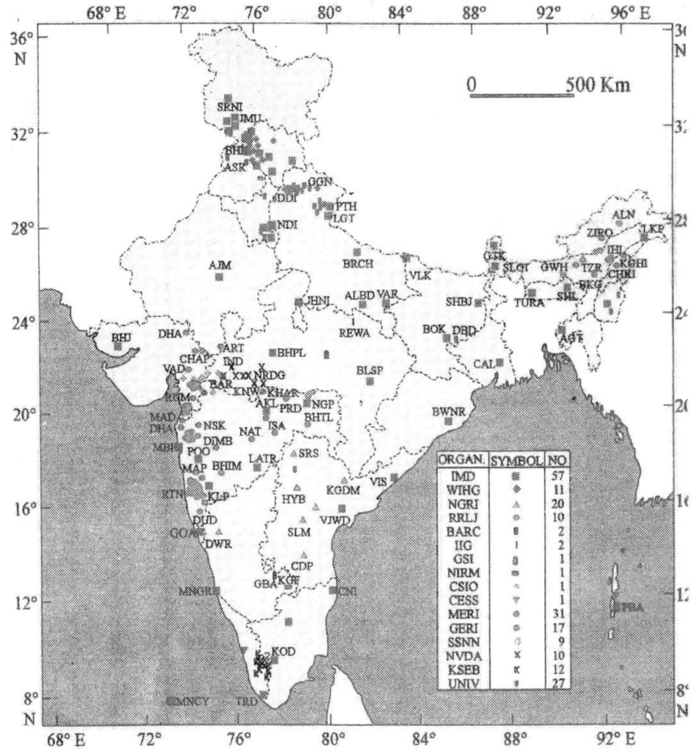

Fig. 2.18 National network showing permanent observatories run by different organisations. Seismological Data Centre and the IMD Headquarter at Delhi (IMD website).

2.9.2 Other Permanent Seismological Stations/Networks

There are more than 150 permanent stand-alone seismological stations around the country, which are being maintained by various agencies other than the IMD. Some important agencies and their seismic stations are described below. However, these information need to be updated.

Bhaba Atomic Research Centre (BARC), Bombay

A most significant addition to the national network of seismological stations was the establishment of a special seismological array at Gauribidanur near Bangalore by the BARC since 1965 in collaboration with U.K. The array is capable of recording smaller events compared to conventional seismographs due to its high sensitivity and low noise. The main purpose of this array is, however, to detect the underground nuclear explosions. In early 1988, the BARC also commissioned an indigenously-built analog telemetered seismic network of eight stations in and around Bhatsa dam, Maharashtra state, for monitoring Reservoir Induced Seismicity (RIS); the network was in operation till late 1990s.

National Geophysical Research Institute (NGRI), Hyderabad

Since 1970s the NGRI is maintaining a well equipped seismological observatory in Hyderabad. It has an equipment-set of three-component short period and a three-component long period analog seismographs similar to the WWSSN. In early 1990s a broadband digital seismograph has been installed in the observatory in collaboration with the GEOSCOPE (France). Further, the NGRI was running a six to ten-station analog network since 1980 in the Koyna dam area to monitor the RIS (Gupta, 1992; Talwani, 1997; Mandal et al., 1998); the network is upgraded to digital telemetry in late 1990s (Rai et al., 1999a). In addition, since 1988 the NGRI is running a six-station analog/16 bit digital telemetered/stand alone network in Tezpur area of Assam state, northeast India. The 16-bit digital system now is upgraded to 24-bit resolution in early 2000.

Regional Research Laboratory (RRL), Jorhat, Assam

In northeast India, the RRL (J) has also established a seismological network consisting of 14 vertical-component short-period seismograph stations progressively since 1980s. Out of these, six stations were running on analog/16-bit digital telemetric system. The 16-bit digital system was upgraded to 24-bit resolution, and the eight stand-alone stations were replaced by 24-bit digital/broadband systems in early 2000. Further, three more broadband stations were established in the epicentre area of the 1897 great earthquake in the Shillong Plateau.

Wadia Institute of Himalayan Geology (WIHG), Dehradun

The WIHG is operating a network of 11 seismological stations with analog recording system since early 1990s. These stations are mostly in the Himachal Himalaya and a few in the Garhwal Himalaya. At five stations three-component short-period seismographs are working and at other six stations only vertical component short-period seismographs are working. The network was upgraded with 24-bit digital/broadband instruments in early 2000.

Gujarat Engineering Research Institute (GERI), Vadodara, Gujarat

The GERI is running a permanent seismic station at Vadodara, Gujarat state. The GERI is also associated with most of the major and medium irrigation projects in Gujarat. These projects run five-to-six station networks with short-period seismographs and digital accelerographs.

Institute of Seismological Research (ISR), Gandhinagar, Gujarat

After the January 2001 devastating earthquake (M_w 7.7) in Bhuj, Gujarat State Government has established the ISR for monitoring earthquakes with much close-spaced network. About 22 broadband seismic stations are in operation in the region with a central recording station in Gandhinagar, the capital city of the State. About 10 strong motions instruments are also installed.

Maharashtra Engineering Research Institute (MERI), Nasik, Maharashtra

The MERI is running about 30 seismological stations in different irrigation projects in Maharashtra state. A seismic station is also functioning at its headquarter in Nasik. Most of the stations have vertical component short-period smoked paper seismographs.

Kerala State Electricity Board (KSEB), Kottayam, Kerala

The KSEB is operating 12 seismic stations in and around Idduki dam site in northern part of the Kerala state. Each station has short-period vertical analog seismograph.

Central Water and Power Research Station (CWPRS), Pune

The CWPRS looks after two to three permanent analog microearthquake networks consisting of three to four seismic stations in each network in different dam sites in the Eastern Himalaya including Sikkim and Bhutan areas.

Centre for Earth Science Studies (CESS), Trivandrum

The CESS is running one permanent observatory at the Centre using an analog microearthquake instrument. The station is equipped with a digital broadband instrument since 1998.

Indian Institute of Technology (IIT), Roorkee

The Department of Earthquake Engineering (DEE), IIT Roorkee, runs a six-station telemetric analog microearthquake network around the Tehri dam, Garhwal Himalaya since early 1990s. The network was upgraded with digital instruments. In addition, the DEE runs several strong-motion arrays in the western Himalaya and in northeast India region. Each array consists of 40 to 50 digital strong-motion seismographs.

University of Delhi, Delhi

The Centre of Georesources, University of Delhi, is running seismic station in the University campus with a digital short-period 3-component seismograph and a short-period vertical analog seismograph since 1990s. The university is operating two more seismic stations in the adjoining Haryana state with short-period vertical seismographs. A broadband three-component digital seismograph has started working at the University campus since early 1998.

Manipur University, Imphal

A short period analog seismograph and a triaxial strong-motion accelerograph are working at the Manipur University campus, Imphal since 1994. During 1996 the University has set up three more seismic stations in the Manipur state with digital triaxial short-period seismographs. These stations are now upgraded with broadband instruments.

Indian Institute of Technology (IIT), Kharagpur

The *Indian Institute of Technology* (Kharagpur) established an eight-station semi-permanent strong-motion array in the Sikkim Himalaya in 1998. The institute has also established a broadband seismic observatory in the institute complex, Kharagpur, in 2004.

Other Universities

The *Kurukshetra University* is running seismic station with a short-period seismograph in its campus since 1970. The *Andhra University* and *Osmania University* in Andhra Pradesh state, the *Indian School of Mines,* Dhanbad in Bihar state and the *Banaras Hindu University* in Uttar Pradesh state established short-period seismograph stations at the respective campus since 1998, and most of them were upgraded to broadband seismic stations in 2003. The Gauhati University and the Tezpur University in Assam state and the Mizoram University in Mizoram state started broadband seismic stations, one each in their campus, since 2000. The University projects are mostly funded by the Department of Science and Technology (DST), New Delhi.

2.10 INDIAN EARTHQUAKE CATALOGUES

A number of catalogues of past Indian earthquakes are available, but none of them is upto date and comprehensive. The earliest publication is by Oldham (1883), which gives a list of significant Indian earthquakes upto 1869. Kelkar (1968) published a list of earthquakes recorded in Maharashtra during the past two hundred years. Gubin (1968) published a list of significant earthquakes in peninsular India upto 1968. Tandon and Srivastava (1974) published a catalogue of earthquakes in India of magnitude 5.0 and above based on available historical data prior to 1900, and instrumental and macroseismic data upto 1970. A study of these catalogues, however, reveals that earthquakes which occurred before 1900 have not been properly listed. The only source for historical earthquakes are the files of old newspapers. Files of the *Times of India* (1839-1983), the *Statesman* (1875-1983) and the *Hindu* (1907-1983) were scanned for the earthquake reports. Srivastava and Ramachandran (1985) have published a catalogue of earthquakes after assigning seismic intensity (MM scale) from these newspaper reports for the period 1839-1900. Bapat et al. (1983) have prepared an earthquake catalogue based on data from the past Indian records and USGS data. More recently, Chandra (1992) has compiled a catalogue of Indian earthquakes in Peninsular and the Himalaya region. In this catalogue, earthquakes with maximum intensity V and above have been included. Since the inception of the WWSSN in 1964 and the GSN stations in 1990s more reliable epicentral data are available from the ISC (International Seismological Centre) Bulletin. The India Meteorological Department brings out National Seismological Bulletin for the local and regional earthquakes using the National Network and Global Network data. In addition, a Seismological Bulletin is being published since 1986 by the RRL(J) and NGRI exclusively for the earthquakes of northeast India, which are being recorded by the permanent local networks in the region.

2.11 SEISMIC ZONING MAP OF INDIA

Seismic zoning map of a country is a guide to the seismic status and susceptibility of a region. Some areas get earthquakes frequently, and some at long interval of time. An effort is, therefore, made in most countries to divide the areas into seismic zones with respect to the severity of expected ground motion. Unlike other countries e.g. China, Japan, Italy, USSR etc., where historical records are available for 1000+ years, the historical records in India are available only for about 200 years. The records of the 19th century as mentioned above are mostly newspaper descriptions of what happened at different places during a particular earthquake. The 1897 great Shillong earthquake provided the first instrumental record that was recorded outside the country. This is true also for the first quarter of the 20th century when instrumental data for the strong earthquakes are available from the instruments in other countries.

Based on the available data, in the first zoning exercise (1959-1962) it was thought that since a map is being presented for the first time to the general public, it should conform to historical indications only. It was thus a useful historical picture of the severity of seismicity of a region. In this zoning map, the Himalaya and northeast India were graded into zones VII-IV, and the Deccan Plateau was marked as zone zero. After the 1967 Koyna earthquake, M 6.7, necessity arose to review the zoning, particularly in the Deccan Plateau. Several updated geological, geophysical and seismological information were considered. The revised map was published in 1970 publication of BIS 1893-1970. In the 1996 revision exercise, the geological and geophysical data were considered. The tectonic features were accounted for in modifying the zoning lines. The main zones in the north were made more or less parallel to the Himalayan tectonic trends, and the Deccan Plateau was still showing as a 'safe zone'. Two major changes were made in this map; 'zero zone' in the Deccan Plateau was abolished, and merging of most of the zones V and VI into one zone (V). Total number of zones were thus reduced from 7 to 5 (Fig. 2.19).

However, occurrence of the 1993 Killari earthquake, m_b 6.3, in the so-called 'safe zone' of Deccan Plateau, and subsequently the 1997 Jabalpur earthquake (m_b 6.0), and the 2001 Bhuj earthquake (M_w 7.7) necessitated a review of the seismic zoning map of the country. With more and more seismological data that are now available from the upgraded observatories and new networks, the map is upgraded and is shown in Fig. 2.20. In this revised seismic zoning map of India (BIS, 2004), the zone I or the so-called least seismic zone or stable land is removed; the zones V to II are demarcated. The zone V is assigned peak ground acceleration (PGA) 0.40g, zone IV 0.25g, III 0.20g and zone II 0.10g.

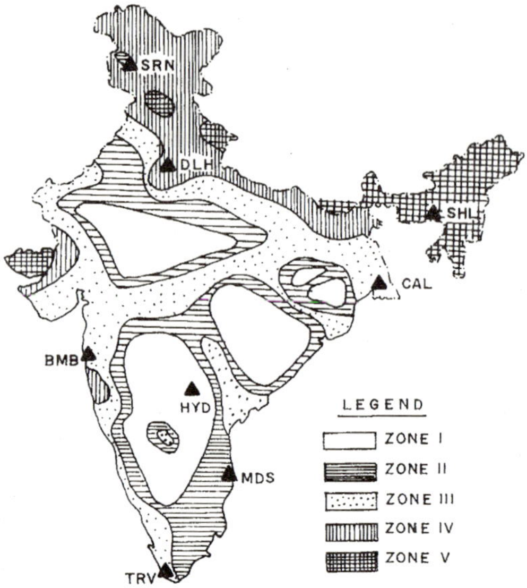

Fig. 2.19 Revised seismic zoning map of India (BIS, 1893-1970) (after Jai Krishna, 1992). Solid triangles indicate WWSSN stations established in 1964. SRN: Srinagar, DLH: Delhi, SHL: Shillong, BMB: Bombay, MDS: Madras, TRV: Trivandrum, HYD: Hyderabad observatory is run by the NGRI/GEOSCOPE.

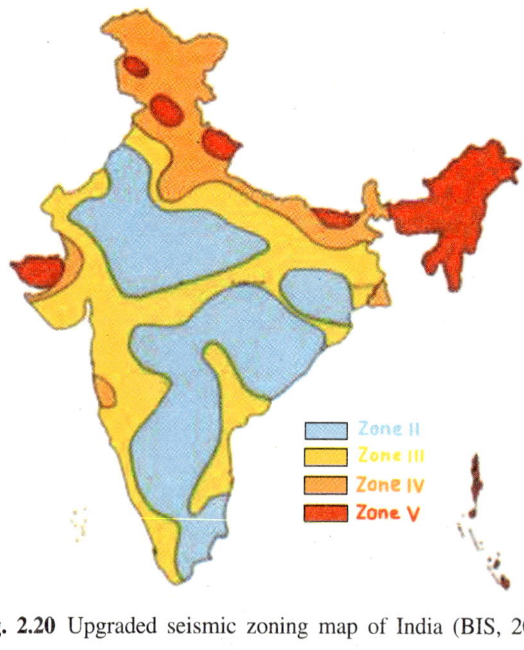

Fig. 2.20 Upgraded seismic zoning map of India (BIS, 2004).

SEISMIC WAVES

2.12 TYPES OF SEISMIC WAVES

Two basic types of elastic waves or seismic waves are generated by an earthquake; these are *body waves* and *surface waves*. These waves cause shaking that is felt, and cause damage in various ways. These waves are similar in many important ways to the familiar waves in air generated by a hand clap or in water generated by a stone thrown into water.

2.12.1 Body Waves

The body waves propagate within a body of rock. The faster of these body waves is called *Primary wave (P-wave)*, or *longitudinal wave* or *compressional wave*, and the slower one is called *Secondary wave (S-wave)* or *shear wave.*

P-wave

The P-wave motion, same as that of sound wave in air, alternately pushes (compresses) and pulls (dilates) the rock (Fig. 2.21). The motion of the particles is always in the direction of propagation. The P-wave, just like sound wave, travels through both solid rock, such as granite, and liquid material, such as volcanic magma or water. It may be mentioned that, because of sound like nature, when P-wave emerges from deep in the Earth to the surface, a fraction of it is transmitted into atmosphere as sound waves. Such sounds, if frequency is greater than 15 cycles per second, are audible to animals or human beings. These are known as *earthquake sound* as discussed in Section 2.6.

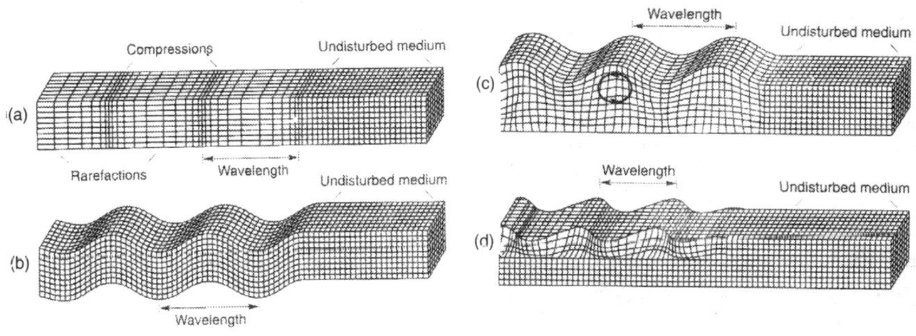

Fig. 2.21 Diagram illustrating ground motion for four types of earthquake waves (a) P-wave, (b) S-wave, (c) Rayleigh wave and (d) Love wave (Bolt, 1999).

The relation between compressional or P-wave velocity (V_p) and the elastic constants E (Young's modulus), σ (Poisson's ratio), K (bulk modulus), μ (rigidity modulus), λ (Lame's constant) and density ρ is given as follows:

$$V_p = \left\{\frac{(\lambda+2\mu)}{\rho}\right\}^{\frac{1}{2}} \tag{2.23}$$

Although Lame's constants are convenient, other elastic constants are also used. From Hooke's law we can obtain the following relations:

$$E = \frac{\mu(3\lambda+2\mu)}{(\lambda+\mu)}$$

$$\sigma = \frac{\lambda}{2(\lambda+\mu)} \tag{2.24}$$

and, $$K = \frac{3\lambda+2\mu}{3}$$

Thus V_p can be expressed as:

$$V_p = \left(\frac{K+\frac{4}{3}\mu}{\rho}\right)^{\frac{1}{2}} \tag{2.25}$$

$$= \left\{\frac{E}{\rho}\left(1+\frac{2\sigma^2}{1-\sigma-2\sigma^2}\right)\right\}^{\frac{1}{2}}$$

$$= \left\{\frac{E}{\rho}\cdot\frac{1-\sigma}{(1-2\sigma)(1+\sigma)}\right\}^{\frac{1}{2}}$$

The above equations (eq. 2.23-2.25) show that the P-wave velocity in a homogeneous solid is a function only of elastic constants and density. One might expect that the elastic constants would be relatively insensitive to pressure, whereas density would increase with pressure. This would mean that the velocity should decrease with depth of burial in the Earth. In fact, this is contrary to actual observations. An explanation for such paradoxial observation is that with increase in density the elastic constants increase much more, which cause higher velocity with higher density. The variation of velocity with depth is reasonably systematic as we go to greater depth.

Polarization of P-wave, when propagating in a homogeneous and isotropic medium, is linear. In inhomogeneous Earth higher frequency waves are, however, affected; they show irregular particle motion.

S-wave

It is known that the S-wave or the *shear wave* shears the rock sideways at right angle to the direction of propagation (Fig. 2.21). As shear deformation cannot be sustained in liquid, shear waves cannot propagate through liquid materials at all. The outer portion of Earth's core is assumed to be liquid because it does not transmit shear waves from earthquakes. The particle motion of the S-wave is perpendicular (transverse) to the propagation. In Fig. 2.21 the particle motion of the S-wave is up and down in vertical plane; it is named S_V wave. However, S-wave may also oscillate in horizontal plane, which is called S_H wave.

The relation between S-wave velocity V_s, the elastic constants and density is given as:

$$V_S = \left(\frac{\mu}{\rho}\right)^{\frac{1}{2}} \tag{2.26}$$

An alternative expression is

$$V_S = \left\{\frac{E}{\rho}\cdot\frac{1}{2(1+\sigma)}\right\}^{\frac{1}{2}} \tag{2.27}$$

Motion of the medium in P-waves being longitudinal, there is no polarization of a P-wave, it is linear, but S-waves being transverse are polarized. A horizontally travelling S-wave if so polarized that the particle motion is all vertical, then it is called an S_V wave, and if particle motion is all horizontal, it is called S_H wave.

The velocity ratio V_P/V_S

Comparing the equations (2.25) and (2.26), we find that the ratio of compressional to shear wave velocity is

$$\frac{V_p}{V_s} = \left(\frac{K}{\mu}+\frac{4}{3}\right)^{\frac{1}{2}} = \left(\frac{1-\sigma}{\frac{1}{2}-\sigma}\right)^{\frac{1}{2}} \tag{2.28}$$

Either expression tells that the P-wave velocity is always greater than the S-wave velocity. The ratio is always greater than 1; first, because K and μ are always positive, second, because σ cannot be greater than 1/2 in an ideal solid. For most consolidated rock V_p/V_s ranges between 1.5 and 2.0. As we see in the later chapters, this is an important parameter in seismological studies.

S-wave Splitting

S-waves are also linearly polarized when propagating in homogeneous isotropic medium. In anisotropic medium they, however, split into fast and slow components. These split waves propagate with different velocity that causes time delay and related phase shift. The two split S-wave components superimposed to an elliptical polarization. The orientation of major axis and degree of ellipticity are controlled by fast and slow velocity directions and the degree of anisotropy. The *S-wave splitting* is often used to study S-wave velocity anisotropy of the Earth (e.g. Silver, 1996; Savage, 1999).

2.12.2 Surface Waves

The second general type of earthquake wave is called *surface wave*, because its motion is restricted to near the ground surface. Such waves correspond to ripples of water that travel across a lake. The wave motion is located at the outside surface itself, and as the depth below this surface increases, wave displacement becomes less and less. Surface waves in earthquakes can be divided into two types: *Love waves* and *Rayliegh waves.* The Love waves are denoted as LQ (or G) and the *Rayleigh* waves as LR (or R). While Rayleigh waves exist at any free surface, Love waves require some kind of wave guide formed by velocity gradient. Both conditions are fulfilled in the real Earth.

Love Wave (LQ)

The British mathematician A.E.H. Love demonstrated that if an S_H ray strikes a reflecting horizon near surface at post critical angle, all the energy is 'trapped' within the wave guide (Love, 1911). These waves propagate by multiple reflections between the top and bottom surfaces of the low speed layer near the surface. The waves are called *Love waves,* and denoted as LQ or G. Its motion is same as that of the S-waves that have no vertical displacement. It moves the ground from side to side in a horizontal plane parallel to Earth's surface, but at right angle to the direction of propagation (Fig. 2.21); so the wave motion is horizontal and transverse.

The Love wave velocity (V_L) is equal to that of shear waves in the upper layer (V_{S1}) for very short wave lengths, and to the velocity of shear waves in the lower layer (V_{S2}) for very long wave-lengths, i.e.

$$\text{Velocity } V_{S1} < V_L < V_{S2} \tag{2.29}$$

The effects of Love waves are result of the horizontal shaking, which produces damage to the foundation of structures. Love waves do not propagate through water, it affects surface water only. It causes the sides of the lakes and ocean bays to move backwards and forwards, pushing the water sideways like the sides of a vibrating tank.

Rayleigh Wave (LR)

Rayleigh (1885) demonstrated that the surface boundary condition can be satisfied leading the existence of a 'coupled' and 'trapped' P-S_V wave travelling along the surface, such as the Earth-air interface, with a velocity lower than shear velocity, and with an amplitude decaying exponentially away from the surface. This second type of surface wave is known as *Rayleigh wave*. In general the surface waves with periods 3 to 60s are denoted R or LR. Like rolling ocean waves, the *Rayleigh waves* develop the particle motion both vertically and horizontally in a vertical plane pointed in the direction of wave propagation (Fig. 2.21). Since Rayleigh waves generate from coupled P and S_V waves the particle motion is always in vertical plane, and due to phase shift between P and S_V the particle motion is elliptical and retrograde (counter clockwise) with respect to the direction of propagation. The amplitude of the motion decreases exponentially with depth below the surface.

The speed of *Rayleigh wave* (V_R) is about 9/10th of shear wave (V_S) in the same medium; the relation is given as

$$V_R < 0.92\ V_S \tag{2.30}$$

For short wave-lengths, V_{Rg} corresponds to 9/10 V_S of the material comprising the surface layer. For very long wave-lengths, the V_{RL} corresponds to 9/10 V_S of the substratum material since effect of the surface layer is negligible when most of the waves travel in the zone below it. As seen from the above equations, S-wave is slower than P-wave, and *Rayleigh wave* is slower than Love wave. Thus as the waves radiate outwards from an earthquake source, the different types of waves separate out from one another in a predictable pattern. An illustration of the pattern seen at a distant place is exemplified in Fig. 2.22.

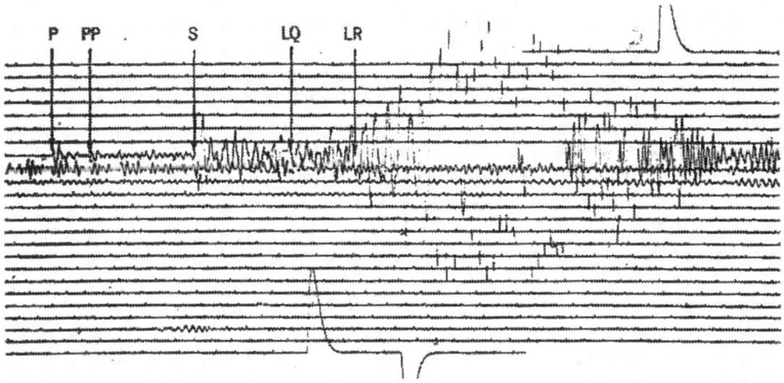

Fig. 2.22 Seismogram showing P, PP, S, LQ and LR phases (Kuthánek, 1990).

2.12.3 Guided Waves

Channel Waves

One of the most important new fields of seismological investigation is the observation of *guided waves*, which include *channel waves* and many types of surface waves. The reader will find details in the book by Ewing et al. (1957).

Channel waves exist in a layered medium such that its properties change only vertically and not horizontally. It is most significant when there is a channel or layer between two discontinuities. For example, coal seam which is a low velocity layer, if sandwiched between two high velocity layers, say sandstones, can act as a channel through which body waves will travel as guided waves or channel waves, which may also include surface waves (Evison, 1955). There is a great deal of literature on channel wave seismology for coal exploration (e.g. Krey, 1963; Dresen and Freystatter, 1976).

L_g Waves

L_g *waves* are one kind of guided waves in the continental crust. These are essentially high frequency Love waves at regional distances of thick continental crust. It travels over long continental paths with relatively little loss of energy, but is cut off abruptly when the path has even a small oceanic segment. The subscript L_g refers to granitic layer. L_g waves being critically incident on the Mohorovičić (Moho) discontinuity propagate by multiple reflections within the crust with a typical velocity 3.5 km/s. These waves dominate the seismograms, specially the horizontal channels. L_g is usually recorded at epicentral distances of about 5° and larger. Ewing et al. (1957) identified it and defined it as a tool for finding boundary of continental structure. An example of L_g wave is shown in Fig. 2.23a.

(a)

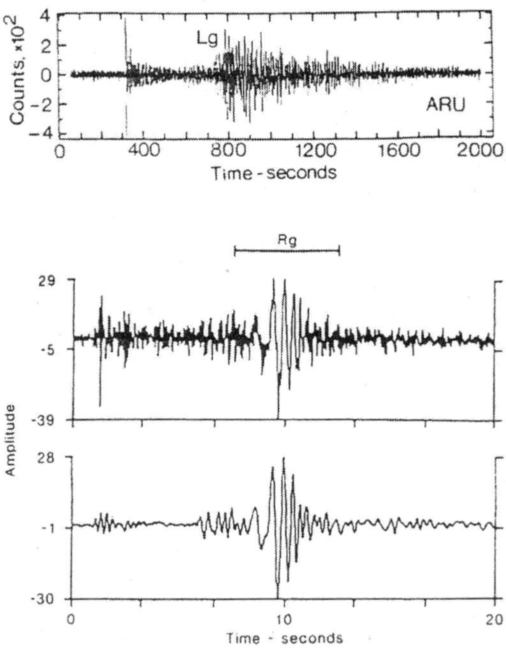

(b)

Fig. 2.23 Seismograms showing L_g and R_g. (a) L_g at epicentral distance 24°. (b) R_g at 39 km, upper trace is the raw seismogram, lower trace in low-pass filtered with a cutoff at 4 Hz (Lay and Wallace, 1995).

R_g wave

At local or regional distances high frequency fundamental mode of *Rayleigh waves* are labeled as R_g. The presence of short period R_g in the seismogram is a reliable indicator of very shallow event (like earthquake, nuclear explosion, mine burst etc.). In other words, absence of R_g *indicates* deeper natural event. The short period R_g waves travel as guided waves through continental crust with velocity 3 km/s; their range of propagation is limited to 600 km or less. An example of the R_g phase is shown in Fig. 2.23b.

T-waves

The term T-wave in seismology literally means Tertiary wave or third wave. These are slow waves, which arrive long after the faster Primary (P) wave or Secondary (S) wave. These slow waves are observed at coastal seismic stations that record sub-ocean earthquakes at regional distances.

Similar to low velocity layer in the upper mantle, there is also low velocity channel in deep ocean. Depending on the salinity and temperature of sea water, the sound velocity decreases to a minimum of about 1.5 km/s from sea surface to 700-1300 m depth, and again increases below this depth. This depth region of low velocity in ocean is called SOFAR (Sound Fixing and Ranging channel). Propagation of T-waves or Tertiary waves generally by SOFAR channel (Bullen and Bolt, 1985) or by multiple reflections between the sea floor and sea surface (Båth and Sahidi, 1974) is every efficient at distances as large as 80°. They show a small gradual beginning, and a gradual amplitude increase, arriving after P and S. The duration can be explained by dispersion. They exhibit rather monochromatic oscillations (Fig. 2.24). These waves are recorded by SOFAR *hydrophones, ocean bottom seismometers* (OBS) or by seismographs, which are located very near to the coast. Unlike P-waves, there is no sharp onset for T-wave group. The appearance of T-phases on seismograms depends largely on the ocean bottom topography, water and land conversion and transmission properties. Ewing et al. (1952) demonstrated that T phase propagates as sound waves through the water and as body waves over the land path, i.e., from the coast or from the continental shelf to the seismic station. T phases were first reported by Linehan (1940). Later, Milne (1959), Johnson et al. (1963) and Adams (1979) reported T-phases from nuclear explosions recorded by hydrophones. Gibowicz et al. (1974) and Chadha (1994) studied T-phases recorded on conventional seismographs for events occurring in the ocean. The T-waves for the large Bhuj earthquake (M_w 7.7) of January 26, 2001 in western India are reported by Pulli and Upton (2002), and for the December 26, 2004 Sumatra mega thrust event (M_w 9.3) by Park et al. (2005).

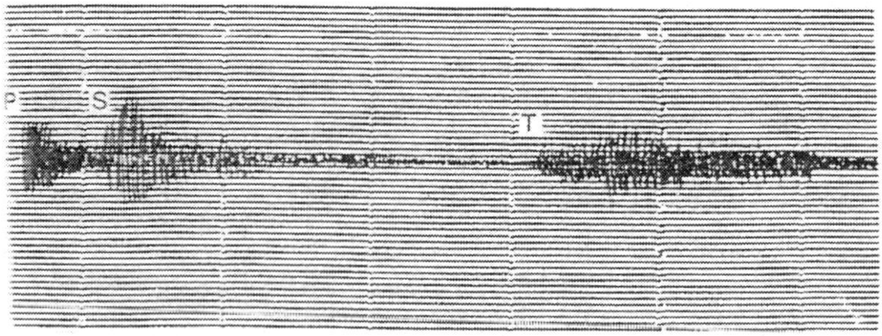

Fig. 2.24 Seismogram showing P, S, and T-phases recorded at an island station (Kuthánek, 1990).

2.13 SEISMIC WAVES AND GROUND SHAKING

The body waves (P and S-waves), when move through the layers of rock in the crust, are reflected and/or refracted at the interfaces between the rock types or layers. When P and S-waves reach the surface of the ground, most of their energy is reflected back into the crust. Thus, the surface is affected simultaneously by upward and downward moving waves. After a few shakes, a combination of two kinds of waves is felt in ground shaking. A considerable amplification of shaking occurs near the surface. This surface amplification enhances the shaking at the surface of the Earth. On the other hand, earthquake shaking below ground surface, say in the mine, is much less. Again combination of two kinds of waves in shaking is not felt at sea. The only motion felt on ship is from the P-waves, because S-waves cannot travel through water beneath the ship. A similar effect occurs as sand layers liquify in earthquake shaking, which is appropriately known as *liquefaction*. There is progressive decrease in the amount of S-wave energy that is able to propagate through liquefied layers; ultimately only P-wave can pass through it.

The above description is not adequate to explain the heavy shaking due to a large earthquake. The effect of surface waves (Love wave and Rayleigh wave), and various kinds of mixed seismic waves including converted and reflected seismic phases at the rock interfaces complicate the matter, and type of ground shaking is further muddled together. The horizontal and transverse motion of the Love waves, and elliptical and retrograde motion of the Rayleigh waves cause severe damage to the foundations of engineering structures and buildings.

The ground shaking is also much affected by soil conditions and topography. For example, in weathered surface rocks, in alluvium and water filled soil, the amplification of seismic waves may increase or decrease as the waves reach the surface from the more rigid basement rocks. Also at the top or bottom of a ridge, shaking may intensify, depending on the direction from which waves are coming and whether the wavelengths are short or long. The site amplifications play an important role in *microzonation* study (e.g. Field and Jacob, 1995; Hartzell, 1992), that identifies sites vulnerable for more damage by seismic waves.

2.14 SEISMIC PHASES AT THE ROCK BOUNDARIES

The body waves (the P and S-waves) are reflected or refracted at the interfaces between rock types. In addition to reflection or refraction of one type, the seismic waves are also converted to other types. As illustrated in Fig. 2.25, P-wave travels upwards and strikes the bottom of a layer of different rock type; part of its energy will pass through the upper layer as P-wave and part as converted S-wave, which is known as P to S conversion (or P_S phase), and part of energy will be reflected back downwards as P and S waves. Similar reflection, refraction and conversion may occur with S-wave. All these *converted phases* are useful for velocity and geological structure study.

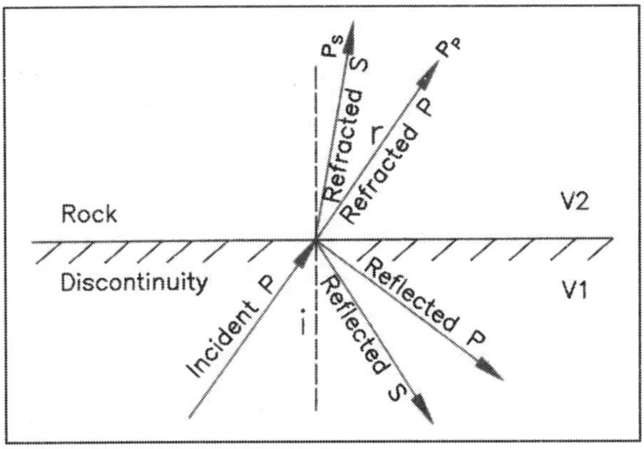

Fig. 2.25 Seismic waves generated by an incident P-wave at the boundary between two rocks.

Huygens' Principle

The new direction of a ray-path can be inferred from *Huygens' principle*. It states that an incident ray, say on a rock boundary, may be treated as new source, about which new hemispherical wave fronts expand on each side of the boundary (Fig. 2.26). Since each of these elementary wave fronts corresponds to only an infinitesimal amount of energy, a physically realistic wave front consists of a surface to which an infinite number of them are tangent. Figure 2.26 illustrates an incoming plane wave. It strikes the rock boundary at point A at time t_1, and becomes active as an infinitesimal source. By the time the incident wave front reaches B at time t_2, wave fronts from A have spread hemispherically into both media. It may be noted that while the radius of the hemisphere is $V_1 (t_2 - t_1)$ in the upper layer, it is $V_2 (t_2 - t_1)$

in the lower medium. The figure further suggests that proportionately smaller hemispheres exist about all points between A and B. The slopping planes, which are tangential to these, physically represent real wave fronts. The new wave directions, reflected or refracted, are normal to these planes.

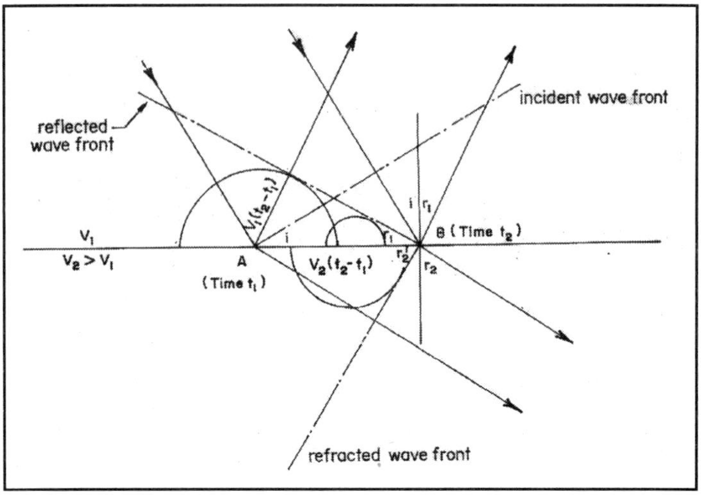

Fig. 2.26 Reflected and refracted wave fronts at a boundary by Huygens' principle.

Snell's law

The upper part of the diagram (Fig. 2.26) demonstrates the law of reflection, i.e. angle of incidence i is equal to angle of reflection r_1. The lower part of the diagram demonstrates the law of refraction, or Snell's law. It is easily deduced that

$$\frac{\sin i}{\sin r_2} = \frac{V_1}{V_2} \tag{2.31}$$

where r_2 is the angle of refraction, V_1 and V_2 are the velocity of the upper and lower layer respectively. This formula can be extended to the case of reflection or refraction of a wave of different type, e.g. reflected or refracted S from an incident P, leading to a generalised form of Snell's law:

$$\frac{V}{\sin i} = \text{constant } (p) \tag{2.32}$$

where V stands for either V_P or V_S on either side of the boundary, and i is the angle between the corresponding ray (incident, reflected or refracted) and the normal on the same side, and p is called *seismic parameter* or *ray parameter.*

The ray parameter is constant for the entire travel path of a ray. The consequence of a ray traversing material of changing velocity, V, is a change in incidence angle, i, with respect to a reference plane. As the ray enters material of increasing velocity, the ray is deflected toward the horizontal. Conversely, as it enters material of decreasing velocity, it is deflected towards the vertical. If the ray is traversing vertically, then $p = 0$, and the ray will have no deflection as velocity changes.

2.15 SEISMIC PHASES AT SHORT DISTANCES

P and S arrivals

The seismic phases generated by an earthquake at a shallow depth ($d < 10$ km) and recorded by a station within a few km of the epicentre, are generally identified as direct P and S arrivals. These are also called $\bar{P}$, $\bar{S}$ or Pg, Sg. A three-component seismogram is illustrated in Fig. 2.27. The apparent velocities of P and S, as determined from time-distance curves, are about 5.5 and 3.2 km/s respectively in the shallow crust.

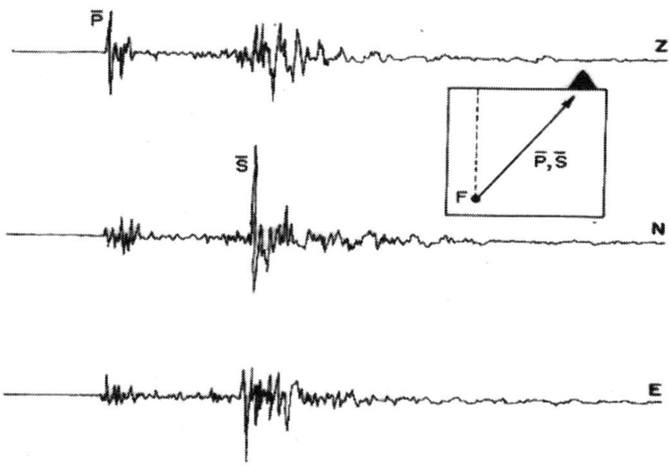

Fig. 2.27 Three-component seismogram showing direct P and S arrivals for a local earthquake (Kuthánek, 1990).

P_n and S_n arrivals

In 1909, A. Mohorovičić identified refracted P_n and S_n arrivals from the interface between the crust and mantle, the layers of lower and higher velocity. He made this discovery on seismograms of an earthquake with epicentre not far from his seismograph station. He found that there is a *critical distance*, generally in the range 100-150 km, beyond which P and S waves are refracted, and arrive with small and long period motion. He designated these phases as P_n and S_n respectively. These are also called *head waves.* The head wave travels with faster apparent velocity along the refractor, and it becomes the *first* arrival. The incidence angle at which the ray is refracted at 90° along the refractor is called the *critical angle.* The ray-geometry is illustrated in

Fig. 2.28. In a standard three-layer crustal model if $V_3 > V_2 > V_1$, three primary travel paths exist between the source and the receiver: (i) direct arrival, which travels in a straight line connecting the source and receiver, (ii) reflected arrival and (iii) a head wave. Additional rays involving multiple reflections will also exist, and make the seismogram complicated. The head waves are followed by the larger and sharper impulse of short period direct P and S arrivals. An example of P_n, P, S_n and S phases is illustrated in Fig. 2.29. The apparent velocities of P_n and S_n as read from their time-distance curves, are about 8.0 and 4.6 km/s respectively, which arc the upper mantle velocities. The surface of separation between the crust and upper mantle is termed as *Moho discontinuity.*

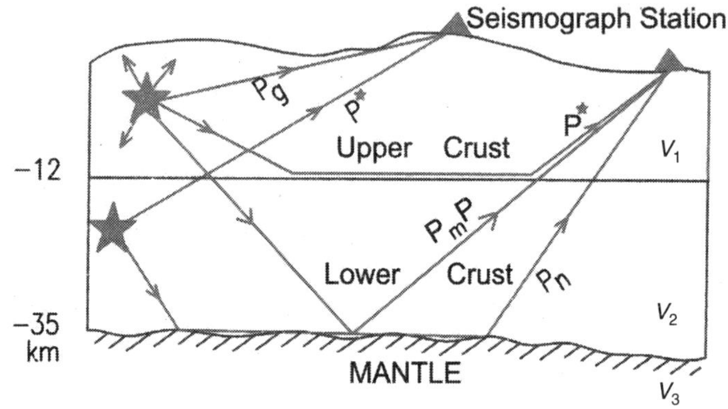

Fig. 2.28 Raypath of direct, reflected and refracted seismic phases and a standard crustal model.

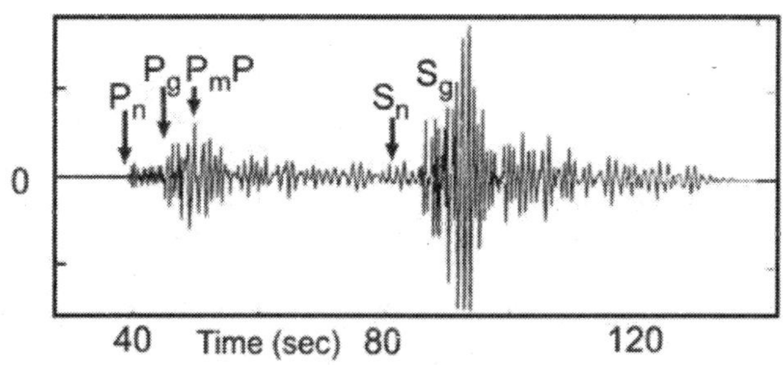

Fig. 2.29 Seismogram showing P_n, P_g, P_mP and S phases recorded for a local earthquake.

P_g, S_g and P^*, S^* arrivals

Later study of seismograms by many investigators revealed further complexity. Conrad observed a small sharp impulse between P_n and P, which he named P^*, and attributed it to refraction through an intermediate layer with a velocity of about 6.5 km/s (Fig. 2.28). The upper boundary of this layer had been called *Conrad discontinuity.* Seismologists have, more or less, accepted the

Conrad discontinuity as separating predominantly *granitic layer* above it and *basaltic layer* below, within the crust, and proposed notations as P_g, S_g and P^*, S^*. The P_g and S_g are practically the same as P and S. A seismogram showing different phases P_n, P_g and P^* is illustrated in Fig. 2.30.

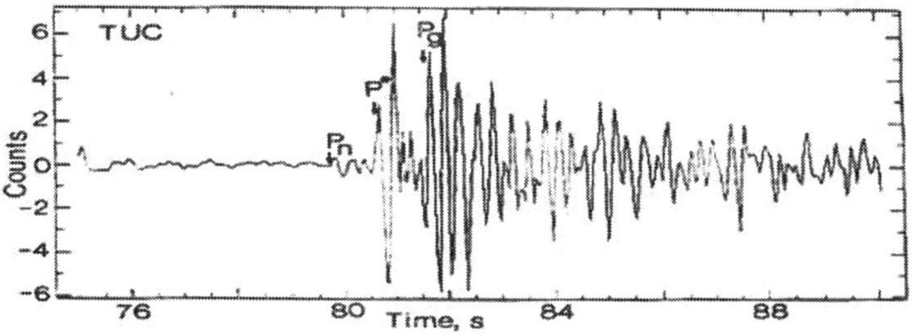

Fig. 2.30 Seismogram showing P_n, P* and P_g (Lay and Wallace, 1995).

Converted Phases

Although conversions of seismic phases at the Conrad discontinuity and at the Moho discontinuity are well illustrated, there is still controversy about the Conrad discontinuity. The conversion of seismic phases P to S (P_s) or S to P (S_p) at the Moho discontinuity is well established (Figs 2.25 and 2.31), (e.g. Cook et al., 1962; Snoke et al., 1977; Kayal, 1986). The reflected rays from the Moho discontinuity are labeled as P_p (or PmP) and S_s (or SmS) (Fig. 2.28).

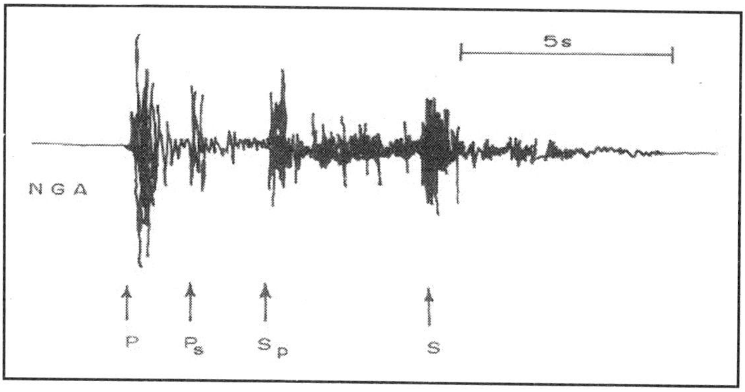

Fig. 2.31 Seismogram showing P_s and S_p phases generated at the Moho discontinuity New Zealand subduction zone (Kayal, 1986).

In rare instances, additional phases between P_n and P_g are present on the seismograms. The P and S leave the focus, travel upward and get reflected as P and S at the free surface and continue further as P_n and S_n. Notations for these phases are given as ${}_pP_n$ and ${}_sS_n$ respectively. The ray paths are illustrated in Fig. 2.32. For a more detailed description of seismic phases of local, regional and teleseismic events readers are referred to Anatomy of Seismogram by Kuthánek (1990). A list of nomenclature of seismic phases for local, regional and teleseismic events is given in Table 2.4.

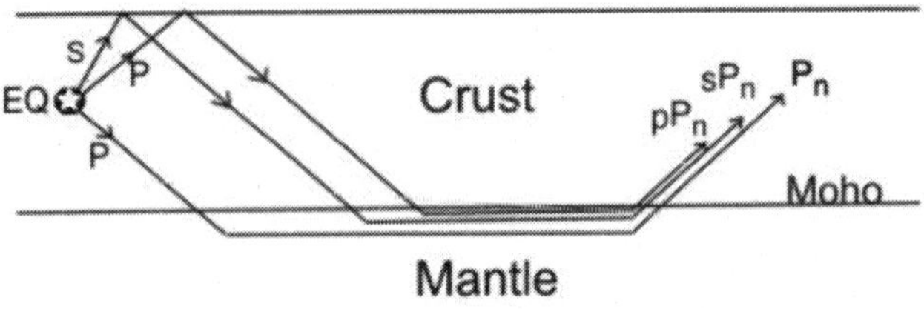

Fig. 2.32 Ray paths of P_n, sP_n and pP_n phases.

2.16 TRAVEL TIMES IN A LAYERED EARTH

2.16.1 Two-layer case

As illustrated in Fig. 2.33 the travel time for the direct arrival is given by

$$T = \frac{X}{V_1} \tag{2.33}$$

The travel time for the reflected arrival is given by

$$T = \frac{2h}{\cos i} \cdot \frac{1}{V_1} \tag{2.34}$$

where h = layer thickness.

The travel time for the head wave is given by

$$T = \frac{r}{V_2} + \frac{2d}{\cos i_c} \cdot \frac{1}{V_1} \tag{2.35}$$

These equations are, however, for a surface source; slight modifications are required for sources within the layer.

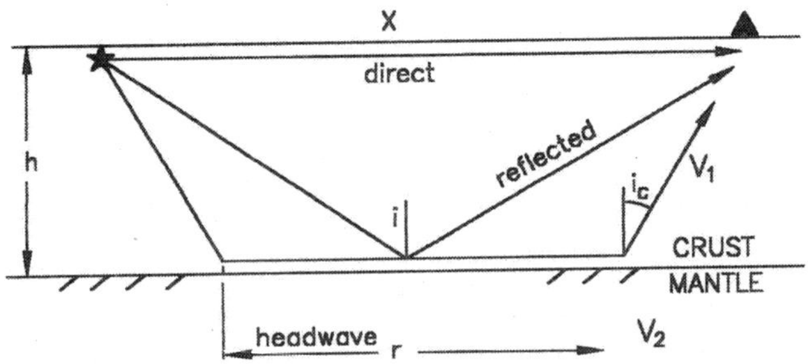

Fig. 2.33 Three principal rays and a crustal layer over a half-space. X = epicentral distance, h = thickness, i_c = critical angle of incidence.

The above equations (eqs 2.33-2.35) determine *travel-time curves.* Figure 2.34 shows the travel-time curves for principal rays for the layered structure in Fig. 2.33. The direct and reflected arrivals exist only at short distances. The direct arrival is described by a straight line with a slope $dT/dx = p = 1/V_1$. The reflected arrival time is described by a hyperbola. The intercept, at $X = 0$, has a travel time $2h/V_1$. At large distances, the branch of the reflected travel-time curve becomes asymptotic to the direct arrival.

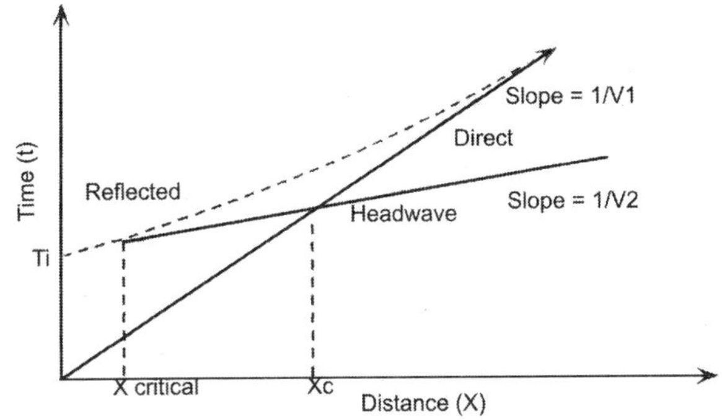

Fig. 2.34 Travel-time curves for the three primary rays in a two-layer velocity structure (Fig. 2.33).

The head wave travel-time curve represents a straight line with a slope $dT/dx = p = 1/V_2$. The head wave travels with a faster velocity, and eventually becomes the first arrival. The direct arrival is the first arrival until the *cross over distance*, X_c, after which the head wave becomes the first arrival. The cross over distance or critical distance, X_c, can be estimated by visualizing that at X_c the travel times of the direct wave and head wave are equal, i.e.

$$T_{\text{direct}} = T_{\text{head}}$$

or,

$$\frac{X_c}{V_1} = \frac{X_c}{V_2} + \frac{2h}{\cos i_c} \cdot \frac{1}{V_1} \tag{2.36}$$

or,

$$X_c = 2h \left[\frac{V_2 + V_1}{V_2 - V_1}\right]^{1/2}$$

The travel time curve in Fig. 2.34 is, however, not complete because we have only considered three rays. There will be many rays, and the seismograms are mixtures of P and S waves, converted phases and multiple reflections. The importance of travel-time curve is its interpretative power, and we can determine the layer velocities and layer thicknesses.

The depth to the interface or crustal thickness (h) can be calculated from the cross over distance (X_c) as shown in equation 2.36. The crustal thickness can also be estimated from the intercept time (T_i):

$$h = \frac{T_i}{2} \cdot \frac{V_1 V_2}{\left(V_2^2 - V_2^2\right)^{1/2}} \tag{2.37}$$

The intercept time T_i can be determined graphically.

2.16.2 Multi-layer Case

The time-depth relation derived for a two-layered media can readily be extrapolated to apply to a large number of layers as long as the velocity in each layer is higher than in the one just above it. The general equation may be written as

$$T = px + 2\sum_{i=1}^{n} h_i n_i \tag{2.38}$$

Figure 2.35 illustrates ray paths and time-distance curves for each media; the lower most designated for generality as n^{th}. The travel-time curve shows many branches due to head waves at each interface.

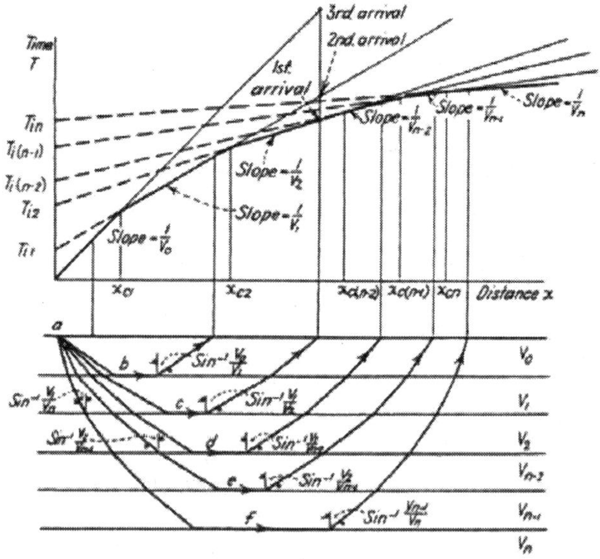

Fig. 2.35 Ray paths and travel-time curves for finely layered Earth (Lay and Wallace, 1995).

Low Velocity Layer and Blind Zone

We may mention two special cases, which complicate the interpretation of travel-time curves. First of these is the presence of a low velocity layer. Figure 2.36 shows that the velocity of layer 2 is less than that of layer 1 and the half space. In such case no *head wave* occurs along the interface between the layers 1 and 2. One would estimate a *pseudo thickness* of a single layer with velocity V_1 over a half space of velocity V_3. This would result in an over estimate of the actual depth to the half space.

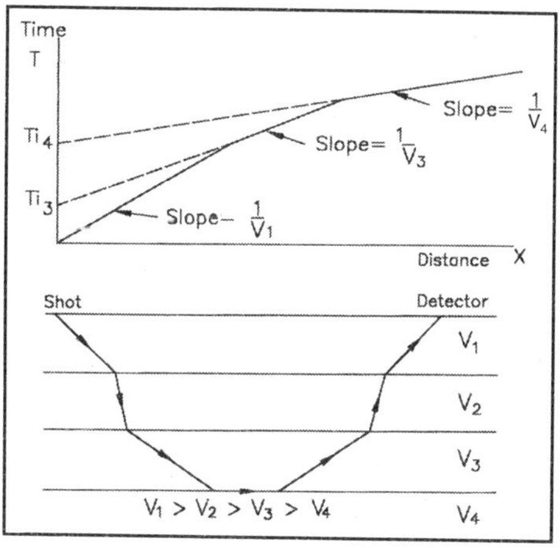

Fig. 2.36 Ray paths for a structure with a low velocity layer, no head wave at the interface V_1 and V_2.

The second special case is the *blind zone,* which is caused by a thin layer or low velocity contrast. The layer may be so thin that the head wave from it is never a first arrival. This may also happen if V_2/V_1 is not much larger than 1. One of the practical examples of blind zone is the *Conrad discontinuity* in the crust.

Increase of Velocity with Depth

Travel-time curves in a continuous medium, i.e. with continuous velocity distribution, shows subtle differences in the character of the travel-time curves (Fig. 2.37). The slowness ($1/V$) observed at a distance X can be found by taking the slope $p = dt/dx$ or tangent of the travel-time curve.

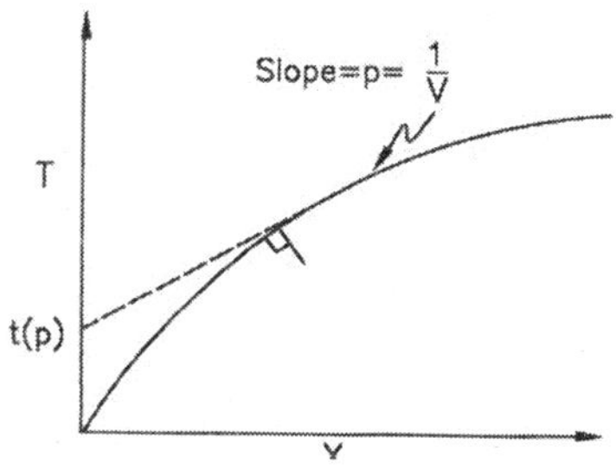

Fig. 2.37 Travel-time curve in a continuous velocity structure.

Dipping Layer

Travel-time curve for *dipping layer* is again complicated; it depends whether the rays are travelling up-dip or down-dip; Fig. 2.38 shows ray-geometry and corresponding travel-time curves. The direct arrivals have the same slowness, but the head waves have different apparent velocities where:

$$\frac{1}{V_u} = \frac{\sin(i_c - \theta)}{V_1} \tag{2.39}$$

and $$\frac{1}{V_d} = \frac{\sin(i_c + \theta)}{V_1}$$

The resulting travel-time equation can be written as

$$T_u = \frac{2h_2 \cos i_c \cos\theta}{V_1} + \frac{X \sin(i_c - \theta)}{V_1} \tag{2.40}$$

$$T_d = \frac{2h_1 \cos i_c \cos\theta}{V_1} + \frac{X \sin(i_c + \theta)}{V_1} \tag{2.41}$$

where *u* and *d* represent up-dip and down-dip observations respectively. It may be noted that the cross over distance is larger for up-dip travel paths. The total travel time for the source-receiver geometry is, however, the same because of reciprocity for interchanged source and receiver locations.

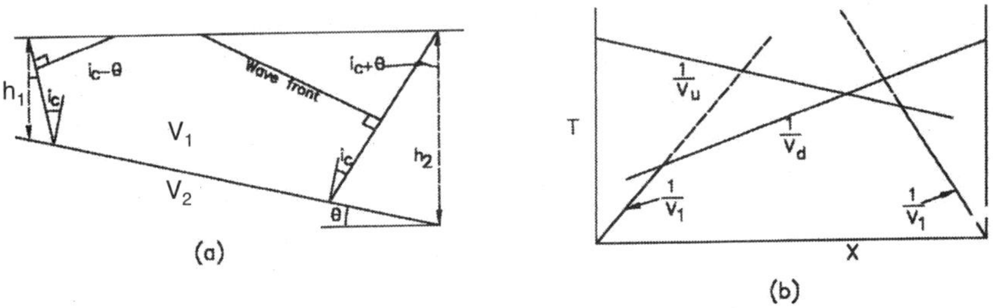

Fig. 2.38 (a) Ray geometry for head wave along a dipping interface. (b) Travel-time curves (up-dip) and down-dip for the dipping structure.

2.17 TELESEISMIC WAVES AND INTERIOR OF THE EARTH

Virtually all our direct informations about the interior of the Earth are derived from the observations of earthquake generated seismic waves only. Since much of the Earth is an elastic solid, two kinds of body waves can propagate through the Earth. The path of the waves are affected by two main discontinuities in the Earth; one at a depth of 30-60 km, and the other at a depth of 2900 km. The first one is called the *Moho discontinuity,* which is of special importance in interpreting seismograms at short epicentral distances upto a few hundred kilometres as discussed above. Evidence of the second discontinuity was discovered by Wiechert and by Oldham in 1906, but precise identification and determination of depth of this discontinuity was made by Gutenberg (1914), hence called *Gutenberg discontinuity*. These two discontinuities divide the Earth internally into an outer shell, called the *crust*, an intermediate shell, called the *mantle*, and a central *core*. The major divisions of the Earth's interior are shown in Fig. 2.39. The mantle is subdivided into upper mantle and lower mantle, and core is subdivided into outer core and inner core.

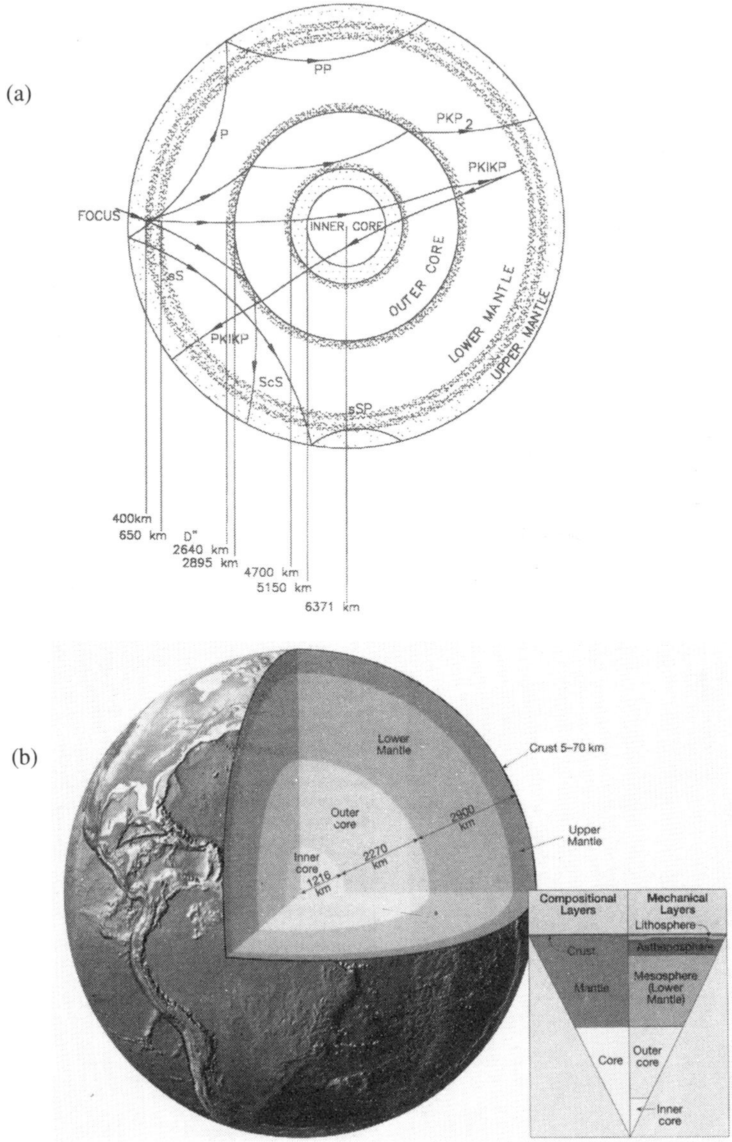

Fig. 2.39 (a) Interior of the Earth and seismic waves which penetrate mantle, core or inner core (modified from Garland, 1979). (b) 3-D view of the interior of the Earth (Courtesy: Roger Masson, U.K.).

Seismic waves arriving at a distance beyond 10° up to about 30° mainly travel through the upper mantle (Moho to 410 km) and through the transition zone to the lower mantle (410-660 km). The strong discontinuities in the transition zone have strong *seismic impedance* (i.e. the quantity $V \times \rho$, that increases with the increase of velocity V and density ρ). This results triplications of the travel-time curve for P waves and S waves, that gives rise to complicated short-period waveforms consisting of sequence of successive onsets with different amplitudes. The base of the transition zone, between the lower mantle and outer core, is labelled as D″ region (Bullen, 1965) (Fig. 2.39a), which may be caused by either chemical differentiation in the mantle or by chemical reactions between core and mantle (e.g. Lay, 1989; Jeanloz, 1990).

At epicentral distances between about 30° and 100°, the P and S waves travel through the lower mantle that is characterized by a rather smooth positive velocity and density gradient. The seismograms are clearly structured with P and S arrivals, followed by multiple surface and core-mantle boundary (CMB) reflections on conversions (Fig. 2.39a). The existence of the great velocity reduction across the CMB causes seismic wave energy to diffract into the geometric shadow zone at distances greater than 100°.

Beyond 100°, only P-wave enters outer core, and reaches surface. There is a dramatic reduction of P-wave velocity, from 13.7 km/s in the lower most mantle to 8.0 km/s in the upper outer core. This P-wave forms a core shadow. Oldham (1906) first observed that a P-wave arriving at the antipodes of an earthquake was late, in comparison with the expected arrival time. He proposed the existence of core of lower velocity than outer region, and predicted the presence of a *shadow zone*. Gutenberg (1914) verified that there was a shadow zone for P between $\Delta = 103°$ and $\Delta = 142°$ with strong arrivals just beyond 142° (Fig. 2.40). Gutenberg estimated the depth to the core boundary as 2900 km, which stood unchallenged until very recently. The shadow zone of the core is not complete, there being arrivals of P-waves of small amplitude through the entire zone. Lehmann (1935) suggested that these arise from *inner core* of higher velocity within the main core. Her proposal of inner core became widely accepted.

The teleseismic arrivals, corresponding to different paths, are illustrated in Fig. 2.39a. The letters P and S are used to designate reflected or refracted waves outside the core. The subscripts **c** and **i** denote reflections from the

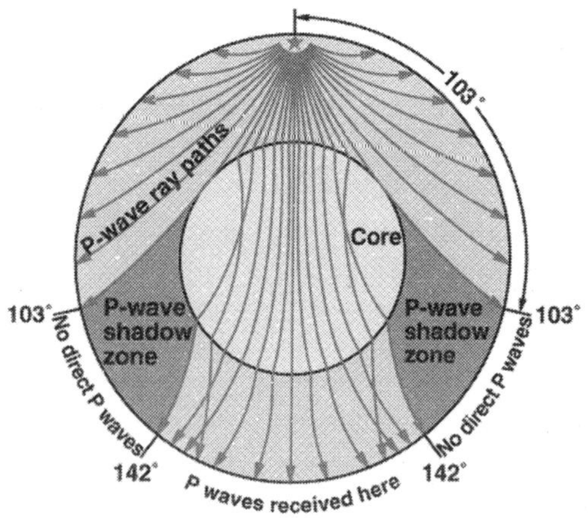

Fig. 2.40 Ray paths of the P-wave through interior of the Earth, showing the shadow zone.

outer and inner core boundaries respectively, while the letters **K** and **I** indicate P waves through outer and inner core. A wave of S type through the outer core has not been observed. We have also phases such as PP, reflected from the Earth's surface. Table 2.4 gives details of different local, regional and teleseismic phases. For global earthquakes and global seismology, readers may refer to Lay and Wallace (1995).

Table 2.4: Nomenclature of Seismic Phases (Kuthánek, 1990)

Symbol	*Meaning*
Local and regional events	
P, S	Direct compressional or shear wave travelling through the upper crust (observed only at very short epicentral distances)
P_g, S_g	Compressional or shear wave in the granitic layer of the crust
P_mP, S_mS	Compressional or shear wave reflected at Moho
P_n, S_n	Compressional or shear wave travelling along (just beneath) the Moho discontinuity, so-called head wave
P^*, S^* (or P_b, S_b)	Compressional or shear wave travelling along (just beneath) the Conrad discontinuity
pP_n	Depth phase that leaves the focus upward as *P*, is reflected as *P* at the free surface and continues further as P_n
sP_n	Depth phase that leaves the focus upward as *S*, is reflected and converted into *P* at the free surface and continues further as S_n
R_g	Short-period crustal surface wave of Rayleigh type
L_g	Guided crustal wave traversing large distances along continental paths
T	Compressional wave propagating through the ocean (Tertiary wave). *T* phases are occasionally observed even at larger teleseismic distances
TP_g (TS_g, TR_g)	Wave that travels the ocean and land portion of the total transmission path as *T* and P_g (S_g, R_g), respectively
Teleseismic events	
P, S	Direct compressional or shear wave, so-called elementary or main wave
PP, PPP, SS, SSS	*P* or *S* wave reflected once or twice at the Earth's surface
SP	*S* wave converted into P upon reflection at the Earth's surface
PPS, PSP, PSS	*P* wave twice reflected/converted at the Earth's surface.
PcP, ScS	*P* or *S* wave reflected at the core-mantle boundary
PcS, ScP	*P* or *S* wave converted respectively into *S* or *P* upon reflection at the core-mantle boundary
pP, pS, pPP, pPS, etc.	Depth phase that leaves the focus upward as *P*(*p* leg), is reflected/converted at the free surface and continues further as *P, S, PP, PS*, etc.

(*Contd.*)

Table 2.4 (*Contd.*)

Symbol	*Meaning*
sP, sS, sPP, sPS, etc.	Depth phase that leaves the focus upward as *S*(*s* leg), is reflected/converted at the free surface and continues further as *P*, *S*, *PP*, *PS*, etc.
pmP	*P* wave reflected at the underside of Moho
pwP	*P* wave reflected at the water surface
PdP	*P* wave reflected at the underside of a discontinuity at depth *d* in the upper part of the Earth. *d* is given in kilometres, e.g., *P*400*P*
Pc, Sc or *Pdif, Sdif*	*P* or *S* wave that is diffracted around the core-mantle boundary
PKP (or *P′*)	*P* wave traversing the outer core
PKS	*P* wave traversing the outer core as *P* and converted back into *S* when again entering the mantle
SKS	*S* wave converted into *S* on refraction when leaving the core
SKP	*S* wave converted into *P* on refraction into the outer core
PKP1, PKP2 or PKP_{BC}, PKP_{AB}	Different branches of *PKP*
PKIKP or P^2, $PKPr_{DF}$	*P* wave traversing the outer and inner core
PKiKP	*P* wave reflected at the boundary of the inner core
PKIIKP	*P* wave reflected from the inside of the inner-core boundary
PKKP	*P* wave reflected from the inside of the core-mantle boundary
PmKP (m = 3, 4,...*)*	*P* wave reflected m – 1 times from the inside of the core-mantle boundary
SmKS (m = 3, 4,...*)*	*S* wave converted into *P* on refraction at the outer core, reflected m – 1 times from the inside of the core-mantle boundary and finally converted back into *S* when again entering the mantle
PKPPKP (or *P′ P′*)	*PKP* wave reflected from the free surface, passing twice through the core
P′dP′	*PKP* reflected at the underside of the discontinuity at depth *d* in the upper part of the Earth, *d* is given in kilometers
LR	Surface wave of Rayleigh type
LQ	Surface wave of Love type
G	Mantle wave of Love type
R	Mantle wave of Rayleigh type
G1, G2	*LQ*-type mantle wave that travels the direct and anticentre routes. Waves that have, in addition, travelled once or several times around the Earth are denoted *G3*, *G4*, *G5*, *G6*, etc.
R1, R2	*LR*-type mantle wave that travels the direct and anticentre routes. Waves that have, in addition, travelled once or several times around the Earth are denoted *R3*, *R4*, *R5*, *R6*, etc.

Global Travel-time Models

The velocities of P and S waves provide information about the Earth's interior. When an earthquake occurs, waves from it are recorded at known locations and at known times, although the latter is not free of error. The travel-time curves are always plotted against the arc distance (Δ). The time interval S-P measured at any station gives a unique indication of arc distance to the earthquake, and distances determined from three or more stations locate the source uniquely within limited errors. The weakness in this method is, however, a time-distance curve must be established before the source can be located, but this location is essential for construction of the curve. So, in the early history of seismology time-distance curve was constructed for the earthquakes, which were precisely located by direct observations or by very close seismographs (Wiechert and Zöppritz, 1907).

Revision of time-distance curves was accomplished by Jeffreys (1935) by least-square technique. In this method, the residuals i.e. the differences between the observed arrival times of P or S and their computed arrival times based on preliminary location and preliminary travel-time curves, are minimized by simultaneous adjustment of epicentre coordinates, origin time and travel-time curve. Number of equations become very great as more earthquakes and stations are utilised. However, Jeffreys devised a method of treating the observations in groups. This analysis in collaboration with Bullen produced the Jeffreys-Bullen (J-B) Tables (1935, 1940). This gives adjusted travel times for P and S waves and also for reflected and refracted waves as functions of the arc distances Δ (Fig. 2.41). Bullen (1965) has described the production of these tables, including the derivation of the correction for the elliptical shape of the Earth. This has provided basic information of body wave seismology, which in turn yielded Earth models that are still taken as standards for reference. Improvement in observational technique, specially timing and availability of large artificial known explosions, have permitted correction to the J-B tables; and the new tables for P-waves were published by Herrin et al. (1968).

In 1987 the International Association of Seismology and Physics of the Earth's Interior (IASPEI) initiated a major international effort to construct new global travel-time tables for earthquake location and phase identification. As a result of this, two models were developed: IASP 91 (Kennett and Engdahl, 1991) and SP6 (Morelli and Dziewonski, 1993). The most significant differences between these new models and the older J-B travel-time model are in the upper mantle and core. The IASP 91 model has been adopted as the global reference model for International Data Centre in Vienna.

Subsequently Kennett et al. (1995) began with the best characteristics of the IASP 91 and SP 6 models, and further improved the global reference model to AK 135. The AK 135 model differs from IASP 91 only in velocity gradient for the D″ layer and in the baseline for S-wave travel times. The

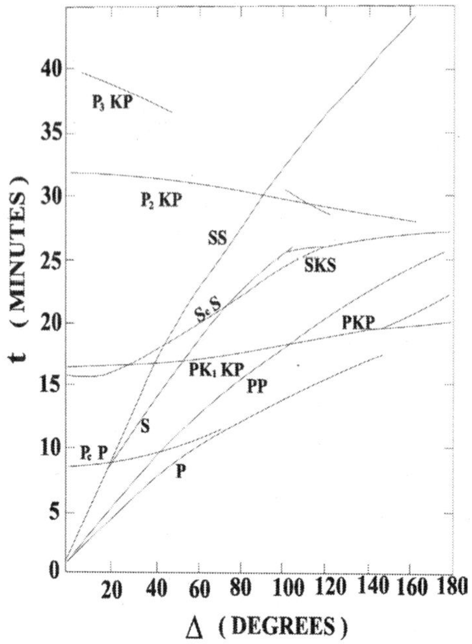

Fig. 2.41 Time-distance curves for some important seismic waves (Garland, 1979).

model AK 135, the body wave travel-time tables and plots, are available via http: // rses.anu.edu.au/seismology/ak135/intro.html, and can be downloaded and printed in postscript.

2.18 WAVE ATTENUATION

Seismic attenuation is caused by either intrinsic anelasticity or scattering effects. The intrinsic anelasticity is associated with small scale crystal dislocations, friction and movement of interstitial fluids. This effect is called *intrinsic attenuation*. The scattering effect, on the other hand, is an elastic process of redistributing wave energy by reflection, refraction and conversion that occur at discontinuities in the medium. This scattering of energy at small scale heterogeneities along travel paths reduce amplitudes of the seismic waves. These effects are called *scattering attenuation*.

The wave attenuation is usually expressed in terms of dimensionless *quality factor* Q, where

$$Q = 2\pi E/\Delta E \tag{2.42}$$

The ΔE is the dissipated energy per cycle. Large energy loss means low Q and vice versa, i.e. Q is inversely proportional to the attenuation.

We can also write an equation for amplitude as a function of distance travelled (Aki and Richards, 1980):

$$A(\chi) = A_o e^{-(f\pi/Q)\,\chi} \tag{2.43}$$

It is obvious from eq. 2.43 that for a constant value of Q a high frequency wave attenuates more rapidly than low frequency wave. Q is observed to be dependent on frequency. At higher frequency, in general, Q increases with frequency. In the Earth, Q is a function of depth, with the lowest Q values occurring in the upper mantle. In general, Q increases with *seismic impedance*, i.e. with increased material density and velocity.

The Q for P-waves in the Earth is systematically larger than Q for S-waves and thus refer to corresponding Q_p and Q_s respectively; $Q_p \simeq 9/4\ Q_s$. Table 2.5 gives approximate values of Q for different rocks.

Table 2.5: Q for different rock types

Rock Type	Q_p	Q_s
Shale	30	10
Sandstone	58	30
Granite	250	70-150
Peridotite	1200	520
Outer Core	8000	0

The mechanisms of intrinsic attenuation are very sensitive to pressure and temperature. High heat flow areas have more attenuation than colder regions. A schematic distribution of Q in a subduction zone is illustrated in Fig. 2.10.

In practice, it is difficult to separate intrinsic attenuation and scattering attenuation. Particularly, local earthquake records are strongly affected by scattering for local crustal heterogeneities, and scattering Q dominates. Scattering Q is usually determined from the decay of coda waves following S_g onsets, and is accordingly called Q_c. Unlike Q defined for intrinsic attenuation, Q_c is not a measure of energy loss per cycle, but rather, a measure of energy redistribution. Q_c is thus very much path dependent, and also strongly frequency dependent. A full discussion on this topic can be found in Aki and Richards (1980).

In this context, it may be mentioned that amplitudes of S-waves are generally five times larger than those of P-waves. This follows from the far field term of Green's function when modeling earthquake shear sources taking into account $V_p \simeq \sqrt{3}\ V_s$. Also, the periods of S-waves are longer, at least by a factor of $\sqrt{3}$, than those of P-waves due to differences in wave propagation velocity.

2.19 SEISMIC DIFFRACTION

Seismic diffraction is defined as the transmission of energy by non- geometric ray paths. In optics, the classic example of diffraction is light "leaking" around the edge of an opaque screen. In seismology, diffraction occurs whenever the radius of curvature of a reflecting interface is less than a few wavelengths of the propagating wave. Figure 2.42a shows a plane wave incident upon an opaque (acoustic impedance is infinite) boundary. Ray theory requires that waves arriving at seismometers at points *F* and *G* have identical amplitudes; no energy is transmitted to the right of point *G*. In fact, the edge of the boundary acts like a secondary source (Huygens' principle) and radiates energy forward in all directions.

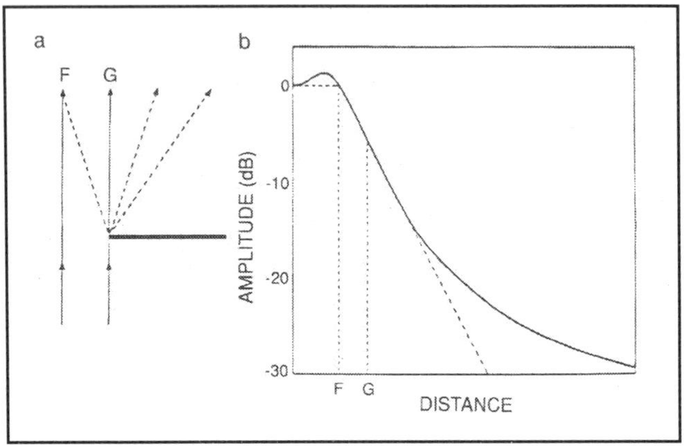

Fig. 2.42 (a) Rays incident on a grating, energy get diffracted around the edge. (b) Amplitude of energy as a function of distance into the diffraction zone (Doorbos, 1989).

These diffractions can be understood from the standpoint of *Fresnel zones*, a concept that states that waves reflect from a large region rather than just a point. Thus, the Fresnel zone causes the ray travelling to *F* to "see" the edge of the reflector, although the geometric ray path clearly misses the boundary. The first Fresnel zone may be thought to as a cone with the edge of the reflector as its apex. For a receiver, that is at a distance *d* beyond the reflector, the cone's radius is given by:

$$r = d + \frac{1}{2}\lambda \qquad (2.42)$$

where λ is the wavelength of the seismic wave. Figure 2.42b shows the amplitude variation predicted for the experiment given in Fig. 2.42a.

The real Earth significantly deviate from simplified global one-dimensional models, thus scattering and diffraction affect not only amplitudes but also travel times. This has to be taken into account in recent development of automated travel-time measurement techniques. Marquering et al. (1999) showed that for an SS wave observed at an epicentral distance of 80°, near surface heterogeneities can influence upon the SS wave travel time.

APPENDIX 2.1

Modified Mercalli (MM) Intensity Scale (1956 version)
(Richter, 1958)

Intensity Description

I. Not felt. Marginal and long-period effects of large earthquakes.

II. Felt by persons at rest, on upper floors, or favourably placed.

III. Felt indoors. Hanging objects swing. Vibration like passing of light trucks. Duration estimated. May not be recognized as an earthquake.

IV. Hanging objects swing. Vibration like passing of heavy trucks; or sensation of a jolt like a heavy ball striking the walls. Standing cars rock. Windows, dishes, doors rattle. Glasses clink. Crockery clashes. In the upper range of IV, wooden walls and frame creak.

V. Felt outdoors; direction estimated. Sleepers wakened. Liquids disturbed, some spilled. Small unstable objects displaced or upset. Doors swing, close, open. Shutters, pictures move. Pendulum clocks stop, start, change rate.

VI. Felt by all. Many frightened and run outdoors. Persons walk unsteadily. Windows, dishes, glassware broken. Knick-knacks, books, etc. off shelves. Pictures off walls. Furniture moved or overturned. Weak plaster and masonry *D* cracked. Small bells ring (church, school). Trees, bushes shaken visibly, or heard to rustle.

VII. Difficult to stand. Noticed by drivers. Hanging objects quiver. Furniture broken. Damage to masonry *D*, including cracks. Weak chimneys broken at roof line. Fall of plaster, loose bricks, stones, tiles, cornices also unbraced parapets and architectural ornaments. Some cracks in masonry *C*. Waves on ponds, water turbid with mud. Small slides and caving in along sand or gravel banks. Large bells ring. Concrete irrigation ditches damaged.

VIII. Steering of cars affected. Damaged to masonry *C*; partial collapse. Some damage to masonry *B*; none to masonry *A*. Fall of stucco and some masonry walls. Twisting, fall of chimneys, factory stacks, monuments, towers, elevated tanks. Frame houses moved on foundations if not bolted down; loose panel walls thrown out. Decayed piling broken off. Branches broken from trees. Changes in flow or temperature of springs and wells. Cracks in wet ground and on steep slopes.

IX. General panic. Masonry *D* destroyed; masonry *C* heavily damaged, sometimes with complete collapse; masonry *B* seriously damaged. General damage to foundations. Frame structures, if not bolted, shifted off foundations. Frames racked. Serious damage to reservoirs. Underground pipes broken. Conspicuous cracks in ground. In alluviated areas sand and mud ejected, earthquake fountains, sand craters.

X. Most masonry and frame structures destroyed with their foundations. Some well-built wooden structures and bridges destroyed. Serious damage to dams, dykes, embankments. Large landslides. Water thrown on banks of canals, rivers, lakes, etc. Sand and mud shifted horizontally on beaches and flat land. Rails bent slightly.

XI. Rails bent greatly. Underground pipelines completely out of service.

XII. Damage nearly total. Large rock masses displaced. Lines of sight and level distorted. Objects thrown into the air.

Masonry A, B, C, D. To avoid ambiguity of language, the quality of masonry, brick or otherwise, is specified by the following lettering.

Masonry A. Good workmanship, mortar, and design; reinforced, especially laterally, and bound together by using steel, concrete, etc.; designed to resist lateral forces.

Masonry B. Good workmanship and mortar, reinforced, but not designed in detail to resist lateral forces.

Masonry C. Ordinary workmanship and mortar; no extreme weaknesses like failing to tie in at corners, but neither reinforced nor designed against horizontal forces.

Masonry D. Weak materials, such as adobe; poor mortar; low standards of workmanship; weak horizontally.

3

Microearthquake Recording and Data Analysis

Introduction

Since 1980s the Geological Survey of India (GSI) has taken a lead in microearthquake investigation for detailed seismotectonics and active fault studies in various parts of India. Although this work also had important beginnings at the NGRI (National Geophysical Research Institute), IMD (India Meteorological Department), University of Roorkee (presently Indian Institute of Technology, Roorkee), WIHG (Wadia Institute of Himalayan Geology) and at the other national research institutes and universities, the work has been pursued most vigorously by the GSI in different parts of the Himalaya, Northeast India, Andaman Islands and Peninsular shield area in the country.

Choice of the seismic system for microearthquake network is a critical issue. The choice must depend on the primary uses of data. As mentioned earlier, a *temporary network* is suitable for quick data collection for seismological mapping or aftershock investigation, whereas a *permanent network* is designed for continuous monitoring of seismicity. Before we discuss on networks, we would first try to understand the earthquake recording instruments, which are called *seismographs*. The seismographs essentially have two components: *seismometer* and recorder. A recorder is an instrument which gives record of the ground displacement or velocity. The records can be analog or digital. The records are called *seismogram*. The part of the seismograph, which is sensing the ground motion, is called the seismometer. It picks up seismic vibrations of the ground and converts them into signals that can be recorded in one of a number of alternative ways.

3.1 SEISMOMETERS: THEORY AND PRACTICE

Commonly we are so much involved in analysing the seismograms and understanding the generating process of the earthquakes, that we do not give much importance on the seismometers. We do not realise on what shaky ground the seismometers are, and they don't do what the manufacturers claim they do. So, it is important that we first understand what a seismometer is, and what it can do, and what it can't do.

The motion of the ground due to an earthquake will have components in all the three cardinal directions. Usually different designs of seismometers are used for picking up different components of ground motion. Further, one group of seismologists is interested to record high frequencies, as high as 100 Hz, while another group expect to record signals with period as low as 100 second down to the prevailing noise level. More recently there is an effort with very sophisticated *long period* instruments, which record with very high *dynamic range* to bridge this gap. Their benefit to *short period* seismology is, however, questionable. Again, the amplitude of ground motion may vary over a wide range of values. This calls for different designs of seismometers, for example, for recording *strong motions* caused by a nearby earthquake, or for recording the motions due to distant earthquakes.

3.1.1 Development of Seismometers

The first known attempts to simply record the occurrence of ground motion were conducted by the Chinese. Figure 2.11 is a picture of the first Chinese seismologist, Chang Heng, and a vertical mechanical seismoscope, which he built as early as 320 A.D. It was one of the first instruments designed specially to record the occurrence of ground motion, but does not produce recording as a function of time. Significant further development of ground-motion sensors was not persued until the 1700s. The Italians developed numerous seismoscopes in the early eighteenth century. The first true seismograph, which recorded the relative motion of a pendulum and the Earth as a function of time, was built by Filippo Cecchi in Italy in 1875. The oldest known seismic record produced by this system is dated February 1887. A group of British seismologists, the best known being John Milne, James Ewing and Thomas Gray, developed first vertical component and first relatively long period systems. By 1900, the first global array of 40 photographically recording seismographs provided initial seismogram data-base to begin to understand the Earth vibrations.

Mechanical System

The principle of a mechanical sesimograph is shown in Fig. 3.1. Here a heavy mass M is hanging through a spring from a rod S, which is fixed to the ground. A pen P is attached to the downward portion of the mass M. A recorder R, which is also fixed to the ground, is rotating and a paper is covered over the recorder. Now, if the ground moves up, the rod as well as the recorder moves up. The mass M, however, does not move initially due to its inertia, and the spring extends. Thus relative to the recorder the mass M apparently moves downward and the pen writes forward motion on the recording paper as shown in the figure. Afterwards the mass will move up and down for some time like a pendulum even if the ground is not moving. The system clearly involves the phenomena of *resonance.* The amplitude of relative motion between mass and the frame may get highly magnified with respect to the ground motion if the frequency of motion is close to the resonant frequency and the seismometer is not adequately damped. Therefore, it is important to choose properly the resonant frequency and the damping of the seismometer. A damping arrangement D is attached to the mass M so that it moves relatively only once when ground moves. We note that upward motion of the ground makes the pen move downward. However, in actual recording we adopt a device to record this as an upward motion. The seismograph shown in the Fig. 3.1 records vertical motion of the ground; so this is called a *vertical seismograph. Horizontal seismograph* can also be made with similar principle as shown in Fig. 3.2, where the pendulum moves like a two-way sewing door. A spring H, called hinge, is used, and this allows the mass M to oscillate about an equilibrium position. The arrow shows the movement of the ground, and relative movement is seen on the recorder R. In order to get a three-dimensional picture of the ground motion, it is necessary to record in three orthogonal (perpendicular) directions: generally, in vertical (Z), north-south (N) and in east-west (E) directions. In Figs 3.1 and 3.2, the part without recorder is the seismometer.

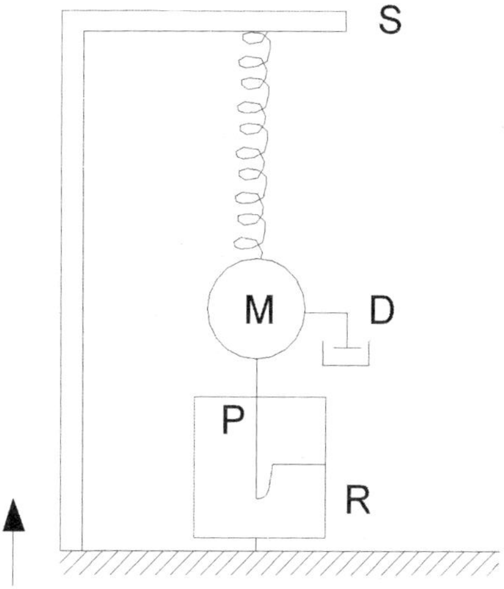

Fig. 3.1 Principle of a vertical seismograph, the arrow shows upward motion of the ground, and the pen (P) records on the recorder (R) (see text).

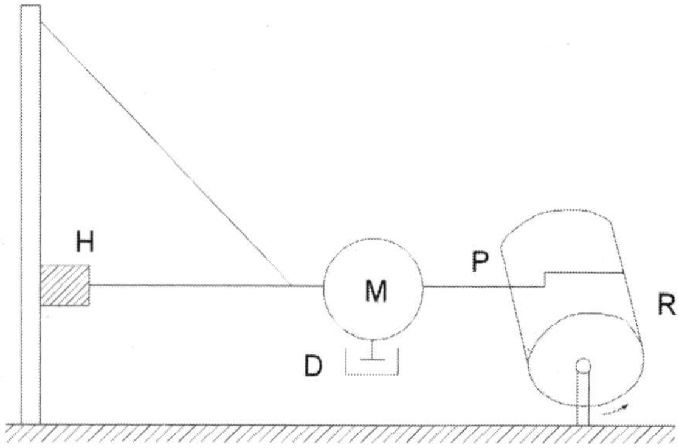

Fig. 3.2 Principle of a horizontal seismograph, the arrow shows the horizontal motion of the ground (see text).

As stated above, the motion of the mass of the seismometer behaves like a forced motion of a damped pendulum. Figure 3.3 demonstrates that if the damping constant (h) of the seismometer is too small or under damped ($h < 1$), there would be undesirable 'ringing' of the signal near the *natural frequency* of the seismometer. If, on the other hand, the damping is excessive or over damped ($h > 1$), the response of the seismometer would be too sluggish. In practice, a compromise between these two extremes is done by keeping the damping constant close to unity. When the damping constant is unity, the seismometer is called *critically damped* ($h = 1$). After moving in one direction, the mass comes back to the equilibrium position without crossing the other direction, and the time taken to come back from extreme position to equilibrium position is half the free period of the pendulum. The actual damping constant in a seismometer is generally kept at about 70% critical damping.

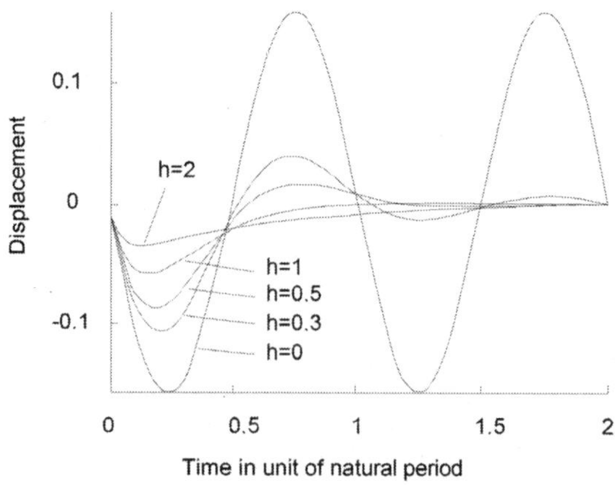

Fig. 3.3 Relative motion of mass (displacement) as a function of time for different values of damping constant (h).

Optical System

The mechanical amplification of ground motion was superseded by galvanometric recording, which was introduced by a Russian seismologist, Boris Galistin (1900-1915). By using a fairly heavy mass in the seismometer and a light mass in the galvanometer, one could get additional amplification. This instrument leapfrogged from magnification of the order of 100 for mechanical seismograph to magnification of the order of 100,000. It came both in horizontal and vertical configurations. An important example of this type of seismometer is *Wood Anderson seismograph.* This is a horizontal seismograph, and was developed in 1920s by the two scientists, Wood, H.O and Anderson, J. A. (Anderson and Wood, 1925). Here the pendulum consists of a metal cylinder (generally made of copper), which is attached to a vertical suspension wire (Fig. 3.4). During ground motion, the cylinder rotates around the wire and a mirror attached to the cylinder reflects a light beam towards the photographic paper on a recorder drum. A horse-shoe magnet surrounds the copper cylinder to give the damping.

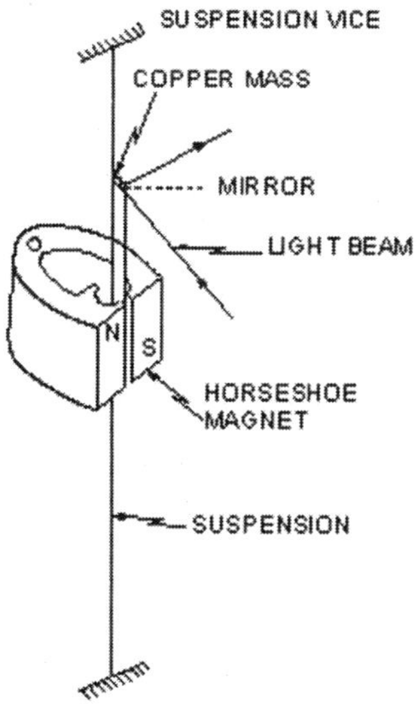

Fig. 3.4 Principle of an optical system, Wood-Anderson seismometer.

Figure 3.5 shows the *response curves* of a few seismometers. The curve 1 with high peak in the upper left corner is 100 kg Benioff. This instrument also involves galvanometric recording; it has magnification 100,000. For short-period seismology, this instrument dominated until it was superseded by electronic seismograph. The Wood Anderson instrument, curve 3, shows lowest magnification among the short-period seismographs shown in Fig. 3.5. For long period, varieties of instruments have been built with peak magnifications of the order of 3000 or so.

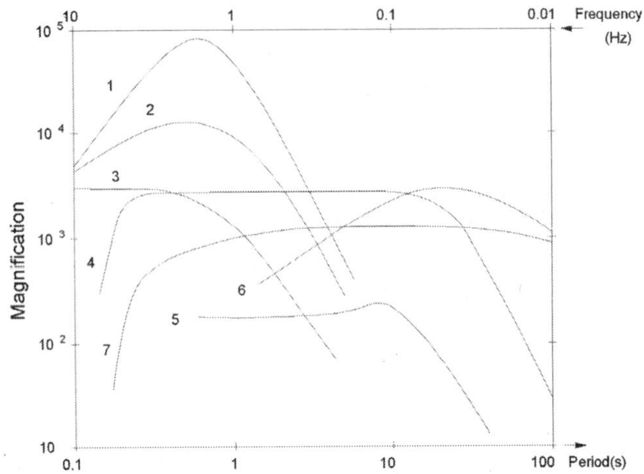

Fig. 3.5 Response curves of a few seismometers (Eaton, 1992): (1) Benioff (SP-WWSSN), (2) Grenet-Coulomb, (3) Wood-Anderson, (4) Kirons, (5) Wiechert, (6) Press Ewing (LP-WWSSN) and (7) Broad-band. (SP and LP stand for short period and long period, respectively).

Electronic System

In the higher frequencies the Earth is quite a bit quieter, and we thus need to run to higher magnifications. The electronically amplified seismometers are different from that of the Benioff; the electronic instruments have a much wider range over which its magnification is high. A few modern electronic seismometers and accelerometers are shown in Fig. 3.6.

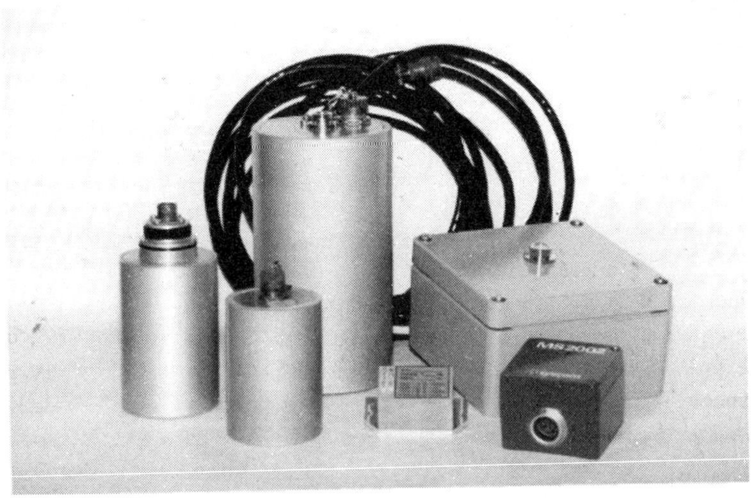

Fig. 3.6 Photograph showing a few electronic seismometers and accelerometers (courtesy Dr. L. Zimakov, Reftek, USA).

Figure 3.7 shows a schematic view of an electronic (or moving coil) seismometer, equivalent to Mark Products L-4 seismometer. It has a free period of one second, and it does not have any mechanical output. As shown, the seismometer mass is suspended by a spring; the mass is surrounded by a magnet and constrained to move vertically. When there is a relative motion between the seismometer mass and the frame, an output voltage is developed. L_0 is the unloaded length of the spring, and L is the loaded length. The heart of a moving coil seismometer is the coil unit, which is called the *transducer*. When the seismometer is disturbed, the mass is displaced, say by an amount Z. The function of the transducer is to convert the relative motion between the coil and the frame into an output voltage. The induced emf produced by the motion is proportional to a motor-constant times the relative velocity of the mass with respect to the frame. When a current is fed through the coil, it reacts with the magnetic field to produce a force, which moves the mass with respect to the frame.

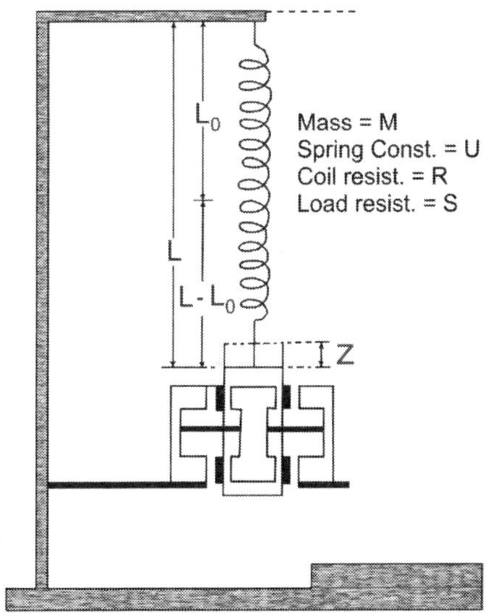

Fig. 3.7 Schematic view of a moving coil seismometer.

One of the critical aspects of the electronic sensors is that they employ force-feedback systems. This involves a negative feedback loop in which a force proportional to the inertial mass displacement is applied to the mass to cancel its relative motion. The force-feedback strategy greatly extends the *bandwidth* and linearity of a seismometer. The force-feedback systems of various types have actually existed at least since 1926. Much of the challenge in designing *broadband* seismometers has been in the development of stable force-generating systems that can respond to the whole range of motions of the Earth.

3.1.2 Equations of Motion for Seismometer

The equations of motion for the seismometer involve gravity, the force due to the spring, damping, and the driving force that comes from the earthquake waves. The force on the mass due to gravity is given by M_g. The restoring force developed by the spring is given by $[(L - L_o) - Z]\, U$, where Z is the disturbance to the system and U is the spring constant (Fig. 3.7). This force turns out to be $M_g - ZU$.

There are two forms of damping; these are:

$$\textit{open circuit damping} = -\Delta\dot{Z}$$

and, $$\textit{electromagnetic damping} = -GI = -\frac{G^2}{R+S}\dot{Z}$$

So, $$\text{the } \textit{total damping} = -\left(\Delta+\frac{G^2}{R+S}\right)\dot{Z} \tag{3.1}$$

The *inertial force* is given by:

$$\textit{inertia} = -M\ddot{\chi}$$

Finally, there is *virtual force* developed by movement of the Earth relative to the χ frame, which is given by:

$$\textit{virtual force} = -M\ddot{\chi}$$

Summing the forces, we get the equation of motion:

$$-M\ddot{Z}-\left(\Delta+\frac{G^2}{R+S}\right)\dot{Z}+M_g-ZU-M_g-M\ddot{\chi} = 0 \tag{3.2}$$

Simplifying we get

$$\ddot{Z}+\frac{1}{M}\left(\Delta+\frac{G^2}{R+S}\right)\dot{Z}+\frac{U}{M}Z-\ddot{\chi} = 0 \tag{3.3}$$

3.1.3 Seismometer Calibration

Calibration of a seismometer establishes knowledge of the relationship between its input (ground motion) and its output (an electrical signal). Since precisely known ground motions are difficult to generate, electromagnetic force is generated in a calibration coil. If the factor of proportionality between the current in the coil and the equivalent ground motion (velocity/acceleration) is known, then calibration is purely electrical measurement.

Figure 3.8 is a schematic diagram of external circuit of a seismometer. The seismometer is characterised by a motor constant G and an internal resistance R. When there is a relative motion between the mass and the frame, an output voltage is developed across the seismometer leads, and it is given by

$$E(t) = G\dot{Z}(t) \tag{3.4}$$

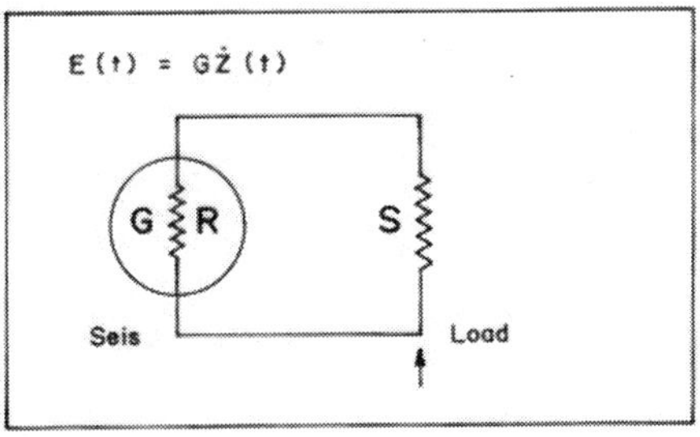

Fig. 3.8 External circuit of a seismometer.

Normally the seismometer is driving some other unit like the input of an amplifier, which has an internal resistance *S*. So, the voltage that actually appears is given by *S*/(*R*+*S*). This is the relation between the output voltage and velocity.

Generally the manufacturer specifies the values of the motor constant, the mass of the seismometer, the free period of the seismometer, the open circuit damping and the damping corresponding to some particular load value. This makes to believe that this is all true, but there are varieties of factors, which may cause the seismometer to change with time.

Testing a closed seismometer, like L-4, is called the *release test*. One way of doing this test is to apply a voltage to the coil which offsets the mass, and calculate what the offset is. If suddenly the voltage is cut off, the seismometer returns to zero while recording its output voltage. It turns out to be a very useful test which produces a *calibration curve* for the instrument. This may be done every day while changing records. An example of calibration curve is shown in Fig. 3.9. This is useful not only to determine the constants of the instrument, but also, in a crude form, to know whether the instrument is centered or not. As we actually carry out the test, we send a current *I* through the seismometer and it produces an offset equal to the motor constant times *I* divided by the spring constant *U*. With a simple experiment in the laboratory, one can determine the instrument constants.

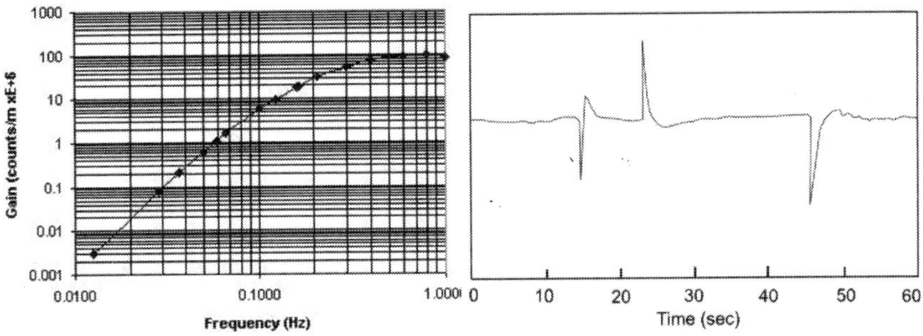

(a) (b)

Fig. 3.9 (a) Calibration curve of a seismometer, (b) calibration test in the field.

For three component (triaxial) seismometers calibration current is split into three adjustable and invertible x,y,z components. These are then adjusted so that test signal appears only at the desired output of the seismometer. Using a *shake table,* mechanical vibration, is the most direct way of obtaining absolute calibration. For details readers may refer to IASPEI manual on Seismological Observatory Practice.

3.1.4 Frequency Response and Dynamic Range

In general, *frequency response* and *dynamic range* describe the expected performance of a seismic system.

Frequency Response

The frequency response describes the relative amplitude at which the system responds to the ground motion as a function of frequency. Pendulums have a special property, namely that after being excited by an arbitrary motion of the frame and subsequently left alone, they swing at a characteristic frequency, f_n (or period T_n), called the *natural frequency, eigen frequency* (German word eigen means own) or *free frequency.* Within certain limits, f_n is amplitude invariable. It is obvious that this property has a direct consequence for the behaviour of pendulum seismometers. When the ground moves forth and back very slowly at a frequency much smaller than f_n, the pendulum's mass follows exactly the movement of the ground. There is no relative movement between the mass and the frame. In other words, when the frequency of the Earth vibrations decreases towards very small values, the amplitude A or the relative movement (pendulum vs frame) tends towards zero. If the frequency f increases, the amplitude A also increases and at the resonant frequency f_n the amplitude A reaches its maximum value. If we further increase the frequency f above f_n the pendulum's mass lags behind the ground vibrations and A again decreases. At very high frequencies of the ground motion, the mass does not move at all which means that the relative movement (in absolute value) between the mass and the frame is equal to that of the frame and $A = 1$. A simple sketch showing the dependence of A upon the frequency of the ground vibrations is displayed in Fig. 3.10.

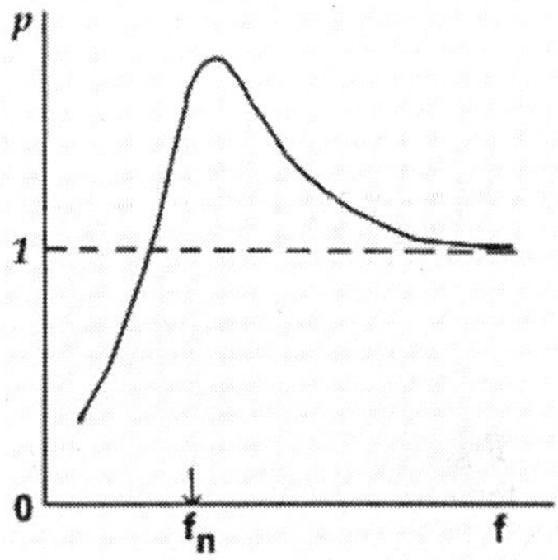

Fig. 3.10 Amplitude-frequency response curve of a seismometer, f_n is a natural frequency.

The frequency content of earthquake waveforms varies with magnitude, from 1 to 100 Hz for small local events to periods of 50 min or more for great earthquakes (Armstrong, 1969; Bacon, 1975). Within this wide spectrum, the earthquakes recorded by a typical microearthquake network show predominant spectral content between 1 and 20 Hz (Oliver et al., 1966).

Corner frequency of a seismograph may be defined as the periods where magnification drops by 3db of peak value i.e. 0.707 of peak magnification. One corner frequency is at the lower side of the period and the other is at the higher side. The range is defined as *bandwidth*. This gives the width of the period within which the seismograph is sensitive to ground motion.

Some interesting relations are inferred from the combined displacement response curves (Fig. 3.11). For earthquakes in the magnitude range slightly greater than 1 to about 4, the peak of the combined displacement response curve, i.e. the frequency of the maximum amplitude that would be seen on the seismogram varies with the magnitude of the earthquake. For the assumed source model this frequency corresponds to the *corner frequency* f_c. For earthquakes more than magnitude 4, the frequency of maximum amplitude will always be 1 Hz. This is because of (i) the rapid roll-off of the seismometer displacement-response below its natural frequency (1 Hz) and (ii) the corner frequency being less than 1 Hz. For earthquakes magnitude less than 1, the frequency of maximum amplitude will be always 30 Hz. This also happens due to (i) roll-off response above 30 Hz and (ii) the corner frequency being greater than 30 Hz.

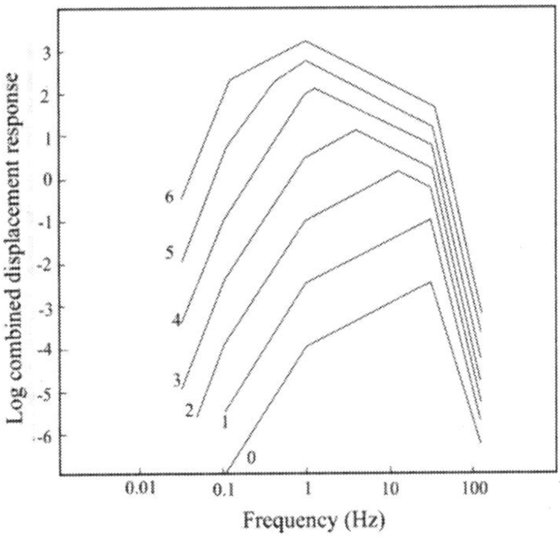

Fig. 3.11 Theoretical combined displacement response as a function of frequency for earthquake ranging in magnitude 0-6. Corner frequencies are shown by arrows, magnitudes are annotated in each curve (Lee and Stewart, 1980).

Dynamic Range

The dynamic range, on the other hand, is the ratio of the largest to smallest input signal that the system can measure without distortion. It is expressed as:

$$\text{Dynamic range (db)} = 20 \log_{10} \frac{\text{clipping level}}{\text{lowest detectable or noise level}}$$

For example, in MEQ–800 analog recorder smallest readable signal amplitude = 0.2 mm and maximum per deflection = 25 mm. Thus,

$$\begin{aligned}\text{Dynamic range} &= 20 \log_{10} \frac{25.0}{0.2} \\ &= 20 \log_{10} (125) \\ &= 45 \text{ db.}\end{aligned}$$

3.1.5 Choice of Sensors

There are several types of seismometers like *Long Period (LP)*, *Broad Band (BB)* and *Short Period (SP)*. The SP seismometers cover a band roughly from one to 50 Hz. Typical examples are Kinemetrics SS-1, Geotech S-13 and the Mark Product L-4C. These are designed for recording local earthquakes. The LP seismometers respond frequencies from about 0.01 second to 1.0 second. These are good for recording teleseismic events. The BB seismometers, on the other hand, span the response spectrum covered by these two instruments. They provide complete seismic information from about 0.01 Hz to 50 Hz, and therefore, allow much broader range of studies than the SP records. They are relatively expensive, and are used mostly for research. Typical examples are the Guralp CMG40T with frequency range 0.03 to 50 Hz and the Wieland-Streckeisen STS2 with frequency range 0.008 to 40 Hz.

There now also exist very broadband (VBB) seismometers, which resolve lowest frequencies resulting from Earth's tides and free oscillations of the Earth. Their important advantage over BB is their ability to record frequencies below 0.001 Hz. Their primary purpose is the research of deep interior of the Earth. The data from a VBB are very important for international seismological community. One or two such stations are recommended in each country depending on the area. Typical example of VBB is the Wieland-Streckeisen STS-1.

Figure 3.12 illustrates the merits of the broadband seismograph relative to WWSSN and GDSN stations with short period or long period seismometers. The broadband system avoids artificial separation of signal energy into separate short-period and long-period channels as done in the WWSSN or GDSN instrumentation.

Further, there is also a choice for recording *velocity* versus *acceleration*. The general rule of thumb is that the seismometer or velocity sensor is used to record weak motion in the magnitude range 0 to 6, and the acceleration sensor, or the *accelerometer,* for strong-motion, magnitude 5 and larger. The acceleration sensors are not sensitive enough to record the smaller events. In practice, accelerometers are used in strong-motion seismograph array for estimation of ground acceleration of large earthquakes for engineering construction purposes. Acceleration of ground motion is expressed in terms

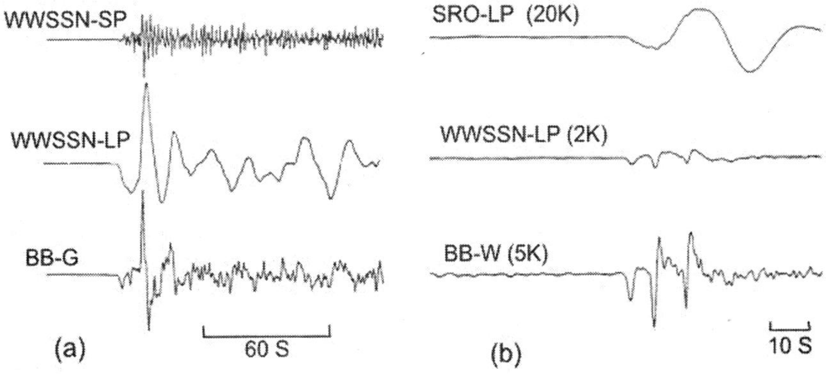

Fig. 3.12 Comparison of seismograms with varying instrument responses for the same ground motion. (a) Teleseismic P-wave of an earthquake recorded by WWSSN short- and long-period seismograph and with broadband seismograph. (b) Teleseismic P-wave recorded by GDSN (SRO-LP), WWSSN-LP and broadband seismographs at an observatory, Germany (Lay and Wallace, 1995). Broadband records contain much more information.

of g (g = acceleration due to gravity = 980 cm/s^2). Accelerometers with 0.25, 0.5, 1, 2 and 4 g full scale sensitivity are available today. In high seismic risk area, a few accelerometers may be kept in a permanent local or microearthquake network to have some on-scale records of large local earthquakes, because the trace amplitudes of such earthquakes, recorded with seismometers close to the epicentres, are generally clipped by the conventional recorders. Although there is significant differences in strong and weak motion network designs, today both type of sensors are integrated into a single system.

Six channel data loggers with three weak and three strong motion channels are cost effective and are in current state-of-the-art. They are capable of covering the whole dynamic range of seismic events, from lowest seismic noise to largest damaging events. In terms of signal processing there is no difference in using a seismometer or an accelerometer.

There is also a question whether to use *vertical component sensors* or use *3-component sensors*. Obviously one would like to use all 3-component sensors to observe ground motions in all three directions towards North, East and Vertical. It would, however, cost about three times as much as the equivalent network using vertical component sensors. Vertical component sensors are really good only for P-wave. It is, however, not easy to recognise the S-wave on a vertical waveform trace. The 3-component seismometers respond to both P- and S-waves distinctly. S-waves are easily identified in horizontal components of ground motions. Basically, the choice of sensor depends on one's objective and the money that can be spent.

Till early 1980s, many permanent seismic stations and temporary microearthquake networks used single component sensors, usually vertical SP seismometers. This was the case because the equipment was analog and the recording was on paper. If three components had been used, neither it would have been cost effective nor easy to generate ground motion vector from three separate paper seismograms.

Today in the era of digital seismology, the situation is different. The price/performance ratio is much favourable for three component systems. Since modern seismological analyses require all seismic information of ground motion vector, one component systems are no longer a desirable choice for new installations. However, single component (vertical SP) systems are still a choice where communication, capability and economy are limiting factors.

3.2 SEISMOGRAPHS: RECORDING SYSTEMS

3.2.1 Analog Recorder

The analog seismographs, till the other day, were very popular as they provide visual records. The complete set is portable in a carrying case except the seismometer. The principal components of the recording system consist of a filter, amplifier, calibration unit, recording media, timing system and a power supply. The recording is made by ink on a photographic paper or by a stylus on a smoked paper or heat sensitive paper. In the heat sensitive recording mode, the pen/stylus remains hot and by scratching it removes the white chemical cover of the recording paper to bring out the black background.

In portable seismograph the record size is approximately 30 × 60 cm. To achieve a timing resolution better than ± 0.1 s the records may be changed daily keeping the drum speed 120 mm/min. Details of the technical descriptions of the seismographs are seldom published. However, early versions of portable seismographs were published by Oliver et al. (1966). The standard portable seismographs are commercially available. Instruments of at least three commercial companies of USA are still widely used in many countries of the world including India. These companies are: *Kinemetrics Inc.*, *Sprengnether Instrumnet Co. Inc.* and *Teledyne Geotech.* The recording system of a portable smoked-paper seismograph is exemplified in Fig. 3.13 and a smoked-paper seismogram is shown in Fig. 3.14.

The main advantage with the analog recording is that if one works in the field he would immediately get a feel about the general level of seismicity just by looking at an analog seismogram. He can right away know how many earthquakes are there, and whether these were large or small. He can immediately know whether the instrument didn't work at all for whatever reason. This is an advantage that the author always liked in the field. Digital recording, however, has its own advantages (see Section 3.2.3). So, a best system is that which produces both analog and digital seismograms. A world example of such mixed network is given in Section 3.3.2.

Fig. 3.13 Recording system of portable smoked-paper seismographs in the field; (a) Sprengnether MEQ-600 and (b) Kinemetrics PS-2 (used by the author in New Zealand and in India respectively).

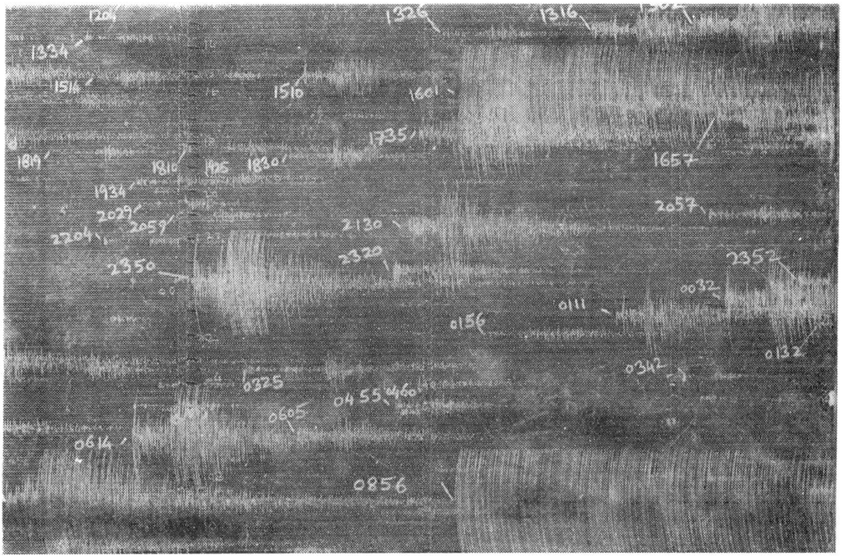

Fig. 3.14 Part of a smoked-paper seismogram showing earthquakes recorded by a PS-2 seismograph. The seismogram showing aftershocks recorded on April 7, 1999 of the March 28, 1999, Chamoli earthquake in the Garhwal Himalaya tectonic zone (courtesy, Sagina Ram). Time marks are one minute apart. The events are clearly identified and annotated. Note that the amplitudes of the higher magnitude earthquakes are clipped. S-waves are difficult to read.

3.2.2 Magnetic-Tape Recorder

The magnetic-tape seismographs are analogous to the analog instruments excepting that the recording is made on a magnetic tape instead of recording on a paper. A magnetic-tape recorder, used by the author, is shown in Fig. 3.15, and a seismogram is shown in Fig. 3.16. It has four-channel recording system; two channels for signal (low gain and high gain), one channel for time-mark and one channel for radio signal. It is a continuous mode of recording system, and one spool (tape-deck) can run for about two weeks at a speed of 0.4 mm/s. Thus it does not need daily servicing. The tape can be read with the help of a play back system and an oscilloscope. A recorded microearthquake on the tape seismograph is shown in Fig. 3.16. The time marks are recorded and the arrival times of the seismic phases can directly be read on the oscilloscope. The reading of the seismic phases can be made precisely on the oscilloscope, but searching of the events by audio-visual (playback and oscilloscope) system is very tedious, and without the help of an analog record this job is difficult. Although recording system provides better timing/reading accuracy and requires less manpower for fieldwork, it has also become obsolete with the advent of the digital system.

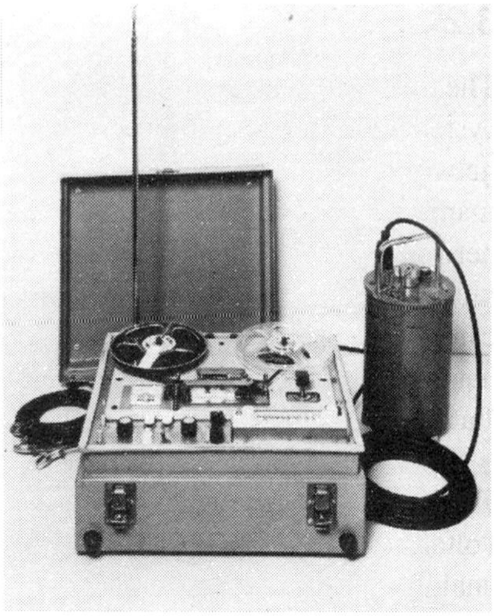

Fig. 3.15 Magnetic-tape recording system (courtesy, R.R. Dibble, 1979), which was used by the author in New Zealand).

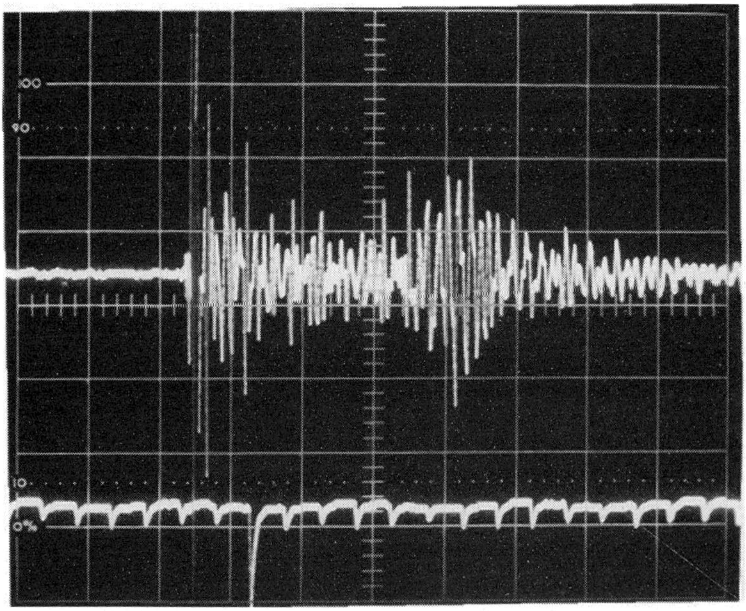

Fig. 3.16 Magnetic-tape seismogram of a microearthquake (photograph from the oscilloscope display). Time marks are shown below; larger mark indicates minute and smaller ones indicate seconds (Kayal, 1983). P- and S-waves are clearly identified.

3.2.3 Digital Recorder

The development of improved digital data acquisition and data analysis system during the last decade has now tipped the balance in favour of a network with digital instruments. This system has not only reduced the manpower requirement, but it is the state-of-the-art technique for running a network system; particularly for a permanent network this system is now a must. The system provides high precision data, and much more information can be extracted from the records compared to the analog system. It is more handy and more light to carry to the field for making a temporary seismic station or a network.

In this system, the seismometer is analog, all other equipments are digital. The amplifier output is connected to a digital recorder, which converts the voltage to counts, say L counts/mv instead of mm/mv as recorded by the analog seismographs. The counts are recorded at equal interval of time, *sample per second* (SPS). A low-pass filter is set according to SPS; its frequency is generally 0.4 times the SPS. For example, if SPS = 100, the low pass filter is 40 Hz. The analog to digital converter (ADC) turns the value into numbers of quantized values. The minimum digitisation step is called digit or count. The smallest unit of a digital value is byte (= 8 bits). A 16-bit ADC can count values from –32768 to 32768. Thus,

$$\text{Dynamic range} \doteq 20 \log_{10} (32768/1) = 20 \times 4.5 = 90 \text{ db.}$$

On the other hand, 24-bit count from – 8,338,608, and this gives dynamic range as high as 140 db. Increase of dynamic range in digital recording compared to analog recording gives the advantage to record the ground motion for a very small magnitude earthquake as well as for large magnitude earthquake without saturation.

Continuous Versus Trigger Mode of Data Acquisition

In *continuous mode* recording, a three-component seismograph with 100 SPS in each component and 16 bits (2 bytes) recording requires in one day a storage of 3 (comp.) × 100 (SPS) × 2 (byte) × 24 (hrs) × 3600 (sec) = 51,840,000 = 50 Mb. This would produce huge volume of data, that is implausible to store for any length of time. Although many large storage devices are coming up recently but it is necessary to keep the storing area to a minimum. To achieve this goal, the network users operate the system in *triggered* mode. In trigger mode, the recording is continuous and in real time, but only stores signal associated with the triggered seismic events. Such systems do not store continuous time histories of seismic signals, rather produce 'event files'. A digital recorder has a few streams of recording; for example, one stream can record at 20 SPS continuously and other stream can record at 100 SPS in trigger mode, i.e. when amplitude of the record goes above the normal level. In trigger mode, recording is done only when the ratio of amplitude of *short term average* (STA) to that of *long term average* (LTA) is more than a given limit. When the signal to noise ratio crosses this limit the recorder starts recording from a few seconds earlier to this time, and records for some time as set by the operator. The digital recorder also maintains the log of this trigger and signal to noise ratio at the time of trigger. Further, for reduction of storing area, compressed techniques are available and this reduces the storing area by half or even less. Main parameters of a digital seismograph are bandwidth, dynamic range and bits (i.e. 16-bit or 24-bit) used in recording.

Advantages of digital recording

Digital recording allows us to do much more quantitative analysis than possible with the analog recording. The main advantages of the digital system are:

1. Large dynamic range due to which amplitude does not saturate in case of large ground amplitude.
2. Easy to store and disseminate the database for more scientific research.
3. Picking up of arrival time of phases on computer screen in an interactive manner by moving the cursor manually. Phases are identified clearly by zooming the portion of interest of the seismogram, and are read very precisely in microsecond. Figure 3.17 illustrates an example of zooming of digital seismogram. This allows one to pick first-motion time and polarity much precisely.

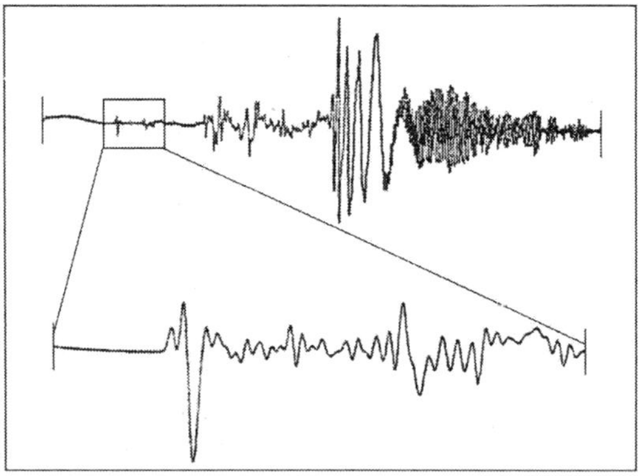

Fig. 3.17 Zooming of a part of the digital seismogram to read P-arrival time and first-motion.

4. Filtering of the trace can be done with the digital seismograms. It is a straight forward matter to convert the time domain into frequency domain, and see how the earthquake motion is distributed by frequency. Figure 3.18 illustrates an example of generation of seismograms at different frequency band within the recording bandwidth. The filtering is applied to a broadband seismogram. One can emphasize surface waves by low-pass filtering, and body waves by high-pass filtering.

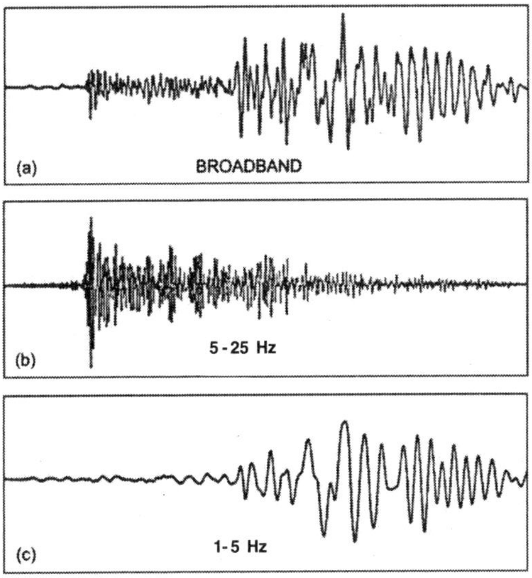

Fig. 3.18 (a) Broad band seismogram (vertical displacement component), (b) short period component of the seismogram, and (c) long period component of the seismogram after filtering.

5. Frequency domain analysis or spectrum analysis by Fourier Transform can be done, and many source parameters can be obtained. The reason for performing this kind of analysis is for modelling earthquake source, site response studies etc. Brune (1970) developed a model for how *the far field spectrum* should behave for an ideal *double couple* earthquake source. In Brune's model there are two parameters which describe the far field spectrum. These are level of low frequency flat part of the spectrum and the corner frequency at which the spectrum begins to decay (Fig. 3.19).

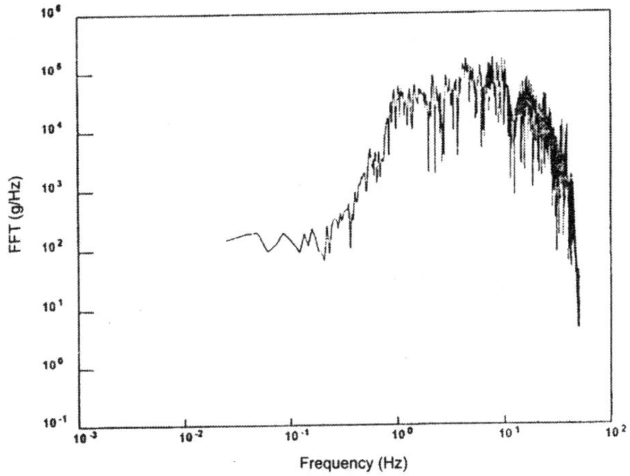

Fig. 3.19 Power spectrum (FFT) of a digital seismogram of an aftershock of the 1999 Chamoli earthquake (Kayal et al., 2002).

6. Ground motion along any direction can be obtained using the three-component digital data. In many studies it is necessary to obtain radial and transverse component by vectorial rotation of the two horizontal components. Further, S-wave is very clear in transverse component. In addition, P-wave group is absent in this component and this can be used to find the direction of the epicentre from the observatory (see Section 3.8.2).
7. Further, digital records allow to estimate *wavelet* parameters. Such parameters provide much deeper insight into the seismic source processes and seismic moment release. The duration of a true ground displacement pulse t_W and the rise time T_r tc its maximum amplitude contain information about size of the source, stress drop and attenuation of the pulse while propagating through the Earth. Integrating over the area underneath a displacement pulse allows to determine its signal moment m_s (Fig. 3.20). In addition, the attenuation and scattering properties along the wave path can be drawn from the analysis of wavelet envelopes (Fig. 3.20).

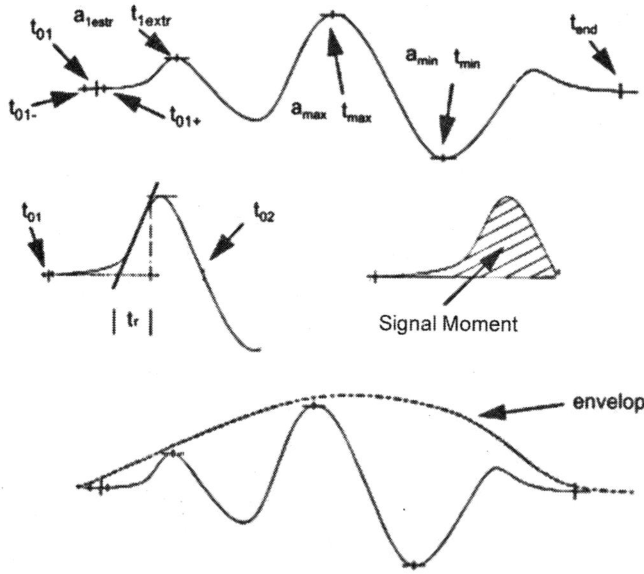

Fig. 3.20 A schematic diagram showing time (t) and amplitude (a) function, signal moment and wavelet envelope (IASPEI, 2002).

3.2.4 Timing System

The *timing system* of a seismograph is an important component. Radio signal (broadcasting at 15.5 KHz) is taken as the primary time base, particularly for the analog systems. A built-in quartz clock of the instrument is synchronised with the radio signal by external connection. The *time signal* is serial binary code at one pulse/second. The *time marks* (hr, min and sec) are automatically recorded by the system on the seismogram. For clarity, only hour and minute marks are generally recorded on analog records (Fig. 3.14). The minute marks can be separated by 60 mm or 120 mm as desired by adjusting the recording-drum speed. The 60 mm separation is a general convention, but 120 mm separation gives better precision in reading the arrival times of the seismic waves, particularly the S-waves. First-motions can also be read more precisely with 120 mm/m drum speed. The only disadvantage with the 120 mm/m speed is that the record lines become very close which makes it difficult for the reader to identify the lines while processing. Overall timing accuracy (including reading error, clock drift etc.) in this system is ± 0.1s for P arrivals, and ± 0.2-0.5s for S-arrivals (Kayal, 1983).

With the advancement of technology, the GPS (Ground Positioning System) sattelite timing are now synchronised with the recording system. This system provides much better precision in timing; accuracy in microsecond is easily obtained. This system is very economically and fruitfully used in digital recording. Synchronisation of the GPS timing into analog system is very expensive. The conversion system from GPS to analog may be as expensive as the seismograph, the whole recording system. Thus timing accuracy in analog system is still maintained by radio-signal synchronisation and also by using higher drum speed. With a little practice in the field, the radio-signal can easily be synchronised with the quartz clock in the analog system, and a high precision timing can be maintained as stated above.

3.2.5 Power Supply

For *power supply* two rechargeable batteries are supplied with each instrument by the manufacturers. These batteries, however, do not run for long time. A built-in charger is provided with the instrument but it requires electricity facility at the recording site for charging, which is, in most cases, not available for temporary network in the field. The batteries can be of course charged at the base camp where electricity is available. A 12 volt, 13 plate car battery can safely be used for power supply. These batteries can be charged with a 2 Amp home-charger at the base camp, and a fully charged car battery can run the instrument for about a month. The batteries should be replaced when the voltage drops to 11 v. For permanent telemetric network, however, Uninterrupted Power Supply (UPS) system should be used in the Central Recording Station (CRS).

3.3 DEVELOPMENT OF MODERN SEISMIC SYSTEM

Initially seismic data were all recorded on drums, using either photographic paper, ordinary paper with ink and pen, or thermo-sensitive paper and a hot stylus. These drum-based systems have been around since about 1900, and are still in use to some extent today. With the development of microearthquake recording instruments, smoked-paper instruments became very popular since 1960s for making temporary networks for aftershock investigation or for any short-term field investigation, and they are still used till date. The main disadvantage with this system is that each record has its own timing system, and each earthquake may be recorded in several different types of recorders. As a result, data processing become very tedious.

The next step in the evolution of seismic data acquisition was the development of the 20-channel *Develocorder system*. This system records seismic data on a 16 mm film, and were very popular during 1970-1980s. Many permanent networks were established with this system in USA, Japan and in New Zealand. The advantage with this system is that all the signals from different seismograph stations are recorded on a common time-base. An example of such record is shown in Fig. 3.21.

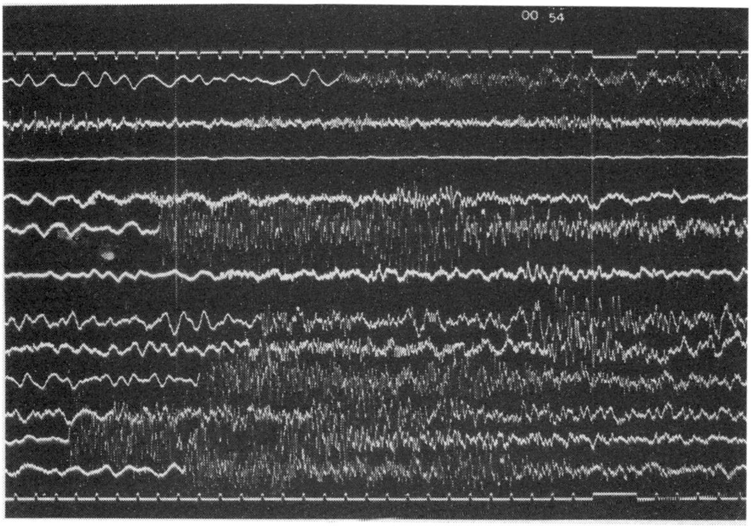

Fig. 3.21 An example of a develocorder-film record for a local earthquake from the permanent Wellington telemetric network (Kayal, 1983). Time marks are shown above and below; larger mark indicates minute and smaller ones indicate seconds.

In 1970-1980s mini computer-based systems were developed in USA and in other advanced countries. They work quite well but they cost up to millions of dollars.

In 1980s several groups of seismologists in USA, Japan, France, South Africa, China and Spain began to develop PC-based seismic systems. These systems have the advantage of being inexpensive compared to the above system, and relatively simple to implement. In 1990 the PC-based system was extended to 128 channels by means of an external *multiplexer*, and in 1991 the system was expanded to provide 16-bit A/D (Analog to Digital) conversion. The system provided far more data, which required powerful software to keep up with the analysis. In 1990, the IASPEI (International Association of Seismology and Physics of the Earth's Interior) in collaboration with the Seismological Society of America published the IASPEI software volumes 1 and 2, which are essentially the network processing softwares; volume 3 came out in 1991, which is a waveform analysis software. Now, all the volumes, i.e. up to volume 6, are available, which provide wide spectrum of data analysis softwares.

It is important to know how many bits should be used to encode the data. A 12-bit system will give a dynamic range of about 2,048 (counts). A 16-bit system has a dynamic range =32,000; this much dynamic range is very good. The research people, however, would like 24 bit systems. These systems have dynamic range =8,388,608. With such systems one can record earthquakes magnitude 1 to 8 on scale. Of course, 24 bit systems cost lots of money. Various institutes in India like NGRI, RRL(J), IMD, GSI, WIHG, started using these systems since 1998.

3.3.1 Permanent Telemetry Network

With the available technology, there exists three types of telemetric data acquisition systems: analog, mixed and digital.

Analog Telemetry

The analog telemetry consists of sensors (which is always analog), analog signal conditioning, frequency modulated (FM) telemetry through Radio (RF) or telephone lines, analog de-multiplexers and analog drum or film recorders. Paper or film records are the final result (Fig. 3.21). There are, however, two shortcomings with such system:

(i) Low dynamic range (40-45 db) and resolution of the acquired data; many events are recorded with too low amplitudes and many with clipped amplitudes.
(ii) Incompatibility of paper or film records with computer analysis.

For these reasons analog telemetry systems are no longer being built.

Mixed Telemetry

The mixed telemetry systems have analog sensors, analog signal conditioning, usually FM telemetry, analog de-multiplexers but digital data acquisition at the central recording site, digital processing and digital data archiving (Fig. 3.22).

The systems have low dynamic range and same disadvantages as the analog systems regarding data completeness and quality. However, one can use modern computer analysis methods except those that require high resolution raw data.

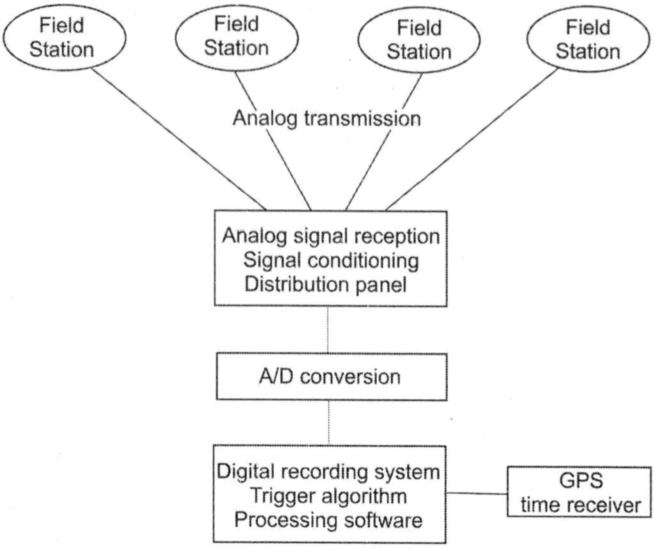

Fig. 3.22 A block diagram of a mixed telemetry system (IASPEI, 2002).

Digital Telemetry

In digital telemetry systems, only the sensors are analog. All other equipments are digital. The digital data are transmitted to the central recording station over fixed digital communication channels (Fig. 3.23). At reception the signals enter the recorder directly. The dynamic range and the resolution are much higher than that of analog or mixed systems. 12-24 bit A/D converters are available today, that correspond to dynamic ranges of approximately 70 to 140 db.

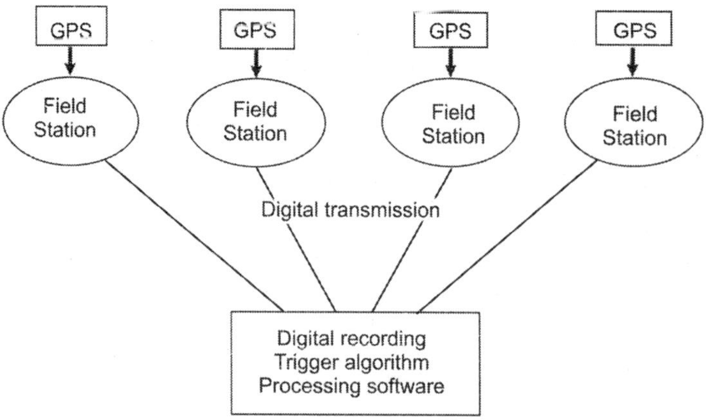

Fig. 3.23 A block diagram of digital telemetry system (IASPEI, 2002).

There are two known design principles that can further increase the dynamic range and/or resolution. The gain ranging method automatically adjusts the analog gain according to the amplitude of the seismic signal and thus prevents clipping of the strongest events. In this the dynamic range is dramatically increased; the resolution, however, remains roughly unchanged. The over sampling principle is another approach which helps the dynamic range and resolution. The data are sampled at much higher rate; the value of each sample of the final output data stream is calculated by statistical model. The increase in the resolution is significant. The resolution, however, depends on the ratio of the over sampling rate to final sampling rate.

3.3.2 World Examples of Network

There are many large permanent telemetric networks in many countries around the world including USA, Japan, China, New Zealand, France, UK, Russia etc. Short descriptions of California and Japan networks are given below, that give an idea of permanent networks.

California Seismic Networks

There are two large networks of more than 350 seismic stations each in northern and southern California. The southern California Seismic Network (SCSN) is one of the largest and most automated telemetric networks in the world. It consists of triggered as well as continuous digital and analog systems. The central recording station is at Caltech. The SCSN as well as the Northern California Seismic Network (NCSN) have been the pioneers in local permanent telemetric system in USA (Lee and Stewart, 1981). It is interesting to note that despite highest technology being adopted, there are still some robust analog stations in the SCSN (Fig. 3.24). The analog stations have its own advantage as stated above.

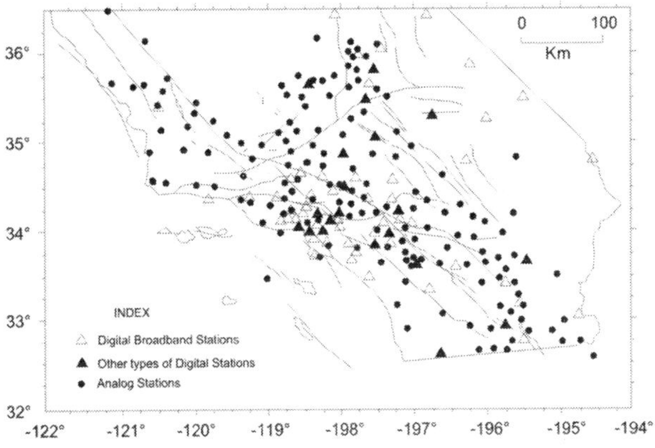

Fig. 3.24 The map shows Southern California Seismic Network (SCSN) with different types of seismometers (SCSN home page http://www.trinet.org.scsn.html).

Japanese Seismic Networks

Japan is seismically very active and it has dense and most modern networks. There are three seismic networks that are run by the National Research Institute for Earth Science and Disaster Prevention (NIED). The first network is named *Hi-net*, it consists of 600 short period (SP) seismographs. The SP sensors are installed in deep (100-200 m) boreholes. Along with the SP sensors strong motion accelerometers are also installed in the deep boreholes. The second network, *F-net*, consists of 70 broadband (BB) seismographs. The BB sensors together with strong-motion sensors are installed in deep (~50 m) vaults. The third one, *K-net*, consists of 1000 strong-motion seismographs, the accelerometers are installed on the ground surface. Any user in the world has open access to the data from these networks via internet (http://www.hinet.basai.go.jp).

3.3.3 Telemetric Networks in India

In India, though operation of digital telemetric network has started recently, a few analog telemetry systems were developed in early 1980s. An analog telemetry network was in operation for about 15 years since early 1980s in Bhatsa (near Mumbai). The network was indigenously developed by the Bhaba Atomic Research Centre (BARC) (Kolvankar, 1988). The BARC is also running a mixed telemetric array, specially designed network, in Gouribidanur in collaboration with UK since 1965 for detection of nuclear explosion.

In late 1980s, the NGRI (Hyderabad) and the RRL-J (Regional Research Laboratory, Jorhat, Assam) jointly started two six-station analog telemetric networks in Assam; one in Tejpur and the other in Jorhat respectively. The Teledyne Geotech Company installed the networks. The seismic data were published in the yearly NGRI-RRL(J) Seismological Bulletins. These two networks are upgraded to 24-bit digital telemetric networks since 2001. The Tejpur network is running with six BB stations, and the Jorhat telemetry network with 14 BB stations.

The NGRI was running a seven-station digital (16 bits) telemetric network in Kudankulam, Tamil Nadu state, in early 1990s, which was established by the Project authority for a nuclear power plant project. No research work or data were published on this network. The University of Roorkee (UR) attempted to install one digital (16 bits) telemetric network in 1980s in Tehri area, Garhwal Himalaya, but it did not perform well due to various technical problems (Kumar et al., 1997). The Tehri telemetric network was redesigned by the UR (presently IIT Roorkee) in 1995 with the support of Tehri dam project authority, which was running with SP analog system. This network is upgraded to 24-bit digital system since 2000. The IMD (India Meteorological Department) is operating a six-station 24-bit digital SP telemetric network in New Delhi since 2002.

After the 1993 Killari earthquake, the DST (Department of Science and Technology, New Delhi) proposed three 14-station digital (24 bits) radio-linked telemetric networks, one at Koyna-Warna, one at Latur in Maharashtra state and the other at Khandwa area in the Narmada-Son Lineament (NSL) zone in Madhya Pradesh state. The first two networks are proposed to be run by the NGRI, and the third one by the GSI. Two telemetric networks, the Koyna and the Khandwa, were established in 2002 and 2003 respectively. The Wadia Institute of Himalayan Geology (WIHG) is running a eight-station digital (24 bits) telemetric network in Western Himalya on a DST project also.

3.3.4 Temporary Seismic Network

A *temporary network* is established for quick investigation as well as for reconnaissance survey. Portable smoked-paper seismographs (analog recorders) were very popular till early 1990s for such surveys because they are reliable, and they produce visual seismograms. A few commercial companies in the United States manufacture the portable units. Indigenous analog instruments are now commercially available in India. Magnetic-tape recorders may also be used in temporary networks. These are, however, expensive and require a playback facility (Kayal, 1983). Temporary microearthquake networks with analog recorders or magnetic-tape recorders require a large amount of manual labour to collect and to process the data. With the advancement of technology, temporary networks are now-a-days preferably made by stand-alone digital seismographs. These require less manpower for data acquisition and data processing. Further, the digital instruments provide much precise GPS timing, and more information of the seismic sources can be extracted from the waveforms.

As mentioned in the beginning, the GSI takes a leading role to carry out microearthquake surveys routinely by temporary networks in different parts of the country for monitoring seismicity and delineating active faults. Other organisations, in addition to the GSI, like the NGRI, IMD, WIHG make temporary networks for special investigations, aftershock study etc. In subsequent chapters we shall be discussing the results of temporary seismic networks that were run in various parts of the country for various investigations.

3.4 STATION DISTRIBUTION AND SITE SELECTION

3.4.1 Station Distribution

For an effective microearthquake network (permanent or temporary), which can provide reliable data for earthquake location, the station sites should be evenly distributed by azimuth and distance. The maximum azimuthal gap between the stations should be less than 180°, and the distance between the stations should not be more than twice the average focal depth of the earthquakes. For example, the focal depths of the earthquakes are typically within 20-30 km in the Shillong Plateau. So, it is desirable that the station spacing is not greater than 40-50 km. Optimal distribution of stations has been studied by Sato and Skoko (1965) and Uhrhammer (1980), which suggest that if the earthquakes are uniformly distributed over a region A, then a network of approximately A/S^2 stations is needed, where S is the station spacing. If earthquakes are concentrated along a fault zone, the total number of stations required could be less.

3.4.2 Site Selection

There are various constraints in site selection, which affect station distribution. If seismic noise is high, benefit of modern instrument is lost and poor detection will result. If stations are situated on soft ground, even BB or VBB records will be useless and SP responses will not be interpretable due to local noises. For example, Quaternary alluvium causes higher background noise compared to bedrock or competent rock. Inaccessible terrain or large bodies of water may also put serious constraints on the station selection. Similarly, cultural noise e.g. near large cities, highways, farming lands or the livestock and wind noise may cause poor quality of records. One way to reduce cultural noise is to place the seismometers at some depth. For example, Steeples (1979) placed seismometers in steel cased boreholes, 50-60 m deep, for a telemetered network, which reduced the cultural noises like those of vehicle and livestock by as much as 10 db. Takahashi and Hamada (1975) have described the results of placing seismometers in deep boreholes in the vicinity of Tokyo. A seismometer is also placed in a 60-70 ft deep borehole in Jorhat (Assam) at the Central Recording Station of the RRL (J) telemetered network (Sitaram M.V., Pers. Comm., 1999). The author also gained similar experience in the fieldwork; when no bedrock exposure was available in the field the seismometers were kept on 5-8 ft thick concrete floors in 6-10 ft deep dug wells. This effort enhanced the gain of the recording instruments by at least 6 db.

Further, a few more pertinent points may be considered for making temporary microearthquake networks. First, office studies (at headquarter) should be made in selecting potential sites. The exercise of preliminary design of the network may be done with the help of a good seismo-geological map. The bedrock having higher *seismic impedance* (which is the product of density and velocity of the rock) produces less noise and provides high gain seismic station. It should be away from man made cultural or traffic noise. Second, the sites should have easy accessibility for daily servicing and changing of records and batteries, particularly for an analog network. With the availability of GPS, the station coordinates and elevation should be obtained more precisely, with ± 10 m accuracy. Maintaining of timing accuracy in analog system is very tricky; the radio signal should be regularly synchronized with the built-in quartz clock of the system.

3.5 ANALOG RECORD PROCESSING AND RECORD KEEPING

Record processing as well as *record keeping* is a procedure for the ultimate goal of data analysis and preparing seismicity map. It includes collecting, storing and cataloging all information associated with the microearthquake surveys.

3.5.1 Record Processing

First, a *scan list* may be made, which is a tabulation of the event-time in chronological order at each station, particularly for analog station. The regional and teleseismic events may also be included in this list with remarks. In addition to these, approximate S-P times are noted in the scan list for each earthquake at each seismograph station. The scan-list may be prepared for each station in the field itself. All these information are useful to save considerable time and to select the events for reading the phase-data for computer analysis at the head quarter.

For large data volume, systematic checking of the measured phase data is needed. Although it sounds easy to read the phase data visually on the computer for digital records, or using illuminated glass-top and an illuminated 10 × magnifier with graduated fine scale for analog records, different analysts may read seismograms differently. Errors may be introduced in measuring, writing and key punching. Like the fieldwork, it is a more tedious job to read the analog seismograms, and finally to prepare a seismicity map after several computer runs by checking and rechecking the phase data many times.

3.5.2 Record Keeping

Experiences show that unless the pertinent information is noted on the permanent record or register, or on a computer file, these are soon forgotten or lost. The following steps may be followed for analogs in record keeping:

Operational Detail

Instrument with station code, operational details like frequency, gain, speed, date, time etc. may be noted on the top left-corner of the smoked-paper itself before the operation is started. Similarly, while changing the record, the time and date may be noted at the end of the record. Immediately after the record is brought to the base camp, label/sticker with all the above details may be pasted on the bottom right corner of the seismogram. This not only helps to provide the necessary information about the record, but also helps to trace out any seismogram from the record-keeping shelf at any time.

Earthquake Annotation

First, the hour signals are annotated on the seismogram using white marker-pen or pencil, and then all the detected events, including the local, regional and teleseismic earthquakes, are annotated with the hour-minute number, e.g. 0320, which means 03 hr 20 min. (Fig. 3.14). This helps in finding the event in other records, and then for reading the phase data and for checking/rechecking the reading later.

Data Sheet

Data sheets are prepared by entering the phase data like P and S-arrival times, P-wave first-motion and signal duration on a sheet. There may be a remark column also for each event. The data sheet may be prepared in the Hypo-71 (USGS) format for easy entry of the data on the computer. This format is acceptable not only by the Hypo-71 computer program but also by other earthquake location programs that may require little modification.

Earthquake List

Earthquake list is prepared with origin-time, epicentre coordinates, focal depth, magnitude and pertinent information of the computing errors. The earthquake list and the phase data are filed on computer as well as on the diskette. These are published in the form of a bulletin.

Scientific Reports

Scientific reports describing seismicity, seismicity-rate, frequency-magnitude relation, focal-mechanism, velocity structure etc. may be published, which are useful for the study of *seismotectonics, risk evaluation, hazard mitigation, microzonation* etc., and for the other users.

3.6 DIGITAL DATA MANAGEMENT

In field experiments with continuous recording systems, one collects about 30-50 MBytes of data per day at a 3-component seismic station with 100 Hz sampling rate. In recent years, hard disks have overcome other mass storage technologies in field recording. They are reliable, robust and cheap. For connecting the hard disk to the recorder, many systems support Small Computer System Interface (SCSI). This system is an accepted and well defined industry standard, and available on most platforms. It is possible to change the hard disk without interrupting the recording process.

The hard disk, like a tape drive, can be used as sequential block device. The data stream is subdivided into blocks, including the header. The header defines the type of data, date, time stamp of the first sample. Besides raw data streams, state of health and status information are generated. Additional channels containing information about internal voltages, temperatures, information on the synchronization against the external clock are recorded. Each subsequent block of data is written to the hard disk. Steim (1986) introduced a widely used algorithm to compress integer time series without loss of information. Sampling three channels at a range 100 Hz, a 4-GByte hard disk will provide recording capacity for about 80 days.

All the data describing the instrumentation and site location related to the stored waveform are called metadata. For archiving and exchange of waveform data, there are several data formats in use, mainly pure ASCII code which is independent of the hardware platform, but widely accepted standard is the SEED format. SEED is the only standard format that holds the most important metadata and the waveform data within one single file.

3.7 BASIC DATA FOR ANALYSIS

Once the scan-list is prepared the following data are read from the seismograms for the selected events for hypocentre location and magnitude determination:

(i) P-wave *arrival time* (direct, reflected, refracted P_n etc.),
(ii) P-wave *first-motion* (up or down),
(iii) Converted-phase arrival times if any (like S_p P_s etc.)
(iv) S-wave arrival time (direct, reflected, refracted S_n etc.)
(v) *Signal duration* (i.e. the time taken after the first onset for the earthquake coda to become indistinguishable from the background noise). An example of a seismogram with clear P, S, P-wave first-motion and signal duration is shown in Fig. 3.25.

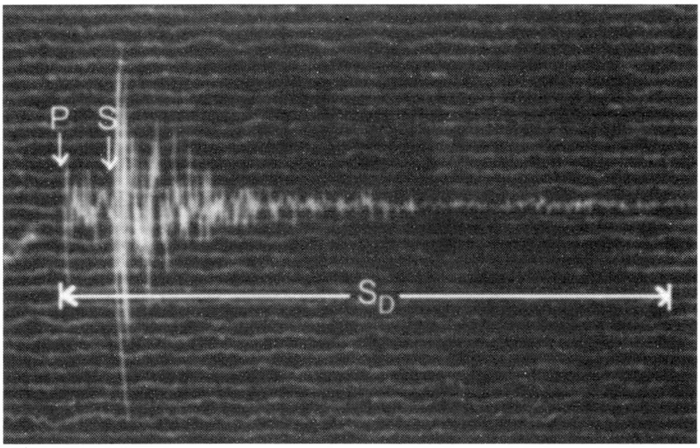

Fig. 3.25 Identification of P- and S-arrivals, P-wave first-motion (down) and signal duration (S_D) of a microearthquake recorded by a PS-2 analog system in Arunachal Himalaya (Kayal et al., 1993).

Precise measurement of phase data, first P-arrival and first S-arrival in particular, is most important as these data are used to compute origin-time, hypocentre parameters and finally to determine focal mechanism. The first onset of P-wave with impulsive arrivals in an analog record should be read with a precision of ± 0.05 second, while for emergent arrivals the uncertainty might be as high as ± 0.1 second. This much reading precision is possible using a 10x graduated magnifier with an illuminated glass top. For S-wave arrivals the uncertainty is larger, ± 0.1 to 0.5 second. The reading accuracy of the digital seismograms is, however, obtained in microseconds (0.001 s) for the P-arrivals as well as for the S-arrivals. Apart from these, the other converted phase data can be more easily read from the digital seismograms compared to analog records. Signal duration is used for estimation of duration magnitude (M_D). The signal duration may be read with an accuracy of ± 1 second or so. The list containing these data for an earthquake, which are read from the seismograms of different stations, is called the *phase list.*

With digital multi-station data and advanced seismogram analysis software, signal detection, signal filtering, polarization analysis and spectral analysis are routine task in digital data analysis. Digital records allow versatile signal processing so that additional *wavelet* parameters can also be measured. Figure 3.20 depicts signal parameters in relation to rise time (t_r) of the displacement pulse, signal moment m_s and wavelet envelope.

Data Reduction or Correction

Two types of corrections are needed in preparing the phase list, one is *station correction* and the other is clock-drift correction or *time correction.*

Station Correction

The station correction is needed when the seismometer is situated on sediments and/or on higher elevation. The sedimentary thickness and the elevation should be considered for the delay arrivals of the phases, which is called *station-delay.* The station-delay for sedimentary thickness can be calculated using a simple formula:

$$T = t\left(\frac{1}{V'} - \frac{1}{V}\right) \tag{3.5}$$

where T is the station delay, t is sedimentary thickness, V_p' is the velocity of sediments and V_p is the model velocity (Fig. 3.26). Similarly, corrections are needed for those stations situated on the higher elevation.

Time Correction

The *time-correction* is needed when there is a drift in the quartz clock of the analog recording instrument. A typical clock drift of a MEQ-600 instrument

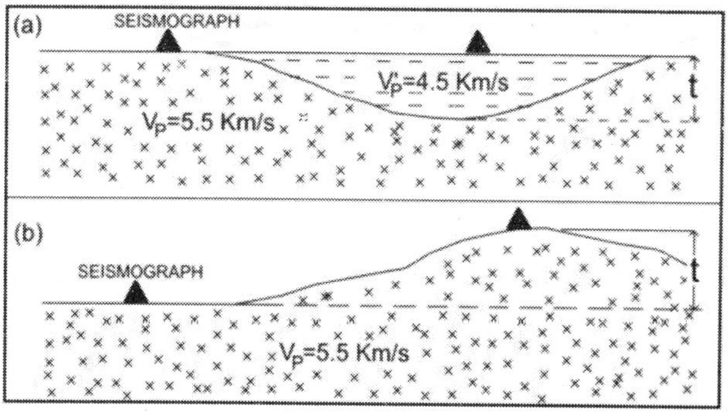

Fig. 3.26 Diagram showing station-delay due to: (a) sedimentary thickness and (b) higher elevation.

is shown in Fig. 3.27. A drift of the clock can be read from the difference in time-marks of the radio-signal and the clock signal, and it can be plotted with time. Necessary corrections can be read from this curve, which may be applied as and where necessary. The GPS timing in the digital system, however, takes automatic care for the drift, and no such graphical correction is needed.

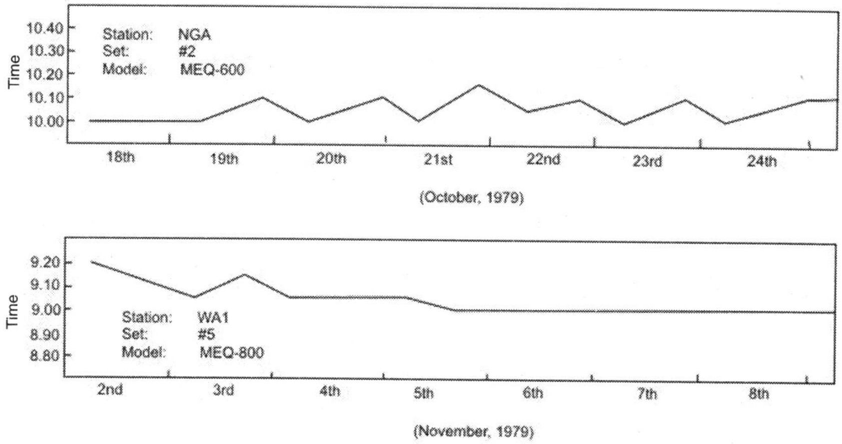

Fig. 3.27 Time-correction graph due to clock drift of a MEQ-600 instrument while using in a temporary network (Kayal, 1983).

3.8 EARTHQUAKE LOCATION

3.8.1 Manual Method

It is possible to get an approximate location of an earthquake without restoring to a computer. If the seismic network is dense and the earthquake is originated inside the network, the station with the earliest arrival can be approximated as the epicentre. This is called *earliest station method,* and it gives an estimate with a probable error of the same order as the station spacing.

Another approach is to contour the P-wave arrival times. This is an extension of the earliest station method. This may give fairly good estimate of the epicentre location if the network is dense enough. If the network is not dense and it consists of a few stations only, then the method is not effective.

A pretty good approximation of the epicentre can be made by *arc method* or *circle method* with three or more S-P times. Figure 3.28(a) shows simple geometry of the epicentral distance *X*, hypocentral distance *D* and depth of focus (*H*) of an earthquake. If T_p is arrival time of the P wave and T_s is that of the S wave, then we can write:

$$T_P = \frac{D}{V_p} \quad \text{and } T_S = \frac{D}{V_S}$$

where V_P and V_s are the velocity of the P-wave and S-wave respectively. Thus we get

$$T_S - T_P = D\left(\frac{1}{V_s} - \frac{1}{V_p}\right) \tag{3.6}$$

Assuming $V_p/V_s = 1.73$, the hypocentre distance can be estimated as

$$D = 8(T_s - T_p) \tag{3.7}$$

where *D* is in kilometres, and S-P is in second.

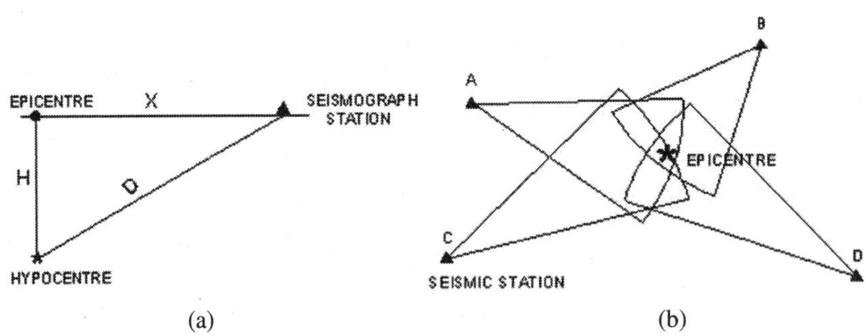

Fig. 3.28 (a) Diagram showing hypocentre, epicentre, and H: focal depth, D: hypocentral distance, X: epicentral distance. (b) Locating epicentre by drawing arcs using S-P interval as radius from the seismograph stations.

After determining *D* at three or more stations, arc with radius *D* can be drawn with each arc centred at the respective station. As shown in the Fig. 3.28b, the epicentre lies in the region where the arcs overlap. This method was commonly used as a quick means of locating earthquakes. The *focal depth* (*H*) can be obtained as

$$H = (D^2 - X^2)^{1/2} \tag{3.8}$$

Adding more observations will give additional intersections that theoretically should pass through the epicentre. In practice, error is always present, both in data and in assumptions, that the ray-paths are straight and the Earth is homogeneous; so scatter in the intersection usually occurs.

The *origin time* of an earthquake can be determined with a very simple graphical technique, called *Wadati plot* (Wadati, 1933). The time separation of the P and S phase ($t_s - t_p$) is plotted against the arrival time of the P-wave. Since $t_s - t_p = 0$ at the hypocentre, a straight line fit on the Wadati diagram gives the origin time at the intercept with the P-arrival-time axis (Fig. 3.29).

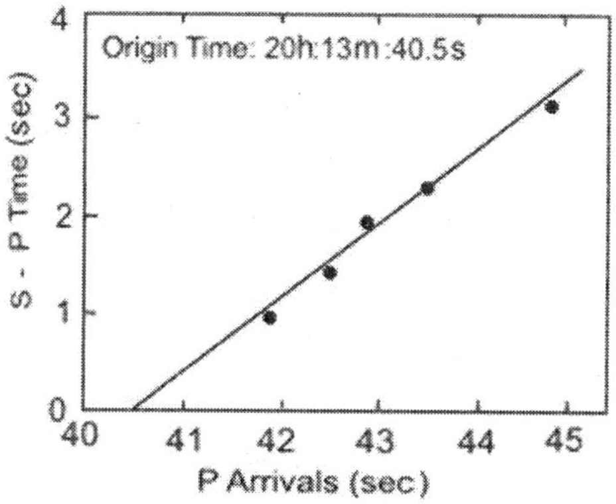

Fig. 3.29 Wadati-plot for determining origin time. The origin time is given by the intercept with the P-arrival time axis.

3.8.2 Single-Station Location

In general, the arrival times of seismic phases at several stations are required to determine earthquake hypocentre and its origin time (Section. 3.8.3). It is, however, possible to use a single seismic station to obtain a crude estimate of earthquake location (Roberts et al., 1989). Single-station method requires three-component recordings of ground motion. Since P waves are vertically and radially polarized, the vector P-wave motion and the amplitude ratio of horizontal components (AE/AN) can be used to infer the back azimuth (against North) from station to the source.

Figure 3.30 displays the nature of P-wave polarization. If the vertical motion of the P-wave is upward, its radial component is directed away from the epicentre. If the vertical component of the P-wave is downward, the radial component is directed toward the epicentre. With suitable software, the ratio of the amplitude on the two horizontal components (AE/AN) can be used to find the vector projection of the P-wave along the azimuth to the seismic source. This has been implemented in the SEISAN software (Havaskov and Ottemoller, 2000). An example is illustrated in Fig. 3.31. The distance (D) to the seismic source may be obtained from $D = 8\ (t_s - t_p)$. Knowing the hypocentre distance, one can estimate the P-travel time, and thereby origin time.

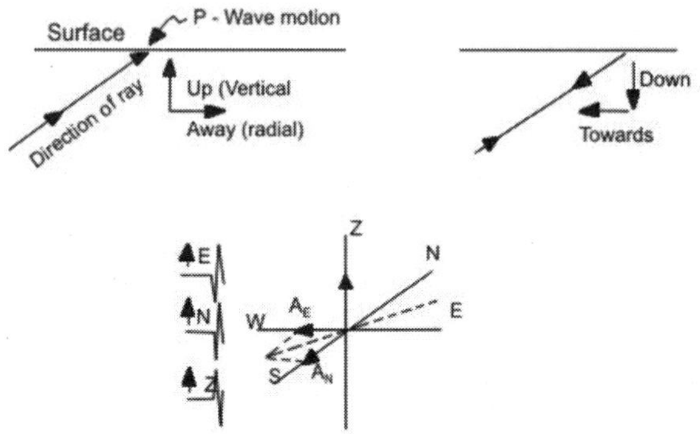

Fig. 3.30 P-wave three-component vector ground motion and determination of azimuth to the source (Båth, 1979).

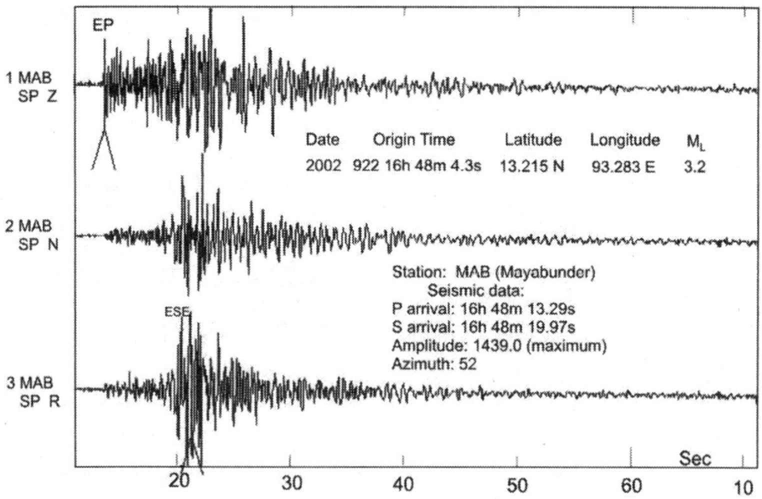

Fig. 3.31 An example of single-station location using Seisan program (Kayal et al., 2004).

For teleseismic events, Borman (1971) showed that in the absence of a sufficiently strong P-wave arrival, the back azimuth can be determined from horizontal components of any later seismic phase which is polarized in the

the Earth model differs significantly from the true velocity structure of the Earth. The question arises, what is meant by the "best" solution? This is the solution that minimizes the *root mean square* (RMS) residual. That is, if we define R_i as the difference between the ith observed and the ith computed arrival time, and if we define RMS as:

$$RMS = \sqrt{\frac{\sum_{i=1}^{N} R_i^2}{N-1}} \tag{3.14}$$

then the "best" solution is the one which minimizes RMS.

There are several ways in which the minimum RMS can be determined. The simplest method might be to perform an exhaustive search, which implies that the solution space must be discretized in some manner to reduce the number of points to be searched to a finite value. One could grid the volume of space, which is thought to enclose the hypocentre, and then evaluate RMS at every point of a grid within that space. The point, which has the minimum value of RMS, would then be returned as the solution. This method relatively works well as long as the volume of space encloses the solution, and as long as the grid spacing is fine enough for the required precision. The method is, however, time consuming. If there are M grid points per dimension, then there must be M^3 evaluations of RMS to find the solution. However, as computers become faster, this method or a combination of it with the method described next, becomes feasible.

A much faster method is to start at a trial hypocentre, linearize the equations, and compute an improved solution. The improved solution becomes the next trial hypocentre and the process is repeated until changes between trial hypocentres are small. The principle problem with this method is in distinguishing a *local minimum* from a *global minimum*. That is, for many earthquake data sets, the RMS does not increase monotonically away from the global minimum. Rather, the RMS has one or more local minima in addition to the global minimum, which we seek. If the first trial hypocentre location is within the "valley" of a local minimum, the solution will converge to that local minimum rather than the global minimum. This problem is illustrated later in more detail (see Section 3.10.2).

3.9.1 Hypo-71 Program

There are many earthquake location programs. We will discuss the Hypo-71 program of the USGS (Lee and Lahr, 1975), the one which is widely used and we are most familiar with. The basic inputs to Hypo-71 are the *station list* (station coordinates), *velocity model*, *control parameters*, and of course the *arrival times* (Annexure 3.1).

Station List

The station names used by Hypo-71 may be up to four characters long e.g. SHIL (Shillong). Elevation is not used by the program even though it is specified in the station list. P-delay (*station correction*) may be specified, and S-delay is calculated from the P-delay as

$$\text{S-delay} = \text{P-delay}\left(\frac{V_p}{V_s}\right) \tag{3.15}$$

A positive residual would indicate that the observed time is larger than the computed time, and would be reduced by a positive delay.

P-wave Velocity Model

The computer program needs a reasonable velocity model to compute the travel-times. A plane layered P-wave velocity model based on available geological and geophysical information may be used for preliminary location of the earthquakes. The model then may be modified by trial and error method. The velocity model can be constrained from the observed arrival data of the recorded earthquakes. Time-distance plot of the direct P-arrivals (P_g and P^*) may be used to estimate the velocity of the crustal layers, and P_n-arrivals may be used for the mantle-velocity estimate. The velocity model may be further modified by 1-D and 3-D inversion. A detailed study of the crust and mantle velocities in the Shillong Plateau is given in Chapter 6. Converted seismic phases may be used to constrain the discontinuities within the crust, say at the boundary of the upper crust and lower crust, and/or at the mantle depth. Readers are referred to some global examples on the study of converted phases at the crustal discontinuities (e.g. Snoke et al., 1977; Calhaem et al., 1977; Kayal, 1986).

Velocity Ratio Vp/Vs

The computer program further needs a constant ratio of P to S-wave velocities (Vp/Vs) for computation of S-wave travel times. For preliminary location, an average value of Vp/Vs = 1.73 of the Earth's crust is used. However, Vp/Vs can be determined with a fair degree of accuracy by the *Wadati-plot* method (Wadati, 1933), or by the *station-by-station* method (Smith, 1983).

In *Wadati-plot* method, the S-P interval time is plotted against the P-arrival time (Fig. 3.29). The technique was originally suggested to estimate the origin time of an earthquake (Wadati, 1933), as discussed above, and later used by many investigators for velocity-ratio estimates (e.g. Semenov, 1969; Utsu, 1969; Nersesov et al., 1971). If the wave paths from the hypocentre of an earthquake to an array of seismograph stations lie in a region of constant Vp/Vs, then the Wadati plot is linear and has a gradient $\left(\frac{Vp}{Vs}\right)-1$. This technique is useful because it yields a fairly accurate estimate of Vp/Vs without the knowledge of source parameters.

Smith (1983) developed a *station-by-station* method to estimate Vp/Vs. In Wadati-plot method, if the number of data per event is small and if there is any significant change in the value of Vp/Vs from station to station, the slope of regression line will change according to the stations, which provide the data for the earthquake. To overcome this difficulty, Smith (1983) used P- and S- arrival data for a number of events and developed the following relation:

$$\mathrm{Vp}\ P_{ij} - \mathrm{Vs}\ S_{ij} - (\mathrm{Vp} - \mathrm{Vs})\ t_i = \varepsilon_{ij};\ i = 1, m;\ j = 1, n \qquad (3.16)$$

Vp and Vs are average P- and S-velocities for the region containing the events and stations. P_{ij} and S_{ij} are P- and S-arrival times at the jth station of a network of n seismographs from the ith event of a set of m events, t_i is the origin time of the ith event, and ε_{ij} is the sum of the model error and reading error. The method does not depend on the location of the event either. The equation (3.16) is a system not more than $n \times m$ equations in m+1 unknowns, Vp/Vs and t_i. The least-squares technique is used to estimate the unknowns (Smith, 1983). De and Kayal (1990) used this technique to evaluate Vp/Vs of the crust in northeast India (see Chapter 6).

Control Parameters

The control parameters for Hypo-71 include first trial location and the distance weighting limits. First trial location may be assumed to the station with earliest arrival. The purpose of distance weighting is to give greater weight to the close stations. The way it works is that for stations less than lower distance limit (XNEAR), the weight given to the data is one. Beyond XNEAR the weight decreases linearly to zero at the outer limit (XFAR).

Additional control over Hypo-71 is achieved through the use of *reset list* (see Annexure 3.1). This is an optional series of up to 13 lines. Each line is used to reset one of 13 variables. If no reset list is present, Hypo-71 will use a set of default values, which are suitable for earthquakes recorded by the California network. The 13 variables on the reset list are as follows:

TEST (01) is the cut off value for RMS, below which Jeffreys' weighting of residuals is not used. It should be set to a value approximately equal to overall timing accuracy of P-arrivals in seconds. For small networks, less than 30 arrival times per earthquake, Jeffreys' weighting will not work well and should not be used. Further, Jeffreys' weighting is designed to catch large arrival-time errors. Therefore it is useful for preliminary runs. After arrival-time errors have been corrected, Jeffreys' weighting is not recommended.

TEST (02) limits simultaneous changes in epicentre and depth. If the epicentre adjustment is greater than this amount, the adjustment is recalculated with focal depth fixed. Test (02) should be set to a value approximately equal to station spacing in km.

TEST (03) is the critical F-value for the stepwise multiple regression (Draper and Smith, 1966). It should be set according to the number and quality of P- and S-arrivals. If Test (03) is set to zero, then simple multiple regression is performed. If Test (03) is set to 2 or greater, then the iterative process may be terminated prematurely. A value between 0.5 and 2 is recommended.

TEST (04) is used to terminate the iterative location process. If the hypocentral adjustment is less than (04) the solution is terminated.

TEST (05) is used to control the maximum focal depth adjustment. If the adjustment (DZ) at any step is greater than Test (05) then DZ is reset to DZ/(K+1) where K = DZ/test (05). Test (05) should be set to a value approximately equal to half the range of focal depth expected.

TEST (06): If no significant variable is found in the stepwise multiple regression, the critical value F is reduced to Test (03)/Test (06), and the regression is repeated. If the Test (03) is set to less than 2, then Test (06) should be set to 1.

TEST (07), TEST (08) and TEST (09) are the coefficients used for calculating duration magnitude (Lee et al., 1972). According to the formula (using default values):

$$\text{FMAG} = -0.87 + 2\log T + 0.0035D \tag{3.17}$$

where T = signal duration in second, and D = epicentral distance in km.

TEST (10) is used to control the maximum adjustment in latitude or longitude. If either adjustment exceeds Test (10) at some iteration, then a new reduced adjustment is calculated using the Test (10).

TEST (11) is the maximum number of iterations in the solution process.

TEST (12) is used if the focal depth adjustment would put the hypocentre in the air. In that case, a new adjustment is calculated utilizing Test (12).

TEST (13) is used to control the radius of a sphere centered on the hypocentre location estimate. The RMS residual is calculated at ten points on the sphere to see if there is a nearby location with lower RMS than the final location (Fig. 3.32). This will not happen if the best location has been found.

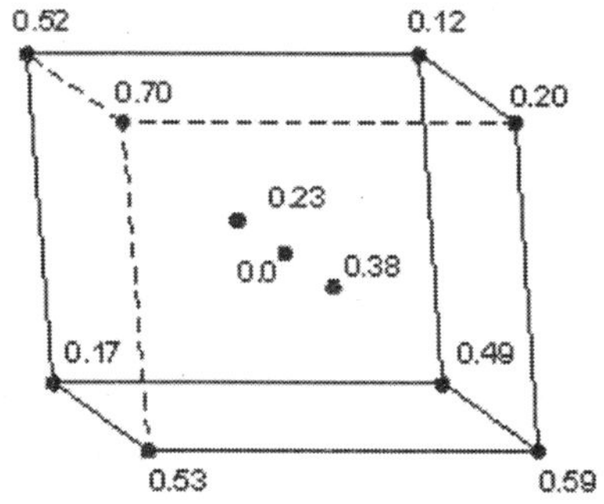

Fig. 3.32 Hypo-71 RMS on ten points on a sphere of 1 km.

3.9.2 HYPOELLIPSE Program

HYPOELLIPSE is modern version of Hypo-71. Mainly the differences between this program and Hypo-71 are discussed here briefly. HYPOELLIPSE is complex and has a lot of options. Depending on the needs, this may or may not be an advantage. Compared to Hypo-71, HYPOELLIPSE is not very stable. Hypo-71 has not changed in the past 30 years, but HYPOELLIPSE probably changes every two or three months. HYPOELLIPSE has lots of options that are useful for research as well as for routine processing of data, but having many options makes the program more difficult to learn.

Instead of just one velocity model, HYPOELLIPSE allows up to ten velocity models to be specified. The Vp/Vs ratio may be varied in each layer. If detailed crustal velocity information is available, this flexibility is quite helpful. In addition to constant velocity layers over a half space, one can also specify a linear increase over a half space. Also, in addition to calculating travel times using a model, up to three different travel-time models may be used. Elevation corrections may be made based on station elevations, which is not done by Hypo-71.

HYPOELLIPSE also has options which allows the variation of RMS with depth to be computed and to make repeated runs of either artificial or real data with added random errors. These options provide a means of testing location stability.

HYPOELLIPSE optionally compute the Vp/Vs ratio for each earthquake processed. The computed Vp/Vs ratio may be used in parts as a diagnostic tool. If the ratio is out of bounds, then a phase interpretation error is indicated. Spatial or temporal variations in Vp/Vs may also be studied.

In HYPOELLIPSE the *station list* contains a history of the station polarity, the station coordinates, the telemetry delay, and the gain as a function of time. Because of this, an arbitrarily long sequence of earthquakes may be proposed even if station parameters have changed during the sequence. As time progresses, the program automatically updates station parameters from the information in the station list. The station list allows independent P- and S-delays instead of having to use a fixed Vp/Vs ratio to compute S-delay, as in Hypo-71.

Location error limits are specified by HYPOELLIPSE in terms of the error ellipsoid. This is the region in space in which there is 66% confidence that the earthquake occurred. Figure 3.33 illustrates an error ellipsoid and its projection onto map view. The ellipsoid can also be projected onto a vertical cross section. The ellipsoid shape is determined entirely by the velocity model and the geometric relationship of the stations to the hypocentre. The scaling of the ellipsoid is based on an estimate of the probable reading errors of the arrival times. The error ellipsoid estimates the precision of the location but does not include any information on potential systematic error that might bias the location.

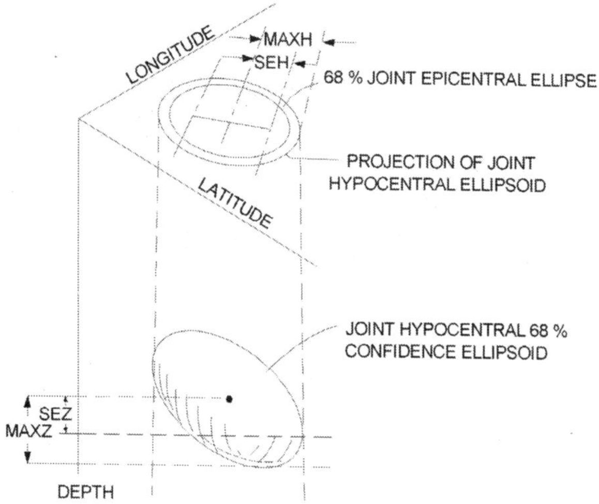

Fig. 3.33 HYPOELLIPSE error ellipsoid on to a map (Lahr, 1992).

3.9.3 SEISAN Program

The SEISAN program is now the most recent and becoming most popular seismic analysis system; this provides a complete set of program for analysing earthquake data from analog as well as from digital records. The SEISAN system is built on a collaborative effort by the British Geological Survey, U.K. and by the University of Bergen, Norway (Havaskov and Ottemoller, 2000). With this system it is possible to enter phase readings manually, or pick them from digital records to locate events, plot epicentres etc. The program is very versatile; it can be used for local as well as for global events.

The routine processing normally produces magnitudes and hypocentres. Event statistics, depth section, b-value etc. can be plotted using this program. Fault-plane solution for individual event and composite solution for a group of events can be made by this program using first-motion data. Fault-plane solution by *waveform modelling* of the digital seismograms is an added advantage with this program. In addition, *azimuth determination, spectral parameters*, *seismic moment*, *stress drop*, *seismic source radius* etc. can be determined by spectral analysis of the P and S waves for local earthquakes. Further, SEISAN contains some integrated research type programs like *coda Q, synthetic modelling,* 1-D *velocity inversion, joint hypocentre* location and a complete system for *site response* and *seismic hazard calculation.* In one word, SEISAN is the most versatile computer program now available for earthquake (analog + digital) data analysis. The program is made available on the website. www.ifif.uib.no/seismo/seisan.html. It has almost replaced the popular Hypo-71 program.

3.10 LIMITATIONS OF ACCURACY IN LOCATION

3.10.1 Depth and Origin Time

There is always certain amount of uncertainty in the observed arrival times and the Earth model used to compute the travel times. How these errors are mapped into spatial location errors depends on the geometric relationship of the stations to the hypocentre and on the phases that are used. Unless an earthquake is shallow and within a dense network, it is essential to have S-phase readings in addition to P-phase readings. The combination of P- and S-phases gives greater constraint on the origin time, and limits what could otherwise be an unbounded trade off between origin time and location.

Let us consider an earthquake at the centre of a circular array of stations (Fig. 3.34). The P-phases will arrive simultaneously at all of the stations. This fixes the earthquake epicentre but places no constraint on depth. We can see why this is so by looking at a cross section of the problem. There is a trade-off between hypocentre depth and origin time. As shown in Fig. 3.34, for any depth, all the P-phases will arrive simultaneously. As depth is increased, the only effect is that the origin time must be earlier. Since there is no information about the origin time, there is no control over the depth. Just one S-phase reading would, however, constrain the path length between the hypocentre and that station. This would serve to fix the origin time and hence the depth.

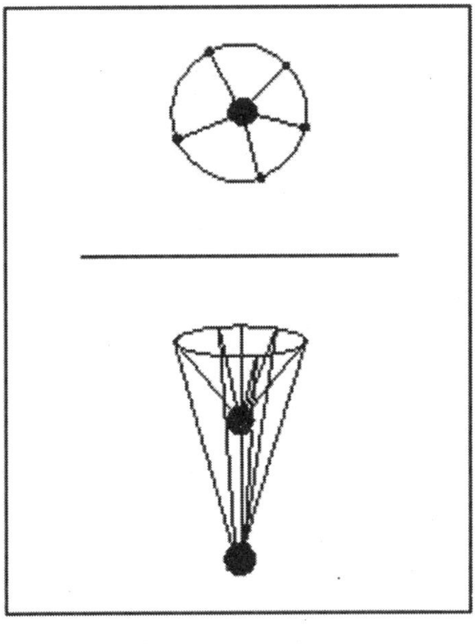

Fig. 3.34 Trade off between depth and origin time (Lahr, 1992).

If an earthquake occurs outside the network and only P-wave arrivals are available, the information that is used to get the distance to the epicentre is the curvature of the P-waves as they pass through the net. If the epicentre is very close to the net, as for the source at A in Fig. 3.35, the curvature is significant. The source B, on the other hand, is further away and the curvature, as measured within the net, is less than that of the source A. If the source is farther away than about ten times the diameter of the net, then the curvature will be nearly plane as they pass through the net, and for all sources which are very far outside the net, the curvature is essentially the same. Therefore, as events get farther outside a network, S-phases become more important to constrain the distance to the source.

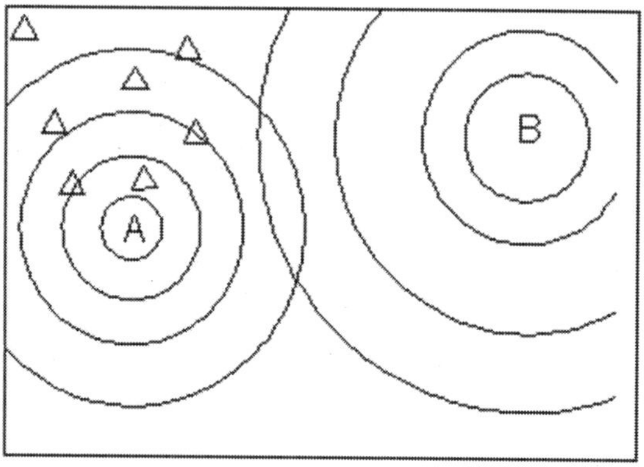

Fig. 3.35 Variation of wave front curvature with epicentral distance. Triangles show the seismograph stations; A and B are the epicentres (Lahr, 1992)

The depth of hypocentre inside a network may be difficult to constrain, even apart from those near the centre of a circular array. The problem is analogous to that of constraining the distance from the network to a distant earthquake. If the earthquake is shallow, relative to the station spacing, then the curvature is significant and the depth is well constrained. However, for deeper earthquakes it becomes progressively more difficult to constrain the hypocentre depth without S-wave readings. The conclusion is that minimum one or two reliable S-arrival data should be used to obtain a well constrained hypocentre solution.

3.10.2 Multiple Minima Problem

Figure 3.36 illustrates a multiple minima problem. In the figure, the vertical axis is depth, increasing downward, and the horizontal axis is root mean square residual (RMS). The curve I shows a smooth decrease with depth to the minimum RMS and a smooth increase as depth increases past the minimum point. Not only is there just one minimum, but the minimum is very narrow. This is a well constrained solution.

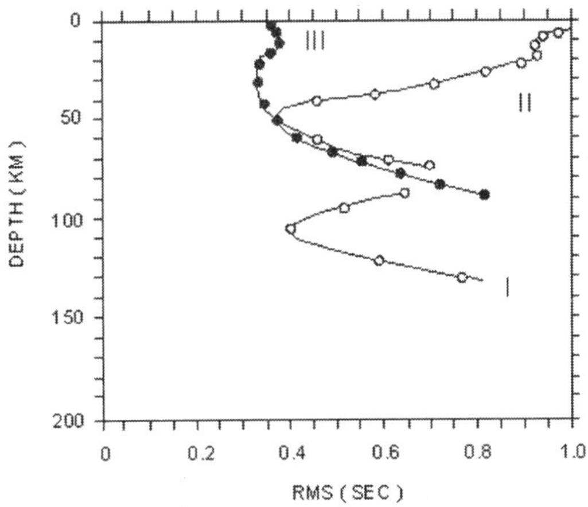

Fig. 3.36 Multiple minima problem in earthquake location (Lee, 1992).

The curve II presents more of a problem. There is a minimum at a depth of about 13 km followed by a small peak, and then another minimum at a depth of about 50 km. In this case, if the location program starts with a trial location near the surface, it will converge at the shallow minimum, even though the minimum at 50 km is actually the deepest or *global minimum.* The error bars for the shallow depth solution may indicate a well constrained minimum. The RMS will, however, indicate that there might be a problem with the solution. On the other hand, if the program started the solution at a depth below 20 km, the algorithm would converge at the global minimum of 50 km. The RMS returned would be much lower than for the shallow solution, but it is possible that the error bars would be wider if the slope near the global minimum is less than the slope near the shallow minimum.

The curve III has a minimum at the surface and another at a fairly shallow depth. The error bars returned for the 20 km depth solution will be rather large since the slope is so around that solution. One might note that the minimum near 20 km is less well constrained than the minimum at the surface, and choose the surface focus as the best solution. This is a mistake, the solution with the lowest RMS should be accepted.

3.10.3 Station Subset Problem

If we have a network with lots of stations and we locate a group of earthquakes, that group may often have a spatial distribution with a distinct trend. This can lead to believe that the distribution is due to some geological control such as a fault.

If the stations used in each location are reviewed it may be found that one or more of the stations was used for only a fraction of the earthquakes. It may be useful then to plot all the earthquakes for which a given station was used, and then plot all the earthquakes for which the station was not used.

The two data sets may produce two distinct clusters which, when combined, give the trend seen in the complete data set. What we have then, is not a trend, rather an artifact; a more localized distribution whose computed location depends on the stations used. Various approaches for relocation of the earthquakes can be made to determine the true trend of seismicity.

3.11 EARTHQUAKE RELOCATION

To determine true trend of seismicity several techniques are applied to relocate the events. Principles of these techniques are briefly discussed.

3.11.1 Master Event Method

One approach in finding a true pattern in a set of earthquakes is called the *master event method*. We may start with one earthquake that is exceptionally well located and has readings for every possible phase at all the stations. It could be a main shock with very clear arrivals. Now, the *station delays* are so adjusted that the earthquake location has zero RMS. Using these adjusted station delays/corrections, another earthquake with nearly the same pattern of arrivals will locate near the master event, even if a different subset of stations is used.

3.11.2 Average Residual Method

Another approach is called the *average residual method*. Here we compute the average residual at each station for the entire set of earthquakes. Then use the averages as station delays and recompute the locations. The process can be iterated a few times. This will reduce the relative errors for this set of earthquakes.

3.11.3 Homogeneous Station Method

There could be scatter introduced into the solutions because of different combinations of stations being used in the solutions. An approach that is often useful in such situations, is to find a subset of phase readings that were used for locating most of the events. Then those events may be relocated that have arrival times for all of these phase readings. All other events and phases may be excluded. This can be a very effective means of eliminating scatter, and may reveal a clear picture of the true hypocentral distribution. The method is also referred to as the *common phase method* (Lahr, 1992). In a small area, the epicentre locations by this method are also independent of the velocity model error (Kayal, 1984, Kayal et al., 2003a). This is greatest advantage in this method, that brings out the true seismicity pattern by precise relative location of the events.

Neither of the above methods, however, can improve the absolute error in source location for earthquakes in the group, but they reduce the relative location errors so that a pattern that would not have been apparent otherwise becomes clear. Figure 3.37 illustrates an example where a clear trend of hypocentres is revealed by homogeneous station method.

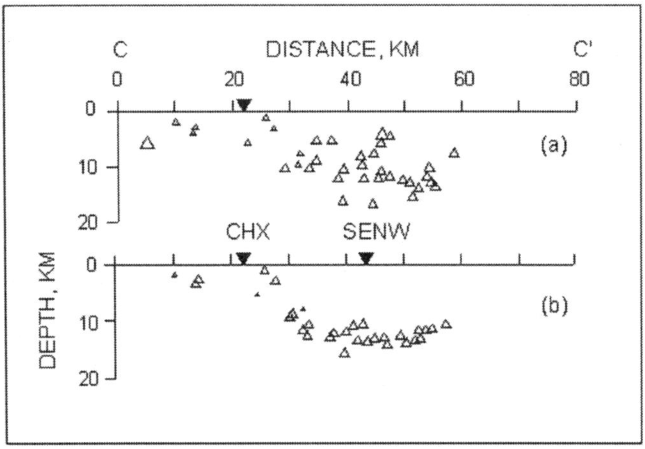

Fig. 3.37 (a) Hypocentres showing no particular trend. (b) Hypocentres showing true trend after improved location by homogenous station method (Lee, 1992).

3.11.4 Double-Difference (DD) Method

The double-difference (DD) technique, a theoretically based recent technique, also belongs to the relative earthquake location method (Got et al., 1994). This technique is based on the fact that if the hypocentral separation between two earthquakes is small enough compared to event-station distance, then the ray paths can be considered identical along their entire length. With this assumption, the differences in travel times for two earthquakes recorded at the same station may be attributed to the differences in spatial separation of the hypocentres. Thus the velocity-model errors are minimized without the use of station corrections.

The DD technique and the relocation algorithm (Hypo-DD) are thoroughly discussed by Waldhauser and Ellsworth (2000). Theoretical travel time differences are calculated based on the assumed 1D velocity model, and the residuals between the theoretical and observed times are minimized by an iterative procedure. When two events are close to each other compared to the event-station distance and to the scale of velocity heterogeneity in the broader hypocentral area, the residual between observed and calculated differential travel time for the two events is calculated as:

$$dr_{\mathrm{k}}^{\mathrm{ij}} = (t_{\mathrm{k}}^{\mathrm{i}} - t_{\mathrm{k}}^{\mathrm{j}})^{\mathrm{obs}} - (t_{\mathrm{k}}^{\mathrm{i}} - t_{k}^{j})^{\mathrm{cal}} \qquad (3.18)$$

where t is the travel time, superscripts i and j correspond to the two different events and the subscript k corresponds to a particular observation (i.e. one particular phase to one common station).

The double difference equations are derived from differentiating Geiger's equation for earthquake location (Geiger, 1912), and have the form:

$$\frac{\partial t_k^i}{\partial m}\Delta m^i - \frac{\partial t_k^j}{\partial m}\Delta m^j = dr_k^{ij} \tag{3.19}$$

where Δm (Δx, Δy, Δz and $\Delta\tau$) is the change in the four hypocentral parameters of the events. The above equation can be expanded as

$$\frac{\partial t_k^i}{\partial x}\Delta x^i + \frac{\partial t_k^i}{\partial y}\Delta y^i + \frac{\partial t_k^i}{\partial z}\Delta z^i + \Delta\tau^i - \frac{\partial t_k^j}{\partial x}\Delta x^j - \frac{\partial t_k^j}{\partial y}\Delta y^j - \frac{\partial t_k^j}{\partial z}\Delta z^j - \Delta\tau^j = dr_k^{ij} \tag{3.20}$$

This equation relates the residual, dr, between observed and theoretical travel time difference between the two events (Eq. 3.18) to changes in each event's locations (x, y, z) and origin time (τ).

In the double-difference earthquake location method, the partial derivatives in Eq. (3.20) are estimated for each event pair-station combination, and the resulting equations build up a system of linear equations of the form:

$$\mathbf{WGm} = \mathbf{Wd}$$

where $\mathbf{G}$ is the matrix of the partial derivatives, $\mathbf{m}$ $(\Delta x, \Delta y, \Delta z \text{ and } \Delta\tau)^T$ the matrix containing the changes in the hypocentral parameters required to improve the model fit to the data, $\mathbf{d}$ the data vector containing the double differences estimated through Eq. 3.18 and $\mathbf{W}$ a diagonal matrix used to weight equations.

The above equations are finally solved by the conjugate gradient algorithm LSQR (Paige and Saunders, 1982) or by Singular Value Decomposition (SVD) method depending on the large or small amount of data respectively.

3.11.5 Joint Hypocentre Method

It is understood that for a given seismic station, the error in the predicted travel time is due to inaccuracies of the assumed velocity model. These deviations may occur anywhere along the travel paths, which may be divided into three categories: (i) deviations from the velocity structure near the source, (ii) deviations near the station, and (iii) deviations along the deeper travel path. For a single event-station pair, it is not possible to isolate the effects of these errors. On the other hand, if a *cluster* of earthquakes, with approximately same location, occurs it is possible to determine a *station correction* that accounts for the inaccuracies of the model structure along the travel path and beneath the station. In this we include a correction term for each of the N stations and recast the problem as

$$r_{ij} = dT_j + \frac{\partial t}{\partial x}dx_j + \frac{\partial t}{\partial y}dy_j + \frac{\partial t}{\partial z}dz_j + \frac{\partial t}{\partial s}ds_i \tag{3.21}$$

where r_{ij} is the residual or error at the ith station for the jth earthquake, i.e.

$$r_{ij} = t_{ij} - (\tau_{ij} + S_i)$$

where t_{ij} is the observed arrival time, τ_{ij} is the computed travel time and S_i is the station correction term. dT_j is the perturbation of the origin time for the jth event. In matrix form, it may be written as

$$\mathbf{r}_j = \mathbf{A}_j \, \mathbf{dx}_j + \mathbf{ds} \qquad (3.22)$$

where $\mathbf{r}_j$ is the vector of residuals, $\mathbf{A}_j$ is matrix of partial derivatives, $\mathbf{dx}_j$ is the vector of origin time and hypocentre adjustment, and **ds** is the vector of station correction term. Solution of this system of equations is Joint Hypocentre Determination (JHD) and was proposed by Douglas (1967). Many authors (e.g. Herrmann et al., 1981; Pavlis and Booker, 1983; Pujol, 1988) proposed efficient inversion schemes for solving the Eq. 3.22; nearly all the schemes involve *singular value decomposition*. The relative locations obtained by JHD are better than those determined by conventional inversion; an example is shown in Fig. 3.38. This method is applied to relocate the earthquakes in northest India region, and the results are discussed in chapter 6.

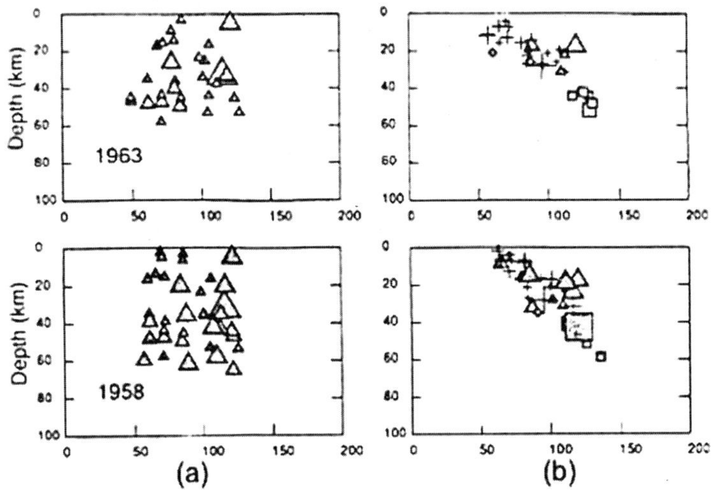

Fig. 3.38 Comparison of earthquake locations: (a) conventional procedures of ISC, and (b) JHD relocations (Schwartz et al., 1989).

3.12 SEISMIC TOMOGRAPHY AND EARTHQUAKE LOCATION

So far we have discussed about one dimensional plane layered velocity model for routine earthquake location and relocation by different techniques. We, however, know that lateral heterogeneity exists at every depth in the

Earth. For the last 50 years seismologists have been mapping gross velocity differences near the surface associated with variations between continental and oceanic crust and upper mantle. In the past 20 years, a concerted effort has emerged to map or image the three dimensional velocity structure inside the Earth using a method called *seismic tomography* (Fig. 3.39).

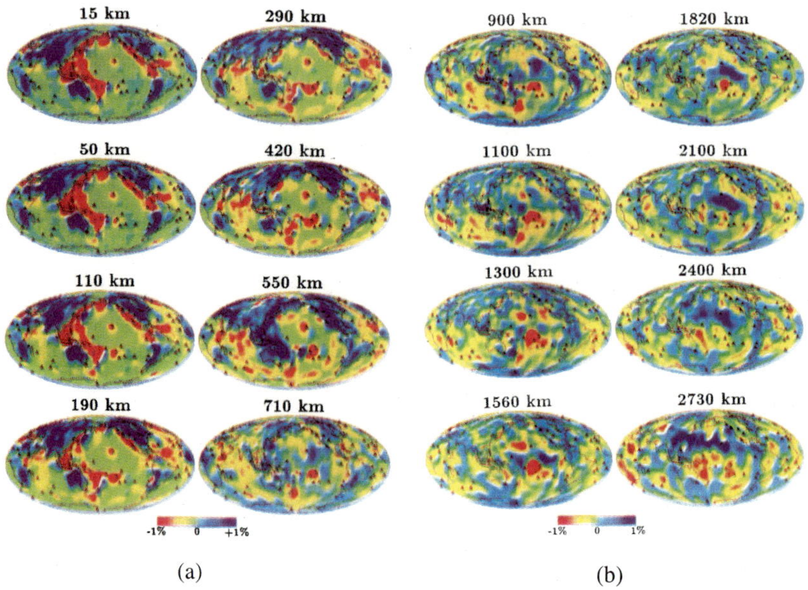

Fig. 3.39 Seismic tomography of the Earth's internal structure (Zhao, 2001).

'Tomo' is from a Greek word, which means "slice". In tomography, we examine a series of two-dimensional slices from a three-dimensional object, and use the information to infer the internal structure of the three dimensional object. The beginning of tomography may be traced back to the discovery of x-rays by W. Roentgen in 1895.

Medical tomography and seismic tomography share some similarities, but they also have significant differences. In radiology, we measure the intensity of received x-rays, which is a function of the *absorption coefficient* of the travel path. In seismology, we measure a travel time, which is a function of the *slowness* (i.e. inverse of velocity) along the ray path. In medical tomography, the x-ray path is a straight line. In seismology, the seismic-ray path is almost never a straight line. Also, in medical tomography, the source and detector locations are known exactly at all times. In seismic tomography, the source locations are unknowns, which must be solved for while inverting for the velocity structure. Figure 3.40 illustrates the basic differences in medical tomography and seismic tomography.

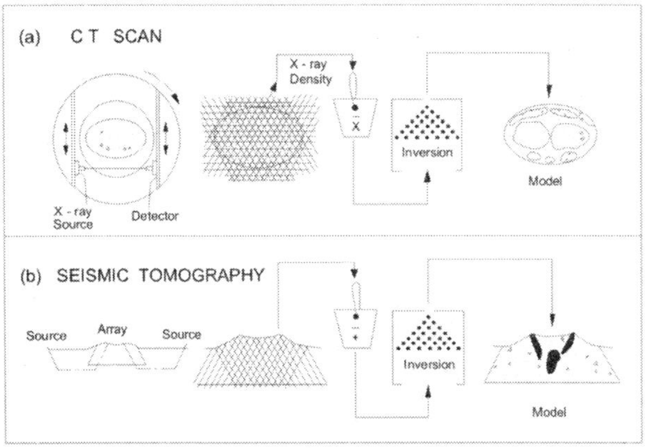

Fig. 3.40 Schematic diagram illustrating the differences between (a) medical tomography, and (b) seismic tomography (Iyer and Hirahara, 1992).

During the early years of the development of seismic tomography, the technique was called *simultaneous inversion*. The idea was to get some information on the velocity structure of the Earth, and at the same time source parameters of the earthquakes were to be determined. In the mid-1970s, it occurred to a number of seismologists that since a seismic network may have hundreds of stations, and since only four stations are required to do a hypocentre inversion, all that extra information might be put to use to invert for the velocity structure of the Earth simultaneously with the hypocentre inversion. Crosson (1976 a, b) worked out the problem for the one-dimensional case, and Aki and Lee (1976) published the results of a three-dimensional inversion. At about the same time, Aki et al. (1977) worked out the three-dimensional inversion method for velocity parameters using teleseismic events.

Examples of seismic tomography geometries are shown in Fig. 3.41, which illustrates application of local and teleseismic earthquakes recorded by an array of surface sensors. While it is possible to locate the events with a one-dimensional velocity model, the crossing-ray coverage allows us to solve for shallow crustal heterogeneity as well. The problem can be formulated for a simultaneous solution of velocity structure and earthquake locations, using the simultaneous inversion technique.

Seismic tomography is not limited to imaging the Earth or part of the Earth using body-wave arrival-time data; it has been extended to include surface waves and waveforms as well (e.g., Geller, 1991; Iyer and Hirahara, 1992). Although inverting for the Earth's velocity structure is most common, other parameters can also be imaged. For example, spatial variations in seismic attenuation from observed S to P wave amplitude ratios was studied by Ho-Liu et al. (1988), and slip velocity by tomographic source imaging of small earthquakes was carried out by Frankel et al. (1986).

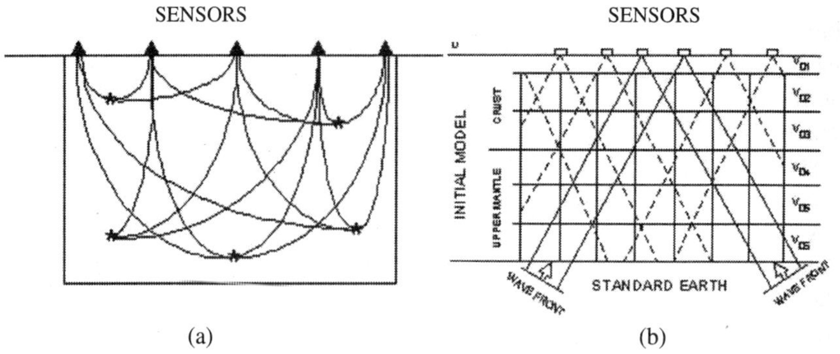

Fig. 3.41 Raypaths for (a) local earthquakes, and (b) for teleseismic earthquakes (Iyer and Hirahara, 1992).

The theoretical aspect of seismic tomography is discussed by many authors (Iyer and Hirahara, 1992; Lee, 1993; Lay and Wallace, 1995). A particular seismic phase with a travel time T is represented by a path integral through the medium as

$$T = \int_S \frac{ds}{v(s)} = \int_S u(s)ds \qquad (3.23)$$

where $u(s)$ is the slowness $[1/v(s)]$ along the path. The *travel-time residual* relative to the reference Earth model may be caused by a velocity or slowness perturbation anywhere along the path. A change in velocity along the ray must perturb the ray path. The path integral through the medium perturbations should equal the observed travel-time residual:

$$\int_s \Delta u(s)\, ds = \Delta T = T_{obs} - T_{com} \qquad (3.24)$$

where $\Delta u(s)$ is the slowness perturbation to be determined. If the medium is subdivided into blocks, we can calculate the path length l_j in the jth block and discretize (3.24) to give

$$\Delta T = \sum_j l_j \Delta u_j \qquad (3.25)$$

Obviously, a single observation is inadequate for partitioning the slowness perturbations along the path. However, if there are many event-station pairs, each with the ith ray path, we can develop a system of **i** equations

$$\Delta T = \sum_j l_{ij} \Delta u_j \qquad (3.26)$$

The equation (3.26) is in the form of a linear system, which can be solved by the *matrix inversion methods*. In this case, the path length of each ray in a block, l_{ij} , is the partial derivative $\partial T_i / \partial u_j$ of the travel time with respect to the slowness of that block. The generalized inverse solution of (3.26) is provided by

$$\mathbf{m} = [\mathbf{G}^T\, \mathbf{G}]^{-1}\, \mathbf{G}^T\, \mathbf{d} \qquad (3.27)$$

where $\mathbf{d} = \Delta T_i$, $\mathbf{G} = l_{ij}$, and $\mathbf{m} = \Delta u_j$. If the generalized inversion is unstable, we resort to damping or singular-value truncation to obtain a solution. The resolution matrix can be computed to reveal how well the model can be reconstructed if the data and model parameterization are perfect. Usually the resolution matrix reveals streaking between adjacent blocks where the ray coverage is inadequate to isolate the anomaly uniquely in each block.

A system like Eq. (3.27) is determined by calculating ray path segments through the model, and this is solved by *damped-least-squares, singular-value decomposition* (SVD), or by iterative procedures called *back projection* that iteratively solve the matrix equations without having to invert huge matrices.

The above mathematical techniques became popular in the 1980s as increasingly large data sets and model parameterizations were adopted. One scheme involves solution of (3.27) in the least squares form:

$$\mathbf{G}^T \mathbf{G}(\mathrm{m}) = \mathbf{G}^T \mathbf{d} \tag{3.28}$$

Rather than solving for the generalized inverse of $\mathbf{G}^T \mathbf{G}$, one can simply use the diagonal of this matrix to approximate the solution, giving

$$\Delta u_j = -\frac{\sum_i \Delta t_i l_{ij}}{\sum_i l^2_{ij}} \tag{3.29}$$

The two sums are computed separately and updated ray by ray. Then each slowness perturbation is calculated, and the model is perturbed. In a procedure called *Simultaneous Iterative Reconstruction Technique* (SIRT), this new model is used to compute new residuals, and the procedure repeats for many iterations. Because no large matrix inversions are performed, huge models can be examined but at the cost of not computing any resolution matrices. This iterative back-projection approach was developed in medical imaging tomography (Swindell and Barrett, 1977).

Simultaneous Inversion

In the following discussion, we will concentrate on only one-type of seismic tomography, the simultaneous inversion of earthquake hypocentral parameters and velocity structure from local earthquakes. This approach is commonly used in seismology, and the basic idea can be extended for inverting for other parameters.

The simultaneous inversion problem can be posed in the following manner. There is an array of (m) stations indexed k =1, 2,..., m. The array is used to record (n) earthquakes indexed j =1, 2,..., n. The hypocentre parameters for the jth earthquake are $(t_j^0, x_j^0, y_j^0, z_j^0)$. There are n earthquakes, so the complete set of hypocentre parameters is $(t_1^0, x_1^0, y_1^0, z_1^0, t_2^0, x_2^0, y_2^0, z_2^0, ..., t_n^0, x_n^0, y_n^0, z_n^0)$. There are a total of $4n$ unknown parameters in this vector.

To parameterize the velocity structure, we superimpose a three-dimensional grid on the region of the crust containing the earthquake sources and the seismic stations. Figure 3.42 shows a schematic sketch of the system. Each of the cells within the grid system is assumed to have constant velocity but the velocity can vary from cell to cell. In practice, the dimensions of each block can be different. For example, around a fault, we probably want to have a finer grid than at great distance from the fault. However, for the purposes of this discussion, let us assume the cells are all equi-dimensional. Further, we assume that there are (L) blocks indexed l =1, 2,..., L. The velocity parameters of the problem are indexed (v_1, v_2,..., v_L). The complete set of unknown for the problem consists of the set of hypocentre parameters and the set of velocities, so the total number of unknowns for the problem is $4n + L$. Mathematical consideration of the simultaneous inversion given by Lee and Pereyra (1993) is briefly presented here.

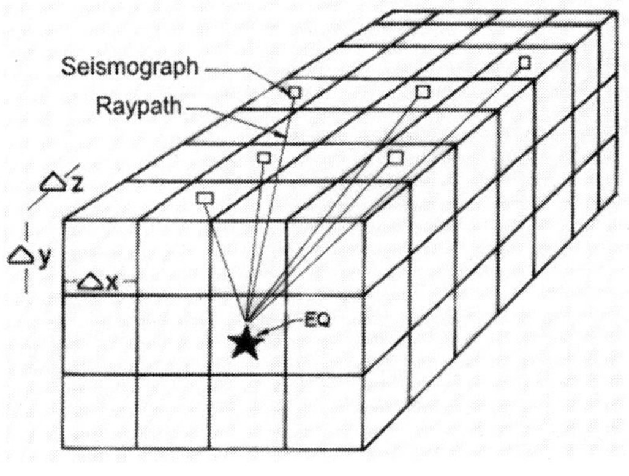

Fig. 3.42 Schematic diagram showing 3-D ray-tracing grid.

It is mathematically convenient to put all the unknowns in one parameter vector, which can be expressed as:

$$\zeta = (t_1^0, x_1^0, y_1^0, z_1^0, ..., t_n^0, x_n^0, y_n^0, z_n^0, v_1^0, ...v_L^0)^T \tag{3.30}$$

We need to define a trial parameter vector, which will be given as:

$$\zeta^* = (t_1^*, x_1^*, y_1^*, z_1^*, ..., t_n^*, x_n^*, y_n^*, z_n^*, v_1^*, ...v_L^*)^T \tag{3.31}$$

The trial vector is essentially a guess, which will be improved during the solution process. For any trial vector, the residual associated with the jth earthquake at the kth station is:

$$\mathbf{r}_{jk}(\zeta^*) = \mathbf{T}_{jk} - \mathbf{T}_{jk}(\zeta^*) - \mathbf{t}^*_j \tag{3.32}$$

Here $\mathbf{T}_{jk}$ is the measured arrival time for the jth earthquake at the kth station. $\mathbf{T}_{jk}(\zeta^*)$ is the theoretical travel time, and $\mathbf{t}^*_j$ is the theoretical origin time. There are (mn) of these residuals and in order to have a solution, we must have (mn) much larger than $(4n+L)$. Given these conditions, the problem is to minimize:

$$F(\zeta^*) = \sum_{j=1}^{n} \sum_{k=1}^{m} \left[r_{jk}(\zeta^*) \right]^2 \qquad (3.33)$$

Equation (3.33) is just the sum of the squared residuals over all the stations and all the earthquakes for (mn) observations.

This is a generalization of the earthquake location problem. In the earthquake location problem, we are given a set of m stations with m corresponding arrival times, and we want to find the "best" origin time and hypocentre location, assuming a velocity model. We make a starting guess for the desired parameters, and through an iterative procedure, we improve the guess until it meets some stopping criterion. In simultaneous inversion, we do this but we also solve for the velocity model at the same time.

Just as in the earthquake location, we define a residual vector:

$$\mathbf{r} = (r_{11}, r_{12}, \ldots, r_{1m} \;\ldots\; r_{n1}, r_{n2}, \ldots, r_{nm})^T \qquad (3.34)$$

Then we define the adjustment vector:

$$\delta\zeta = (\delta t_1, \delta X_1, \delta Y_1, \delta Z_1, \ldots \delta t_n, \delta X_n, \delta Y_n, \delta Z_n, \delta v_1, \ldots \delta v_L)^T \qquad (3.35)$$

At each step of the process, we must solve a set of simultaneous equations for the adjustment vector in order to improve the estimate of the parameter vector. In matrix notation, the system of equations is:

$$\mathbf{B}\,\delta\zeta = -\,\mathbf{r} \qquad (3.36)$$

Here $\mathbf{B}$ is the Jacobian matrix generalized to include the n earthquakes and L velocity parameters, i.e.,

$$B = \left[\begin{array}{ccccc|c} A_1 & 0 & 0 & \cdots & 0 & \\ 0 & A_2 & 0 & \cdots & 0 & \\ \cdots & \cdots & \cdots & \cdots & \cdots & C \\ 0 & 0 & 0 & \cdots & A_n & \end{array}\right] \qquad (3.37)$$

Here the $\mathbf{0}$'s are (m by 4) matrices with zero elements, and the $\mathbf{A_j}$'s are (m by 4) Jacobian matrices of the form:

$$\mathbf{A_j} = -\left(\begin{array}{cccc} 1 & \frac{\partial T_{j1}}{\partial x} & \frac{\partial T_{j1}}{\partial y} & \frac{\partial T_{j1}}{\partial z} \\ 1 & \frac{\partial T_{j2}}{\partial x} & \frac{\partial T_{j2}}{\partial y} & \frac{\partial T_{j2}}{\partial z} \\ . & \cdots & \cdots & \cdots \\ . & \cdots & \cdots & \cdots \\ \cdots & \cdots & \cdots & \cdots \\ 1 & \frac{\partial T_{jm}}{\partial x} & \frac{\partial T_{jm}}{\partial y} & \frac{\partial T_{jm}}{\partial z} \end{array}\right)\Bigg|_{\zeta^*} \qquad (3.38)$$

The **C** element of matrix **B** is a (*mn* by *L*) matrix where each row of the matrix has **L** elements and describes the sampling of the velocity blocks of a particular ray path. If Γ_{jk} is the ray path connecting the $\mathbf{k}^{th}$ station to the $\mathbf{j}^{th}$ earthquake hypocentre, then the elements of the ith row $[i = K + (j - 1)m]$ of **C** are given by:

$$\mathbf{C}_{il} = -\Pi_{jkl}\left(\frac{\partial T_{jk}}{\partial v_{l}}\right)\Bigg|_{\zeta^*} \quad \text{for } l = 1, 2, ..., L \tag{3.39}$$

where Π_{jkl} is defined by

$$\Pi_{jkl} = \begin{cases} 1 \ \textit{if}\ l^{th}\ \textit{block penetrated} \\ 0\ \textit{otherwise} \end{cases} \tag{3.40}$$

It is advantageous to work with slowness rather than with the velocity, where we define the slowness in the lth block as $u_l = 1/v_l$. From this, we have:

$$\delta_{ul} = -\frac{\delta v_l}{v_l} \tag{3.41}$$

To approximate the travel time derivative with respect to velocity to first-order accuracy, we make the approximation:

$$\frac{\partial T_{jk}}{\partial u_l} = \frac{\partial T_{jkl}}{\partial u_l} \tag{3.42}$$

where T_{jkl} is the travel time for the ray Γ_{jk} in the lth block. Also, $T_{jk} = u_l\, S_{jkl,}$ where S_{jkl} is the length of the raypath Γ_{jk} in the lth block. Assuming that the dependence of S_{jkl} on u_l is of second order,

$$\frac{\partial T_{jkl}}{\partial u_l} \simeq S_{jkl} = v_l T_{jkl} \tag{3.43}$$

Thus Equation (3.39) becomes:

$$C_{il} = \Pi_{jkl}\frac{T_{jkl}\left(\zeta^*\right)}{v_l^*} \quad \text{for } l = 1, 2, ..., L \tag{3.44}$$

Since each raypath samples only a small number of blocks, most elements of matrix **C** are zero. Therefore, the matrix **B** is sparse, with most of its elements being zero. Equation (3.32) is a set of (*mn*) equations written in matrix form which may be rewritten as:

$$\delta t_j + \frac{\partial T_{jk}}{\partial x}\Bigg|_{\zeta^*}\delta x_j + \frac{\partial T_{jk}}{\partial Y}\Bigg|_{\zeta^*}\delta y_j + \frac{\partial T_{jk}}{\partial z}\Bigg|_{\zeta^*}\delta z_j - \sum_{l=1}^{L}\frac{\pi_{jkl}T_{jkl}}{V_l^*}\delta v_l = r_{jk}\left(\zeta^*\right)$$

For $K = 1, 2, ... m$, $J = 1, 2, ... n$ (3.45)

Equation (3.45) is a generalization of the equation used for hypocentre determination. The only difference is that the simultaneous inversion equation has the summation term, which is an approximation to the sums of partial derivatives of the travel times with respect to velocity in each block.

To summarize the solution, first we guess a trial parameter vector. Essentially, one must have some estimates of where all the earthquakes are, and have some initial approximation for the velocity. Next, we calculate the theoretical travel times and the spatial derivatives for the trial vector (all stations and all earthquakes). This information is used to construct the **B** matrix. Also, we compute the residual vector, which is essentially the difference between the observed travel times and the theoretical travel times. Then we solve a system of (n by m) linear equations. This gives the adjustment vector, which is added to the trial vector. This is repeated until the solution converges to some pre-defined limit or is terminated after a pre-defined number of iterations. The process is analogous to hypocentre location, but instead of doing it for just one earthquake, we use many earthquakes, and instead of using an assumed velocity model, we solve for that as well.

The central idea of seismic tomography is not that complicated. Yet, it took nearly two decades to develop the method to its current state. It all started with the realization that local networks provided far more data than were required to locate earthquakes. From that point, the method has been refined over the years, and now there are many mathematical variations in the implementation of simultaneous inversion algorithms. However, all of them share the essential theoretical formulation of the problem that has been outlined here.

It is not possible to cover all details of the process in this brief discussion of seismic tomography. Instead, it is attempted to give a basic understanding of how the method works. If one is interested in working with seismic tomography, there are many excellent references available, such as Lanczos (1961), Forsythe et al. (1977), and Gill et al. (1981). For readers not familiar with these mathematical topics, a careful study of the above mentioned references and texts books (e.g., Menke, 1984; Tarantola, 1987) will be of great help. A collection of classic papers on inversion of geophysical data may be found in Lines (1988).

For seismic tomography, a general review by Thurber and Aki (1987) is a good start. In addition, Claerbout (1985), Nolet (1987), Lee and Pereyra (1993) and Iyer and Hirahara (1992) treat this subject in some detail. For medical tomography, Swindell and Barrett (1977) is a good introduction. In addition, there are classic papers by Cormack (1963) and by Houndsfield (1973), and a good text book by Herman (1980).

Many authors have developed software for seismic tomography study (e.g. Aki and Lee, 1976; Thurber, 1981; Paige and Saunders, 1982; Um and Thurber, 1987). A state-of-the-art software is developed by Zhao et al. (1992). Although the conceptual approach of this software parallels that of Aki and

Lee (1976), it has many additional features: (i) It is adaptable to a velocity structure that includes several complex shaped velocity discontinuities like the Moho discontinuity, dipping slab boundary etc., and images 3-D velocity variations in the model. (ii) It uses pseudo-bending technique of Um and Thurber (1987) and Snell's law to search the fastest ray. (iii) The LSQR algorithm (Paige and Saunders, 1982) with damping regulation is used to solve the large and sparse system of observation equations, allowing a great number of data to be used to solve a large tomography problem. (iv) The nonlinear tomographic problem is solved by iteratively conducting linear inversions. In each iteration, perturbations to hypocentral parameters and velocity structure are determined simultaneously to determine the crustal and mantle structure in a region. The method has been used very successfully in Japan (Zhao et al., 1992 and 2004), in Southern California (Zhao and Kanamori, 1993 and 1995), in northeast India (Kayal and Zhao, 1998), in western India, Bhuj 2001 earthquake source area (Kayal et al., 2002b) and in many subduction and volcanic zones (Zhao and Kayal, 2000).

3.13 RECEIVER FUNCTION

Although detailed investigations of the Earth structure utilize multiple seismic recordings as discussed above, a few techniques have been developed to image crustal layering or discontinuity beneath isolated three-component stations. The most widely used method is called *receiver function analysis*. The teleseismic P waves that are incident upon the crustal section below a station produce P to S conversions at crustal boundaries as well as multiple reflections in the shallow layering. It is observed that the P to S conversion has much stronger amplitude on the longitudinal component than on the vertical component. By *deconvolving* the vertical-component signal from the longitudinal component, the obscuring effects of source function and instrument response are removed, leaving a signal composed of primarily S-wave conversions and reverberations below the station. These deconvolved horizontal components are called receiver function traces and are inverted for a model of the shear wave velocity layering in the crust.

Receiver function analysis reveals the crustal interfaces at depth, but the absolute velocities and depths of the boundaries are not well resolved. Using independent determination of the average crustal velocities provided by a surface-wave dispersion analysis can provide a better constrained model. This technique can give stable results if the layers are horizontal. In such a case the receiver function is the same for all azimuths, and the tangential component of the P-wave motion is very small. In regions of three-dimensional structure, the inversion is usually ill posed, and forward modelling of the azimuthally varying receiver functions is used to constrain the crustal and uppermost mantle structure. With the invent of the broadband instruments, the receiver function analysis has become popular in 1980s (e.g. Burdick and Langston, 1977; Owens et al., 1987; Ammon, 1991). Some work is done in this line in India (Kumar et al., 2004; Gupta et al., 2003; Rai et al., 2003). There are about 30 broadband stations in the national network running in the peninsular and extra-peninsular India; the installations started since 1996 in phases. These broadband data have been very useful for this study. In addition, several temporary networks and stand alone stations with broadband instruments are being established under the DST sponsored research projects by the NGRI in peninsular India, RRL-Jorhat in northeast India and by the WIHG in NW Himalaya for this study.

3.14 SEISMIC WAVEFORM INVERSION AND MODELLING

It is mentioned early that one of the goal of a seismologist is to understand every wiggle of a seismogram, which carry information about the source as well as of the medium through which the waves propagate. These waves contain the source effects (magnitude and source time function), the propagation effects (reflectivity, diffraction, etc. in the medium) and the seismograph response effects. Mathematically each of these effects can be modelled to match the waveform of a seismogram. Such mathematical construction of a seismogram is known as *synthetic seismogram*. There are two ways to extract the information about seismic sources and subsurface parameters from seismic data: (i) *Inversion* and (ii) *Modelling*.

In inversion, the model parameters (Earth structure/source characteristics) are updated automatically until the residuals between the *synthetic seismogram* and the *observed seismogram* are minimized. In modelling, the model parameters are changed by human interference until a model is found such that the wave fields generated from the model can be best explained by the observed data. These *waveform modelling* and *inversion method* are very comprehensive compared to travel time methods (Lay and Wallace, 1995). Waveform inversion can produce results of higher resolution than travel time inversion, because it utilizes the information contained in the seismic waveform that are the source and subsurface structures. Waveform inversion and modelling are tightly coupled because inversion requires numerical modelling of wave propagation, and modelling needs a model that is obtained from the inversion.

Theoretical Aspects

Normally for full waveform inversion, earthquake source is assumed as *double couple*. In 2D, a double couple source can be parameterized as

$$s(x, z, t_0, a, R_\theta) \tag{3.46}$$

where x and z are the horizontal and vertical positions respectively, t_0 is the origin time, a is the amplitude of the double couple force, and R_θ is the angle associated with the orientation of the double couple, the local rupture direction.

For a spatially and temporally extended earthquake, the composite source parameters correspond to a series of n double-couples:

$$s(s_1, s_2, s_3, \ldots, s_n) = s(x_1, z_1, t_{01}, a_1, R_{\theta 1}, \ldots x_2, z_2, t_{02}, a_2, R_{\theta 2} \ldots x_n, z_n, t_{0n}, a_n, R_{\theta n}) \quad (3.47)$$

As this is a discrete representation, the total energy release is the integral over the energy associated with all the grid points in the area.

A seismogram or a ground motion at a site can be represented as $u(t)$, i.e. amplitude as function of time. In other words, in seismic waveform modelling, $u(t)$ is represented as the resultant of effects of three basic filters (Lay and Walace, 1995):

$$u(t) = s(t) \times g(t) \times i(t) \quad (3.48)$$

where $u(t)$ is the seismogram, $s(t)$ is the signal from seismic source, $g(t)$ is the propagation filter and $i(t)$ is the seismometer response. Further, the $s(t)$ and $g(t)$ can be divided into a filter that accounts for these multiplicity of arrivals due to reflections and refractions, and a filter that accounts for the attenuation. Similarly, $s(t)$ can be divided into filters that account for source radiation and fault rupture characteristics. For a more theoretical information readers should refer to Helmberger (1983), Wallace and Helmberger (1982), Lay and Wallace (1995) and many other text books.

The linear filter theory is used in constructing synthetic seismogram; the seismogram is treated as the output of a sequence of linear effects. Figure 3.43 illustrates P-wave ground displacement and its records on SP and LP seismometers. The trapzoid shape of ground motion may be considered as output of filters that account for the effects of rupture on a finite fault plane. The ground motion is then distorted by recording characteristics of the seismometers as illustrated in Fig. 3.43.

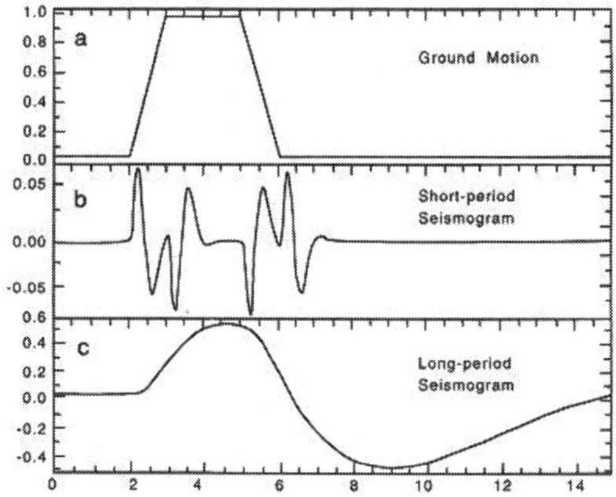

Fig. 3.43 (a) A trapezoid time function corresponding to hypothetical ground motion, (b) Seismogram produced by SP seismometer and (c) Seismogram by LP seismometer (Lay and Wallace, 1995).

For inversion we define the source region to be larger than that which we anticipate exists, and solve for the source parameters of all points in this region. In the result, only the actual source region will correspond to significant energy.

The inherently nonlinear waveform inversion is solved as a sequence of liberalized inversions. It can be represented in three directional components (vertical, radial and transverse) as below:

$$u_V(\Delta, t) = M_0 \; [a \; u_{VDD}(\Delta, t) + b \; u_{VDS}(\Delta, t) + c \; u_{VSS}(\Delta, t)] \quad (3.49)$$

$$u_R(\Delta, t) = M_0 \; [a \; u_{RDD}(\Delta, t) + b \; u_{RDS}(\Delta, t) + c \; u_{RSS}(\Delta, t)] \quad (3.50)$$

$$u_T(\Delta, t) = M_0 \; [a' \; u_{TDS}(\Delta, t) + b' \; u_{TSS}(\Delta, t)] \quad (3.51)$$

where M_0 = seismic moment = $\mu_s \, A \, D$, μ_s = modulus of rigidity at the source, A = area of the fault, D = average slip (displacement) of the rock above the fault and a, b, c, a' and b' are the coefficients and are nonlinear functions of (i) dip δ and (ii) strike direction θ of the fault, and (iii) slip direction λ of the shear dislocation constants

$$a = \sin\lambda \sin 2\delta \quad (3.52)$$
$$b = -\cos\lambda \cos\delta \cos\phi + \sin\lambda \cos 2\delta \sin\phi$$
$$c = 0.5 \sin\lambda \sin 2\delta \cos 2\phi + \cos\lambda \sin\delta \sin 2\phi$$

and

$$a' = \sin\lambda \cos 2\delta \cos\phi + \cos\lambda \cos\delta \sin\phi$$
$$b' = \cos\lambda \sin\delta \cos 2\phi - 0.5 \sin\lambda \sin 2\delta \sin 2\phi$$

and $u(\Delta, t)$ is the response while travelling through the medium (Earth) and also depends on the type of source mechanism, time function and epicentre distance Δ. The solution for δ, θ and λ are obtained by grid search method of nonlinear inversion to get the minimum misfit error between observed and synthetic seismograms.

A general source is represented by a *moment tensor* [M] with six independent elements. However, in most tectonic earthquakes, there is no volume change at the source and thus we have only five independent elements of [M]. A synthetic seismogram is obtained by sum of five terms where each is a product of an element of [M] and the corresponding Green's function (synthetic seismogram). The moment tensor elements can be evaluated by simple linear inversion so as to get the minimum difference between synthetic and observed seismograms. The moment tensor is decomposed into best double couple moment tensor and a remaining CLVD tensor. The resulting double couple moment tensor is then interpreted for dip, strike, slip direction of the shear dislocation and percentage of double couple in the moment tensor solution.

The goal of waveform modelling is to recover seismic source parameters like source depth, time function, fault orientation and seismic moment (*moment tensor solution*) by matching the observed and synthetic waveforms. From mid 1970s through early 1980s many studies were made on the earthquake source parameters using teleseismic waveform modelling, mainly long period WWSSN data. The methodology involved matching long period P and SH waves that are well distributed in azimuth about the source (Fig. 3.44). Subsequently, attempt has been made to recover source orientation by waveform inversion at a single seismic station (Kanamori et al., 1990). Earthquakes at local and regional distances with new generation very broadband high dynamic range instruments make it possible to obtain focal mechanism for moderate size earthquakes. Source parameters can also be obtained from surface wave inversion. Surface wave inversions are best performed at long periods (>100 s). Moment tensor solutions in understanding the focal mechanism or source process are further discussed in Section 4.8.

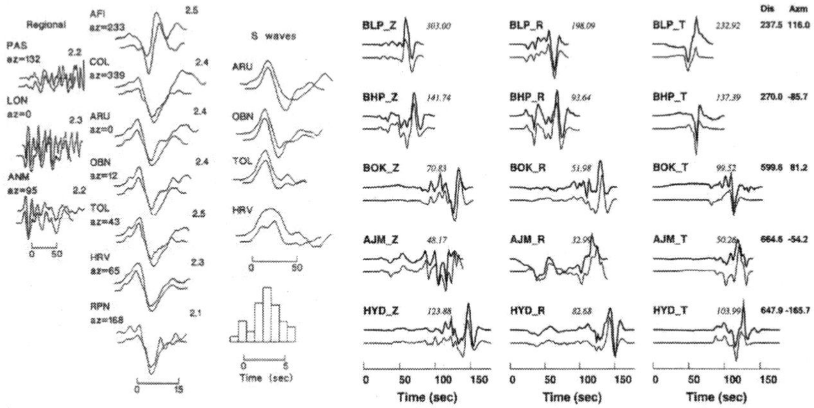

(a) (b)

Fig. 3.44 (a) An example of waveform modelling for the 1989 Loma Prieta earthquake (Wallace et al., 1991). (b) An example of waveform modelling for the 1997 May 21, Jabalpur earthquake (Rao et al., 2002). (The observed (thick) and synthetic (thin) displacement seismograms are shown.)

3.15 MICROEARTHQUAKE MAGNITUDE

Richter magnitude scale is very popular for estimating magnitude for local earthquakes. The Richter magnitude scale is, however, based on Wood Anderson seismograph (see Section 2.1), which seldom records earthquakes with magnitude less than 3.0. So, we need a convenient method for estimating magnitude of the local earthquakes recorded by microearthquake networks with high gain instruments.

An approach was suggested by Brune and Allen (1967) to calculate the ground motion from the recorded maximum amplitude, and from this to compute the response expected from a Wood Anderson seismograph. There are, however, two problems in this approach: (i) Richter (1958) pointed out that the maximum amplitude on the Wood-Anderson record may not correspond to the wave with maximum amplitude on a different instrument's record, and (ii) maximum amplitude are mostly saturated (clipped) by the high gain microearthquake recorders.

The other approach is to use signal duration instead of maximum amplitude. This idea appears to originate from Bisztricsany (1958). He determined the relationship between earthquake magnitudes 5 to 8 and duration of the surface waves of epicentral distance 4°-160°. Solovev (1965) and Tsumura (1967) studied determination of magnitude from total signal duration for local earthquakes in Japan. Since then many authors derived empirical relationships between signal duration and magnitude for local earthquakes.

Lee et al. (1972) established an empirical formula for estimating signal duration magnitude (M_D) for the local earthquakes recorded by the USGS Central California microearthquake network using signal duration. For a set of 351 earthquakes, they computed the M_D equivalent to local magnitudes as defined by Richter (1958). Correlation of the local magnitudes with the signal durations measured by the USGS microearthquake network is shown in Fig. 3.45. They obtained the following relation:

$$M_D = -0.87 + 2.00 \log \tau + 0.0035 \Delta \qquad (3.53)$$

where M_D is duration magnitude equivalent to Richter magnitude, τ is signal duration in seconds and Δ is epicentral distance in kilometres. In an independent work, Crosson (1972) obtained a similar relation in Puget Sound region, Washington, USA.

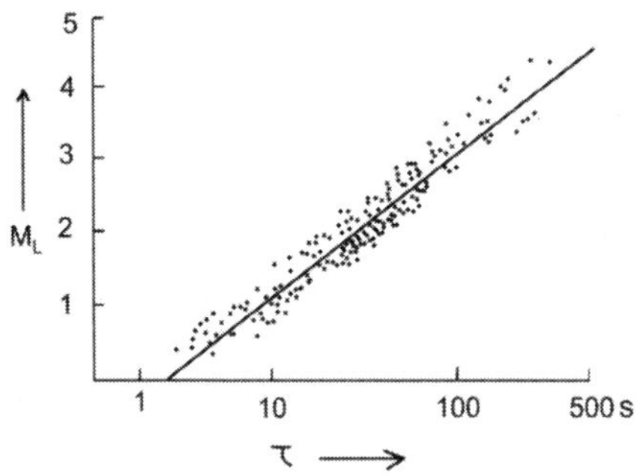

Fig. 3.45 Relation between Richter magnitude and signal duration (Lee, 1992).

Since 1972 many authors (e.g. Hori, 1973; Real and Teng, 1973; Herrmann, 1975; Bakun and Lindh, 1977; Griscom and Arabasz, 1979; Johnson, 1979; Suteau and Whitcomb, 1979) have determined signal duration magnitude and given a relation in the form:

$$M_D = a_1 + a_2 \log \tau + a_3 \Delta + a_4 h \qquad (3.54)$$

where τ is signal duration in seconds, Δ is epicentral distance in kilometres, h is focal depth in kilometres, and a_1, a_2, a_3 and a_4 are empirical constants. These constants are determined by correlating signal duration with the Richter magnitude for a selected set of earthquakes and taking epicentral distance and focal depth into account.

It is, however, important to note that different authors have defined signal duration differently. In practice, it is difficult to measure total signal duration of an earthquake. Lee et al. (1972) defined signal duration to be the time interval in seconds between the onset of the first P-wave and the point where the seismic signal, peak to peak amplitude, no longer exceeds 1 cm as it appears on a Geotech Develocorder viewer with 20 × magnification. Though this definition is arbitrary, it is easy to apply, and the measurements are reproducible by readers within 1-2 s; this much precision is fair enough to compute the M_D. Since the background noise is about 0.5 cm (peak to peak), this definition is equivalent to a cut off point where the signal amplitude is approximately twice that of the noise, an example is illustrated in Fig. 3.25. For a given earthquake, signal duration should be measured for as many stations as possible. Duration magnitude is then computed for each station, and the average of the station magnitudes is taken to be the earthquake magnitude.

Using signal duration to estimate earthquake magnitude has become a common practice in microearthquake networks. However, studies to improve estimation of magnitude of local earthquakes have been carried out. Johnson (1979) presented a method of magnitude estimation based on decay of coda amplitudes using digital seismic traces. Suteau and Whitcomb (1979) studied relationships between local magnitude, seismic moment, coda amplitude and signal duration. The signal duration relation developed by Lee et al. (1972) for estimation of magnitudes of the local earthquakes recorded by the microearthquake networks in California is, however, widely used by many authors. These estimations are found to be compatible with the Richter magnitudes.

3.16 EARTHQUAKE PHENOMENA: POWER LAW RELATIONS

The earthquake phenomena with respect to magnitude, time and space possesses *power-law* relation. The Gutenber-Richter (1944) frequency-magnitude relation, *b-value*, is a power law relation involving magnitude. Similarly *aftershock attenuation (p-value)* follows another power law, Omori Law, involving time (Omori, 1894). Two-point spatial correlation function for earthquake epicentres also displays a power law structure (Kagan and Knopoff (1980), and represents a self similar mathematical construct, the *fractal*; the scaling parameter is called the *fractal dimension* (Mandelbort, 1982). All these relations are also important to understand earthquake process and tectonics.

3.16.1 Frequency-Magnitude Relation b-value

Magnitude of an earthquake is the most commonly used parameter of earthquake size. The statistical distribution of sizes for a group of earthquakes is complicated. Gutenberg and Richter (1944) provided a simplest earthquake occurrence of frequency-magnitude relation, which describes a power law relation:

$$\log_{10}N = a - bM \quad (3.55)$$

where N is the number of earthquakes in a group having magnitude larger than M, a is a constant and b is the slope of the log-linear relation. The estimated slope of the log-linear relation or the coefficient b is known as b-value. The b-value varies from 0.5-1.5 depending on tectonics, structural heterogeneity and stress distribution in space (Mogi, 1962; Scholtz, 1968). It has been shown that the relation also holds for aftershock sequence (Utsu 1961 and 1969).

The b-value should be estimated carefully as the self-similarity may break with the following three stages: smaller events (M<3.0), medium events ($3 \leq M \leq M_{saturate}$), and larger events ($M \geq M_{saturate}$). The smaller events may give lower b-value because of shortage of recorded smaller events, while bigger events may give higher b-value because of the saturation of the magnitude (Scholtz, 1990).

The b-values are estimated using two methods: the *least-square fit method* and the *maximum likelihood method*. In the least-square fit method, the log values of the cumulative number of earthquakes (N) are plotted with magnitude (M). A sample of plot is shown in Fig. 3.46 for the 1999 chamoli earthquake sequence. The b-value is estimated from the slope of the least square fit line, the long-linear relation of the N and M.

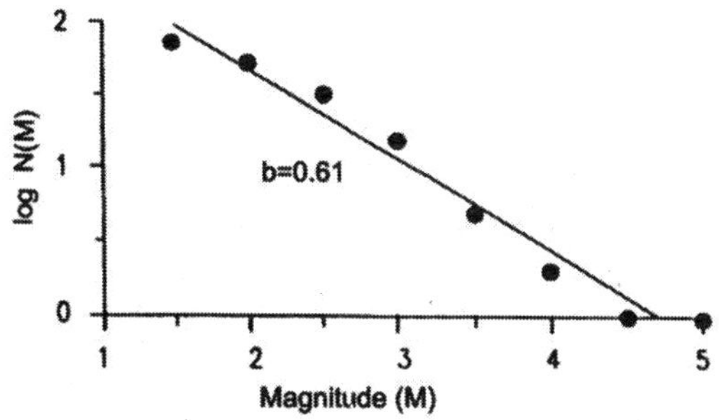

Fig. 3.46 Frequency-magnitude plot, 1999 Chamoli aftershocks, NW Himalaya (Kayal et al., 2003b).

In the maximum likelihood method, Aki (1965), based on theoretical considerations, gave an estimate of b-value as:

$$b = \frac{\log_{10} e}{\overline{M} - M_0} \tag{3.56}$$

where $\overline{M}$ is the average magnitude and M_o is the threshold magnitude. An estimate of error, standard deviation δb of the b-value was given by Aki (1965), then modified formulation was given by Shi and Bolt (1982) as follows:

$$\delta b = 2.3b^2 \sqrt{\frac{\Sigma(M_i - \overline{M})}{n(n-1)}} \tag{3.57}$$

where M_i is the magnitude of ith event, $\overline{M}$ is the average magnitude for a set of earthquakes and n is the number of earthquakes in the set. An estimated b-value by this method for the 2001 Bhuj earthquake sequence is 1.21 ± 0.001; this is comparable with that (1.20) of the least-square fit method (Kayal et al., 2000a).

3.16.2 Aftershock Attenuation (p-value)

Omori (1894) first presented a famous formula about the time dependent attenuation of the aftershock activity, a power law relation, as follows:

$$N(t) = K/(t+c) \tag{3.58}$$

where $N(t)$ is the frequency of aftershocks per unit time interval at time t, K and c are constants.

Utsu (1957) made a modified version of the Omori's formula as:

$$N(t) \propto t^{-p} \tag{3.59}$$

where p is a rate-constant of aftershock decay. Equation (3.59) is called modified Omori formula with exponent p =1, which implies that relaxation function for aftershock activities on frequency shows a temporal fractal property. Several researchers have empirically shown that the p-values of large earthquakes were closed to 1 but ranged from 0.6 to 2.5 (e.g. Mogi, 1962; Utsu 1969; Kisslinger and Jones, 1991; Guo and Ogata, 1995, 1997). It is, however, not clear why each aftershock sequence has a significant different p-value. Hirata (1987) argued that the p-value may be related to *fractal dimension* of pre-existing fault system in an aftershock region.

Figure 3.47a illustrates the aftershock sequence with time for the 2001 Bhuj earthquake in western peninsular of India. The aftershock frequency per day, $N(t)$ day^{-1}, is plotted against time t (days) on double logarithmic scale in Fig. 3.47b. The p-value is estimated from the slope of the regression line fitted to these data by least-square method. The estimated p-value is 0.91, near to normal p=1 (Kayal et al., 2002a).

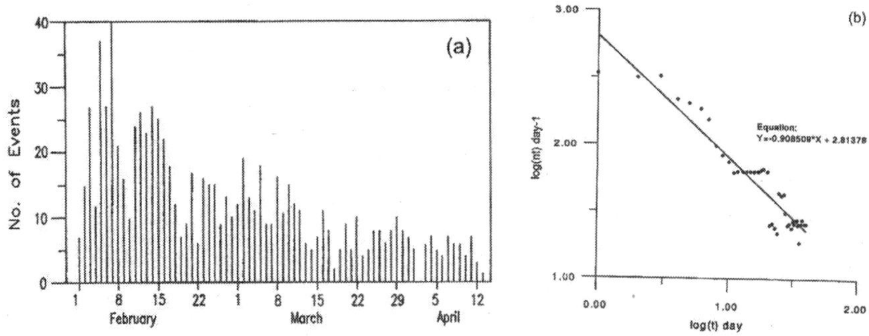

Fig. 3.47 (a) Aftershock decay with time, the January 26, 2001 Bhuj earthquake sequence, and (b) p-value of the sequence from the slope of the log-log plot (Kayal et al., 2002a).

3.16.3 Fractal Dimension

Fractal properties of seismicity, a stochastic self-similar structure in time and space distribution of earthquakes, can be measured by *fractal dimension*, which is introduced as a sophisticated statistical tool to quantify dimensional distribution of seismicity, its randomness and clusterisation (Hirata, 1989; Kagan and Knopoff, 1980; Ogata, 1988).

The most commonly used methods for fractal dimension calculation is the box counting method which measures the *capacity dimension* D_0, and the correlation integral method which measures the *correlation dimension* D_2. In the box counting method an active fault system of a given extent is overlaid with a grid of square boxes. Grids of different size boxes are used. This method considers the number of cubes necessary to accommodate each of a sequence of seismic events. The main disadvantage in this method is that it does not consider the number of seismic events in each box (Xu, 1992). It takes into account only the fact that the boxes are occupied or not. This method is not reliable especially when the number of data point is limited (Hirata, 1989). The correlation dimension is widely applied in seismology, especially to spatial distribution of earthquakes. This technique is preferred to box counting method because of its greater reliability and sensitivity to small changes in clustering properties (Kagan and Knopoff, 1980; Hirata, 1989).

The fractal dimension of the spatial distribution of seismicity is calculated from the correlation integral given by Grassberger and Procaccia (1983):

$$D_{wr} = \lim_{r \to 0} \log (C_r)/\log r \tag{3.60}$$

where (C_r) is the correlation function. The correlation function measures the spacing or clustering of a set of points, which in this case is earthquake epicentres, and is given by the relation:

$$C(r) = \frac{1}{N(N-1)} N_{(R<r)} \tag{3.61}$$

where $N_{(R<r)}$ is the number of pairs (X_i, X_j) with a smaller distance than r. The correlation integral is related to the standard correlation function as given by Kagan and Knopoff (1980):

$$C(\varepsilon) = \frac{1}{N(N-1)_i} \sum_{\substack{i,j=1 \\ i \neq j}}^{N} \delta^F (X_i - X_j - \varepsilon)$$

$$C(r) = \int_0^r d\varepsilon^F c(\varepsilon) \tag{3.62}$$

where F is 2 in the case of two-dimensional space, and 3 in the case of three-dimensional space. F is 2 for the epicentre distribution. If the epicentre distribution has a fractal structure, we obtain the following relation:

$$C(r) \sim r^{D_2} \tag{3.63}$$

where D_2 is a fractal dimension, more strictly, the correlation dimension (Grassberger and Procaccia, 1983). By plotting $C(r)$ against r on a double logarithmic coordinate, we can practically obtain the fractal dimension D_2 from the slope of the graph. The distance (r) between two events, (θ_1, ϕ_1) and (θ_2, ϕ_2), is calculated by using a spherical triangle as given by Hirata (1989):

$$r = \cos^{-1} (\cos \theta_1 \cos \theta_2 + \sin \theta_1 \sin \theta_2 \cos (\phi_1 - \phi_2)) \quad (3.64)$$

The slope is obtained by fitting a least-square line in the scaling region as shown in Fig. 3.48. A detailed study of fractal of dimension is made in the NE India region as discussed in Section 6.7.

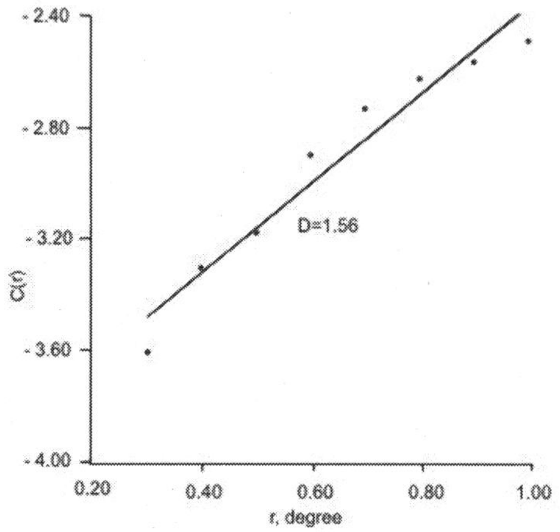

Fig. 3.48 Fractal dimension estimate by plotting $C(r)$ versus r in NE India region (Bhattacharaya et al., 2002).

Tosi (1998) illustrated that possible values of fractal dimension for epicentral distribution are bound to range between 0 and 2, which is dependent on the dimension of the embedding space. Interpretation of such limit values is that a set with D=0 has all events clustered into one point; at the other end of the scale, D=2 indicates that the events are randomly or homogeneously distributed over a two-dimensional embedding space.

Hirata (1989) reported temporal variations in fractal dimension to quantify the seismic process, and Shimazaki and Nagahama (1995) demonstrated that active fault systems in Japan possess self-similarity with fractal dimension of 0.5 to 1.6. Time variation of spatial fractal dimension also suggest that there may be a positive or negative correlation among b-value, p-value and fractal dimension (Ouchi and Uekawa, 1986; Main, 1991; Nanjo et al., 1998). Hence it is important to understand these parameters in assessing earthquake risk of a tectonically active region.

ANNEXURE 3.1

Example of an input data file (Hypo-71 format): Chamoli aftershocks recorded by a temporary network in Western Himalaya

Control Parameter

```
RESET TEST(01)=0.1 sec
RESET TEST(02)=20.00 km
RESET TEST(03)=-0.50
RESET TEST(04)=0.20 km
RESET TEST(05)=5.0 km
RESET TEST(06)=1.0
RESET TEST(07)=-0.87
RESET TEST(08)=2.00
RESET TEST(09)=0.0035
RESET TEST(10)=5 km
RESET TEST(11)=16.0
RESET TEST(12)=0.5 km
RESET TEST(13)=1.0 km
```

Station Coordinates

```
SML 3014.80N 7914.21E
MTN 3023.00N 7919.00E
GHT 3015.00N 7925.00E
CHD 3026.00N 7905.00E
JSM 3030.00N 7934.00E
```

Velocity (Vp) Model

```
5.50 0.000
6.50 12.000
7.10 20.000
8.10 50.000
```

Control parameter, P and S phase details and signal duration

```
  3.   200.   500.  1.70       3 0   4      1      1  0 0 0000   0 0.00 0 0.00
 SML   EPD0  990420202429.60  32.90IS 0   0.                   30.
 MTN   IPD0  990420202427.90  29.20IS 0   0.                   44.
 GHT   IPD0  990420202429.10  32.10IS 0   0.                   41.
       10
 SML   IPU0  990421204821.30  24.30IS 0   0.                   26.
 MTN   IPD0  990421204819.50  21.30IS 0   0.                   40.
 GHT   IPD0  990421204820.80  23.50IS 0   0.                   36.
       10
 SML   IPD0  990421233421.00  24.10IS 0   0.                   40.
 MTN   EPD0  990421233419.20  21.50IS 0   0.                   19.
 GHT   IPD0  990421233420.90  24.50IS 0   0.                   48.
       10
 SML   IPD0  990424071359.30  62.00IS 0   0.                   24.
 MTN   IPD0  990424071358.10  59.90IS 0   0.                   24.
 GHT   IPD0  990424071358.40  60.60IS 0   0.                   41.
```

4

Dynamics of Faulting and Fault Plane Solution

Introduction

Explanation of immediate faulting or mechanism of earthquake is one of the most fascinating and significant problems in seismology. The physical process of *elastic strain* accumulation and the triggering mechanism are the basics to understand the earthquake kinematics. The term *focal mechanism* or *fault-plane solution* conventionally refers to fault orientation, the displacement and stress release patterns and the dynamic process of seismic wave generation.

There is now convincing explanation that some parts of the Earth's crust and the upper mantle gradually come under mechanical stresses due to plate movements. Sudden fractures occur at weak places in the stressed rocks, which release stress and strain simultaneously, thus emitting seismic waves or earthquake waves. The fractures are ones in which blocks of rock on either side of the fracture plane or fault plane move in opposite directions in a motion of shear. Since motions of shear do not involve change in volume, such fractures can occur anywhere in the brittle part of the Earth's outermost layers, even at substantial depths in the upper mantle in some situations where the pressures are very high.

4.1 CLASSIFICATION OF FAULTS

Faults are ruptures along which the opposite walls move past each other. The main feature is differential movement parallel to the surface of the fracture. The faults can be classified as *thrust fault, normal fault* and *strike-slip fault* based on the nature of relative movement along the fault. The block above the fault plane is called *hanging wall*, the block below the fault plane is the *foot wall* (Fig. 4.1). The *dip* is the angle between the horizontal surface and the plane of the fault; *hade* is compliment of the dip. The hanging wall, foot wall, dip and hade of a fault are illustrated in Fig. 4.1.

A standard nomenclature *rake* has evolved for describing slip direction, that is useful to introduce at this point. The actual motion of the two blocks on either side of the fault plane is defined by a slip vector which can have any orientation on the fault plane. The direction of slip vector is given by the angle of slip or *rake* (λ). It is measured in the plane of the fault from the strike direction to the slip vector showing the motion of the hanging wall relative to the footwall (Fig. 4.1b). The magnitude of the slip vector is given by D, the total displacement of the two blocks.

Three basic categories of fault slip are commonly used to characterize motions on the faults relative to the Earth surface. These are described below.

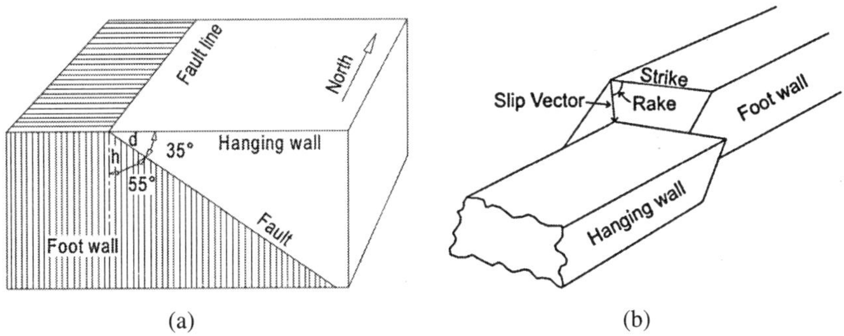

Fig. 4.1 (a) Diagram showing fault plane, dip (d = 35°), hade (h = 55°), hanging wall and footwall, and (b) the slip vector and rake (λ).

Thrust Fault

A *thrust fault* is a fault along which the hanging wall (upper side of the fault) has moved up relative to the foot wall (Fig. 4.2). The thrust is one that dips less than 45°. A *reverse fault* is a thrust that dips more than 45°, and an *overthrust* that dips less than 10°. In pure thrust-faulting the slip vector is parallel to the dip direction of the fault and it is upward, so λ = 90°. Thrust faulting involves crustal shortening, and implies *compression.*

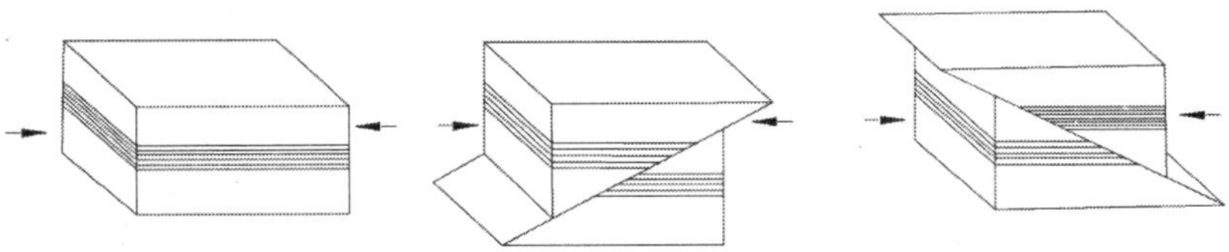

Fig. 4.2 Pure thrust faulting; dip $0° < \delta < 90°$, rake = + 90°. Arrows indicate compression.

Normal Fault

A *normal fault* is a fault along which hanging wall has moved relatively downward (Fig. 4.3). In pure normal faulting the slip vector is also parallel to the dip direction of the fault plane, but it is downward; the $\lambda = -90°$ (270°). Normal faulting involves lengthening of the crust, and implies *tension.* There are many possibilities concerning the actual movement; the footwall may remain stationary and the hanging wall goes down; or the hanging wall remains stationary and the footwall goes up, or both blocks move down, but the hanging wall moves more than the footwall, or both blocks move up; but the footwall moves more than the hanging wall. Some geologists use the term *gravity fault* in preference to normal fault.

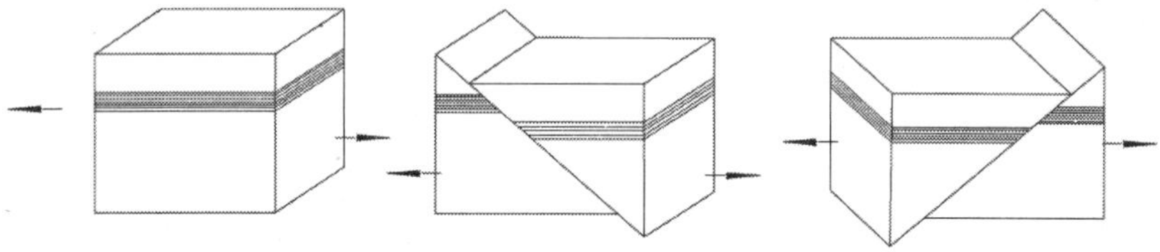

Fig. 4.3 Pure normal faulting; dip $0° < \delta < 90°$, rake = –90°. Arrows indicate tension.

Strike-slip Fault

In contrast to the normal dip-slip (Fig. 4.4a) or reverse dip-slip (Fig. 4.4b), a *strike-slip fault* is a fault along which displacement has been essentially

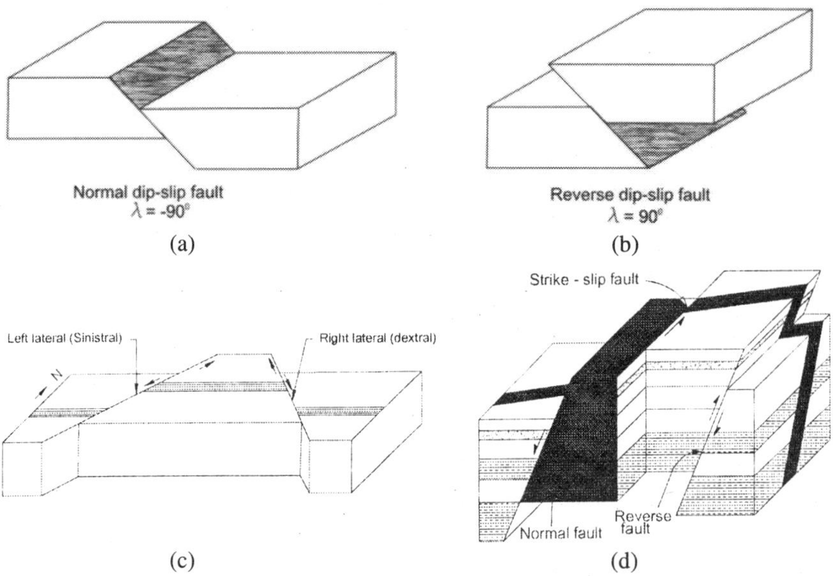

Fig. 4.4 Slip in different faulting (a) Normal dip-slip fault (b) reverse dip-slip fault, (c) strike-slip faulting; left-lateral slip (dip = 90°, $\lambda = 0°$) and right-lateral slip (dip 90°, $\lambda = 180°$), (d) combination of strike-slip, normal and reverse faulting.

parallel to the strike of the fault (Fig. 4.4c, d), i.e., the dip-slip component is less or negligible ($\lambda = 0$ or 180°). For $\lambda = 0$, the hanging wall moves to the right so that the opposite wall, faced by an observer, moves relatively to the left (Fig. 4.4c). This is called *left-lateral slip* or *sinistral* fault. When λ = 180°, the hanging wall moves to the left and the opposite wall, faced by an observer, moves relatively to the right (Fig. 4.4c) This is *right-lateral slip* or *dextral* fault. In general λ will have a value different than these special cases, and the motion is then called *oblique* slip.

We often use term normal fault or thrust fault with strike slip component and so on. A combination of normal, strike-slip and reverse faulting is illustrated in Fig. 4.4d.

4.2 FORCE AND STRESS

Force is defined by Newton's Second Law as that which accelerates a mass, and amount of force per unit area is termed as *stress*. The unit of force, *Newton*, is defined as that force which gives to a mass of one kilogram an acceleration of one metre per second per second. The unit of stress is then Newton per square metre (N/m^2).

A stress is usually resolved into two components (Fig. 4.5), a normal stress (σ) perpendicular to the area, and a shearing stress (τ) acting on the plane of the area.

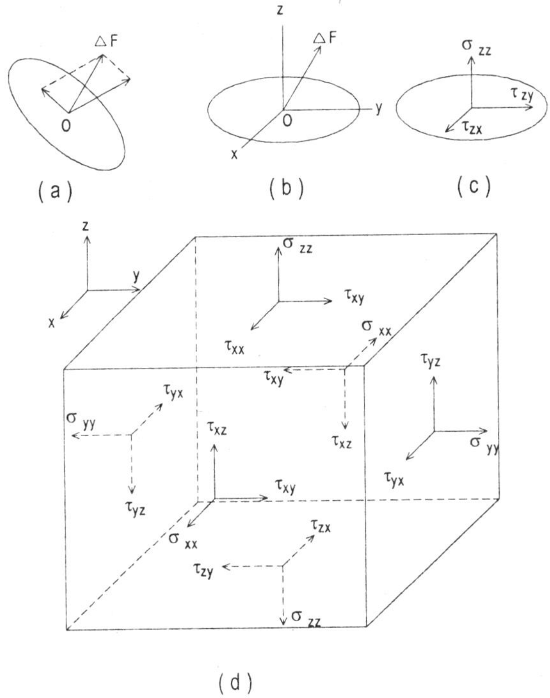

Fig. 4.5 (a-c) Stress acting on an element, (d) stress component in three dimensions.

4.2.1 Geometry of Stress

The geometry of stress will depend on both the nature of the system of applied forces and on the position and orientation of the surface area ΔA with respect to the system. For example, if ΔA lies on the X-Y plane, there will be two shear components (τ_{zy} and τ_{zx}) and one normal component (σ_{zz}) (Fig. 4.5c). The first suffix indicates the direction of normal to ΔA and the second the direction of stress component in which it acts. There will be similar components when ΔA is parallel to other two coordinate planes. In all, nine such stress components can be associated (Fig. 4.5d); these are as follows:

$$\begin{matrix} \sigma_{xx} & \tau_{xy} & \tau_{xz} \\ \tau_{yx} & \sigma_{yy} & \tau_{yz} \\ \tau_{zx} & \tau_{zy} & \sigma_{zz} \end{matrix} \qquad (4.1)$$

Stress is thus not a simple vector, but one whose components are also vectors, which is what is called *tensor.*

4.2.2 Principal Stress

Given the nine components (Eq. 4.1), the stress on any inclined plane can be determined. When all components involving z are zero, i.e. in a state of plane-stress, the nine components reduce to just four (Fig. 4.6a). This is, however, strictly valid for very thin plates. In all other general situations, there will be three mutually perpendicular stresses at each point. These are: σ_1 (maximum principal stress), σ_3 (minimum principal stress) and σ_2 (intermediate principal stress), (Fig. 4.6b). The value of intermediate principal stress (σ_2) generally plays a secondary role in deformation and failure of material, and therefore, it is the σ_1 and σ_3 stresses which are usually of much interest.

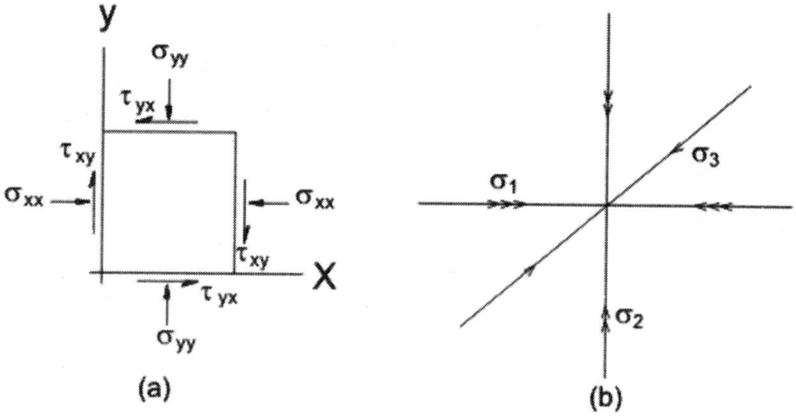

Fig. 4.6 (a) State of plane-stress, (b) three mutually perpendicular stresses.

The stress state in any medium can be described by what is called the *stress ellipsoid* (Fig. 4.7). This is an ellipsoidal surface with its centre at the concerned point in the material. The stress, measured as the force per unit area normal to a specified plane passing through the point, is represented by the length of the radius of this ellipsoid drawn normal to that plane. An ellipsoid has three principal axes. The components of stress along these three axes are designated as the three principal stresses at the point, and are referred to as the maximum (σ_1), intermediate (σ_2) and minimum (σ_3) principal stresses as mentioned above.

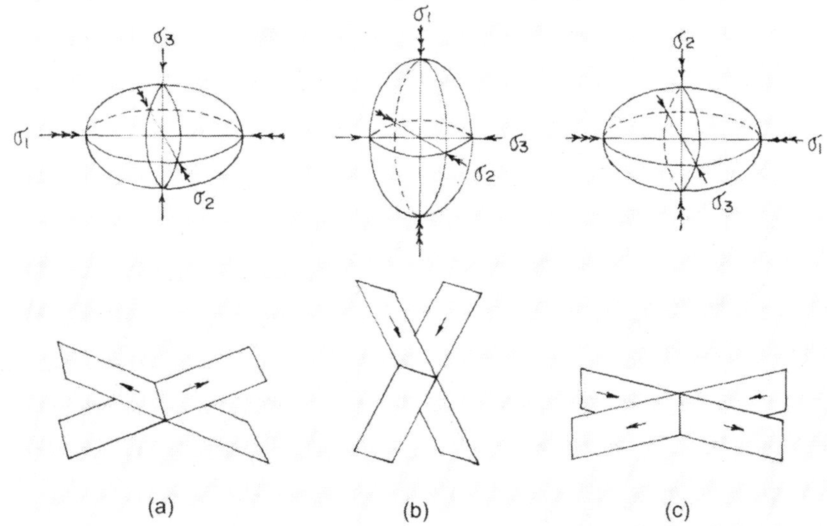

Fig. 4.7 Stress ellipsoids: (a) thrust-faulting, (b) normal faulting and (c) strike-slip faulting.

At any point in a liquid all these three components of principal stress are equal, and the stress ellipsoid becomes a sphere. This is what is expressed by *Pascal's law*. In a solid, however, the maximum principal stress component may be much more than the other two components.

4.3 CRITERIA OF FRACTURE

Brittle solids respond to excessive strain by breaking. The nature of breaking depends on the manner in which the stress is applied. As long as the strain resulting from the stress in the material is within a certain limit, the energy of deformation gets stored up as strain. But when a certain limiting strain is reached, the material can no longer deform further to accommodate stress, and goes into a sudden fracture. We refer to this as *brittle fracture.*

Let us consider the most basic stages of faulting, which are: (i) initiation of rupture and (ii) frictional sliding during the rupture process. Coulomb in

1773 introduced a simple theory for rock failure referred to as the Coulomb failure criterion, which states that the shear stress (τ) of a rock is equal to the initial strength of the rock, plus a constant times normal stress (σ_n) on the plane of failure; this can be expressed as:

$$\tau = c + \mu\sigma_n \tag{4.2}$$

where τ and σ_n are the *shear* and *normal stresses* resolved on the concerned plane within the material, c is the initial strength of the rock called *cohesion,* and μ is a constant called the *coefficient of internal friction.*

Under compressive stress, shear fracture in a homogeneous solid happens only when σ_2 and σ_3 are unequal. The slip plane, the plane in which the shear stress is maximum, contains the intermediate principal stress axis σ_2, and makes an angle θ with the maximum principal axis. This angle is 45° when internal friction is zero, but smaller for all real material.

In the laboratory, the relationship between sliding displacement and applied shear stress is not smooth. In general no slip occurs on the fault surface unless the critical value of τ is reached, then sudden slip occurs followed by a drop in stress. This causes a time interval of "no slip" during which stress again builds upto the critical value for sudden-slip; the episode is repeated. This type of frictional behaviour is known as *stick-slip*, or unstable sliding. For extremely smooth fault surfaces the slip may be continuous, and referred to as stable sliding. Earthquakes are generally thought to be recurring slip episode on pre-existing faults followed by period of 'no-slip' and increasing strain. The difference between the shear stress just before the slip episode and just after the slipping has ceased is known as *stress drop*. The stress drop observed in an earthquake may represent only a fraction of the total stress supported by the rock.

4.4 GRIFFITH'S THEORY OF CRACK PROPAGATION

Although Coulomb failure criteria and the Anderson theory of faulting (see Section 4.5) provide a static framework for understanding the faulting process, it does not consider what happens at the microscopic level inside a rock. Griffith, in 1920, showed that in all real materials failure happens at much lower values of stress than what is expected theoretically for a homogeneous material. It has been found that cracks are always present at a microscopic level. These are referred to as *Griffith cracks.* Under an externally applied stress the concentration of stress at the crack tips can cause the cracks to extend, and thus, it is the existing Griffith cracks which determine the fracture strength of the bulk material.

Confining pressure has very profound effects on the behaviour of rocks under stress. For small values of confining pressure the pre-existing cracks close. At larger pressures, if the confining condition is not adequately met, a rock mass may extend laterally in an inelastic fashion. This property, in which volume dilatation occurs as a result of application of *deviatoric stress*, is known as *dilatancy*. This is interpreted as being due to the development of new microcracks throughout the body of the stressed rock, together with an increase in the total void space in the cracks. The deviatoric stress is defined in Section 4.5.

Higher temperatures also tend to reduce the strength of a rock. In the presence of water, which is ubiquitous in the Earth's lithosphere, the deformation behaviour of rocks under stress can be very different from what may be expected to happen in the dry state. When a rock is subjected to constant stress that is less than its instantaneous fracture strength, it suffers what is called *static fatigue*. This is a process of failure after the lapse of some characteristic time, which depends on the fluid content, temperature, activation energy for the propagation of the microcracks and the applied stress. The rock initially deforms at a slow rate, then faster, and finally goes into brittle fracture. This type of creep behaviour, although macroscopically viscoelastic, involves brittle fracture at microscopic scale throughout the stressed volume. Thus, crack growth can occur at lower stresses, leading finally to a fracture or fault.

Purely brittle behaviour in rocks changes into crystalline *plasticity* at sufficiently high temperatures and pressures. The transition is broad, from semi-brittle, which is essentially a mixture of brittle and plastic processes at microscopic scale, to plastic. The macroscopic rheology is *ductile*. The degree of ductility depends largely on the lithology and temperature. *Ductility* is used to refer to a rheology that allows large macroscopically homogeneous deformation regardless of the micro-mechanisms involved. *Plasticity*, on the other hand, is used to imply an underlying mechanism that is plastic in nature, such as dislocations in the lattice. The term *cataclastic* flow refers to a kind of deformation characterised by ductile stress-strain behaviour, but involving some degree of internal cracking as well as frictional sliding.

4.5 DYNAMICS OF FAULTING

The application of Coulomb failure criterion to geologic materials as well as stress in the Earth to predict newly created fault orientations is understood by *Anderson's Theory of faulting* (Anderson, 1951).

Equation (4.2) can be displayed graphically in a *Mohr diagram*. Figure 4.8 represents the *Mohr* diagram on a σ-τ plot along with the straight line, which gives the condition for failure. The Mohr circle construction consists of a circle with radius $(\sigma_1 - \sigma_3)/2$ units centered on the horizontal axis, which forms the origin. The values of the normal stresses defined by the two points of intersections of the circle with this horizontal axis are the principal stresses σ_1 and σ_3. The direction of failure or orientation of the fault is defined by the perpendicular to the envelope (angle 2θ in Fig. 4.8). The slope of the failure envelop is related to the *internal friction* by $\lambda = \tan \phi$. A fault will form at an angle (90°–θ) from the axis of maximum compressive stress σ_1; $\theta = \pm (45° \pm \phi)$.

For every general state of stress the mean stress $\bar{\sigma}$ can be defined as

$$\bar{\sigma} = \frac{(\sigma_1 + \sigma_2 + \sigma_3)}{3} \tag{4.3}$$

In two dimensions, $\bar{\sigma}$ is represented by the centre of *Mohr circle*. The degree to which the stress system departs from this mean is termed as a *deviatoric stress*. In two dimensions, the radius of Mohr circle is a measure of the deviatoric stress.

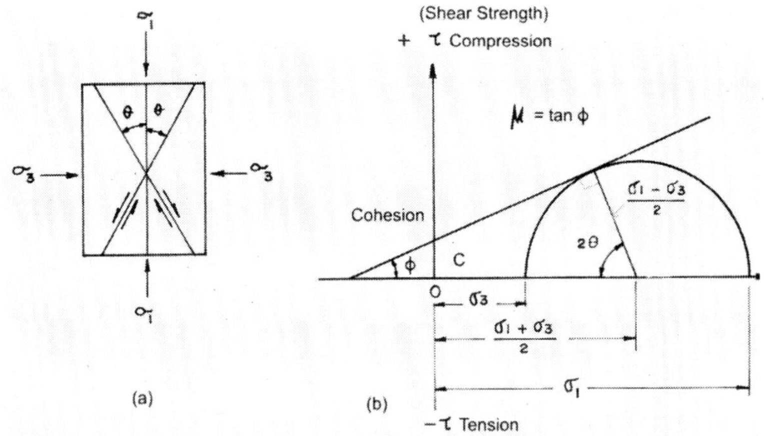

Fig. 4.8 (a) Shear fracture in a homogeneous solid. The slip plane contains the σ_2. (b) *Mohr circle* on a σ-τ plot.

Let us now try to understand what happens if a rock mass is subjected to tectonic stresses. It is convenient to assume that in the absence of such stresses, the stress at a point at depth is nondeviatoric, and its value depends on the overburden. In this hypothetical situation, the mean stress is the total stress. Anderson (1951) termed this as the standard state. With the onset of tectonic stress, the state of stress at the point becomes deviatoric. As the deviatoric stress becomes too large, the material fails and cause faulting. In combination with the geometrical relationship between fracture planes (fault planes) and stress directions the dynamic classification of faulting are given below:

Thrust Faulting

In thrust faulting σ_1 is horizontal and dominant, σ_3 is vertical, σ_2 is horizontal and both are orthogonal to σ_1. There will be two sets of planes across which tangential stress is maximum (Fig. 4.7a). Both sets will have their strike parallel to σ_2 and perpendicular to σ_1. The planes dip at an angle 45° in opposite directions. It is, however, seen that the planes of faulting do not exactly follow the directions of maximum stress. If the planes dip at less than 45° it will produce thrust faulting and if at more than 45° it would be reverse faulting (Fig. 4.9a). Symbolic representation of the fault-plane solution is also shown; fault-plane solutions are discussed in Section 4.7.

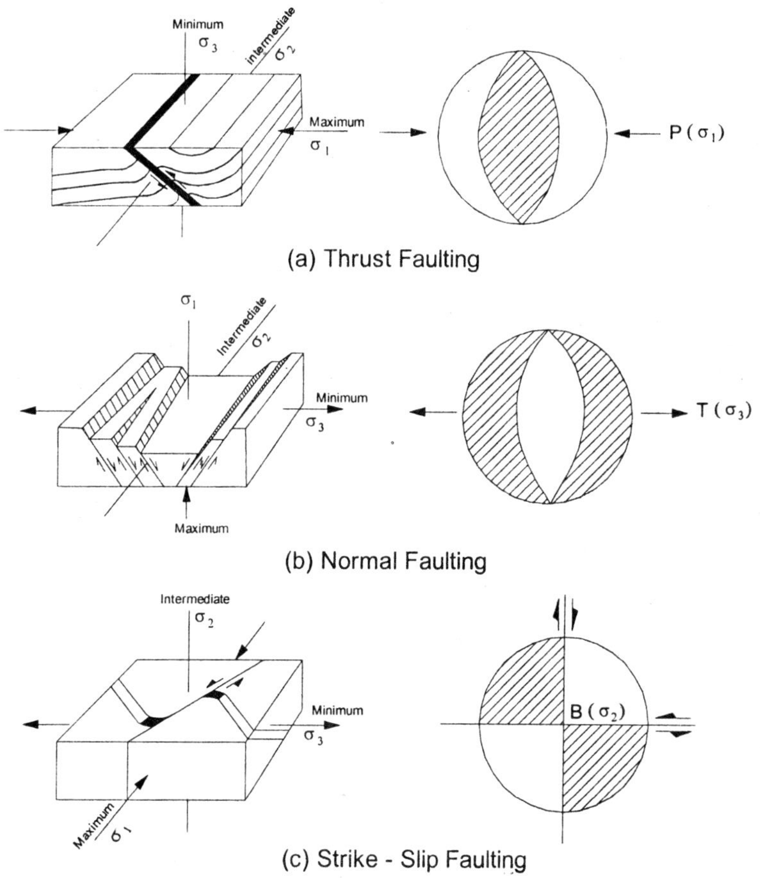

Fig. 4.9 Dynamics of faulting, stress orientations and symbolic representation of fault-plane solutions: (a) thrust faulting, (b) normal faulting, and (c) strike-slip faulting P, T and B represent σ_1, σ_3 and σ_2 respectively.

Normal Faulting

In normal faulting σ_3 is dominant and horizontal, σ_1 is vertical and σ_2 is horizontal and orthogonal to each other. The planes of maximum tangential stress are parallel to σ_2 and perpendicular to σ_3, and they dip in opposite directions at 45° (Fig. 4.7b). The planes of actual faulting will deviate from these positions so as to form smaller angles with σ_1. The result will be normal faulting with the fault planes dipping at angle more than 45°, the expected fault plane solution is shown (Fig. 4.9b).

Strike-slip Faulting

In strike-slip faulting both σ_1 and σ_3 are horizontal and σ_2 is vertical. The planes of maximum tangential stress, in this case, are vertical and at angles 45° with σ_1 (Fig. 4.7c). The planes of actual faulting will, however, deviate from these positions so as to form smaller angles with σ_1. These will be still vertical and motions along these planes will be nearly horizontal. The result will be strike-slip faulting, the fault plane solution is illustrated in Fig. 4.9c.

4.6 ELASTIC REBOUND THEORY

The elastic rebound theory of Reid (1910) is commonly accepted as an explanation of how most earthquakes are generated. Like a watch spring that is wound tighter and tighter, the crustal rocks are elastically strained more and more under a deviatoric stress. The process continues, when the spring breaks and rebounds, and the elastic strain is released suddenly. Similarly, when the accumulated strain in the rock exceeds the strength of the rock, the rock breaks or faulting occurs (Fig. 4.10a). The opposite sides of the fault rebound to a new equilibrium position, and the energy is released partly as heat and partly as elastic waves. These waves are earthquakes. If earthquakes are caused by faulting, it is possible to deduce the nature of faulting from an adequate set of seismograms, and in turn the nature of the tectonic stress that causes the earthquakes.

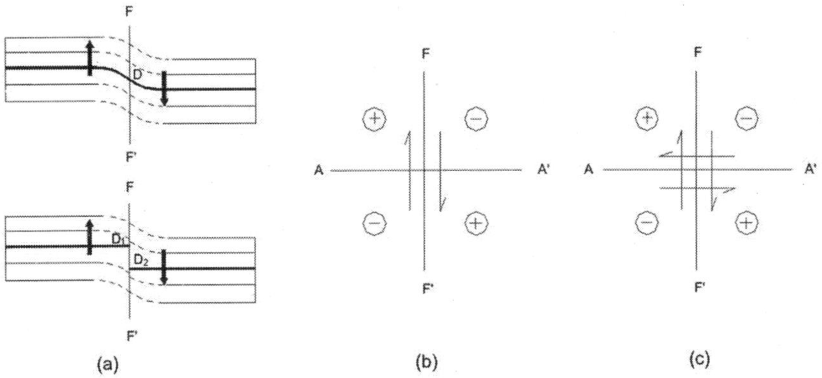

Fig. 4.10 (a) Schematic representation of the elastic rebound theory, (b) single couple hypothesis and (c) double couple hypothesis.

4.6.1 Double-Couple Hypothesis

From 1920s to the 1960s, many authors in USA, Japan and USSR have developed various methods of determining *fault-plane solution*. Readers are referred to the review articles by Honda (1962) and by Stauder (1962) for details.

Let us now try to understand a simple earthquake mechanism as was suggested for the 1906 San Francisco earthquake. Figure 4.10a illustrates the plan view of horizontal motion on a vertical fault F-F′, and the arrows represent the relative motion of the two sides of the fault. Thus, the material behind the arrows is dilated from the source, and the material ahead of the arrow is compressed. Consequently, the area surrounding the earthquake focus can be divided into four quadrants in which the P-wave first-motions are alternatively of compression and dilatation in nature. The quadrants are separated by two orthogonal planes, A-A′ and F-F′. The F-F′ plane is the fault plane here, and the A-A' plane is the auxiliary plane (Fig. 4.10b).

The above example is based on the *single-couple hypothesis* proposed by H. Nakano (as cited by Honda, 1962), i.e. two forces oppositely directed along a fault-plane send alternate compressions and dilatations into quadrants separated by the fault-plane and an orthogonal plane (Fig. 4.10b). A *double couple hypothesis* was then proposed by Honda (1962), which explains the above compressions and dilatations into quadrants separated by two orthogonal planes but with two sets of forces (Fig. 4.10c). With one set of force, F-F′ is fault-plane and A-A' is auxiliary plane, and with the other set of force, A-A′ is fault-plane and F-F′ is auxiliary plane.

There was much debate about the suitability of the single couple versus the double couple model for faulting. Although single-couple model makes less sense physically, the main reason for not ruling it out was that the P-wave radiation from both models is indistinguishable. However, elastodynamic solutions for actual stress and displacement discontinuities in the medium confirm the equivalence of double couple body forces and shear dislocations (Lay and Wallace, 1995). The double-couple force system has no net moment; the strength of the two couples can be represented by the *seismic moment* M_0, which is shown to equal μAD, where μ is the rigidity, A is the fault area, and D is the average displacement on the fault.

4.6.2 Radiation Pattern

Figure 4.11a illustrates the P-wave radiation pattern. The length of a vector from the origin to some point on the radiation pattern represents the relative amplitude of the P-wave at that orientation with respect to the fault plane. The radiation pattern shown in the upper part of the figure is actually just a two dimensional slice through a three dimensional pattern. It may be noticed that the amplitude of the pulses decreases towards the nodal planes, and is shown by decreased size of the polarity symbols in the lower part of Fig. 4.11a. This directly corresponds to the reduced vector length near the nodal planes. Thus decreased amplitude of the P-wave may be noted as its *nodal character*, which can be a guiding factor in drawing nodal planes.

S-waves are radiated with a different pattern, as shown in Fig. 4.11b. For S waves, the maximum amplitude occurs on the fault plane and perpendicular to the fault plane. This is often observed that for a local earthquake reduced amplitude P-wave is followed by larger amplitude S-wave.

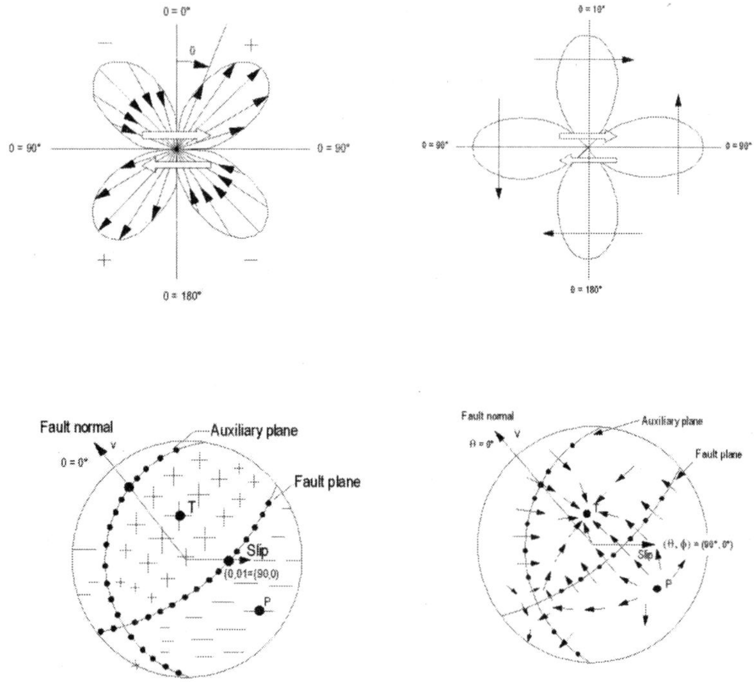

(a) (b)

Fig. 4.11 (a) Radiation pattern of P-wave, and (b) radiation pattern of S-wave.

4.7 EQUAL AREA PROJECTION AND FAULT PLANE SOLUTION

On the basis of the above theory and hypothesis, it is possible to infer orientation of the fault and direction of motion if the directions of P-wave first-motions around the source are known. Thus, it is necessary to trace the observed P-wave first-motions back to a hypothetical focal sphere, i.e. a small sphere enclosing the earthquake source, and to develop techniques to find two orthogonal *nodal-planes* which separate quadrants of compressions and dilatations on the focal sphere. The projection of the P-wave first motions can be made either by the equal-area projection or by stereographic projection. The equal-area projection is very similar to *stereographic projection.* The equal-area projection is preferred for fault-plane solution because the area on focal sphere is preserved in the equal-area Schmidt net compared to the stereographic Wulf net (Fig. 4.12).

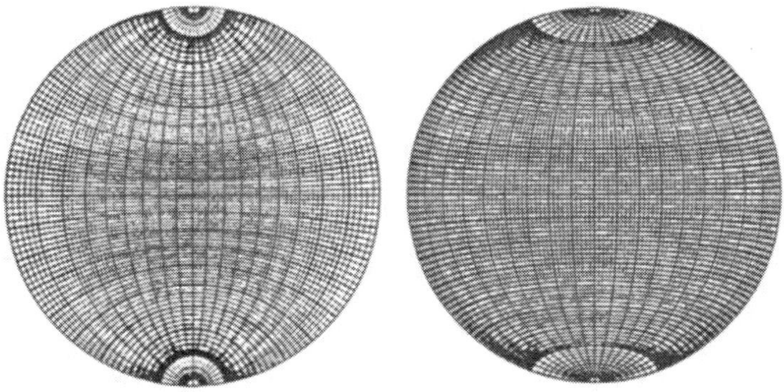

(a) (b)

Fig. 4.12 Comparison of (a) stereographic (Wulff) net and (b) equal-area (Schmidt) net.

Two elementary problems on equal area projection (lower focal hemisphere) are illustrated below. Readers are referred to some elementary texts for further details (e.g. Billings, 1954; Ragan, 1973).

Problem 1

Representation of a fault plane striking N30°E and dipping 40°SE on an equal area projection is given below.

Solution

I. On a piece of tracing paper let us draw a circle making the outer circumference of the equal-area net and mark N, S, E and W.
II. Put a drawing pin through the centre of the equal-area net. Mark the strike counting 30° clockwise from the North, and rotate the overlay tracing-paper 30° counter clockwise to coincide the North of the net with the mark of strike. Now, count 40° along the east-west axis from the circumference for the dip, and trace the great circle-arc from the net on the tracing paper as shown in Fig. 4.13. This represents the required fault plane.
III. If we count 90° from the great circle along the east-west axis, we find a point P which represents the *pole* or *normal axis* to the fault plane.
IV. By reversing the process we can get the strike and dip of the fault plane.

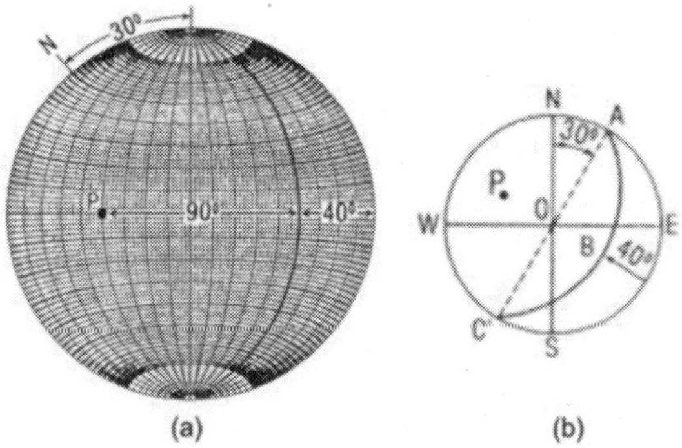

Fig. 4.13 (a) Equal-area plot of a fault-plane strikes N30°E and dips 40°SE. P is the pole of the fault plane, (b) as it appears after the plot (Ragan, 1973).

Problem 2

Representation of a point/axis with 138° azimuth and 40° plunge is illustrated below.

Solution

I. Tracing of the circumference of the equal area net may be done as above.
II. Mark the azimuth on the overlay tracing paper counting 138° clockwise from the North, or 42° anti clockwise from the South. Now, rotate the tracing paper to bring the azimuth-mark on the east-west axis of the net, and count 40° along the east-west axis from the circumference for the plunge. This point represents the required point/axis on the equal area-projection (Fig. 4.14).
III. By reversing the procedure we can get the azimuth and plunge of the point/axis.

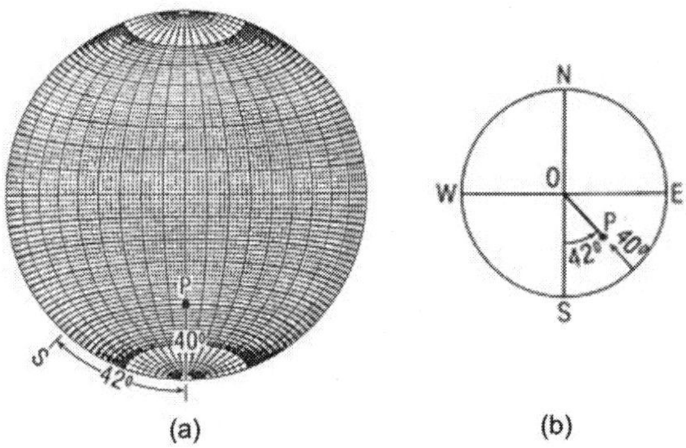

Fig. 4.14 (a) Equal-area plot of a point P, (b) as it appears after the plot (Ragan, 1973).

4.7.1 Plotting of P-wave First-motion Data and Fault-plane Solution

For most of the microearthquake networks, focal mechanisms or fault-plane solutions are based on the P-wave first motions as the first motions of the later phases are difficult to identify. Because most of the seismic rays in a microearthquake array are up going rays, it is convenient to use equal-area projection on the upper hemisphere. The procedure for plotting the data and the theoretical consideration are given below:

1. First arrivals of the P-waves and their corresponding directions of motion for an earthquake are read from the vertical-component seismograms. Normally, the symbols **U** or **C,** or + for up motions, and **D** or – for down motions are used.
2. An illustration for tracing the P-wave first-motion back to the focal sphere (upper hemisphere) is given in Fig. 4.15. The position of the seismic station on the surface of a focal sphere is determined by two angles, ø and θ, where ø is the azimuthal angle (measured clockwise from the North) formed by the line joining the earthquake epicentre to the given station, and θ is the take-off angle with respect to upward vertical for the seismic ray from the earthquake hypocentre to the given station. The former is computed from the coordinates of the hypocentre and the given station, and the latter is determined while computing travel-time derivatives; both are estimated by the Hypo-71 or SEISAN program. It may be mentioned that the program determines the take-off angle or angle of incidence (AIN) with respect to downward vertical. In this case, we must take care of the angle θ for up going rays i.e. when AIN > 90° by substituting θ = 180°–AIN. For shallow (depth <50 km) local earthquakes, seismic rays are mostly recorded as up going rays by a microearthquake network. The down going rays (AIN < 90°), however, may be plotted as refracted rays, i.e. plotting P-wave first-motion data on the opposite quadrant substituting θ = AIN.

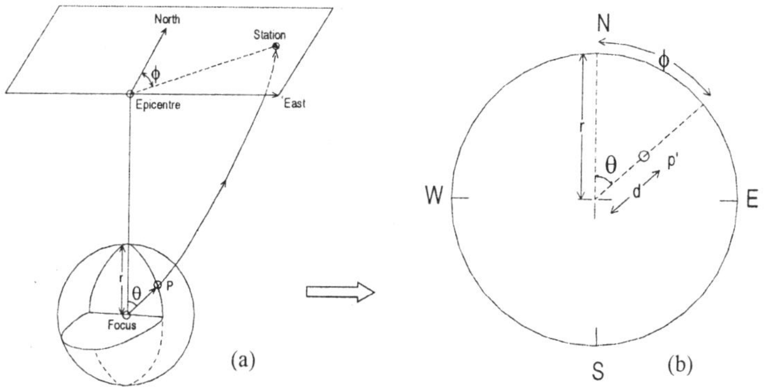

Fig. 4.15 (a) Tracing the P-ray on the upper focal hemisphere, (b) equal-area projection.

A point on the focal sphere may be specified by R, ϕ and θ, where R is the radius, ϕ is the azimuthal angle and θ is the take-off angle with upward vertical. In equal-area projection these parameters (R, ϕ, θ) are transformed to plane polar coordinates (r, α) by the following formulas (Mailing, 1973).

$$r = 2R \sin(\theta/2), \qquad \alpha = \theta$$

Since the radius of the focal sphere R is immaterial and maximum value of r is more conveniently taken as unity, the above eq. is modified to read

$$r = \sqrt{2}\sin\left(\frac{\theta}{2}\right), \qquad \alpha = \theta$$

This type of equal-area projection was introduced by Honda and Emura (1958) in fault-plane solution.

3. Figure 4.16a illustrates a P-wave first-motion plot (upper hemisphere) for a microearthquake recorded by a 20-station network in New Zealand (Kayal, 1984). Fifteen reliable first-motion data are used for the solution. Solid circles are used for compression and open circles for dilatation. The P-wave first-motions on the focal sphere are separated so that the adjacent quadrants have opposite polarities. These quadrants are delineated by two orthogonal great circles, the *nodal planes*. To do this, rotate the plot such that a great-circle arc on the net separates the compressions from dilatations as much as possible. In this example (Fig. 4.16a), the arc MBN represents the projection of a nodal plane, which strikes 38° and dips 58° NW. It may be noted that the concave side of the plane points the dip direction in the upper hemisphere plot. So, it is important to mention about the use of the hemisphere. The pole or normal axis to this plane is the point C_1, which is 90° from the great circle arc MBN. Since the second nodal plane

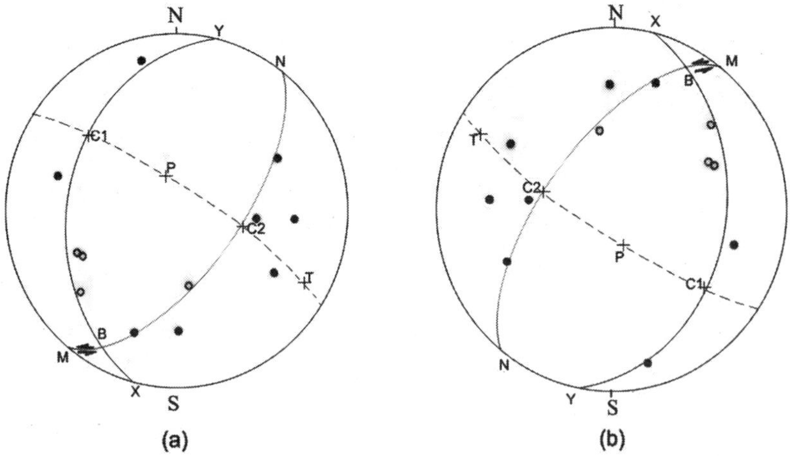

Fig. 4.16 (a) Fault-plane solution on upper hemisphere of an earthquake recorded by a temporary microearthquake network in New Zealand (Kayal, 1984). (b) The solution is illustrated in lower hemisphere.

is orthogonal to the first, its great circle-arc must pass through the point C_1. To find the second plane, rotate the plot so that another great circle-arc passes through the point C_1 and also separates the compressions from dilatations. This arc XBY strikes 14° and dips 32° ESE as measured from the equal-area net. The pole or normal axis of the second plane is the point C_2, and it lies on the first nodal plane.

For a comparison and for a clear understanding, the same data are shown on a lower hemisphere plot (Fig. 4.16b). Note that the compression and dilatation data are just plotted on the opposite quadrants respectively. Substituting $\alpha = 180^0 + \text{AIN}$ when AIN > 90^0. In this plot, the convex sides of the planes point the dip direction. The solution, however, remains the same. Lower hemisphere plot is most common for teleseismic events.

4. The intersection of the two nodal planes is represented by the point B, which is the position of the intermediate stress (σ_2) axis or *null axis*. The plane normal to the null axis is represented by the great circle-arc C_1PC_2T which is called equatorial plane; it contains two important axes: the *P-axis* (σ_1) and the *T-axis* (σ_3). The P axis is 45° from C_1 and C_2, and in the dilatational quadrant, and the T axis is 90° from the P axis and lies in the compressional quadrant. The P-axis represents the direction of *compressional stress*, and the T-axis the direction of *tensional stress*. In fault-plane solutions the stress axes are more popularly expressed as the P, T and B, rather than σ_1, σ_3, and σ_2 respectively. Now, if we select the first nodal plane MBN as the fault plane then the point C_2 represents the *slip vector*. It may be mentioned that we cannot distinguish which nodal plane is the fault plane from the P-wave first-motion plot alone. The geological information of existing faults, the P-wave radiation pattern i.e. nodal character, distribution of the epicentres and the seismic cross sections or depth-sections of the earthquakes are used to infer the fault plane. This is discussed in the following chapters with practical examples.
5. We can summarize our results from the above P-wave first-motion plot as follows:

   (a) Nodal Plane 1 (preferred Fault Plane):
   Strike: 218°, dip: 58° NW, slip angle or rake: 95°
   (b) Nodal Plane 2 (Auxiliary Plane):
   Strike: 14°, dip: 32° ESE, slip angle or rake: 32°
   (c) P-axis: strike: 344°, plunge: 74°
   (d) T-axis: strike: 118°, plunge: 12°
   (e) B-axis: strike: 211°, plunge: 11°

6. The fault-plane solution (Fig. 4.16) represents almost a pure normal faulting with a small left-lateral strike-slip motion as shown by arrows.

4.7.2 Composite Fault-plane Solution

The composite solution is generally used for a small microearthquake network. A temporary microearthquake network of five to six stations may be sufficient for locating an earthquake, that occurred within the network, with a fair degree of precision, but the first-motion data would be too meagre to study its focal mechanism. In such cases, a cluster of earthquakes at a same depth range may be selected for a composite fault-plane solution. The basic assumption in such an attempt is that the events are generated by the same mechanism. The solution may be improved by relocating these selected events by the *Homogeneous Station Method* (Kayal, 1984). The reliable first-motion data of all these events are then plotted on the equal-area net for fault-plane solution. A composite solution of a group of microearthquakes is illustrated in Fig. 4.17. The solution is obtained with a good number of first-motion data, and the nodal planes are well constrained. There are a few inconsistencies; some dilatational data are in the compressional quadrant and vice versa. The overall solution is a pure normal faulting ($\lambda = -90°$).

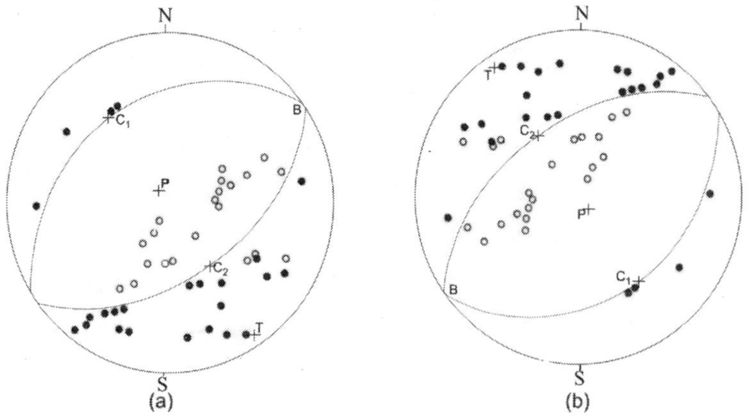

Fig. 4.17 (a) Composite fault-plane solution (upper hemisphere) of a group of microearthquakes recorded in New Zealand (Kayal, 1984). (b) The solution is illustrated in lower hemisphere.

Estimation of Rake in Oblique Slip

The Fig. 4.18 further demonstrates estimation of strike, dip and slip angle or rake in case of oblique slip. In this figure, F1 is the fault plane, F2 auxiliary plane and EP is the equatorial plane. P1, P2 and P3 are poles of F1, F2 and EP respectively. Plunge of the P and T are shown.

Beach-ball presentation

Figure 4.19 illustrates the beach-ball presentations of different fault-plane solutions. The shaded area indicates zone of compression and the open area zone of dilatation. The solid circle represents the P-axis and the open circle the T-axis.

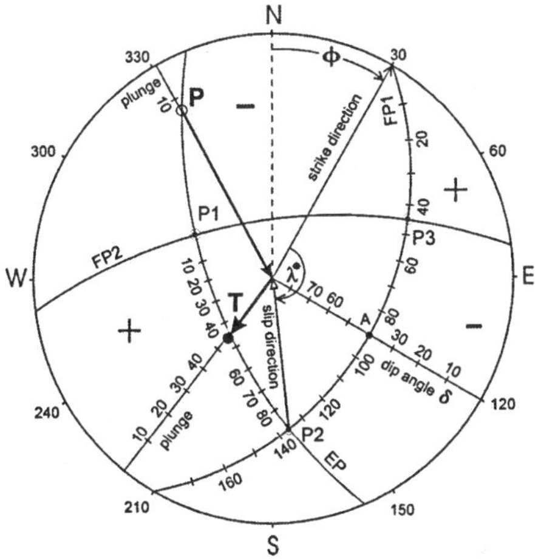

Fig. 4.18 The plot illustrates a solution where rake or $\lambda^* = 180°–\lambda$ in case of thrust faulting component, and $\lambda^* = –\lambda$ in case of normal faulting component (IASPEI, 2002).

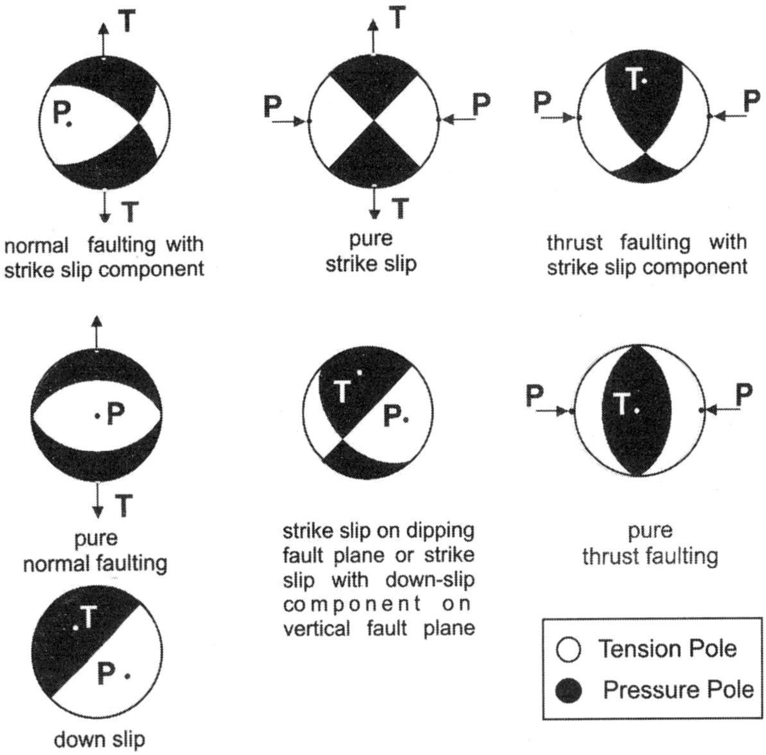

Fig. 4.19 Beach ball presentation of fault plane solutions.

4.7.3 Computer-assisted Fault-plane Solution

There exists a number of softwares for determination of both single and composite fault-plane solutions from first-motion data. For local earthquakes flat-layered velocity models are acceptable. The computer programs (Hypo-71, Hypoellipse, Hypoinverse or Seisan) compute the azimuth AZM and take off angle AIN at which the P-wave leaves the focus and list them in the output files.

The program FPFIT (Reasenberger and Oppenheimer, 1985) determines double-couple fault-plane solution based on P-wave polarity readings. The program accepts the output files of the earthquake location programs Hypo-71, Hypoinverse or Hypoellipse as its input files. The inversion is achieved through a grid-search procedure. It finds the source model by minimizing a normalized weighted sum of first-motion polarity discrepancies. In addition to minimum-misfit solution, FPFIT finds alternative solutions corresponding to significant relative misfit minima. The existence of several minima may be due to insufficient first-motion data, location errors, polarity misreadings or velocity-model errors. The display program FPPLOT shows the final fault-plane solution and the estimated uncertainty.

The FOCMEC program accepts the Seisan output files, and determines best fitting double couple fault-plane solution (Havaskov and Ottemoller, 2000). It is more versatile program that uses P, S_H, S_V polarities and/or S_V/P, S_H/P or S_V/S_H amplitude ratios. The amplitude ratios, however, need to be corrected to the focal sphere taking into account geometrical spreading, attenuation and free surface effects. The FOCMEC solution is obtained by grid search over strike, dip and slip of the double couple source. The program FOCPLOT, provided with the FOCMEC in the Seisan, allows us to plot upper or lower hemisphere projections of the focal sphere. All the required parameters including strike and dip of the nodal planes, P, T, B axes and rakes are estimated. This is a very useful program for fault-plane solution of single or selected cluster of events. This program is extensively used for the aftershock studies in the recent years, and discussed in the following chapters.

4.7.4 Pitfalls in Fault-plane Solution

First-motion Data

Unlike arrival times, which are read with error limits, a P-wave first-motion reading is either correct or wrong. Strictly, the well read first-motion data should be used for fault-plane solution. For example, Lee et al. (1979) have shown how drastically different fault-plane solutions can be obtained if a key station (PYR) is ignored for less reliable data; this is illustrated in Fig. 4.20. So, reliability of data is very important.

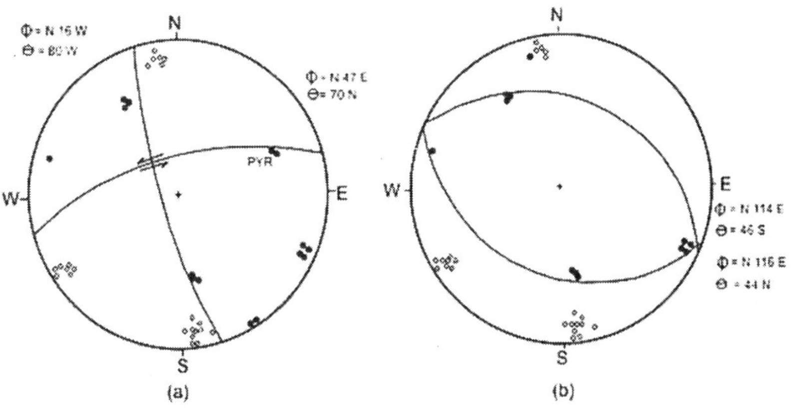

Fig. 4.20 Fault-plane solutions (lower hemisphere plot) are different if two first-motion data of a key station PYR are ignored (Lee et al., 1979).

Station Polarities

Care must be taken to check the instrument/station polarities. Instrument polarity can be checked at the laboratory. One of the most important tasks in operating a microearthquake network is to ensure that the direction of motion recorded by a seismogram is the true direction of ground motion. Large nuclear explosions or teleseismic events may be used to ensure the accuracy of station polarities. Readers are referred to the work of Houck et al. (1976) for details.

Focal Depth and Velocity Model

P-wave first-motion plot depends on the earthquake location and take-off angles; thus a precise location of the focus and a reasonably realistic velocity model of the study region are needed. Engdahl and Lee (1976) showed that proper ray-tracing is required for fault-plane solution. Care must be taken to plot the refracted P-arrivals. They have further emphasized that velocity model has a vital role as it determines the take-off angle, ray path etc.

Station Distribution

Fault-plane solutions may be ambiguous or may not be possible if station distribution is poor. At least 15-20 well distributed and reliable first-motion data should be used for a fault-plane solution.

An example of ambiguous focal mechanism or non-unique solution is illustrated in Fig. 4.21. The figure shows that the first-motion data fit with three totally different solutions. It is warned here that experience is very much needed in judging the nodal planes and data reliability. Knowledge of regional geology and tectonics is needed to judge the fault plane. The beginners should take the help of an experienced seismologist before drawing any conclusion on focal-mechanism solutions.

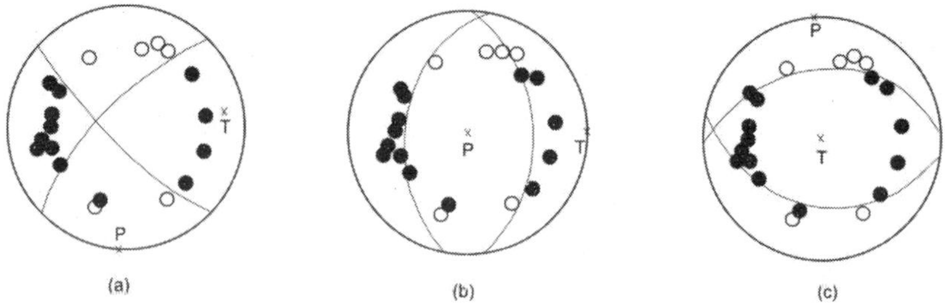

Fig. 4.21 An example of non-unique fault-plane solutions with the same data set (Lahr, 1992).

4.8 MOMENT TENSOR SOLUTION

In recent years, the conventional fault-plane solutions, particularly for medium-size or large earthquakes, are being replaced by *the moment tensor solutions.* This has been possible by the advent of high precision broadband digital data, in conjunction with the high speed computers. It is now a practice to invert the waveforms recorded at seismic stations in various azimuths around the globe to obtain the moment tensor elements and subsequently derive the best double couple solution and the P, B and T axes orientations (see Section 3.14). A moment tensor is an alternative or state-of-the-art description of a seismic source. The displacement across a fault is described in terms of equivalent body forces, which produce the same radiation as the fault slip. A moment tensor is defined as:

$$M = \begin{bmatrix} M_{xx} & M_{xy} & M_{xz} \\ M_{yx} & M_{yy} & M_{yz} \\ M_{zx} & M_{zy} & M_{zz} \end{bmatrix}$$

where M_{xy}, for example, represents a couple whose force is in the x direction and arm in the y direction defined in an appropriate coordinate system with respect to the seismic source. Inversion may be done in time domain or frequency domain. Care must be taken to match the synthetic and observed seismograms. In most inversion schemes, focal depth is assumed to be constant. For details, readers are referred to the book on global seismology by Lay and Wallace (1995).

There are many variations of the algorithm used to obtain a moment tensor solution from the waveforms of body waves or surface waves (e.g. Stump and Johnson, 1977; Patton, 1980; Kanamori, 1980; Ward, 1980). All these methods have a priori assumption of the parameters defining the Earth's structure, the event location and the origin time. An approach, that allows an improvement in the location parameters and simultaneously solves for the moment tensor using the body wave portions of the seismograms, has been developed by Dziewonski and Anderson (1981).

The moment tensor thus obtained is decomposed in terms of its principal axes; the eigen values as well as the directions of the eigen vectors referred to the lower hemisphere of the focal sphere, as is customary in conventional focal mechanism studies of global earthquakes. The first step in decomposition is the calculation of eigen values and eigen vectors of the seismic moment tensor. For this the program MTEIG can be used. It performs rotation of the moment tensor M into principal axis system. The axis associated with the largest positive eigen value is the *tension* or *T-axis*, the largest negative value is the *compressional* or *P-axis*, and the intermediate the *null* or *B-axis*. An illustration of focal mechanism solutions for a pure strike-slip, thrust and normal faulting and the corresponding moment tensor representation is shown in Fig. 4.22.

$$M_1: \begin{bmatrix} 0 & 1 & 0 \\ 1 & 0 & 0 \\ 0 & 0 & 0 \end{bmatrix}; \quad M_2: \begin{bmatrix} 1 & 0 & 0 \\ 0 & -1 & 0 \\ 0 & 0 & 0 \end{bmatrix}; \quad M_3: \begin{bmatrix} 0 & 0 & 0 \\ 0 & 0 & 1 \\ 0 & 1 & 0 \end{bmatrix}$$

$$M_4: \begin{bmatrix} 0 & 0 & 1 \\ 0 & 0 & 0 \\ 1 & 0 & 0 \end{bmatrix}; \quad M_5: \begin{bmatrix} -1 & 0 & 0 \\ 1 & 0 & 0 \\ 0 & 0 & 1 \end{bmatrix}; \quad M_6: \begin{bmatrix} 1 & 0 & 0 \\ 0 & 1 & 0 \\ 0 & 0 & 1 \end{bmatrix}$$

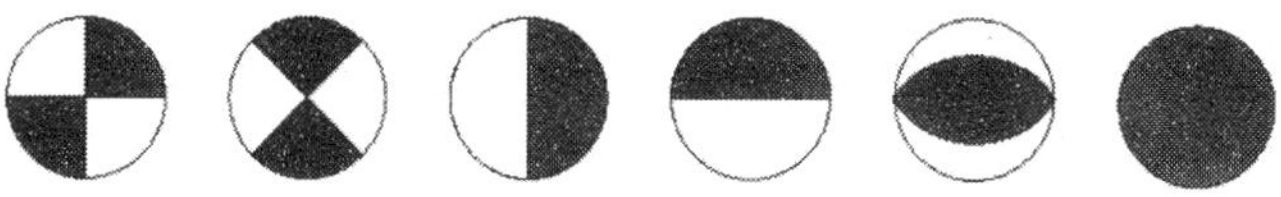

M1 M2 M3 M4 M5 M6

Fig. 4.22 Representation of moment tensor solutions for different faults.

Generally the quality of moment tensor inversion depends, to a large extent, on the number of data available and the azimuthal distribution of stations about the source. In inversion with body waves, Dufumier (1996) gives a systematic overview of the effects caused by differences in azimuthal coverage and effects caused by the use of only P waves, P and S_H waves, or P, S_H and S_V waves. An overview with respect to the effects caused by an erroneous velocity model for the Green function calculation and effects due to erroneous hypocentre coordinates is made by Kravanja et al. (1999).

Moment Tensor Solutions by Global Agencies

NEIC Fast Moment Tensor Solution

The U.S. National Earthquake Information Centre (NEIC, USGS) in co-operation with the IRIS data management centre makes an effort to determine seismic moment tensor for earthquakes with body wave magnitudes $m_b > 5.0$. Teleseismic P waveforms are used to compute seismic moment tensor using a technique based on optimal filter design (Sipkin, 1982). More information is available under http://gldss7.cr.usgs.gov/neis/FM/fast moment.html.

Moment tensor solutions derived from low frequency data reflect the properties of rupture process, and those derived from high frequency data are sensitive to the dynamic part of rupture process during which most of the seismic energy is radiated. Initial constraints on focal mechanism are provided by the polarities from P, pP and PKP and from transversely polarized S waves. The fault-plane solution and depth are then refined by least square fitting of synthetic seismograms.

Harvard CMT Solution

The Harvard University (USA) group routinely estimates the Centroid Moment Tensor (CMT) solutions for the medium to large magnitude ($m_b \geq 5.0$) earthquakes by inverting the data from various digital seismograph stations around the world. These data are regularly published in the *Physics of the Earth and Planetary Interiors* (Dziewonski et al., 1989, for example). In addition, the Harvard CMT catalogs can also be accessed on line through Internet by accessing the home page of the Harvard group (http://www.seismology.harvard.edu/projects/CMT). The Harvard CMT method makes use of both very long period (T > 40 s) body waves and surface wave at T > 135 s.

The Harvard CMT catalogs consist of centroid moment tensor solutions determined over the period from 1976 till present. The CMT inversion seeks a solution for the best point source location of the earthquake. It represents the integral of the moment density over the extended rupture area. This centroid location may, for large earthquakes, significantly differ from the hypocentre location based on P-wave and S-wave first arrivals. The hypocentre location corresponds to the rupture initiation point, whereas centroid location corresponds to the average location of the area of maximum release of seismic moment, fault extent and rupture directivity. These are discussed with practical examples in the subsequent chapters.

4.9 EARTHQUAKE MECHANISMS AND PLATE TECTONICS

Earthquake mechanisms play a major role in development of our understanding of global plate tectonics; their distribution is used to map plate boundaries. The *slip vectors* of the earthquakes provide information about the direction of plate motion at individual boundary.

4.9.1 Oceanic Spreading Centre Mechanisms

Earthquake mechanisms from spreading centre are illustrated in Fig. 4.23. The spreading centre shows a portion of spreading ridge is offset by transform faults. New lithosphere forms at the ridges, then moves away. The relative motion of lithosphere on either side of a transform is in opposing directions. The direction of transform offset determines whether there is right-lateral or left-lateral motion.

Figure 4.23 shows that the spreading centre is composed of north-south trending ridge segments, offset by transform faults, which trend approximately east-west. Both the ridge crest and transform faults are seismically active. The mechanisms show that the relative motion along the transform is right lateral or left lateral. The earthquakes occur exclusively on the active transform fault segments rather than on the inactive extensions, known as *fracture zones*. No relative plate motion occurs on the fracture zones; these are often marked by distinct topographic features due to contrast in lithospheric ages across it. The seismicity is different on the spreading ridges; normal faulting earthquakes are observed on the ridge crest; the nodal planes trend along the ridge axis (Fig. 4.23).

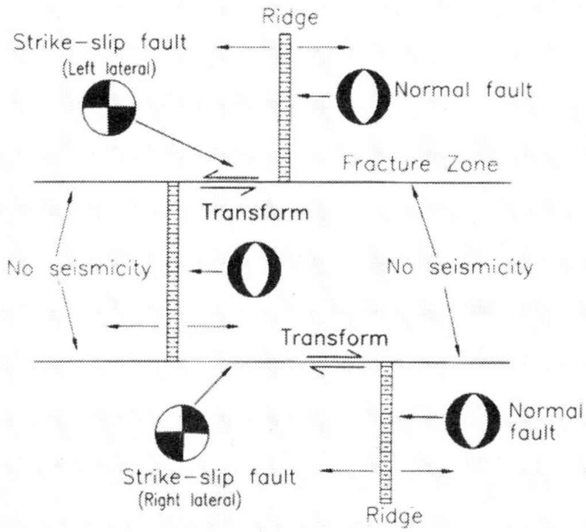

Fig. 4.23 Tectonic setting and earthquake mechanisms at an oceanic spreading centre.

4.9.2 Subduction Zone Focal Mechanisms

Majority of large earthquakes occur in subduction zones; their focal mechanisms reflect various aspect of subduction tectonics. Figure 4.24 illustrates some of the features observed in subduction zones.

Thrust faulting is observed for the shallow earthquakes at the interface between overriding plate and subducting plate. Slip vectors of the focal mechanisms may give the direction of plate motions. The flexural bending of the subducting plate, on the other hand, produces normal faulting earthquakes at shallower depth of ~20-25 km, and thrust faulting earthquakes in the lower part 40-50 km. The observations constrain the position of neutral surface separating the upper extensional zone from the lower flexural zone, thus providing information on the mechanical state of the lithosphere. There has been some controversy; whether the normal faulting earthquakes as "bending" events or "slab pull" events. The deeper earthquakes in the Wadati-Benioff zone go down to 700 km and their mechanism provide information about physics of the subduction process. The essence of the process is the penetration and slow heating of a cold lithosphere in the warmer mantle. The thermal structure, seismic velocity, attenuation characters are diagnostic for the subducting plate.

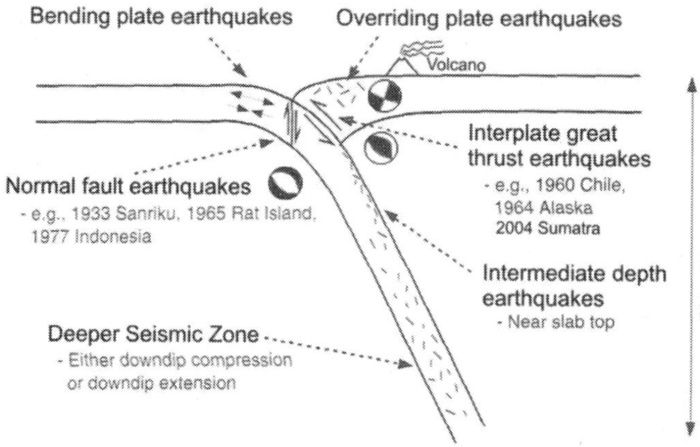

Fig. 4.24 Possible earthquake mechanisms in a subduction zone.

4.9.3 Diffuse Plate-Boundary Focal Mechanisms

The continental crust is much thicker, less dense and has different mechanical properties from oceanic crust. The plate boundaries in continents are often diffuse, and a broad zone of deformation is evident, such as India-Eurasian collision zone in the Himalaya or the Pacific-North America boundary zone in the western United States.

In the Himalayan collision zone the earthquakes are characterized by thrust mechanism and the plate boundary zone is wide (see Chapter 5). The Pacific-North American plate boundary, on the other hand, is extensional, essentially transform along the San Andreas fault system, and convergent in the eastern Aleutians. These changes are well reflected in focal mechanisms (Fig. 4.25).

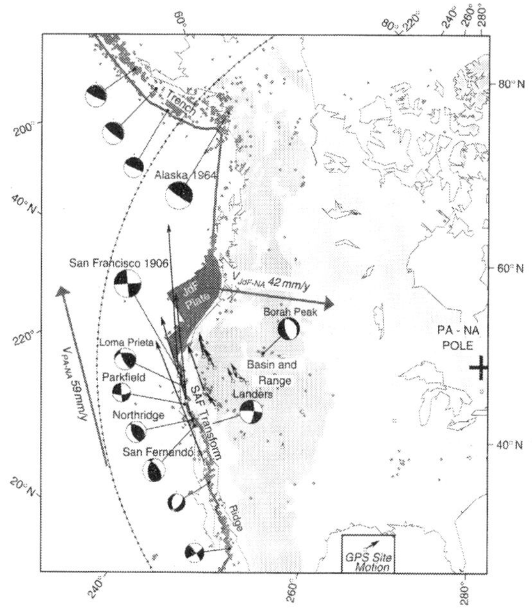

Fig. 4.25 Focal mechanisms of earthquakes in the North America-Pacific plate boundary zone, transform along the San Andreas fault (SAF) and convergence along Aleutian trench (Lee et al., 2002).

4.9.4 Intraplate Focal Mechanisms

The intraplate earthquake mechanism have important use to learn poorly understood tectonic processes of internal deformation of major plates. The seismicity of such regions is generally thought to be due to reactivation of pre-existing faults or weak zones in response to intraplate stresses. Intracontinental earthquakes occur less frequently than plate boundary events, recurrence estimates average 500-1000 yrs. As a result, understanding how these intraplate seismic zones operate is a major challenge. Three such intraplate strong earthquakes occurred during the last decade in peninsular India. A detailed discussion on mechanisms of these events are given in Chapter 7.

5

Himalayas, Pamir-Hindu Kush and Foredeep Region

5.1 INTRODUCTION

The geophysical data in the Indian ocean floor suggest that India got separated from the Antarctica and Australia during Cretaceous (e.g. Le Pichon, 1968; McKenzie and Sclater, 1971; Bowin, 1973; Sclater and Fisher, 1974; Johnson et al., 1976; Curray et al., 1982; Powell et al., 1980). The Indian plate travelled north and north-eastward in between the Ninety East Ridge and Chagos-Laccedive transform fault during Cenozoic (Gansser, 1966), (Fig. 5.1). The Indian continent had to travel 1000 km before achieving its initial contact with the Eurasian plate in late Cretaceous to early Eocene (McKenzie and Sclater, 1971). This contact in the north is demarcated as the *Himalayan arc* and in the east as the *Burmese arc*. Many authors (e.g. Dewey and Bird, 1970; Verma et al., 1977; Seeber and Armbruster, 1981) have suggested that the Himalayan arc is now involved in a continent-continent collision, whereas the Burmese arc is involved in a subduction process (e.g. Fitch, 1970; Verma et al. 1976a; Nandy, 1976; Mukhopadhyay and Dasgupta, 1988; Kayal, 1989 and 1996a; Satyabala, 2003).

The continental collision at the Himalayan arc, between India and Asia, is an example of crustal thickening and mountain building on large scale. The Himalaya is the greatest mountain chain in the world, about 2500 km long with loftiest peaks from west to east, viz, Naga Parbat (8,125 m), Everest (8,848 m) and Namcha Barwa (7,755 m). The width of the belt varies from 250 km to 350 km. The Himalayan arc, convex southward, is possibly the most profound tectonic belt that has occurred in the geological past. It is responsible for the uplift of the Himalaya, Tibetan Plateau and rejuvenating the tectonic architecture of the Karakuram of Kunlun (Valdiya, 1988). The collision caused changes in the Earth's orography and consequent climate changes (Rowley, 1996). The uplifting process is still going on at about one centimetre per year along with continued erosion and denudation, with different rates from time to time and place to place. The eroded materials shed into different depositional settings within the Himalaya to Bay of Bengal and Arabean sea by youthful rivers and drainage network (Sinha and Upadhyay, 1995; Valdiya, 1988).

Early ideas about the architecture of the Himalaya assumed that the Himalayan mountains rest on a rigid substratum. The geodetic measurement showed that observed deflection was much smaller than expected for a rigid model (Prat, 1855). Then the concept of isostacy, 'crustal root', became the dominant geological thinking. However, the gravity measurements deviated from the local isostatic equilibrium. The gravity observations suggest that the Himalaya is not completely supported by buoyancy force, rather it is held up, at least in part, by an upward force.

The concept of under thrusting of continental lithosphere beneath another was first proposed by Argand (1924). With the advent of global tectonics and increasing amount of data, many authors (e.g. Le Fort, 1975; Molnar et al., 1977; Bird, 1978; Seeber et al., 1981; Chen and Molnar, 1981; Barazangi and Ni, 1982; Molnar and Chen, 1982; Ni and Barazangi, 1984; Molnar, 1984) suggested that the entire 2500 km long Himalayan arc, extending from Kashmir in the northwest to Arunachal Pradesh in the northeast, evolved as a consequence of collision of the Asian and Indian continents some 50 m.y. ago. In front of the northward drifting, Indian continent lay a huge prism of continental margin comprising roughly 1600 to 50 m.y. old sediments resting on a hard basement of even older (> 2000 m.y.) crystalline rocks. As the Indian landmass moved northwards this sedimentary piles with its crystalline foundation complexly folded, faulted and thrusted as a consequence of collision of the Asian and Indian continents. These faults/thrusts, from north to south, are: the *Trans-Himadri Fault* (THF), the *Main Central Thrust* (MCT), the *Main Boundary Thrust* (MBT) and *the Himalayan Frontal Thrust* (HFT) (e.g. Nakata, 1972; Valdiya, 1976; Ni, 1989; Schelling, 1992). The junction of the two colliding continents is known as *Indus Suture Thrust* (IST) (Fig. 5.1), (Crawford, 1974; Thakur, 1981; Valdiya, 1984).

All these thrusts/faults in the Himalayan arc are characterised by topographic breaks which divide the entire Himalaya into four physiographic subdivisions; from south to north these are: sub Himalaya or *Outer Himalaya, Lesser Himalaya, Higher Himalaya* or Great Himalaya and Tibetan or *Tethys Himalaya* (Gansser, 1964; Valdiya, 1976). There is a prominent syntaxial bend at each end of the Himalaya; the *western syntaxis* and the *eastern syntaxis* (Fig. 5.1). Our interest is in the part of the Himalaya between these two syntaxes. Geologically, this entire Himalayan belt can be subdivided into different tectonic domains. The southernmost domain, the *Himalayan foredeep,* also called the *Indo-Gangetic Alluvial Plains* (IGAP), is separated from the Outer (Siwalik) Himalaya by the HFT. The Siwalik domain has an average elevation of about 100 m. It comprises of folded, faulted and tilted

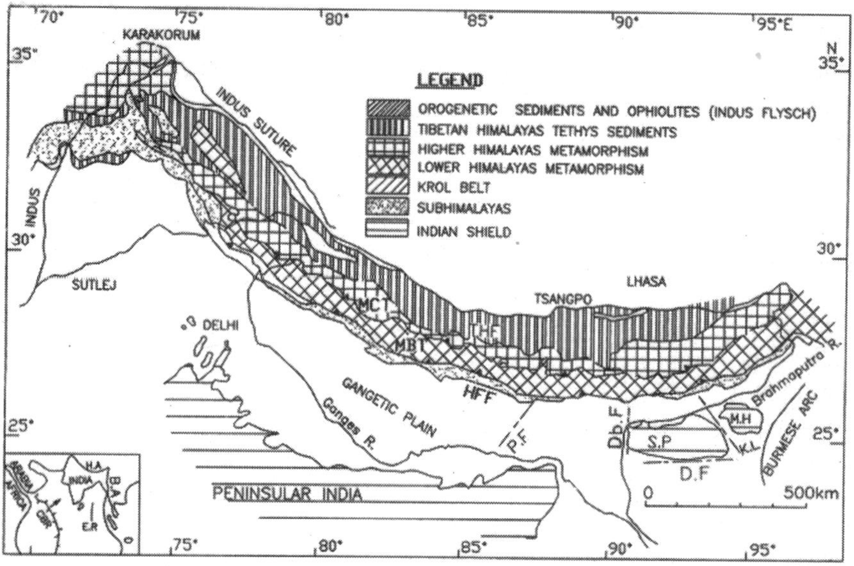

Fig. 5.1 Himalayan arc with the major tectonic features. HFF: Himalayan Frontal Fault, MBT: Main Boundary Thrust, MCT: Main Central Thrust, THF: Trans Himadri Fault, PF: Patna Fault, Db.F: Dhubri Fault, SP: Shillong Plateau, MH: Mikir Hills, KL: Kopili Lineament, DF: Dauki Fault, Inset: sketch map showing north-northeastward motion of the Indian plate, HA: Himalayan Arc, BA: Burmese Arc, CBR: Carlsberg Ridge, ER: East Ridge (Kayal, 2001).

molasses of the Miocene and younger age. The northern boundary of this domain is defined by the MBT that dips northward at 30°-50°. The Lesser Himalaya domain, which follows on the north of MBT, has an average elevation of 2500 m with a slight gradient towards the north. The domain is composed of the metasediments of the Precambrian to Palaeozoic age. The region is characterised by the presence of several nappes of metamorphic rocks brought southwards over long distances by the imbricate thrusting; the rocks are considerably folded and fractured (e.g. Wadia, 1961; Gansser, 1964; Fuchs, 1975; Valdiya, 1976 and 1988). The MCT dips northward at 30°-50°, and separates the Lesser Himalaya domain from the Great Himalaya domain on the north. In the Great Himalaya domain the topography once again rises steeply to about 4500 m beyond which lies the perennial snows. The thrust is characterised by a zone of intense shearing, no master thrust-surface is apparent (Valdiya, 1980; Sinha Roy, 1982). The MCT appears to have developed since mid Tertiary (Valdiya, 1980; Gansser, 1982). Further north, the THF defines the boundary between the Great Himalaya domain and the Tethys Himalaya domain. The Tethys Himalaya domain is composed of the Tethyan sediments of late Precambrian to late Cretaceous. The IST, on the other hand, is represented by intensely deformed and vertically disposed rocks of the ocean-floor and associated sediments that had accumulated in a deep ocean-trench in front of the Asian landmass. Possibly, there was an *island arc* adjacent to this trench which was sandwiched and eventually welded to the colliding continents (e.g. Gansser, 1964; Stocklin, 1980; Allegre et al., 1984).

5.2 GEOLOGICAL PROCESS

Valdiya (1987 and 1988) argued that the Indian continental plate did not slide under the Tibetan plate, instead its leading northern edge encountering considerable resistance, buoyed up giving rise to an elongated up warp all along the front. Its consequences manifest the domal bulges of the basement crystalline rocks, the bending (10°-15°) of the crustal plate to the south of the IST (Lyon-Caen and Molnar, 1983), and splitting up of the synclinorial sedimentary basin into a stack of thrust sheets and fault blocks (Gansser, 1964) (Fig. 5.2). The junction zone, the IST, continues to be a source of high heat flow 125-175 mW/m^3 (Francheteau et al., 1984); this is evident from the belt of numerous hot springs and the occurrence of a large number of plutons of very young granite, 1 m.y. to 15 m.y. in age, which sprang from a shallow source, less than 10 km deep (e.g. Teng and Zhang, 1981; Pham et al., 1986).

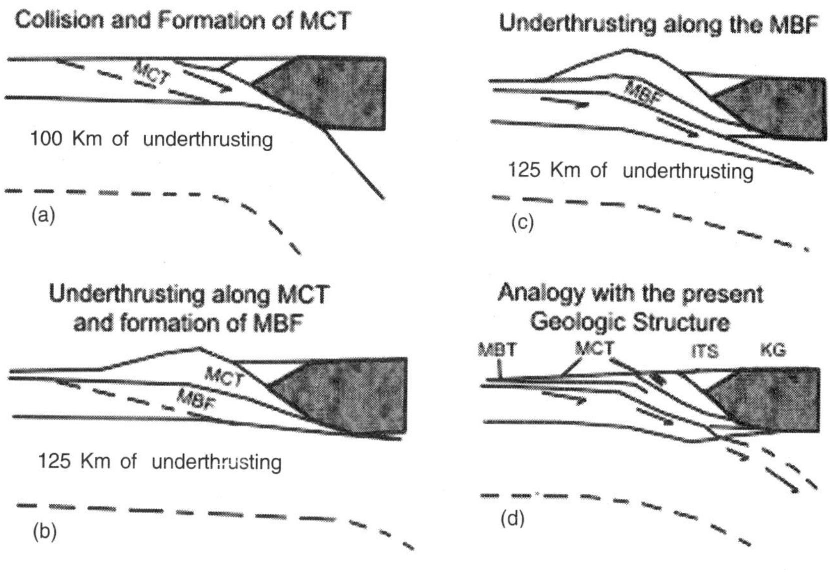

Fig. 5.2 Evolution of the Himalayas : (a) Formation of the MCT (Main Central Thrust), (b) Formation of the MBT (Main Boundary Thrust), (c) Under thrusting along the MBT, (d) Klippe of crystalline rocks transported by the MCT (Lyon-Caen and Molnar, 1983).

The Himalayan convergence speeded up about 20-18 m.y. ago with the breaking up of the leading edge of the Indian crust, giving rise to the MCT and the THF. Continued uplift with southward translation of the basement rocks raised the Great Himalaya, broke up or detached the sedimentary cover from the crystalline basement along the THF, and moved the high-grade metamorphics of the Himadri domain southward. Trampling the Lesser Himalayan sedimentary rocks, the Himadri between the MCT and THF rises like a huge lithotectonic slab in a horst-like manner. This slab is extensively penetrated and injected by 28-20 m.y. old granitic bodies, resulting from differential melting of the high-grade metamorphic rocks of the basement. High frictional heat, generated by strong movements along the MCT, is believed to be responsible for the melting of rocks. (For details see Himalayan Geology volumes published by the Wadia Institute of Himalayan Geology, Dehradun.)

Strong and repeated folding accompanied by deep faulting and uprooting (110-80 km) of rock-masses have brought the basement rocks (2000 m.y.) over the younger (1600-570 m.y.) sedimentary rocks in the Lesser Himalaya. In the Lesser Himalayan terrain there are thus vast and imposing piles of older, highly deformed and metamorphosed rocks resting on comparatively young, less deformed and unmetamorphosed sedimentary rocks. Along the southern flank even the severely folded and repeatedly faulted sedimentary pile is uprooted and thrust southwards over the Siwalik along the MBT.

The highly deformed Siwalik domain is separated from the Indo-Gangetic plains of recent origin by the HFT. Adjacent to the HFT, the floor of the Indo-Gangetic basin has been intermittently subsiding since the very inception, leading to accumulation of enormously thick deposits by the rivers in their flood plains and channels. Between the HFT and the MBT, the Siwalik is not only considerably compressed but also differentially uplifted along a multiplicity of faults, giving rise to small horsts and grabens. Over the past 50 million years, the belt of deformation has successively shifted southward from the IST (50-40 m.y. ago) to MCT in the middle Miocene (20-18 m.y.), to MBT 3.4-1.2 m.y. ago, and to the HFT in the Holocene (1100-4500 yr BP) (Valdiya, 1984). The logical culmination of the continuing northward push of the Indian shield would be corresponding southward advance of the Siwalik over the sediments of the Indo-Gangetic plains, thereby creating a higher degree of tectonic movements in the foothills (Valdiya, 1984).

The processes, which initiated the birth of the Himalaya some 50 m.y. ago, have not ceased but rather continue to influence structural and geomorphic developments of the mountain domain. All the four terrains of the Himalaya experience ceaseless neotectonic movements along the many thrusts and faults that cut the giant body of the Himalaya. Since the total movement has so far been greater and faster along the MCT, the Great Himalaya has attained great heights.

5.3 GEOPHYSICAL DATA IN HIMALAYA

The geophysical data base for the Higher Himalaya is rather poor. The published ground based geophysical data include only a few regional gravity and magnetic traverses in some parts of the Himalaya (Gulatee, 1956; Qureshy, 1969, 1971; Kono, 1974; Marrusi, 1976; Das et al., 1979; Wang et al., 1982). Some of the data were collected during special expeditions like the one led by Norin (1946) and by the Sino-French team in Nepal and Tibetan regions (Allegre et al., 1984).

An *isostatic gravity map* of the Himalaya region is presented by Qureshy and Kumar (1992) based on about 1100 gravity stations data collected by the WIHG (Wadia Institute of Himalayan Geology), ONGC (Oil and Natural Gas Corporation) and SOI (Survey of India). The map shows negative and positive anomalies couple over the foredeep and Himalaya (Fig. 5.3). The Himalayas are associated with a positive isostatic anomaly running between HFT and MCT. Isostatic anomalies of the order of –120 m Gal over the foredeep and +100 m Gal over the Himalaya are observed (Fig. 5.3). The negative anomaly over the foredeep is attributed to the presence of about 5 km thick Quaternary-Tertiary sediments in the Indo Gangetic plains. The positive anomaly +100 m Gal over the Himalaya, on the other hand, is

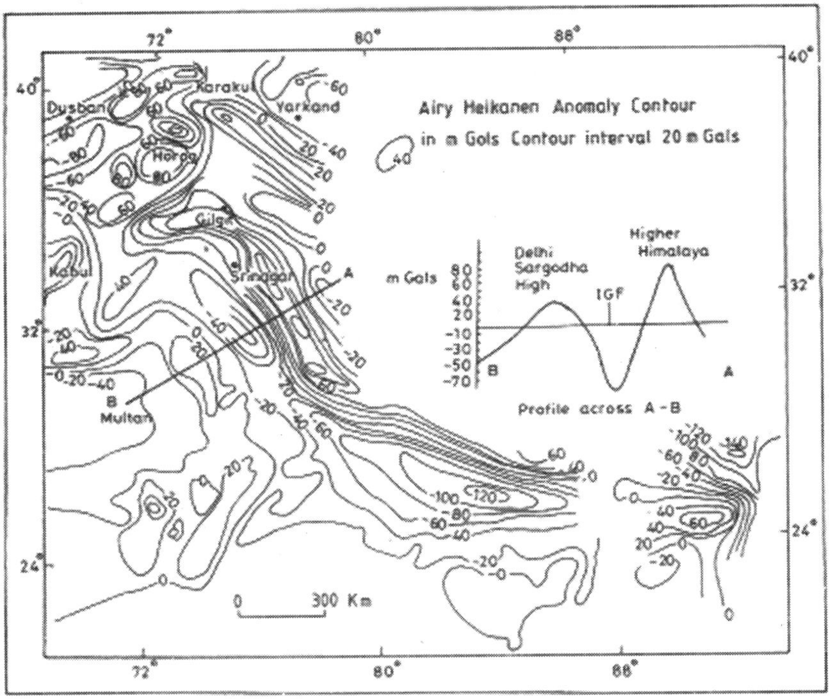

Fig. 5.3 Isostatic anamoly map of northwest Himalaya foredeep region (Qureshy and Kumar, 1992).

attributed to densification as well as thickening of the crust beneath the Himalaya (Qureshy and Midha, 1986).

The *Bouguer anomaly map* shows steep northward negative gradient (Fig. 5.4). The anomalies between –60 m Gal and –500 m Gal over the Himalayas have been explained due to flexure of the crust and crustal thickening along MBT, MCT and IST (Warsi and Molnar, 1977; Quershy and Warsi, 1980; Lyon-Caen and Molnar, 1983; Molnar, 1984; Verma and Prasad, 1987). The Indian lithosphere under thrusts beneath the Ganga basin and Lesser Himalaya at a shallower angle than beneath the High Himalaya implying the *Mohorovičić* (Moho) discontinuity also has different dips. Lyon-Caen and Molnar (1983) showed that the dip of the Moho discontinuity of the Indian

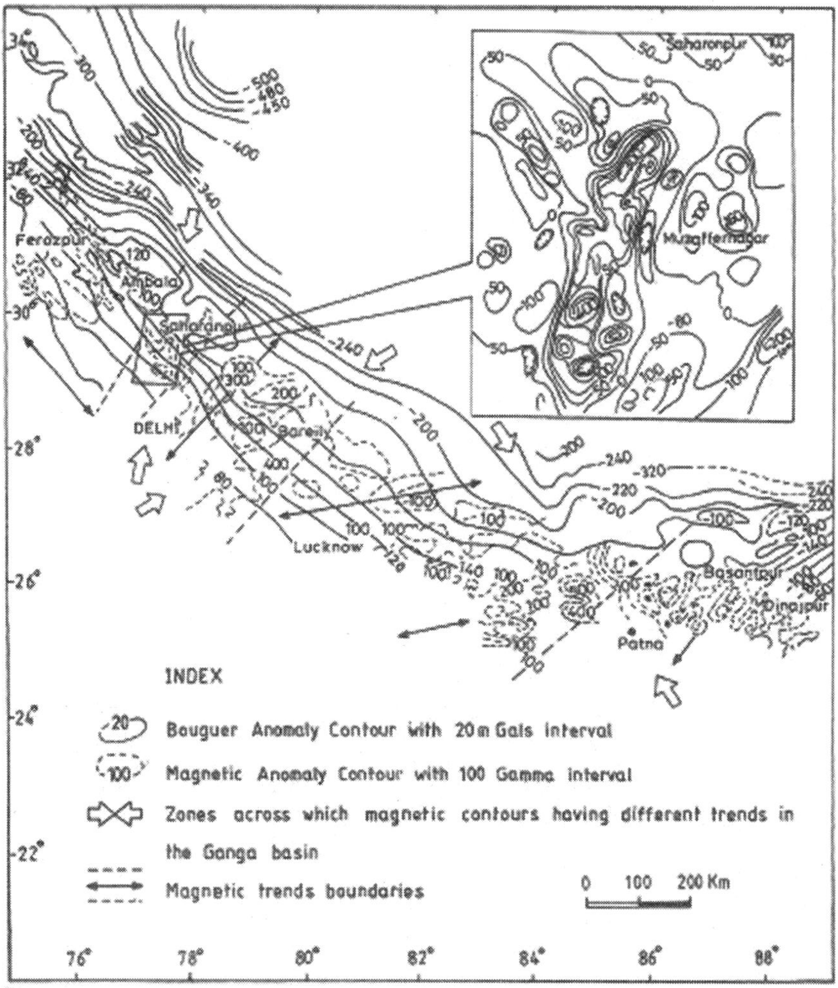

Fig. 5.4 Bouguer anomaly map superimposed on magnetic (VF) anomalies, northwest Himalaya and foredeep region (Choudhury, 1975).

plate increases from about 3°N beneath the Ganga basin and Lesser Himalaya to about 15°N beneath the High Himalaya. The gravity data explain the broad features of the Moho configuration that are consistent with smoothly increasing Moho depth from 38 km below the Ganga basin to about 55 km beneath the High Himalaya (Qureshy and Midha, 1986). The abrupt change of elevation at the foot of Greater Himalaya and the steep dip of the MCT are caused by 'ramping' of the Himalayan crust (e.g. Lyon-Caen and Molnar, 1983; Pandey et al., 1995).

Deficiency and excessive distribution of mass in the Himalaya must be supported by stress differences within lithosphere and mantle. Thus gravity anomaly can be used to examine how the mountains are supported. Lyon-Caen and Molnar (1983) computed Bouguer gravity anomalies across Himalaya using an elastic plate model with different flexure rigidity. Their calculations indicate that the load of the mountain is too great to yield gravity anomalies over the Lesser Himalaya and the Ganga basin, that are comparable with the observed values. A moment must be applied to the end of the elastic plate to properly support the Himalayan mountains. This moment can be related to either horizontal force or vertical force. They ruled out the horizontal force, because it would require an excess horizontal compressive stress of at least 330 Mpa. They suggested that vertical force can be related to phase change of the lower crustal material in the under thrusted Indian lithosphere.

In a recent collaborative work between USA and China, Jin et al. (1996) used Bouguer gravity anomalies to map out the geometry of the Indian and Eurasian plates (Fig. 5.5). Their modelling suggests that the mantle portion of the gravitationally heavy Indian plate is subducted beneath the Eurasian plate for a distance of 500-700 km north of the MBT. They have determined that the mantle suture between the Indian and Eurasian plates lies about 100 km north of the IST (Fig. 5.5). They have further suggested that of the many surface faults, only the MBT is directly connected to a fundamental plate boundary in the mantle. Molnar (1988) suggested that the stress induced by the overly thickened crust itself may be what causes the normal faulting in the High Himalaya. Such stresses within the crust may be responsible for the eastward motion of Tibet; this is marked by slightly higher elevation of the plateau to the west (Dewey et al., 1988).

The main control for mapping the subsurface of the Himalaya consists of Deep Seismic Sounding (DSS) profiles, and of late by receiver function techniques. *Wide angle reflection profiling* in the western Himalaya and in the Nepal-Tibet region have indicated substantial crustal thickening (Beloussov et al., 1980; Hirn et al., 1984; Kaila et al., 1984). The reflection profiles indicate maximum crustal thickness of about 75 km beneath the western as well as central Himalaya. The Moho discontinuity is delineated with steep dips from the Kashmir valley towards the Great Himalaya. The reflection seismograms recorded at the Kathmandu valley from shots in

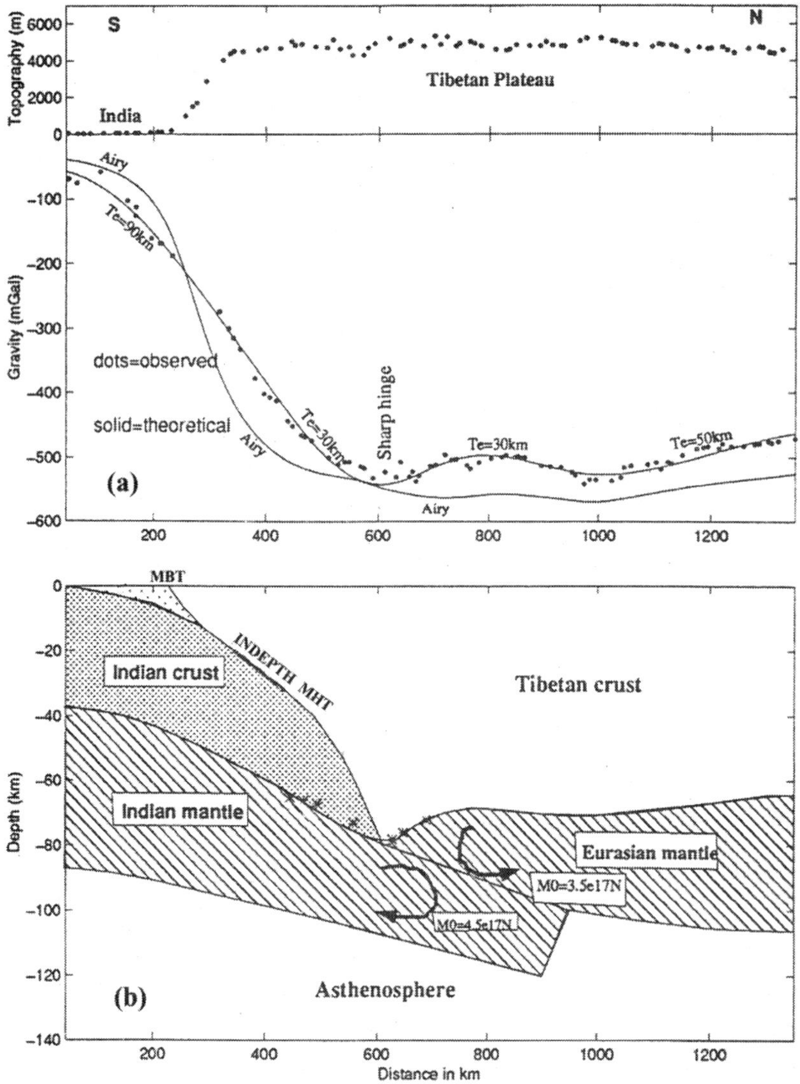

Fig. 5.5 (a) Topography and Bouguer gravity anomaly along a NE-SW profile from Himalayan foredeep to Qaidam Basin (see Fig. 5.7). (b) A model in which the Indian plate has an initial elastic thickness of 90 km, reduces to 30 km after it passes beneath the High Himalaya (Jin et al., 1996).

southern Tibet showed two clear phases separated by about 2s. The later phases were interpreted as P_mP reflections from the Moho discontinuity at about 75 km depth (Lepine et al., 1984). The earlier phase was attributed to reflection from the top surface of the under thrusting Indian plate. A step in the Moho was suggested just north of the High Himalaya (Lepine et al., 1984).

Surface wave dispersion studies have been made by several authors to estimate crustal thickness and upper mantle velocity beneath the Himalaya region (e.g. Gupta and Narin, 1967; Bird and Toksöz, 1977; Chun and Yoshi, 1977; Pines et al., 1980; Patton, 1980). Most of the estimates agree upon an average crustal thickness of 70 km beneath the Tibetan Plateau. Holt and Wallace (1990) mapped the crustal thickness and mantle velocity beneath the Tibetan Plateau, Hindu Kush, Karakuram, southwest China, Burma and northern India regions through a linear least square inversion of 130 individual long period P_{nl} wave forms. They found that the crustal thickness beneath the Tibetan Plateau, Karakuram and Hindu Kush regions range from 63 to 72 km, and the P_n velocity (8.25 km/s) is 0.10-0.20 km/s faster than those beneath northern India (8.1 km/s). Using earthquake travel times, Barazangi and Ni (1982) and Ni and Barazangi (1984) reported high P_n (8.43 km/s) and high S_n (4.73 km/s) velocities beneath the Tibetan Plateau. They concluded that the high P_n and S_n velocities and efficient propagation of high frequency S_n waves were the evidences that the Indian continental shield had under thrust Tibet.

A recent study on S receiver function by Kumar et al. (2005) estimated the crustal thickness 45-65 km in the Tien-Shan and adjacent regions, and it deepens to ~70 km depth in Pamir and Karakuram (Fig. 5.6). The thickness of the lithosphere varies between 90 and 120 km underneath the Tien Shan and increases to 160 km beneath the Tarim basin. The observations indicate a scenario of continental collision and subduction (Kumar et al., 2005). A similar study across the Nepal and Tibet Himalaya imaged the Moho at 40–70 km (Schulte-Pelkum et al., 2005) (Fig. 5.6), and in the northeastern Himalaya at 35-50 km (Ramesh et al., 2005). A detailed study for the northeast region is discussed in Chapter 5.

In a further recent study Kumar et al. (2006) imaged the collision architecture of the Indian and Asian continental plates beneath the Tibetan Plateau (Fig. 5.7). The base of the Indian lithosphere dips northward from a depth of 160 km beneath the Himalayas to a depth of 220 km just south of Bangong suture. The base of the Asian lithosphere is nearly horizontal at the depth of 160-180 km from central to northern Tibet. There is a vertical gap of about 50 km between India and Asian lithosphere. The Moho depth constrained by P and S receiver functions (Kosarev et al., 1999; Kind et al., 2002 and Kumar et al., 2006) indicate penetration of the Indian lower crust several hundred kilometres to the north of the Zangbo suture (Yuan et al., 1997).

The traditional *geodesy*, fundamental technique of triangulation and distance measurement, was started in 19th century in India. Even after 200 years of fieldwork by the Survey of India, there are, however, only a few reports in the open literature on geodetic monitoring of crustal deformation in the Indian section of the Himalaya. The most notable reports are by Middlemiss (1910), Rajal et al. (1986), and by Roy and Hasija (1995). Although no

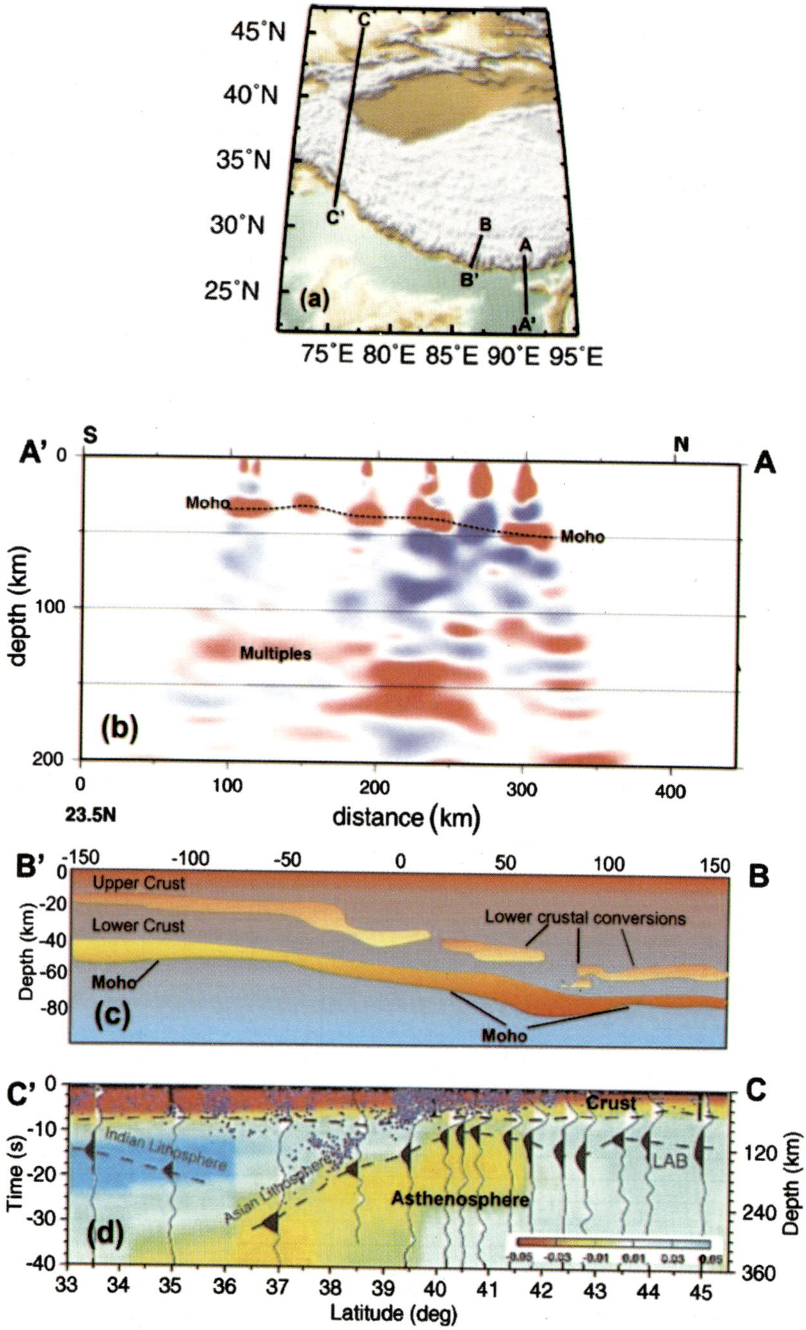

Fig. 5.6 (a) Location of seismic sections across Himalaya. (b) Seismic section along A-A′, eastern Himalaya (Ramesh et al., 2005), (c) Seismic section along B-B′, central Himalaya (Schulte-Pelkum et al., 2005). (d) Seismic section along C-C′, northwestern Himalaya (Kumar et al., 2005) (Copyright @AGU).

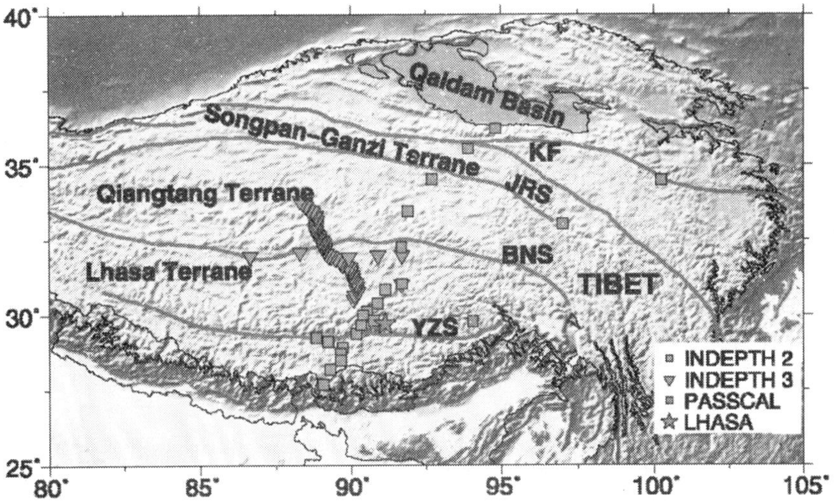

(a)

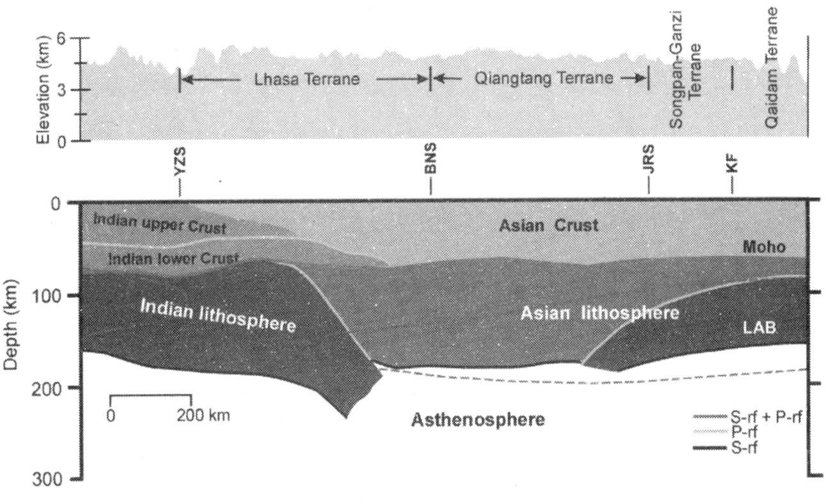

(b)

Fig. 5.7 (a) Topographic map of Tibet and seismic stations used for receiver function analysis; YZS: Yarlung-Zangbo Suture, BNS: Bangong-Nujiang Suture, JRS: Jinsha River System, KF: Kunlun Fault. (b) Architecture of the Indian and Asian lithosphere plates based on S receiver function (Kumar et al., 2006) and P-receiver function results (Kosarev et al., 1999; Kind et al., 2002), after Kumar et al., 2006.

Indian earthquake, including the four great earthquakes in the Himalaya, is well constrained by the geodetic measurement, the coseismic and interseismic elevation changes support the view that the great earthquakes occur through a cyclic process of strain accumulation and release (Gahalaut and Chander, 1997).

The *repeat level surveys* exhibit an uplift as much as 6 mm/yr in the Himalaya (Chugh, 1974; Molnar et al., 1977). Recently *fission-track dating* has shown that the northern edge of the Indian plate underwent 5 km of uplift during last 30-20 m.y., implying uplift of about 1 mm/yr in the western Himalaya (Zeitler et al., 1982; Validya, 1984). The Nanga Parbat, however, has undergone 10 mm/yr uplift over the last 100,000 years, which is about 10 times the average Himalayan rate (Zeitler, 1985). It needs to be examined whether such high rate of uplift is due to isostatic rebound or is caused by local tectonic forces arising from plate tectonic processes. It may be noted that the Bouguer anomalies and crustal thickness in the Himalayas show a correlation with elevation (Qureshy, 1969 and 1971). This implies general prevalence of isostatic equilibrium in the Himalaya.

With the advent of modern technique, *Global Positioning System* (GPS), many shortcomings in traditional studies are removed. Within the last decade, the GPS studies have provided three fundamental constraints concerning the seismogenic framework of the Indian plate: its overall stability (< 0.01 μ strain/year), its velocity of collision with Asia (58 ± 4 mm/year at N 44°E), and its rate of collision with southern Tibet (20.5 ± 2 mm/year) (Freymueller et al., 1996; Larson et al., 1999; Bilham et al., 1998). A 26-station GPS network was established in northwest Himalaya, across the zone of the 1905 Kangra earthquake by Banerjee and Bürgmann (2002). Their estimated slip-rate for the currently locked HFT is of the order of 14 ± 1 mm/yr, which is consistent with the geologic studies (Powers et al., 1998; Wesnousky et al., 1999). This comparison assumes that the elastic strain accumulating inter-seismically along the High Himalaya will ultimately be relieved by slip of the HFT to near its surface trace along the Siwalik foothills. In Nepal Himalaya, it is also reported that the geologic slip-rate estimates along the HFT are in fact very close to those inferred from the GPS deformation field (Larson et al., 1999; Lave and Avonac, 2000).

The geophysical observations made by the China, US, Germany and Canadian collaborative project INDEPTH (International Deep Profiling of Tibet and Himalaya) included *common midpoint (CMP) reflection, wide angle reflection, receiver function* and *magnetotelluric* investigations (e.g. Zhao and Nelson et al., 1993; Nelson and Zhao et al., 1996; Huang et al., 2000; Haines et al., 2003). The transect was about 300 km long, that extend from the crest of the Himalaya (27°43′N, 89°09′E) to roughly centre of the Lhasa block (30°35′N, 91°25′E). The reflection data revealed an undulatory reflection horizon at depths 15-20 km from Sangpo suture to north end of the INDEPTH (Fig. 5.8). The *receiver function* analysis of the INDEPTH teleseismic data shows a marked low velocity zone (LVZ) within the crust in the Tethyan slab, the top of which coincides with the reflection band at depths 15-20 km. The *magnetotelluric* investigation also indicates anomalous electrically conductive crust exists at depths 15-20 km beneath the entire northern two thirds of the INDEPTH survey. These observations suggest that

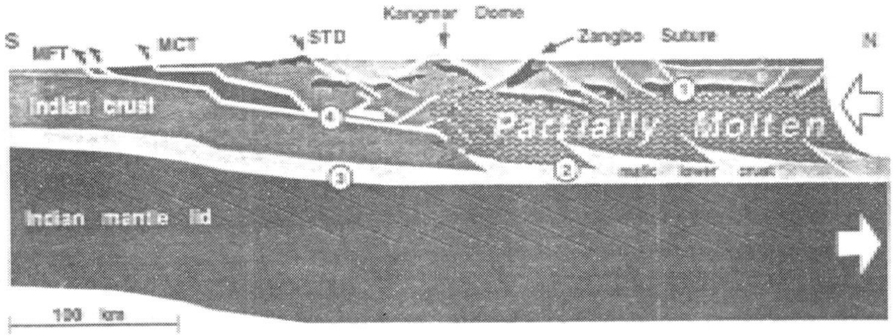

Fig. 5.8 INDEPTH lithosphere scale cross section of the Himalayan collision zone; MFT: Main Frontal Thrust, MCT: Main Central Thrust, STD: South Tibetan Detachment along the Yadong-Gulu rift in southern Tibet (Nelson and Zhao et al., 1996).

the reflection horizon at 15-20 km depth imaged on the reflection profiles marks the top of the mid-crustal partial melt layer underlying the northern part of the INDEPTH survey.

The reflectivity in the south provides information on the geometry of the crustal structure, the thrust plane along which India is currently under thrusting southern Tibet. This plane is traceable as a discrete gently north-dipping reflector up to ~28°45′N, where it disappears beneath Kangmar dome. The surface trace of the plane meets the HFT to the south; they named it Main Frontal Thrust (MFT) (Fig. 5.8). The reflection as well as MT data further suggest that the midcrustal partial melt layer underlies the northern part of the INDEPTH survey, and it dies out southward in the subsurface between the suture and the Kangmar dome (Fig. 5.8). Lack of Neogene to recent mafic volcanism in the region suggests that the warming results from crustal thickening, not intrusion of mantle derived melts into the crust. The Moho is only observed on the INDEPTH CMP profiles near the southern end of the transect where it is found to be at a depth of ~ 75 km (Zhao and Nelson et al., 1993; Brown et al., 1996). To the north the Moho is constrained by the receiver function estimates from broadband earthquake data that estimated depth 70-80 km along the length of the survey. This implies that the crust-mantle boundary is essentially flat beneath the survey region.

Further, twelve broadband Magnetotelluric (MT) soundings were carried out by Lemonnier et al. (1999) across the central Himalaya, Nepal, to investigate a detailed electrical structure. They have reported high conductivity in the foreland basin (~30 Ωm) that contrast with the high resistive Indian basement (>300 Ωm). The basement depth below the Ganga basin is at about 5 km. To the north of MBT the metamorphic rocks of Lesser Himalaya and overlying Klippe are rather resistive (>1000 Ωm). The most remarkable electrical structure is probably the 20 km deep conductor (~30 Ωm) beneath the boundary between the Lesser and Higher Himalayas, at the *ramp/basement thrust* zone (Fig. 5.8), which is characterized by intense microseismic activity

(Pandey et al., 1995). The high conductivity is inferred to probable existence of metamorphic fluids released by under-thrusting of the Indian basement and/or by the microcracks/brittle deformation around the ramp due to interseismic stress build-up. Geomagnetic and geoelectric investigations in the Western Himalaya also revealed conductive zones embedded in a narrow high seismicity zone at the 'ramp'/basement thrust zone (Arora and Singh, 1992).

5.4 CONCEPTUAL TECTONIC MODELS

5.4.1 Western Himalaya

Many authors proposed conceptual tectonic models of the Himalaya (e.g. Le Fort, 1975; Seeber et al., 1981; Ni and Barazangi, 1984; Valdiya, 1984). The main tectonic features of the models are: (i) the *Plane of Detachment*, a thrust with a very shallow dip, that extends beneath the foredeep, the Lesser Himalaya and the sub-Himalaya; the detachment plane separates the Himalayan sedimentary wedge from the Indian shield basement; (ii) *the Basement Thrust*, a more steeply dipping thrust that juxtaposes the basement of the Indian shield with the Tethyan slab, the pre-collisional leading edge of the Indian shield (Powell, 1979), and (iii) the *Basement Thrust Front*, the transition between the Plane of Detachment and the Basement Thrust. These are the major features of the Himalayan tectonics and underlying causes of the prominent surface features, like: (a) the high topographic gradient between the Lesser and High Himalayas and (b) the surface trace of the MCT. Based on these tectonic features and knowledge of the teleseismic events, two seismotectonic models are suggested: *steady state model* and *evolutionary model,* that are most referred by the earth scientists.

Steady State Model

The steady state tectonic model is proposed by Seeber and Armbruster (1981). The model consists of the downgoing Indian shield, overriding Tethyan slab and a sedimentary wedge, which is decoupled from the two converging slabs (Fig. 5.9). In this model the MBT and MCT are two active thrusts, contemporaneous features, which meet the Plane of Detachment at depth. The interface between the down-going slab and the sedimentary wedge is the Plane of Detachment. Farther north lies the Basement Thrust, the interface between the Tethyan slab and the down-going Indian Plate. The *Basement Thrust Front* (BTF), the transition between the Plane of Detachment and the Basement Thrust, coincides with the high topographic gradient and the surface trace of MCT. The BTF is marked as the zone of moderate magnitude thrust earthquakes. Lyon-Caen and Molnar (1983) and Pandey et al. (1995) proposed that the change in elevation as well as the steep dip of the MCT is caused

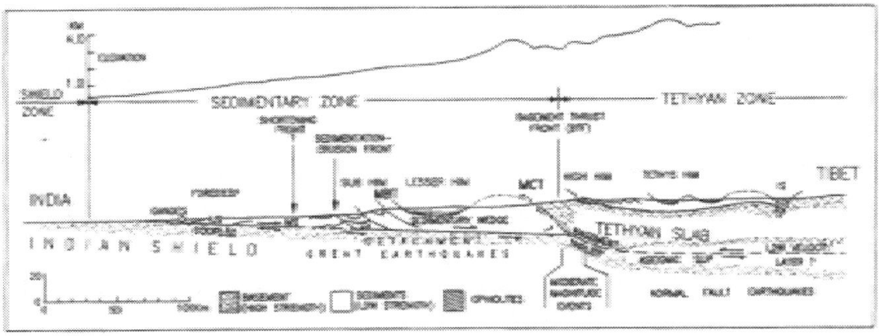

Fig. 5.9 Steady state tectonic model of the Himalaya; Q: Quaternary, US, MS, LS: Upper, Middle and Lower Siwalik, IS: Indus Suture (modified by Kayal, 2001 from Seeber et al., 1981 and Valdiya, 1988).

by *ramping* of the Himalayan crust over the northern edge of the Indian plate, and the 'ramp' represents the BTF to the north of MCT. The model further postulates that the great Himalayan earthquakes are associated with the Plane of Detachment.

Evolutionary Model

Ni and Barazangi (1984), on the other hand, proposed an evolutionary tectonic model (Fig. 5.10). In this model they suggested that ruptures of the great Himalayan earthquakes may have started in the interplate thrust zone and propagated south of the MBT along the Plane of Detachment. The MCT and MBT are assumed to be similar but successive tectonic thrusts. The southern thrust zone, i.e. the MBT becomes new boundary of the continental convergent zone, while the older thrust zone, the MCT, becomes less active and a dormant feature now. The well located earthquakes, as reported in the ISC (International Seismological Centre) Catalog, tend to occur in about 50 km wide zone between the MCT and MBT, and is known as the Main Himalayan

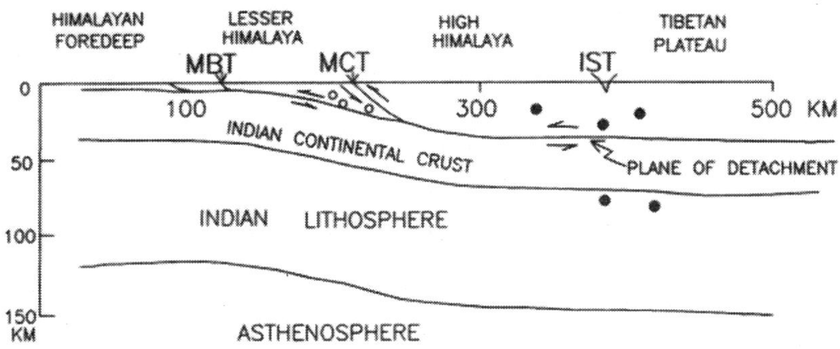

Fig. 5.10 Evolutionary tectonic model of the Himalaya; open circles indicate earthquakes of thrust faulting and solid circles indicate earthquakes of normal faulting zones (Ni and Barazangi, 1984).

Seismic Belt (MHSB). The model postulates that all the Himalayan thrust-fault earthquakes occur at the detachment surface between the under-thrusting Indian plate and the upper Himalayan blocks, whereas all the Tibetan normal-fault earthquakes occur in the shallow crust as well as in the upper most mantle (Fig. 5.10).

5.4.2 Western Syntaxis and Pamir-Hindu Kush

In the Western Syntaxis, the Hazara Syntaxis, northwest end of the Himalaya meets the Pamir-Hindu Kush region (Fig. 5.11). The Pamirs have an average elevation of between 4 and 5 km, and consist of a collage of suture terrains which accreted to the Eurasia during the Triassic – early Cretaceous. To its south lies the Karakuram (also called Karakorum) and Hindu Kush, which are geologically continuous (Rex et al., 1988; Searle, 1991, 1996). The southern boundary of the Hindu Kush and Karakuram is marked by the Shyok suture zone and referred to as the Main Karakuram Thrust (MKT). To the south of MKT lie the Kohistan and Ladak blocks, island arc terrains that are part of the 2500 km long Trans Himalayan batholith. The curvilinear ISZ, MMT and MCT are traced to the south of MKT. The MMT, Main Mantle Thrust, is western extension of the MCT. The present day relative motion of India with respect to Asia, within the Pamir-Hindu Kush region, is approximately due north at 45 mmy^{-1} (DeMets et al., 1990). This convergence is accommodated from the Himalayan thrust front in the south, right through to the Tien Shan in the north (Burtman and Molnar, 1993).

The Chaman Fault to the west is about 1000 km long, an active transform fault constituting the plate suture (Wellman, 1996). It marks the western boundary of the indenting Indian plate. Along the southern part of the Chaman Fault, a slip rate of about 10 mmy^{-1} has been deduced, and to the east of the Chaman Fault, within 80 km, several north-south extending active faults exhibit a left lateral displacement with similar slip rate (Nakata et al., 1980). The Hindu Kush appears to continue geologically eastwards into the Karakuram with no major suture or fault system separating the two mountain ranges. It is suggested that the Karakuram Fault (KaF) to the east marks the eastern limit of the Pamir-Hindu Kush Seismic zone (Pegler and Das, 1998).

The geophysical studies in the Pamir region have favoured subduction in opposite directions beneath Pamir and Hindu Kush (e.g. Hamburger et al., 1992; Burtman and Molnar, 1993; Fan et al., 1994). Chatelain et al. (1980) argued that lack of recent volcanism within the region, combined with the presence of intermediate depth seismicity, indicates a very brief period of subduction in the late Tertiary involving the subduction of two small ocean basins beneath the Hindu Kush and Pamirs. However, there is no geological evidence for the existence of any such basins and many authors suggested that the intermediate depth seismicity represents subduction of continental crust (e.g. Roecker 1982; Hamburger et al., 1992; Burtman and Molnar, 1993; Fan et al., 1994).

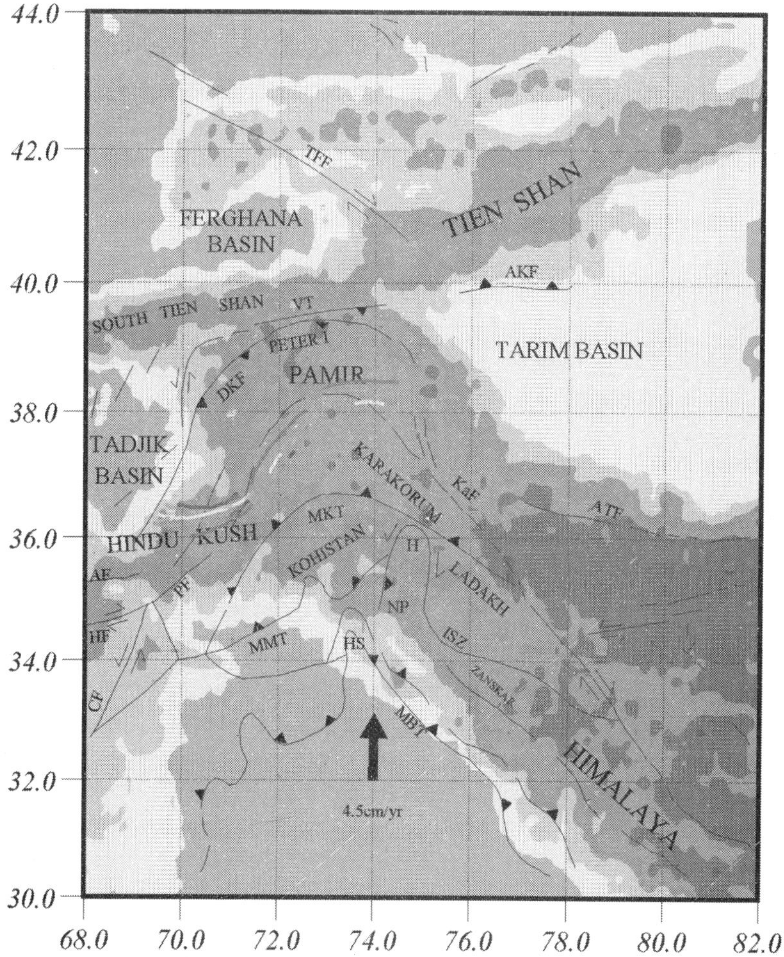

Fig. 5.11 Tectonic map of the Pamir-Hindu Kush region, western syntaxis, VT: VakhshThrust, DKF: Darvaz-Karakul Fault, KaF: Karakuram fault, AF: Andarab Fault, HF: Heart Fault, PF: Panjer Fault, CF: Chaman Fault, ISZ: Indus Suture Zone, MBT: Main Boundary Thrust, MKT: Main Karakuram Thrust, MMT: Main Mantle Thrust, NP: Naga Parbat, H: Haramosh, HS: Hazara Syntaxis, ATF: Altyn Tagh Fault, AKF: Atushi-Keping Fault, TFF: Talas Ferghana Fault (Pegler and Das, 1998). The black arrow shows the direction of plate motion of India relative to Eurasia (DeMets et al., 1990)

Roecker (1982) found evidence for low velocities associated with the Hindu Kush seismic zone, shallower than 200 km, from a local tomography inversion for P and S waves, and suggested that this was indicative of the subduction of continental crust, rather than oceanic crust. Vinik et al. (1977) and Mellors et al. (1995), however, reported high velocity zone at a depth greater than 200 km beneath the Hindu Kush. Mellors et al. (1995) suggested that low velocity at shallower than 200 km and high velocity zone deeper than 200 km depth beneath the Hindu Kush need not be incompatible, but might represent the transition between the subduction of continental and oceanic crust. Beneath the Pamirs there is not much evidence for existence of either unusually high or unusually low velocities. It was suggested that the Pamir seismic zone represents southward subduction of the Asian continental crust down to depth of about 200 km (Hamburger et al., 1992; Burtman and Molnar, 1993; Fan et al., 1994). Fan et al. (1994) gave a seismotectonic model representing steep subduction of the Asian plate, which is bent down by the gently dipping Indian plate (Fig. 5.12).

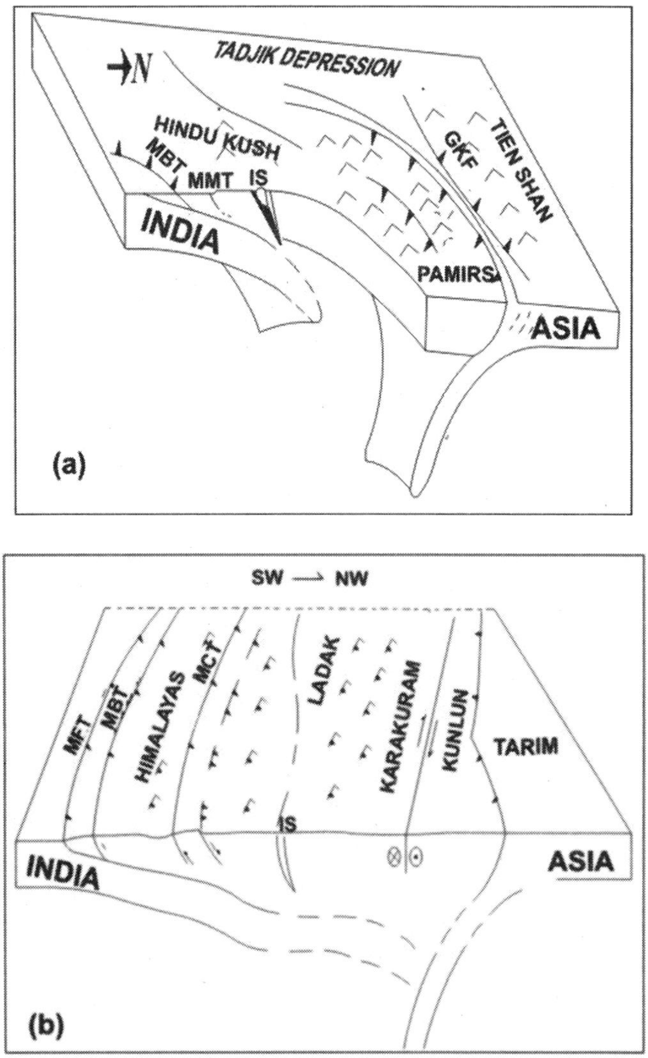

Fig. 5.12 (a) Schematic model of the Pamir region and (b) schematic model of the Karakuram region (Fan et al., 1994).

An isostatic gravity anomaly profile across the region is shown in Fig. 5.13. Three negative anomalies are seen, which correspond to the foreland basin of the Himalayas, Karakuram and Tarim basin. Molnar (1988) suggested that the negative gravity anomaly across south margin of the Tarim basin would require under-thrusting of the basin that extend over 100 km southwest beneath Kunlun. He suggested that the negative anomaly beneath the Karakuram was an evidence of abnormally cold mantle, which is consistent with intense crustal shortening. He further suggested that the down welling of mantle material beneath the Karakuram may be suppressing the Moho in this region, allowing unusually deep seismicity.

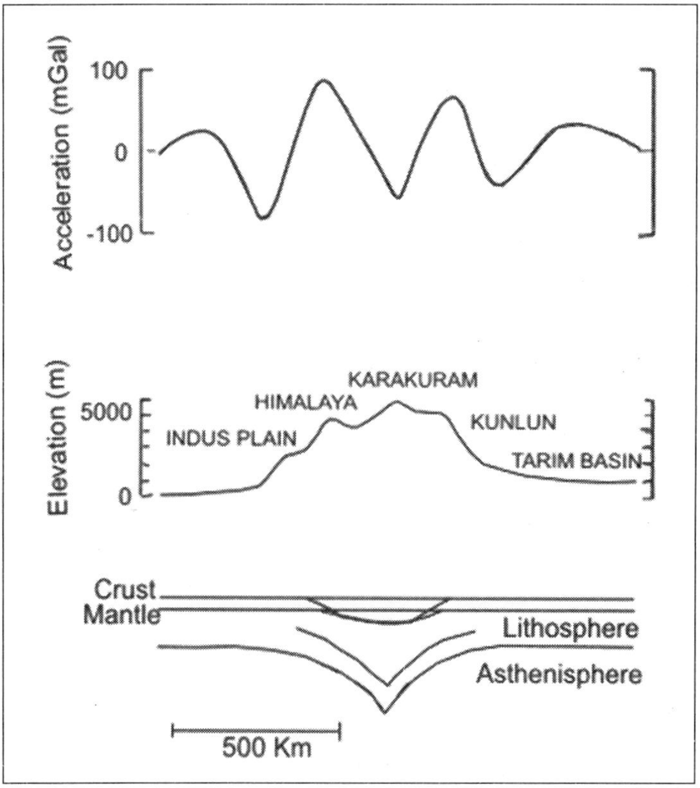

Fig. 5.13 Sketches of isostatic gravity anomaly, topography and interpreted section across the Himalaya, Karakuram and Kunlun (Molnar, 1988).

5.5 HIMALAYAN FOREDEEP

The Himalayan foredeep is a different tectonic domain that separates the Himalayas to the north and the peninsular India to the south. The Himalayan foredeep, the Indo Gangetic Alluvium Plain (IGAP), is an E-W tectonic basin located along the southern margin of the Himalayan fold belt (Fig. 5.1). The basin was formed during Paleogene-Neogene, and is related to the uplift of the Himalaya vis-a-vis flexural down warp of the Indian lithosphere under the supracrustal load of the Himalaya. The axis of the foredeep migrated with time, and the southern limit of the deposition progressively shifted southwards so that along the up dip margin of the basin the youngest Siwalik sediments or even alluvium rest directly over the basement/platform complex. The IGAP contains substantial thickness of Tertiary and Protcrozoic Vindhyan sediments with thickness increasing from south to north reaching up to 5-6 km near the foothills.

In terms of crustal nature and overlying sediments, the Himalayan foredeep widely varies from west to east. The crustal-model studies indicate that to the west of Aravalli ranges the crust is of oceanic affinity (Chun, 1986), and the crust to the east of Malda-Kishenganj fault is also of oceanic nature (Seeber and Armbruster, 1981). In the central part, the Gangetic foredeep is underlain by continental cratonic crust over which the Vindhyan platform sediments were deposited.

The basement contour map of the IGAP is presented by Raiverman et al. (1983), which indicates that the basin is deep along the Himalayan foot-hills. The Bouguer gravity anomalies run almost parallel to the length of the basin with some prominent northeast inflections. These northeast inflections may reflect the extension of the shield elements and subsurface ridges. An average crustal thickness of the IGAP is estimated between 34~37 km by body wave as well as by surface wave study (e.g. Kaila et al., 1968; Tandon and Choudhury, 1967; Gupta et al., 1982). Based on these data Choudhury (1975) prepared the isopatch map (Fig. 5.14).

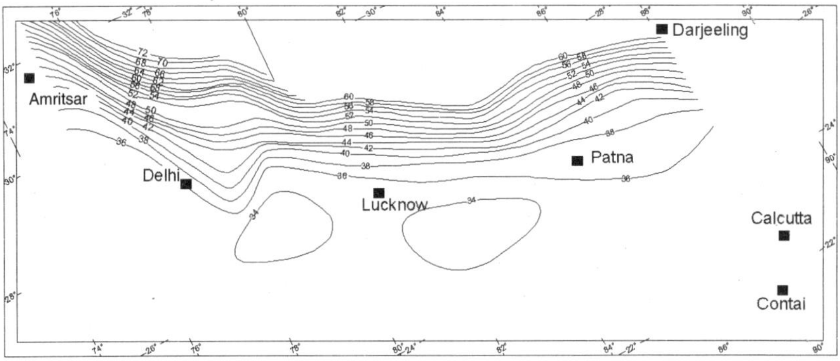

Fig. 5.14 Isopatch map of Indo-Gangetic Plains (Choudhury, 1975).

Geophysical and drilling investigations revealed many basement ridges and depressions defining the basement geometry of the IGAP in relation to the Siwalik foredeep to the north and peninsular shield to the south (Fig. 5.15). The major subsurface ridges beneath IGAP include the Delhi-Aravalli

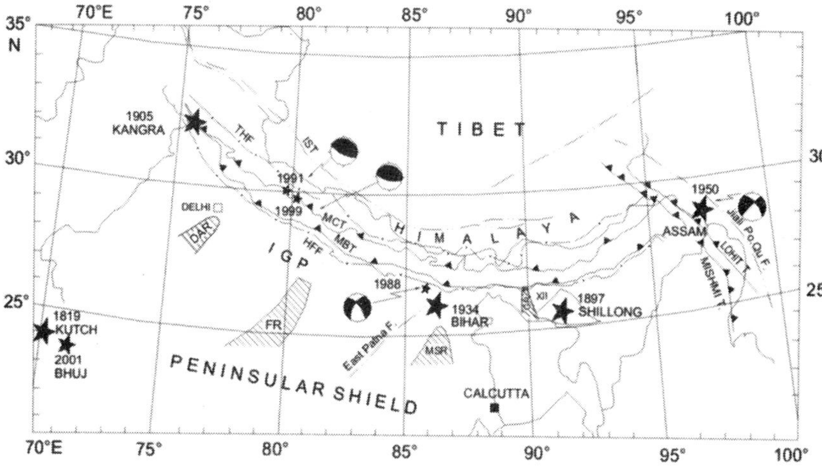

Fig 5.15 Tectonic map of the Himalaya and the foredeep region with the transverse basement ridges; DAR: Delhi-Aravalli Ridge, FR: Faizabad Ridge, MSR: Munger Saharsa Ridge, GR: Goalpara Ridge (modified from Kayal, 2001).

ridge, the Faizabad ridge, the Munger-Saharsa ridge and the Goalpara ridge. The Delhi-Aravalli horst has been redefined and expanded from the original concept of the Delhi-Hardwar ridge (Eremenko and Negi, 1968). Based on structural pattern and gravity observations most of these ridges are found to be bounded by faults. The major depressions in between the ridges from west to east are the Kangra, Subathu, Dehradun, Ramganga, Sarda and Gandak depressions.

The unconformity section of the Himalayan foredeep is dominated by both E-W and N-S trending basement highs and lows. Some of these are *horst* and *graben* structures within the basement, which controlled the thickness of the sediments within individual sub-basins (Fig. 5.15). The geophysical surveys related to oil exploration in the Himalayan foredeep have shown that the foredeep is divided into at least four basins separated by subsurface basement ridges running N-S (Mathur and Evans, 1964; Sastri et al., 1971; Agarwal, 1977). These ridges are in tectonic continuation from the Indian shield. The Aravalli subsurface ridge (Aravalli-Delhi Fold Belt) separates the Punjab basin in the west from the Ganga basin in the east. In the Punjab basin, Neogene foredeep sediments directly overlie the basement. In the Ganga basin, on the other hand, the pre-foredeep sequence is represented by the Vindhyan platform sediments of variable thickness deposited over the Precambrian crystalline basement. The Munger-Saharsa (M-S) ridge limits the deposition of Vindhyan sedimentation to the east. The M-S ridge is a positive landmass since Proterozoic till the initiation of the Neogene foredeep formation. To the east of the ridge, the Gondwana sediments overlie the rifted basement of the Purnea basin. The west and east Patna faults in the east Ganga basin and the Malda-Kishengunj fault in the Purnea basin are the known mapped faults that orient oblique/transverse to the Himalayan belt (see Fig. 5.51). These faults were possibly formed during rifting phase with Permo-Carboniferous Gondwana sedimentation. These faults are deep seated and were active at least up to mid-Cretaceous when Rajmahal trap erupted along some of these faults bordering the M-S ridge and those in the Purnea basin (Mahadevan, 1994). Farther east, beyond the Goalpara wedge of the Shillong Plateau, lies the Cretaceous-Tertiary Assam basin forming the foredeep for the northeastern Himalaya (Chapter 6).

Aeromagnetic, deep seismic and gravity surveys reflect the disposition of the subsurface ridges and depressions underlying the foredeep (Morgan and McIntyre, 1959; Sastri et al., 1971; Farah, 1973; Rao, 1973; Agarwal, 1977). Major faults identified are the Moradabad, Lucknow, west Patna, east Patna, M-S pair faults, Malda Kishenganj (M-K) fault etc. The Moradabad fault outlines the Sarda, and the Patna fault outlines the Gandak depressions. It is believed that most of these faults extend northward, across the Himalayan foothills (Valdiya, 1976), although very little is known about the form of their continuity.

A compiled Bouguer anomaly map of the eastern part of the Himalayan foredeep is shown in Fig. 5.16 (data source: Rao, 1973; Mukhopadhyay et al., 1986; Verma et al., 1988). An overall E-W trend is seen for most part of the foredeep. A value of –40 to –50 mGal is observed in the northern edge of the exposed shield area, which decreases steadily northwards to as low as –200 mGal near Siwalik foothills. The steady fall in gravity is attributed to thicker sediment from south to north as well as for the distant root effect of the Himalaya (Lyon-Caen and Molnar, 1983). The steep gradient and amplitude variation suggest appreciable change in crustal configuration both in thickness and density between the foredeep crust in the west and in the east. The prominent subsurface tectonic features are reflected in the gravity map.

The deep seismic data revealed roughly E-W trending subsurface faults below the Gangetic foredeep (Hari Narain and Kaila, 1982), which were

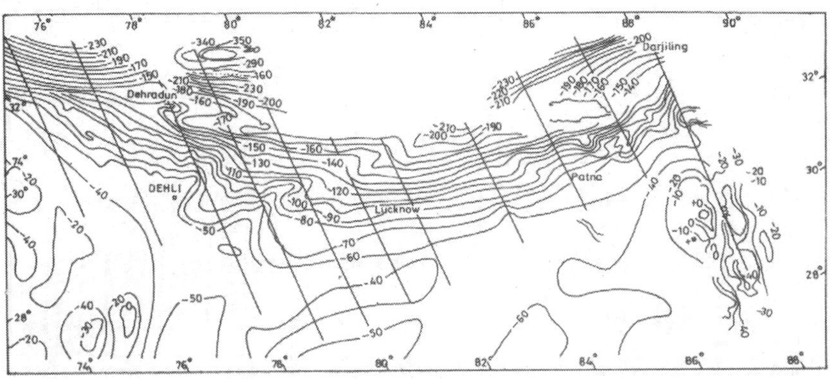

Fig. 5.16 Bouguer anomaly map of the eastern Himalayan foothills and the Indo-Gangetic Plain; contour values are in mGal (Choudhury, 1975).

oldest and originated during the Precambrian as possible passive margin normal faults. The horst and grabens are preserved below the foredeep. The DSS crustal-sections are illustrated in Chapter 7 (Section 7.3.5).

The satellite images indicate that several lineaments are regionally extensive. These extensive lineaments are developed between the Gangetic foredeep and the Indus suture across the whole width of the Himalaya; a few of them even transgress into Tibet. Some of these lineaments parallel the outline of the subsurface ridges underlying the foredeep (Fig. 5.15). It may be noted that several transverse lineaments produce noticeable offsets on the MCT, MBT or on the HFT. A few of these extensive lineaments appear to be seismically active (GSI, 2000).

5.6 LARGE EARTHQUAKES IN THE HIMALAYA

The list of earthquakes in the Himalaya goes back to the first century, and at least from the sixteenth century it seems to be complete for the large earthquakes. The catalogue of instrumentally recorded earthquakes has been made comprehensive since the inception of the World Wide Seismograph Station Network (WWSSN) in 1964. These data are published in the ISC (International Seismological Centre) bulletins. The earthquakes prior to 1964 have been compiled by Chandra (1978), Bapat et al. (1983) and Gupta et al. (1986), and recently Chandra (1992) updated the data base. Figure 5.17 shows a map of epicentres of earthquakes M≥7.0 or Io≥VIII; details are given in Table 2.3 (see Chapter 2). Some of these large (M≥7.0) earthquakes were very destructive and caused loss of many human lives. These include, from west to east, the earthquakes of Mandra 1969 (Io=XI), Kashmir 1828 (Io=X), Kashmir 1885 (Io=X), Kangra 1905 (M 8.0), Deoprayag 1803 (Io=IX), Nepal 1833 (Io=X), Nepal 1869 (Io=X), Bihar-Nepal 1934 (M 8.4), the Shillong 1897 (M_s 8.7) and the Assam 1950 (M_s 8.7). The Delhi earthquakes of 1720 (Io=X), 1883 (Io=X), and 1960 (Io=VII) occurred south of Himalaya. In the recent years, three damaging strong earthquakes occurred in the Himalaya. These are the Bihar-Nepal earthquake (M_s 6.6) of August 20, 1988 in the Himalayan foredeep, and the Uttarkashi earthquake (m_b 6.6) of September 30, 1991 and the Chamoli earthquake (m_b 6.3) of March 28, 1999 in the Garhwal Himalaya. The two Garhwal Himalayan earthquakes (1991 and 1999) are well studied using the large quantity instrumental records of the aftershocks.

Further, another devastating earthquake M_w7.8, Io = X, occurred on October 8, 2005 in the Hazara syntaxis, that caused a huge (~80,000) loss of human lives in the Kashmir Himalaya (Avouce et al., 2006).

The main Himalayan seismic belt (MHSB) and its foredeep region in the Indian Himalaya, between the western syntaxis and the eastern syntaxis,

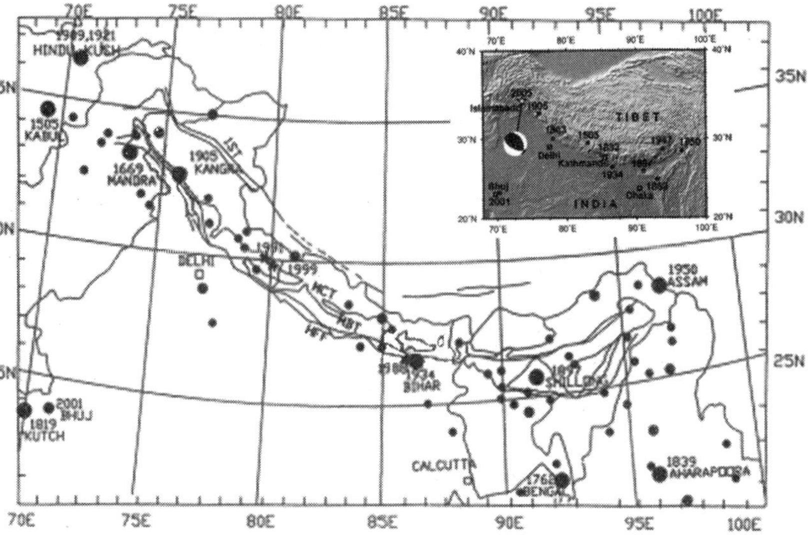

Fig. 5.17 Map showing large earthquakes in the Himalaya and its foredeep region and in the western Peninsular region; epicentres of three damaging earthquakes: 1988 Bihar/Nepal (Ms 6.6), 1991 Uttarkashi (m_b 6.6) and 1999 Chamoli (m_b 6.3) are shown by star symbols (Kayal, 2001). Inset: Map shows the large historical earthquakes and the devastating 2005 Kashmir earthquake (Mw 7.8) with its fault plane solution (Rao et al., 2006).

may be divided into three segments: (i) northwestern and western Himalayas which include Jammu & Kashmir (J&K) Himalaya and Himachal Pradesh in the northwestern part, and Garhwal and Kumaun Himalaya in the western part, between longitude 72°-80°E, (ii) the central and eastern Himalayas which include Nepal Himalaya in the central part and Sikkim and Darjeeling Himalaya in the eastern part, between longitude 80°-90° E, and (iii) the northeastern Himalaya which includes Bhutan and Arunachal Pradesh between longitude 90-98°E. The seismicity and seismotectonics of northeastern Himalayan foothills and its adjoining areas including eastern syntaxis, Assam valley, Shillong Plateau, Tripura fold belt, Indo-Burma ranges and the Andaman Islands are discussed separately in Chapter 6. In this section our interest of study is in the western syntaxis, northwest and western Himalayas and in the central and eastern Himalayas.

5.7 WESTERN SYNTAXIS AND PAMIR-HINDU KUSH SEISMICITY

5.7.1 Seismicity Maps

Recently a detailed study of seismicity in the Western Syntaxis and Pamir-Hindu Kush region has been done by Pegler and Das (1998). The ISC data lists 9127 events within a region bounded by 24-42°N and 68-78°E for the period 1964-1992. About 6000 events were relocated by them using Joint Hypocentre Determination (JHD) method. Out of these, 3260 events were relocated with 90 percent confidence limits of less than 30 km (in latitude, longitude and depth). These relocated events, at different depth ranges, are shown in a series of maps; the reliable fault-plane solutions are also presented in these maps. Brief descriptions of these maps are given below, that shed light on seismotectonics of the western syntaxis zone.

Seismicity: 0-75 km depth

It is observed that the shallow (0-75 km) activity is mostly concentrated along the northern margin of the Pamirs, at the Darvaz-Karakul and Vakhsh fault systems (Fig. 5.18). The NE-SW trending Darvaz fault marks the boundary between northwestern margin of Pamirs and the Tadjik basin (Fig. 5.11), and meets the E-W trending Vakhsh thrust (Burtman and Monlar, 1993). The sinistral strike-slip fault-plane solutions of the events (e.g. 11/01/78 and 10/26/84) are comparable with the Darvaz fault system. The Vakhsh thrust displays thrust faulting (solutions 12/18/77 and 8/12/88); there is also evidence of right-lateral strike-slip faulting (solutions 8/25/83 and 4/17/90). To the north of Pamirs, at the northern margin of South Tien Shan, there is thrusting (solutions 10/13/85 and 5/6/82). There is also evidence of dextral strike-slip faulting along the Talaso-Ferghana fault. The NW-SE trending seismicity running from 77°E 31°N to 73°E 35°N and the depth section suggest northeastward subduction of the Indian plate beneath the Himalaya. To the west, west of 73°E, on the other hand, northwestward subduction is suggested (Pegler and Das, 1998).

The N-S trending cluster of events to the north, between 36.8°N and 37.5°N, exhibits E-W extension (e.g. solution 03/05/90). This cluster is at a depth range 40-60 km, not at the surface. Burtman and Molnar (1993) suggested that this E-W extension might indicate cessation of crustal thickening beneath the Pamirs. A similar observation is made by England and Houseman (1989) for Tibet.

A notable feature of the seismicity in this depth (0-75 km) range is the low seismic activity along the NW-SE trending dextral Karakuram fault. This is a significant tectonic feature and believed to have a high slip-rate, as much as 32 mm/y (Avouac and Tapponier, 1993).

Seismicity: 75-100 km depth

The seismicity in this depth range is restricted to a much smaller area; three prominent clusters (a, b and c) are marked (Fig. 5.19). A large gap exists between the clusters a and b. Three CMT solutions (04/27/85, 04/17/79 and 05/17/90) within the near E-W trending cluster (a) show strike slip solutions with horizontal T-axes. The cluster b shows a near NNE-SSW trend, and the

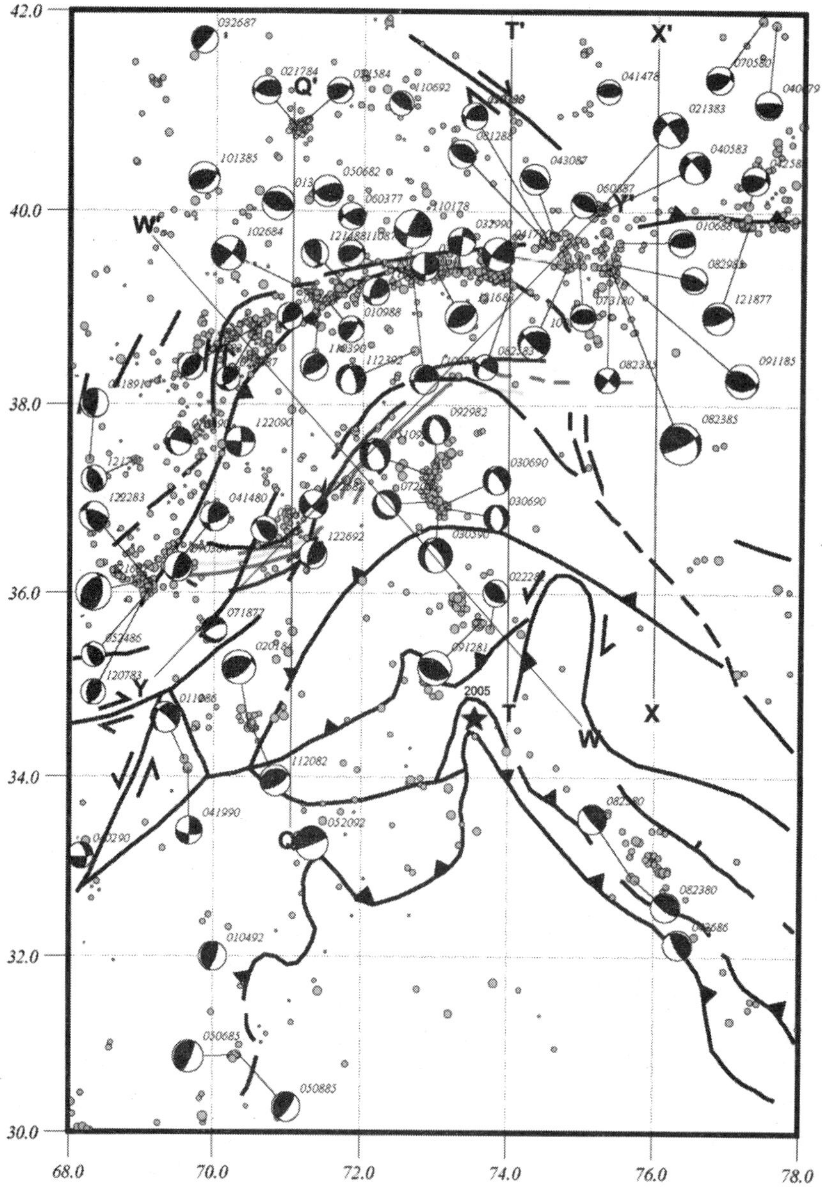

Fig. 5.18 Seismicity (depth 0-75 km) and fault plane solutions in the Pamir-Hindu Kush region; the beach balls indicate usual representation of the fault plane solutions (Pegler and Das, 1998). The epicentre of the October 8, 2005 earthquake Mw 7.8 is shown by solid star.

CMT solutions exhibit mostly reverse faulting with near horizontal N-S trending P-axes. The cluster c has an arcuate trend, varying from E-W at its western end through to NE-SW at its northeastern end. The CMT solutions

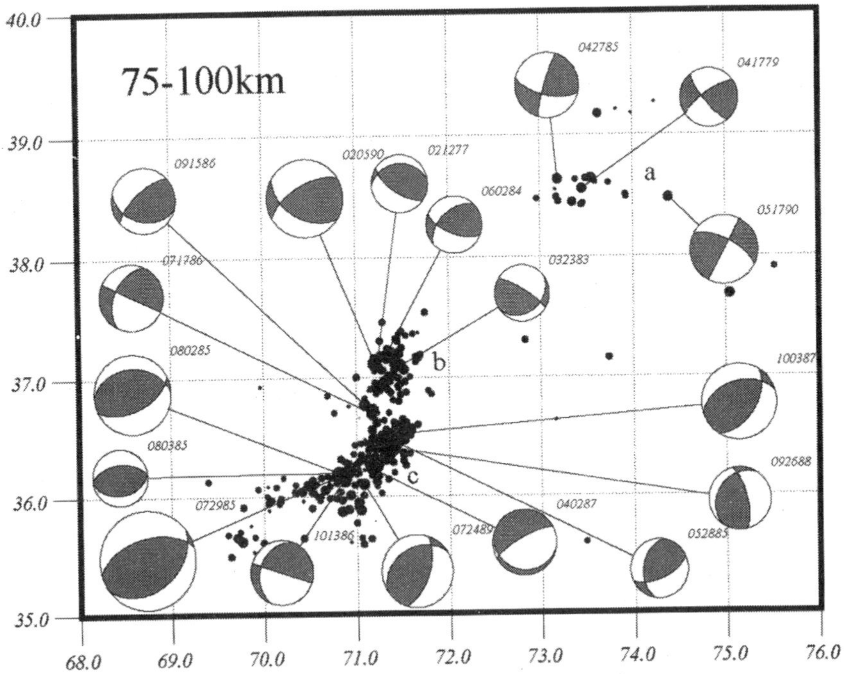

Fig. 5.19 Seismicity (depth 75-100 km) and fault plane solutions, Pamir-Hindu Kush region (Pegler and Das, 1998).

mostly show reverse faulting, with one normal (07/24/89) and one strike-slip mechanism (07/17/85). Both T-axes and B-axes are sub-vertical. It is observed that seismicity south of 36°N is shallower than those of cluster c.

Seismicity: 100-125 km depth

The seismicity at the 100-125 km depth range shows a narrow S-shaped band, approximately 700 km long and 40 km wide (Fig. 5.20). It, however, does not represent a single feature. Two large gaps d and e are marked. The gap d, approximately 40 km long, marks the boundary between the Hindu Kush to the southwest and the Pamirs to the northeast. Similar gaps are also reported by microearthquake surveys (Chatelain et al., 1980; Roecker et al., 1980).

The CMT solutions for the earthquakes, north of 37°N in the Pamirs, mostly show normal and strike slip faulting with near horizontal T-axes (Fig. 5.20). The trend of T-axes is comparable to the trend of seismicity. To the south of 37°N, in the Hindu Kush region, the solutions mostly show reverse faulting with near vertical T-axes, and B-axes are parallel to the strike of the seismic zone; this is a classical pattern for subduction (Pegler and Das, 1998).

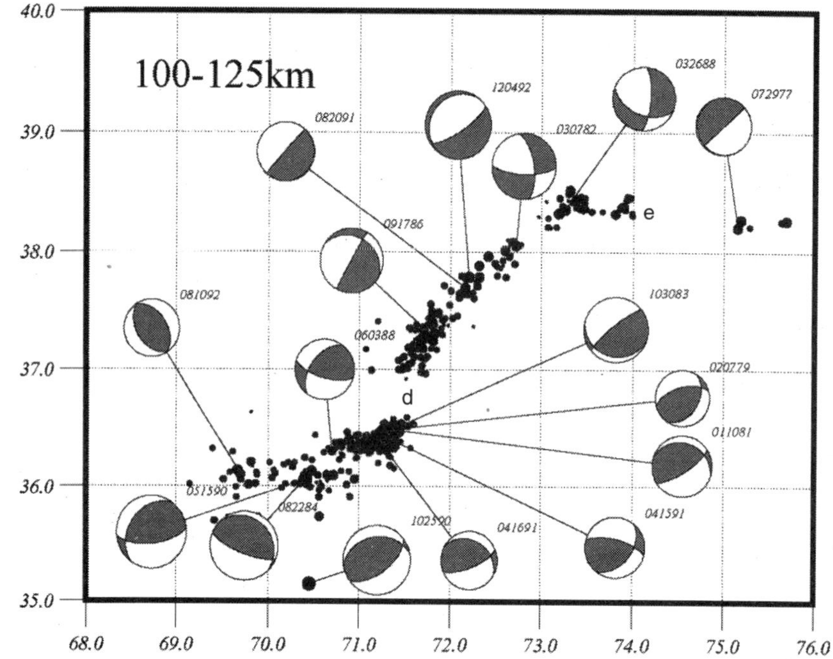

Fig. 5.20 Seismicity (depth 100-125) and fault plane solutions, Pamir-Hindu Kush region (Pegler and Das, 1998).

Seismicity: 125-150 km depth

In this depth range the activity is less (Fig. 5.21), but the S-shaped trend is very much similar to that observed at 100-125 km depth range. The width of the seismic zone is narrowed to 20 km. The seismic gap between the Hindu Kush and Pamir is wider in this depth range. Three CMT solutions are obtained for the earthquakes in the Hindu Kush zone, two show near vertical T-axes as observed at the 100-125 km depth range.

Seismicity: 150-175 km depth

Although low activity is observed in this depth range, the epicentres show four clusters, two beneath the Hindu Kush and two beneath the Pamirs (Fig. 5.22). The continuity of the S-shaped seismicity is also maintained. Only one CMT solution is reported for an event in the Hindu Kush zone; it shows a reverse faulting with near vertical T-axes.

Seismicity: 175-200 km depth

The seismicity in the depth range 175-200 km is more intense than those in the 150-175 km depth range. There are two dense clusters in the Hindu Kush

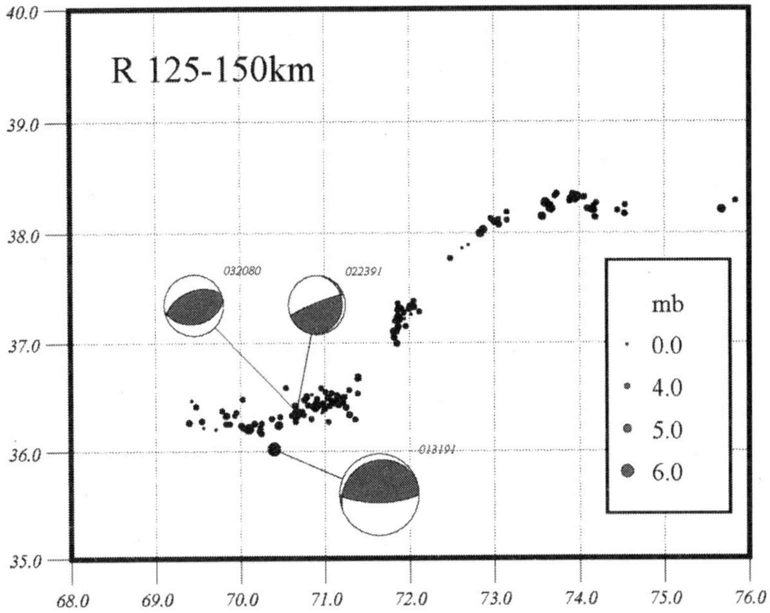

Fig. 5.21 Seismicity (depth 125-150 km) and fault plane solutions, Pamir-Hindu Kush region (Pegler and Das, 1998)

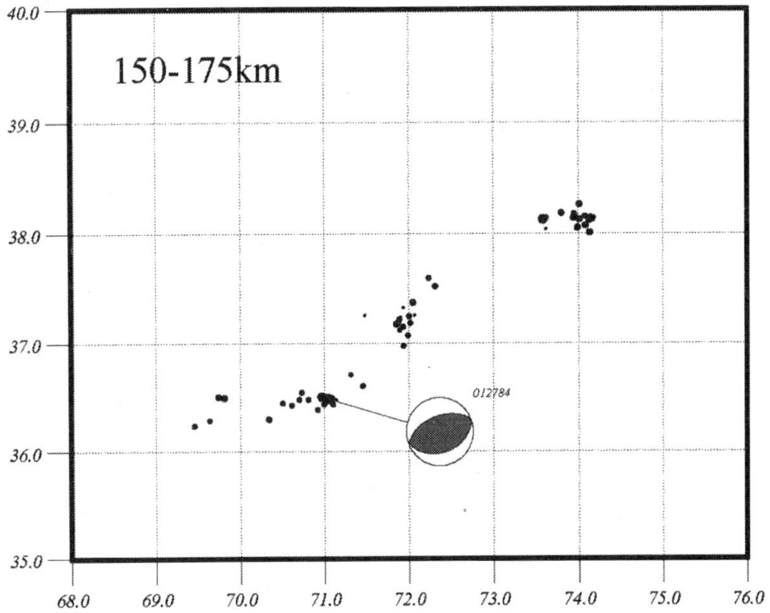

Fig. 5.22 Seismicity (depth 150-175 km) and fault plane solutions, Pamir-Hindu Kush region (Pegler and Das, 1998)

zone, one in the centre and one in the east, and one narrow forked like seismic zone in the west (Fig. 5.23). In the Pamirs, the NE-SW trend of seismicity is maintained. Six CMT solutions are obtained for the central cluster in the Hindu Kush, all show thrust faulting with near vertical T-axes.

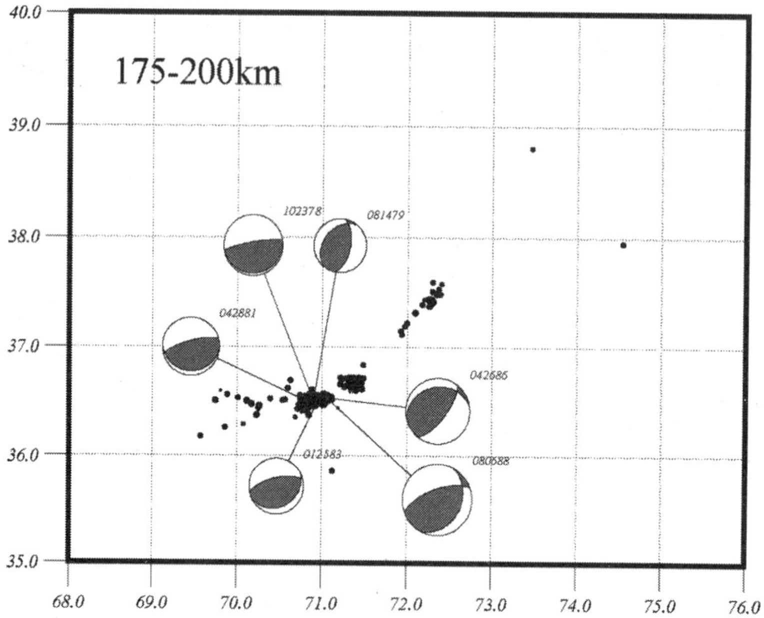

Fig. 5.23 Seismicity (depth 175-200 km) and fault plane solutions, Pamir-Hindu Kush region (Pegler and Das, 1998).

Seismicity: 200-225 km depth

The seismicity in this depth range is intense along the arcuate Hindu Kush seismic zone, but no activity along the Pamir seismic zone (Fig. 5.24). Almost all the CMT solutions show near vertical T-axis, and mostly show northeast plunging rather than southeast plunging as was observed for the 175-200 km depth range (Fig. 5.23). In the western and central parts of the zone, the B-axes are mostly parallel to the seismic zone. In the eastern part of the seismic zone no specific trend of the stress axes is observed. Further, it may be noted that a cluster of events occurred at 36.4°N, 71.2°E, to the south of the main arcuate Hindu Kush seismic zone at its eastern end. This is also observed in the 225-250 km depth range as discussed below.

Seismicity: 225-250 km depth

It is interesting to note that the cluster at 36.4°N, 71.2°E is elongated NE-SW in this depth range into a second zone of intense activity. This zone is

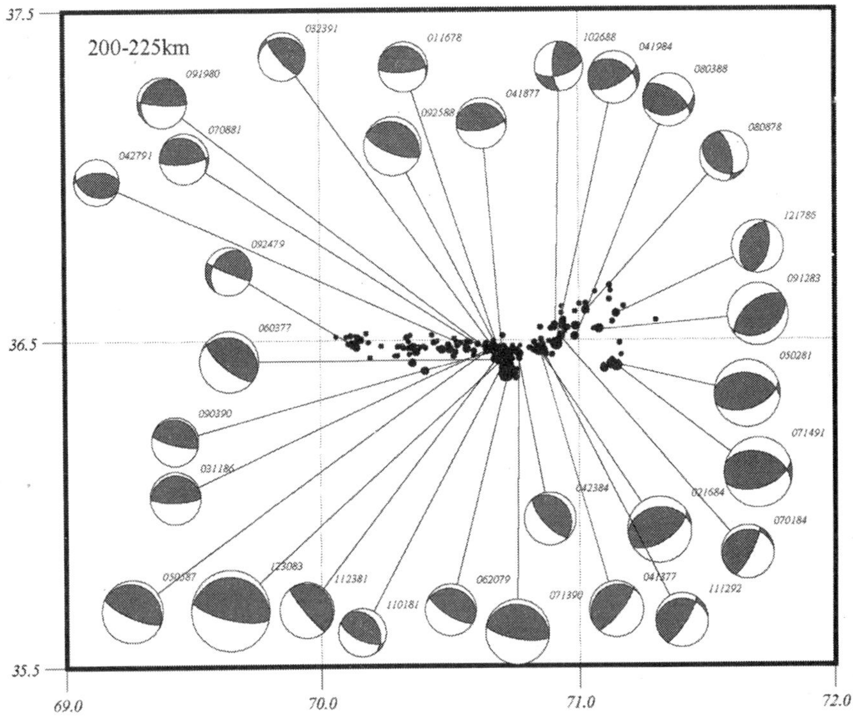

Fig. 5.24 Seismicity (depth 200-225 km) and fault plane solutions, Pamir-Hindu Kush region (Pegler and Das, 1998).

parallel to the main arcuate Hindu Kush seismic zone, about 30 km southeast (Fig. 5.25). Seven CMT solutions are obtained in the eastern part, four show E-W oriented P-axes. One event in the southeast cluster shows normal faulting. One CMT solution in the western seismic zone shows thrust faulting with N-S oriented P-axis.

5.7.2 Seismicity Cross Sections

Cross Section Q-Q′

A representative N-S cross section along Q-Q′ (Fig. 5.18), across the Hindu Kush seismic zone, is illustrated in (Fig. 5.26). The section shows some interesting features. The north dipping Hindu Kush zone is evident; the zone bends sharply at around 100 km depth. Within the dipping zone, there is a gap or low activity at 150-180 km, and again the intense activity at 180-270 km. The shallower part of the zone, at 60-100 km depth, is narrow and dips approximately at 20° to the north. At the intermediate depth range (100-150 km), the zone dips nearly at 50° to the north, and at the deeper range (180-270 km) the seismic zone is nearly vertical.

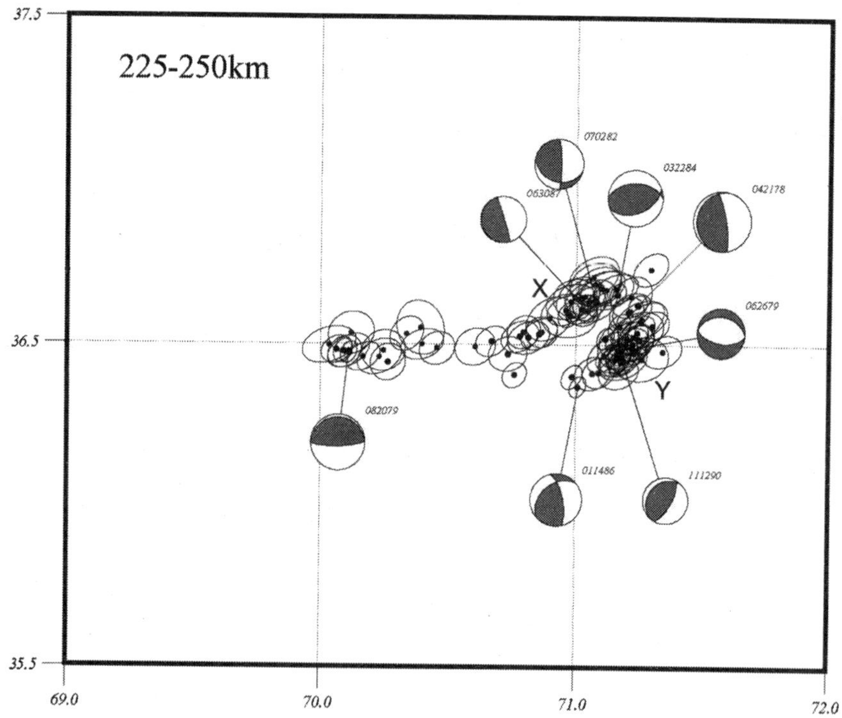

Fig. 5.25 Seismicity (depth 225-250 km) and fault plane solutions, Pamir-Hindu Kush region; 90 per cent confidence ellipsoids are projected (Pegler and Das, 1998).

There is an offset (north up and south down) in the Hindu Kush zone as it bends sharply at around 100 km depth. This section is about 100 km long and 30 km wide dipping at 10-15° north at a depth range 70-100 km below the Pamirs (Fig. 5.26), and links up with the sharply dipping Hindu Kush seismic zone. This zone represents the cluster b in Fig. 5.19, and lies above and to the north of Hindu Kush seismic zone, with focal mechanisms showing N-S shortening.

The cluster of events in the deeper seismic zone at 180-270 km shows an inverted V-shaped feature. The northern prong of the inverted V displays higher level of seismic activity and dips approximately at 80° to the north. This northern prong imaged the cluster of events (X) that are observed in Fig. 5.25, while the southern prong imaged the cluster of events (Y) to the southeast of the main Hindu Kush seismic zone. Below the Tien Shan Plateau a cluster of shallower events (depth 0-50 km) are observed.

Cross Section W-W′

A representative NW-SE cross section, along W-W′ (Fig. 5.18), across the Pamirs is shown in Fig. 5.27. The cluster of seismicity at depth 100-220 km

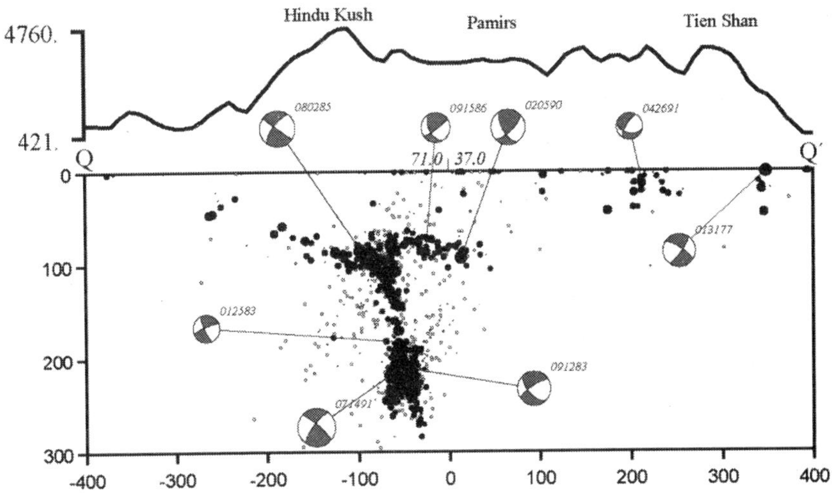

Fig. 5.26 Seismicity cross section Q-Q′, Fig. 5.18 (Pegler and Das, 1998).

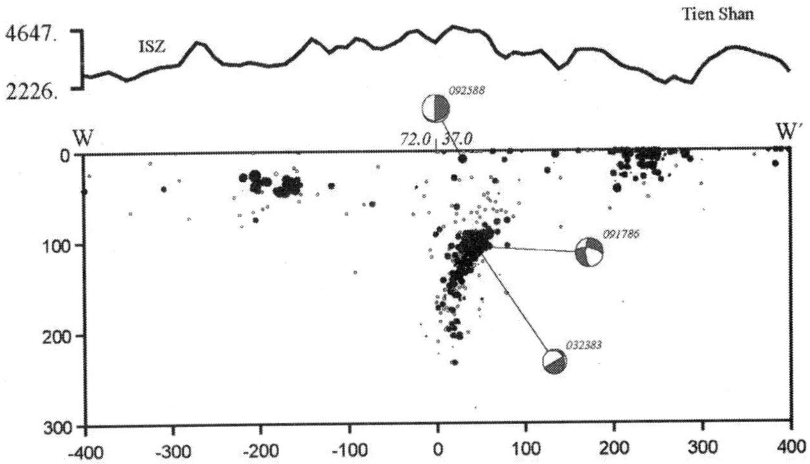

Fig. 5.27 Seismic cross section along W-W′, Fig. 5.18 (Pegler and Das, 1998).

reveals the geometry of the Pamir seismic zone. The shallower cluster 100-120 km dips southeast at approximately 30°. T-axes of the CMT solutions are parallel to the dip of the structure. The deeper cluster of events at 150-220 km suggests a near vertical dip. Several cross sections across the Pamir seismic zone were examined (Pegler and Das, 1998), and in each section it showed that the top of the southeast dipping intermediate depth seismic zone (100-220 km) abruptly terminates at around 90 km depth.

Shallower seismicity (0-50 km) at the northern margin of the Pamirs appears like a south dipping wedge. A two-slab model would require a link between the intermediate depth seismicity at 100-220 km and the shallow cluster that is situated at more than 130 km horizontal distance to the northwest, but there is no seismicity at 70-100 km depth to indicate such a link.

The north dipping trend of seismicity in the 20-60 km depth range, 200 km south of the centre of the section, may be extrapolated to the surface in the vicinity of the MMT. The events (e.g. 09/12/81, 02/22/82) show thrust faulting, near 73.5°E 35.5°N (Fig. 5.18).

Cross Section T-T′

A representative N-S cross section, across the E-W trend of the Pamir seismicity, along T-T′ (Fig. 5.18), is shown in Fig. 5.28. The Pamir seismic zone is reflected in this section, but as a very narrow (20 km) seismic zone dipping at approximately 60° between depths 100-160 km (Fig. 5.28). The cluster of events at 0-80 km depth, about 160 km north of the centre of the section, represents activity beneath the northern margin of the Pamirs. This shallow cluster shows a near vertical structure; the events (e.g. 08/25/83) show thrust faulting (Fig. 5.18). Further to the north, the shallow activity represents thrusting (e.g. 04/14/78; Fig. 5.18) at the northern margin of the Tien Shan. The cluster of events at around 30-50 km depth, about 200 km to the south of the centre of this section, represent thrusting (e.g. 09/12/81; Fig. 5.18) along a gently northward dipping plane (Fig. 5.28). The plane may be extrapolated to the surface around the Nanga Parbat.

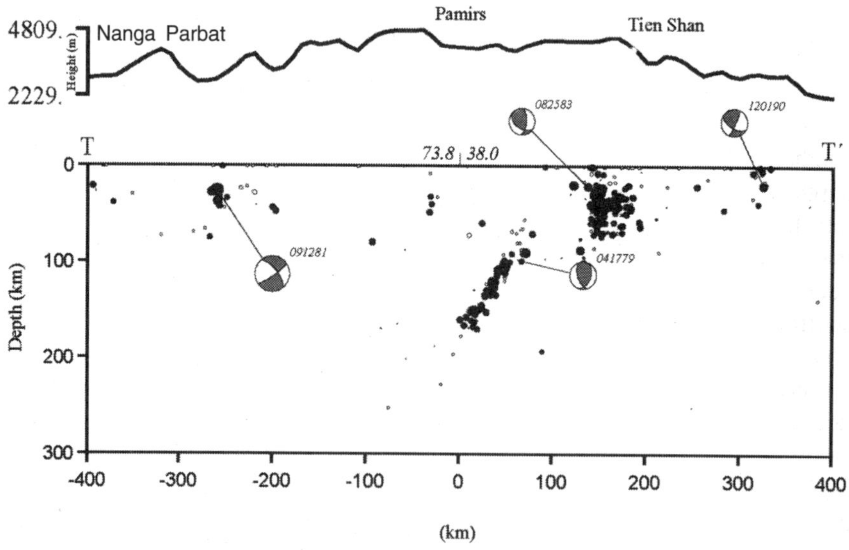

Fig. 5.28 Seismic cross section along T-T′, Fig. 5.18 (Pegler and Das, 1998).

Cross Section X-X′

A representative N-S cross section along X-X′, through the eastern end of the Pamir seismic zone (Fig. 5.18), is illustrated in Fig. 5.29. The activity although is much less at the intermediate depth range, it shows a near vertical trend at depth 100-130 km. Beneath the Tarim basin, intense shallow activity is observed, but there is no clear indication of any clear trend within the seismicity. The events (e.g. 04/25/83) in the cluster show thrust faulting (Fig. 5.18).

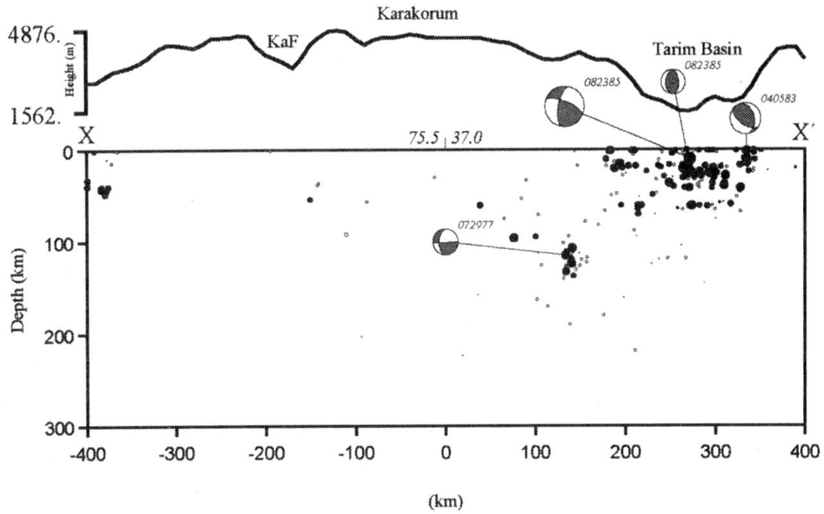

Fig. 5.29 Seismic cross section along X-X′, Fig. 5.18 (Pegler and Das, 1998).

Cross Section Y-Y′

The NE-SW cross section along Y-Y′ (Fig. 5.18), approximately parallel to the trend of the Pamir seismic zone is illustrated in Fig. 5.30. It reveals another interesting view on the seismicity at depth. The top of the intermediate depth seismicity shows an undulating upper surface, which can be divided into three segments. The first form the Hindu Kush zone, where the top of the seismic zone is about at 75 km depth; below this lies a deeper seismic zone at 180-280 km depth. The second segment, from centre of the Y-Y′ section to 100 km further northeast, shows a convex upper surface approximately at a depth of 120 km. The third segment continues until 220 km east of the centre of Y-Y′ where the top of the Pamir seismic zone is at depth of 90 km. Farther to the east, the shallow intense seismicity is observed below the Tarim basin.

5.7.3 Seismotectonic Perspective

There has been much debate in the literature on the nature of the structure that reflects the seismicity in the Pamir–Hindu Kush seismic zone, whether

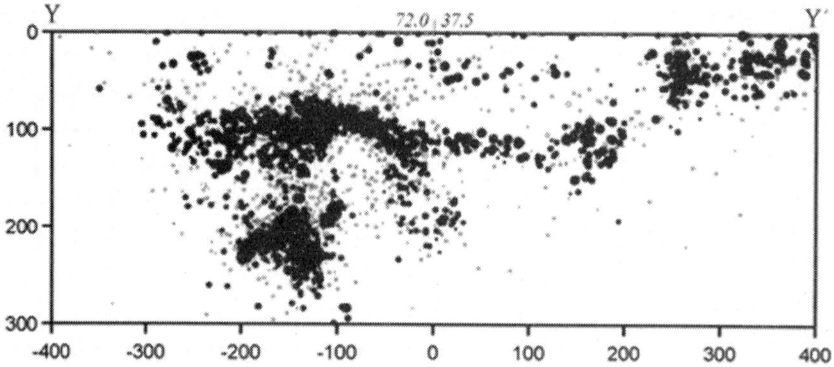

Fig. 5.30 Seismic cross section along Y-Y′, Fig 5.18 (Pegler and Das, 1998).

it is caused by one or two opposing subduction slabs (e.g. Billington et al., 1977; Chatelain et al., 1980; Hamberger et al., 1992; Burtman and Molnar, 1993; Fan et al., 1994; Pegler and Das, 1998). Pegler and Das (1998), based on relocation of 3260 events, argued that the Pamir-Hindu Kush seismic zone is a single contorted zone of seismicity. The different seismotectonic features of the seismic zone are discussed below.

Intermediate Depth Seismicity

The intermediate depth seismicity below Hindu Kush as well as below Pamirs indicate slab-like seismic zones (Figs 5.26 and 5.27). The near vertical T-axes at intermediate depth along Hindu Kush seismic zone are comparable with subduction tectonics of oceanic plates, which descend to intermediate depths (Billington et al., 1977, Roecker et al., 1980; Chatelain et al., 1980). The youngest oceanic rocks within the western syntaxis of the Himalaya, however, represent the closure of Tethys some 50-40 Ma ago. Further, there is no evidence of subduction related volcanism within the Tertiary. Assuming that the Hindu Kush seismic zone reaches the surface in the vicinity of MMT (Main Mantle Thrust), the total subducted slab length down to 300 km depth would be around 500 km. With the present day convergence rate 45 mmy^{-1} between India and Eurasia it should take about 10 Ma to subduct a 500 km long piece of lithosphere. Even with a half of this rate, it would still take around 20 Ma to subduct 500 km of lithosphere. Therefore, it is unlikely that the seismicity at depth beneath the Hindu Kush can really represent the former Tethys oceanic lithosphere; it must represent more recent subduction. Chatelain et al. (1980) argued that the lack of recent volcanism within the region indicates a brief period of subduction of two small oceanic basins beneath the Hindu Kush and Pamirs. There is, however, no geological evidence for the existence of any such basins. On the other hand, many authors suggested that the intermediate depth seismicity represents subduction of continental crust beneath the Hindu Kush (e.g. Roecker, 1982; Hamburger et al., 1982; Burtman and Molnar, 1993; Fan et al., 1994). It was further suggested that the reported low velocity zone shallower than 200 km (Roecker, 1982) and high velocity zone below 200 km (Vinnik et al., 1977) beneath the Hindu Kush represent the transition between the subduction of continental and oceanic crust (e.g. Roecker, 1982; Burtman and Molnar 1993, Mellors et al., 1995).

Beneath the Pamirs, there is little evidence of either unusually high or unusually low velocities. It has been argued that the Pamir seismic zone represents the subduction of Asian continental crust down to a depth of around 200 km (Fig. 5.12) (Hamburger et al., 1992; Burtman and Molnar, 1993; Fan et al., 1994), though such subduction of continental crust down to 200 km is unclear and uncommon (Burtman and Molnar, 1993). Recent receiver function study (Kumar et al., 2005), however, imaged the deeper (~200 km) subducted lithosphere in the region (Fig. 5.6).

Subduction Process

Billington et al. (1977) first described the Pamir-Hindu Kush seismic zone as a single contorted zone; the zone beneath the Pamirs may have become overturned. Roecker et al. (1980) and Chatelain et al. (1980), based on microearthquake surveys, showed the existence of a 50 km gap between the Pamirs and Hindu Kush zones. They suggested that this gap represents two opposing subducted slabs. The most comprehensive study of Pegler and Das (1998) also shows a gap of about 50 km, but they argued that this gap does not suggest two-slab model, rather one contorted subduction zone. The gap in seismicity is interpreted as evidence for a tear in the plate where it has overturned and detached from its shallower continuation.

The Hindu Kush seismic zone bends significantly from an E-W trend in the west to a more NE-SW trend in the east, in an alignment to the Pamir seismic zone. Pegler and Das (1998) argued that if the two zones are of separate origin, one would expect to see a difference in the two trends. The spatial accommodation of the bent slab is possible as the inferred fault extends across and beyond the Hindu Kush seismic zone and forms the inverted V-shaped or forked feature that is observed in the seismic data (Fig. 5.26).

The Hindu Kush seismic zone, if extrapolated to the surface, reaches to the MMT or MKT (Fig. 5.26). It is widely accepted that Hindu Kush seismic zone represents a northward subduction (e.g. Billington et al., 1977; Roecker et al., 1980; Chatelain et al., 1980; Hamberger et al., 1992; Burtman and Molnar, 1993). If we believe subduction of the Asian lithosphere to southeast beneath Pamirs the surface locus of subduction either points to Tien Shan (Hamburger et al., 1992) or to northern margin of the Pamir (Chatelain et al., 1980; Fan et al., 1994) (Fig. 5.12a). The seismicity beneath the northern Pamirs and Tien Shan in section W-W′ shows a zone of shallow seismicity at 0-40 km depth (Fig. 5.27); there is no evidence to suggest that this extends to depths greater than 50 km, and certainly not to 200 km. Pegler and Das (1998) argued that this shallow seismic zone does not link with the intermediate depth seismicity as there is a gap of about 140 km which is aseismic between 40 and 100 km depth. They suggested rather two seismic zones of different origin than a subduction of the Asian lithosphere down to 200 km depth. The intermediate seismic zone does not extend shallower than 90 km, and its top appcars to be marked by an abrupt cut off in seismic activity (Fig. 5.27). In receiver function images (Fig. 5.6d) the lithosphere asthenosphere boundary (LAB) in the Asian plate is well demarcated (Kumar et al., 2005), two seismic zones with a gap are shown, and no clear Benioff zone is observed.

The topography of the region highlights another difference between the Pamir and Hindu Kush seismic zones. The topography is high where the Hindu Kush seismic zone bends sharply at around 100 km depth (Fig. 5.26). The topography decreases to the east and west, it is more gentle, inclines northwards into the Pamirs, and above the Pamir seismic zone the topography is flat (Fig. 5.26). This observation suggests that there is no crustal link between the geometry of the seismic zones and the topographic relief.

Steepening of the Hindu Kush seismic zone from west to east, becoming vertical and then possibly over turning, is explained by a single Benioff zone rather than two opposing Benioff zones (Pegler and Das, 1998). It is argued that the Pamir seismic zone, which dips steeply to the southeast and displays seismicity within the same depth range as the Hindu Kush seismic zone, only 50 km away, as a single feature. Pegler and Das (1998) suggested that the Hindu Kush seismic zone itself overturns to dip to the southeast at its eastern end.

To the west, the Hindu Kush seismic zone is bounded by the sinistral strike ship Darvaz-Karakul and Chaman fault systems (Fig. 5.11). The Karakuram fault marks the eastern limit of the Pamir-Hindu Kush seismic zone. The high level of seismicity beneath Hindu Kush does not continue east of 71.6°E, beneath Karakuram (Fig. 5.18). The dextral movement along the Karakuram fault has probably accommodated the overturning of the north dipping slab, which has torn and overturned beneath the Pamirs.

Pop-up Structure: Karakuram

The relocated seismicity and its cross-section in the Karakuram region is shown in Fig. 5.31. The intermediate depth seismicity (70~220 km) beneath the Pamirs terminates eastwards at around 75.5°E. Some intermediate depth events down to 120 km are, however, recorded to the southeast, beneath the Karakuram (Fig. 5.31). A section KA–KA′ through this region was taken, roughly perpendicular to the Karakuram fault, MKT and the Himalayan thrust. The section defines a triangular wedge bounded by the MKT to the

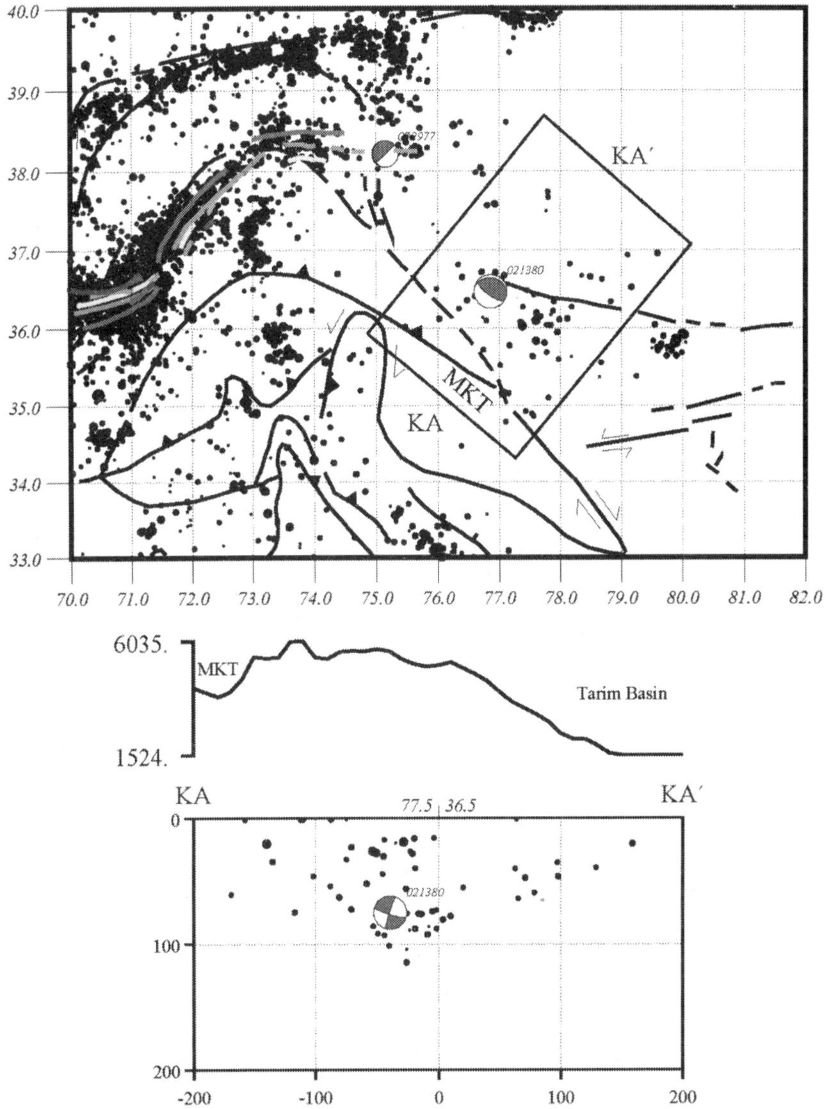

Fig. 5.31 (a) Seismicity in the Karakuram region is shown by a rectangular box, MKT: Main Karakuram Thrust, (b) seismic cross section along KA-KA′ (Pegler and Das, 1998).

southwest and by the southern margin of the Tarim basin to the northeast. It shows a ‘pop-up’ structure for the Karakuram (Pegler and Das, 1998). This observation is comparable with the geological section (Searle and Tirrul, 1991), in which the MKT is interpreted as deep seated break-back thrust, extending down to 100 km beneath the Karakuram. The CMT mechanism of the event 02/13/80 is interpreted as thrusting along the MKT at 77 km depth.

The section KA-KA′ also implies thrusting of the Kunlun over Tarim, by as much as 200 km and down to 100 km depth (Pegler and Das, 1998).

An isostatic gravity profile across the region shows three negative anomalies, which correspond to the foreland basin of the Himalayas, the Karakuram and the Tarim basin (Fig. 5.13). Molnar (1988), based on the gravity data, suggested that the Tarim basin underthrusts the Kunlun, and the thrusting probably extend over 100 km southwest beneath Kunlun. This interpretation is compatible with the seismicity section KA-KA′. Molnar (1988) further suggested down welling of mantle material beneath the Karakuram, which may be a reason for deeper seismicity. Pegler and Das (1998) argued that within the single slab model, the Karakuram marks the region where most down welling of mantle material is expected.

5.8 NORTHWEST AND WESTERN HIMALAYAN SEISMICITY

A regional seismicity map of the northwest and western Himalayas, longitude 76°-80°E, and latitude 29°-36°N for the period 1900-2000, is shown in Fig. 5.32. The epicentres located since 1964 are fairly reliable, which incorporate the global network data as well as the national network data. Earthquakes of magnitude 4.5 and above are shown in the map. A significant aspect of the seismicity is that a belt of high rate of earthquake occurrence follows the arcuate mountain chain, structure and topography. The belt of earthquakes within the zone of MBT and MCT, called the Main Himalayan Seismic Belt (MHSB), mostly lies in the Lesser Himalaya, and is more or less uniform along the entire Himalayan arc except a few very active or cluster zones. The epicentres are close to the surface trace of MCT. These are shallow earthquakes (depth <30 km), and mostly occur above the Plane of Detachment. The activity is much less in the Indo-Gangetic Plains, foredeep region.

To the north of MHSB, the seismicity in the Tethyan zone is wide spread. The IST does not show much seismic activity; there is no record of large earthquake in this zone. Seismicity, surface faulting and topography are more or less uniform over the Tethyan zone, which suggests a uniform tectonic process in this zone.

5.8.1 Damaging Earthquakes in MHSB

Northwestern Himalaya

The 1555 (AD) event was possibly the largest historical event that occurred in the northwestern Himalaya with an intensity XII (MM scale) (Avouac et al., 2006). Since 1800, three large ($M \geq 7.0$) and 25 strong ($6<M<7$) earthquakes occurred in this part of the Himalaya that includes Jammu & Kashmir (J&K), Himachal Pradesh and Kinnur (Fig. 5.32). Isoseismals of four damaging

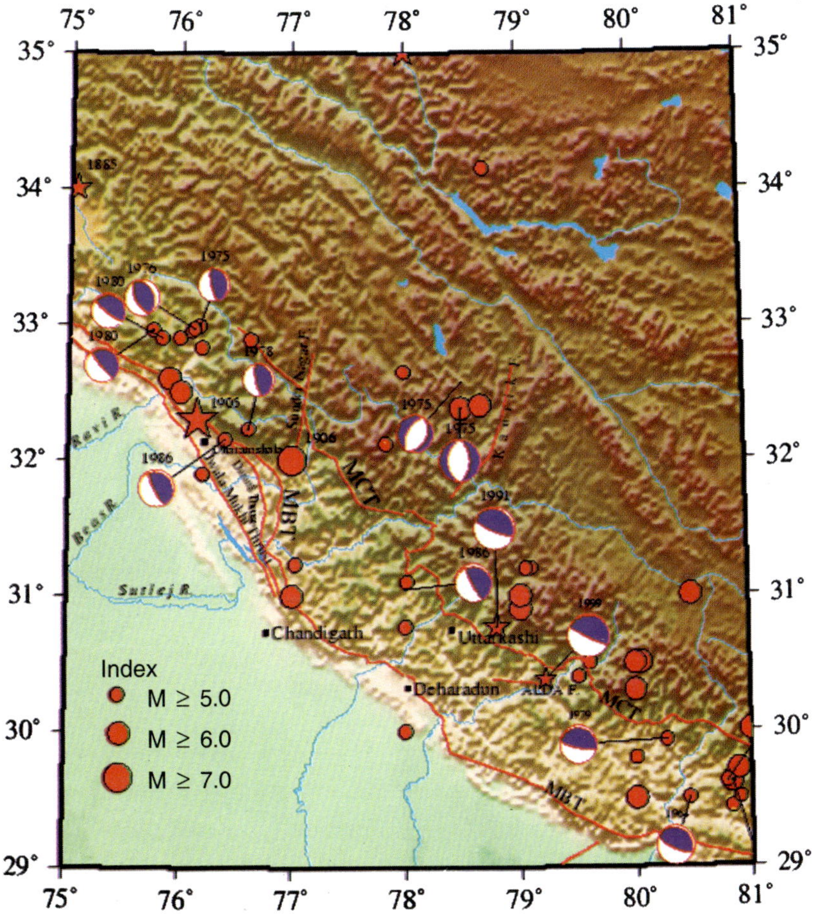

Fig. 5.32 Topographic and seismicity map of the northwest and western Himalaya region. Tectonic features are taken from GSI (2000a); available CMT solutions are also shown. The 1905 great Kangra earthquake and 1991 Uttarkashi and 1999 Chamoli earthquakes are shown by star symbols.

earthquakes of 1885, 1929, 1937 and 1939, M~6.9-7.2, in the northwest Himalaya are shown in Fig. 5.33. Different intensity scales were used to draw the isoseismals. The 1885 Kashmir earthquake, intensity X (M 7.0), took about 3000 lives and caused severe damage around Srinagar.

The other significant or damaging earthquakes that occurred in recent years in the northwest Himalaya are the Chamba earthquake (M 7.0) of June 22, 1945 and the Kinnur earthquake (M 6.2) of January 19, 1975. The Chamba earthquake caused considerable damage, but the 1975 Kinnur event took about 40 human lives and caused collapse of houses in about 20 villages. Maximum intensity reached IX for both the events (GSI, 2000). The N-S Isoseismal trend of the 1975 Kinnur event is comparable with the general seismicity tectonic feature in the area. The event shows a normal fault solution (Fig. 5.32).

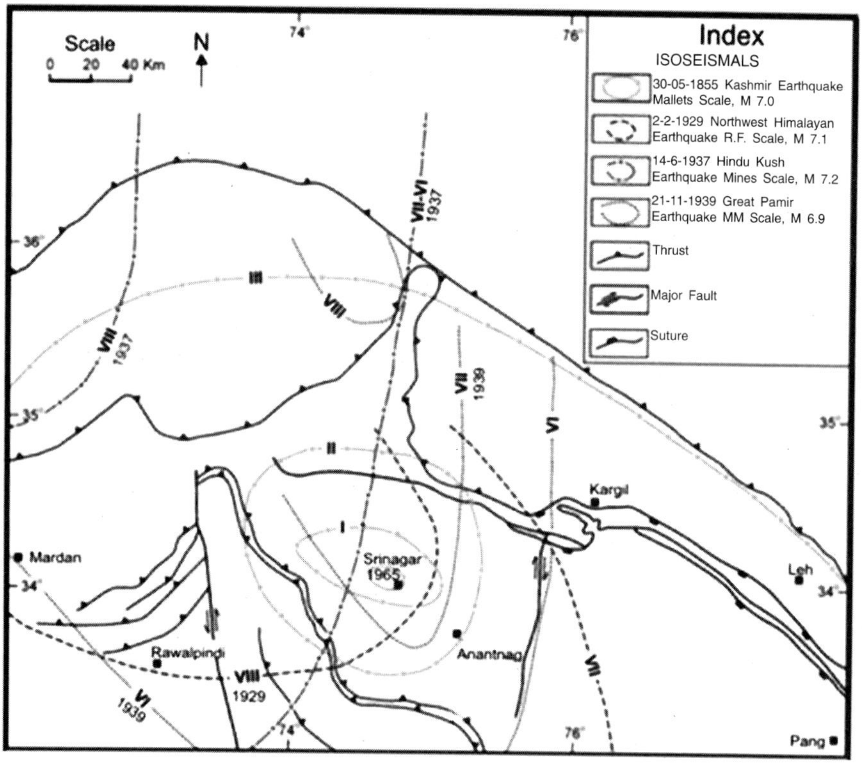

Fig. 5.33 Isoseismals of the four damaging earthquakes in NW Himalaya (GSI, 2000a).

The 1905 great Kangra earthquake M 8.0 (Richter, 1958) occurred in the northwestern Himalaya, Himachal Pradesh. Details of this earthquake is given in Chapter 2 (Section 2.7). The Kangra area continues to show high seismic activity through occurrence of moderate magnitude earthquakes from time to time (Fig. 5.32). In the recent years, three medium magnitude earthquakes of 1968 (M 5.0), 1978 (M 5.4) and 1986 (M 5.7) in this area caused extensive damage near Dharamsala (GSI, unpub. reports). The isoseismals were oriented along the MBT, similar to that of the Kangra earthquake. The great 1905 earthquake occurred to the south of MBT at an estimated depth 25 km. The published fault-plane solutions of the earthquakes in this part of the Himalaya are illustrated in Fig. 5.32. All the solutions in the HSMB show pure thrust faulting with a north to northeast dipping nodal plane comparable with the major thrusts in the area.

The most recent, the October 8, 2005 Kashmir earthquake (M_w 7.6) in the western syntaxis zone caused severe damage and took about 80,000 lives. The earthquake was caused by pure thrust faulting, and the maximum intensity reached X in the MSK scale (Fig. 5.34). The tectonic setting, rupture zone and the epicentre of the 2005 Kashmir earthquake are shown in the map (Fig. 5.34a). The map also shows the rupture zones of the past large earthquakes (1555, 1905, 1413) in northwest and western Himalayas. The aftershocks of the 2005 event show a NW rupture zone (Fig. 5.34b), and the aftershocks mostly occurred by thrust faulting (USGS report). A continuous surface rupture over a distance of ~75 km, that cuts across the Hazara syntaxis reactivating the Tanda and Muzaffarabad faults was reported by Avouac et al. (2006). The rupture terminates at the hairpin turn of the MBT showing a strong structural control on the surface rupture. It may be, however, noted that the aftershock activity was intense further northwest, beyond the abrupt northern termination of the rupture (Fig. 5.34b). Maximum intensity X on the MSK scale was reported by Zare (2005, pers comm.) and by the GSI (GSI 2006, website) (Fig. 5.34c and d).

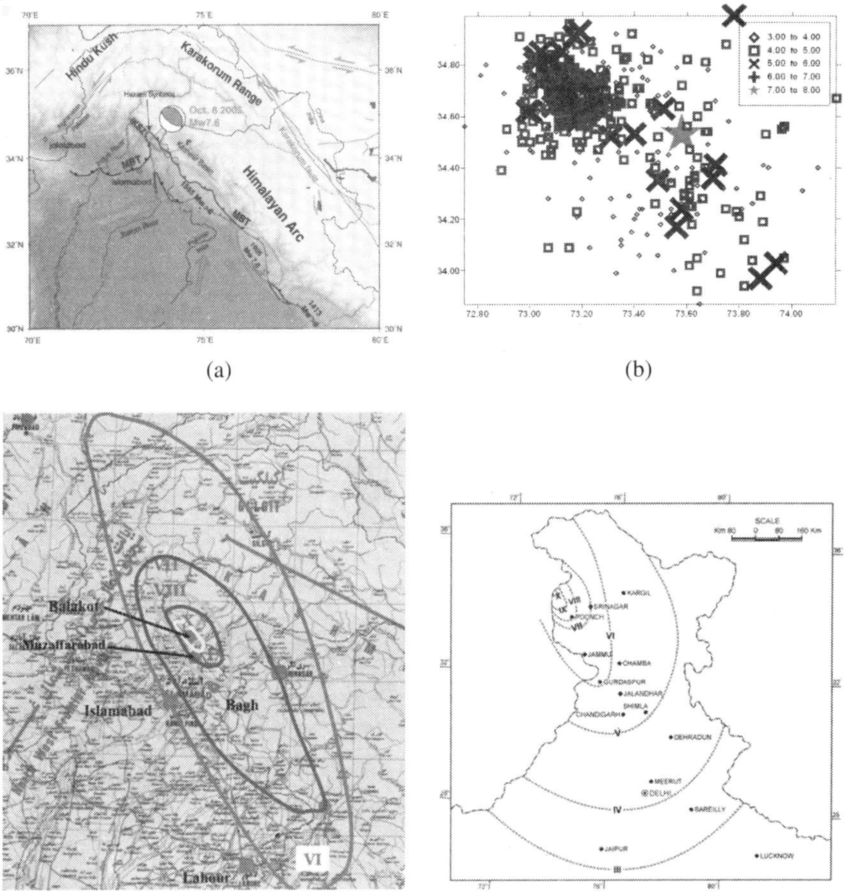

(a) (b)

(c) (d)

Fig. 5.34 (a) Tectonic map and rupture zone of the large earthquakes in the northwest and western Himalayas and epicentre of the October 8, 2005 Kashmir earthquake (Avouac et al., 2006). (b) Aftershocks of the 2005 Kashmir earthquake (USGS report). (c) Isoseismal map of the 2005 Kashmir earthquake prepared by Zare, M. (2005, pers. comm.), and (d) isoseismal map prepared by the GSI (website:http://www.gsi.gov.in).

Western Himalaya

In this part of the Himalaya that includes Garhwal Himalaya and the Kumaun Himalaya, one large earthquake (M>7.0), intensity IX, occurred on September 1, 1803 near Uttarkashi, north of Delhi (Fig. 5.35). Based on historical records, an isoseismal map (Fig. 5.35) is prepared by Rajendran and Rajendran (2005). Most reports mention severe damages in the provinces of Garhwal and Kumaun Himalayas as well as in some locations in the Gangetic Plains (Raper, 1810) (Table 2.3). This earthquake caused severe damages between Delhi and Lucknow. A long sequence of aftershocks was reported. Since 1900 about 12 events M>6.0 occurred in this tectonic block (Fig. 5.32).

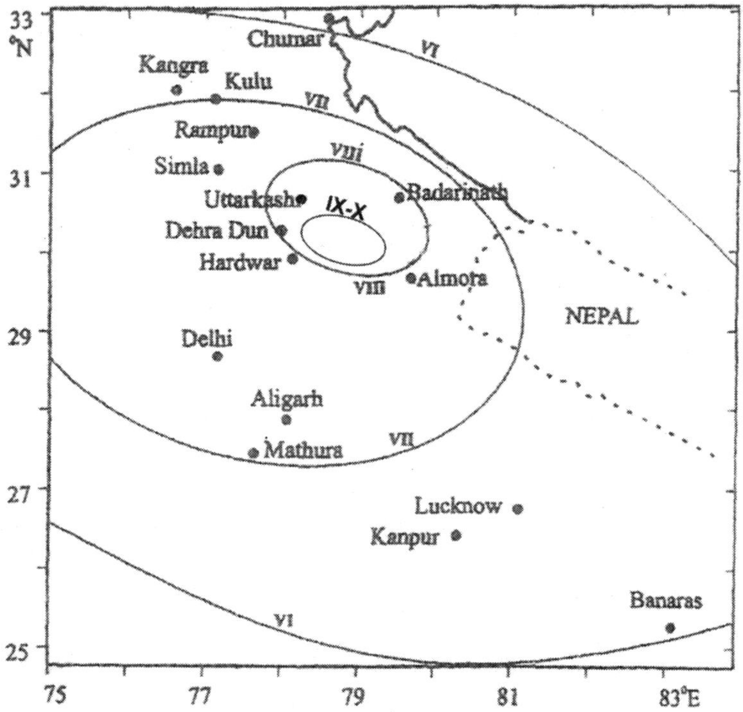

Fig. 5.35 Isoseismal map of the 1803 earthquake (Rajendran and Rajendran, 2005).

The recent two damaging earthquakes, the Uttarkashi earthquake (m_b 6.6) of September 30, 1991 and the Chamoli earthquake (m_b 6.3) of March 28, 1999, are the most well studied earthquakes in the western Himalaya (Fig. 5.32). A brief discussion of these two earthquakes is given below. Temporary networks were deployed for aftershock investigations of these two damaging earthquakes, that provided a better understanding of seismotectonics in this part of the Himalaya.

Uttarkashi Earthquake m_b 6.6, 1991

The Uttarkashi earthquake (m_b 6.6) of October 20, 1991 rocked the Garhwal region of the western Himalaya for about 45 seconds at 02h 55m, Indian Standard Time (IST), and left behind a trail of about 800 deaths, 5000 injured and 100,000 damaged houses with a reported intensity of IX on the MSK scale (GSI, 1992). Numerous landslides occurred in the Uttarkashi and adjoining districts. A detailed report on the macroseismic investigation is given by the GSI (1992), and by Rastogi and Chadha (1995). An isoseismal map is shown in Fig. 5.36. Chandrasekaran and Das (1992) estimated maximum ground acceleration 0.3 g from the analysis of the strong motion accelerograms.

The main shock was located by the USGS (US Geological Survey) at 30° 46′.80 N and 78° 46′.20 E, and centroid depth at 10 km (Fig. 5.32). The IMD estimated the focal depth at 12 km. The intensity IX as well as half of the intensity VIII lies to the south of the surface trace of the MCT; the maximum intensity IX encompasses an area of 20 sq km. The IMD and the USGS locations of the main shock lie outside the meizoseismal area. On the basis of phase data from the close seismograph stations, Rastogi (1995) relocated the main shock within the meizoseismal zone, at 30° 40.20 N and 78° 36.00 E. The relocated epicentre of the main shock seems to be more realistic with the observed aftershock cluster located by the close-spaced

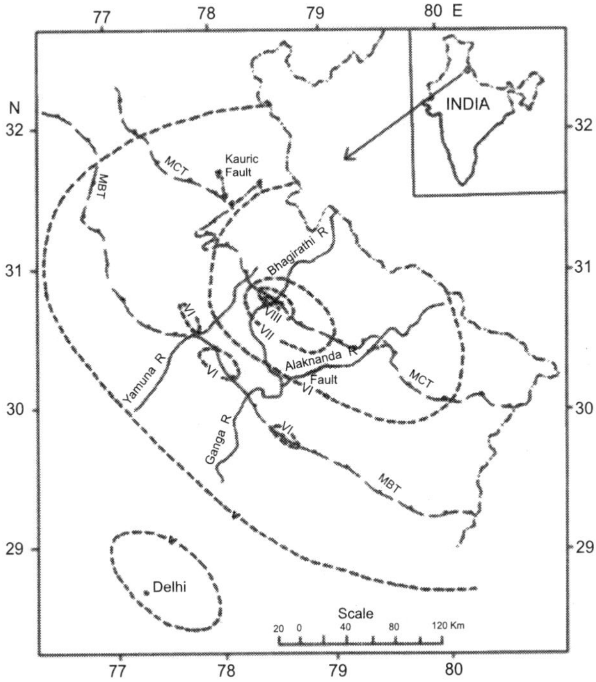

Fig. 5.36 Isoseismal map of the 1991 Uttarkashi earthquake (GSI, 1992).

temporary microearthquake network (Kayal et al., 1995; Kayal, 1996b); it is discussed later (Section 5.8.2).

The main shock was well recorded by a network of accelerometers; six three-component accelerometers within a radius of 60 km of the main shock were triggered. Forward modelling of these records shows that the rupture propagated towards the west (Cotton et al., 1996). Analysis of the teleseismic data, 34 broadband body waves (22 P and 12 S) recorded by the IRIS/GSN, MEDNET and GEOSCOPE, gives a centroid depth 14 km, and seismic moment 1.2×10^{19} Nm (Cotton et al., 1996). Moment tensor solution (USGS) of the main shock shows a thrust faulting (Fig. 5.37). The north dipping nodal plane (strike 295°, dip 5°) is the inferred fault plane. The available fault-plane solutions of the past earthquakes also show similar thrust faulting (Fig. 5.37). Details of the aftershocks M<4.0 recorded by the temporary microearthquake network is discussed later (Section 5.8.2).

Chamoli Earthquake m_b 6.3, 1999

The Chamoli earthquake (M_L 6.8, IMD; m_b 6.3, USGS) of March 28, 1999 (local time: 00h 35m, March 29) occurred in the Chamoli area which falls in the Garhwal Himalaya tectonic zone of the MHSB (Fig. 5.37). The USGS, based on the global network data, immediately reported the epicentre at 30.49° N and 79.29° E and the centroid depth at 12 km. The event was well recorded by the permanent national network including the broadband stations of the IMD and by the permanent local Garhwal network of the WIHG (Wadia Institute of Himalayan Geology). There were 11 permanent seismograph/accelerograph stations in operation in the Garhwal network. These stations provided good quality data for the Chamoli earthquake sequence. The IMD relocated the main shock using a total of 373 seismic phases from the Worldwide and Indian stations, and the epicentre is given at 30.41°N and 79.42°E and focal depth at 21 km IMD (2000). The USGS and the IMD locations of the main shock are shown in Fig. 5.37. The moment tensor solution (USGS) of the main shock revealed a thrust-fault mechanism (Fig. 5.37), similar to that of the Uttarkashi earthquake of 1991 (Fig. 5.37). The IMD also reported analogous fault-plane solution of the main shock using the first-motion data of the global and the national network (IMD, 2000). The north dipping nodal plane (strike 282°, dip 9°N, slip 95°) is the inferred fault plane.

No significant *foreshock* was reported by the IMD (2000). Rastogi (2000), however, argued that a few foreshocks of lower magnitude were recorded by the local seismograph stations in and around the region in the preceding month. It may be mentioned that the GSI made a temporary microearthquake network in the epicentre area about a year before the main shock to monitor seismic activity. A quiescent period of activity was reported during the three-month survey, August-October, 1998; only 12 events M < 3.0 were recorded with S-P interval of 10 seconds (GSI Unpub. Report). A similar quiescent period before the Uttarkashi 1991 earthquake was reported by Kayal (1996b).

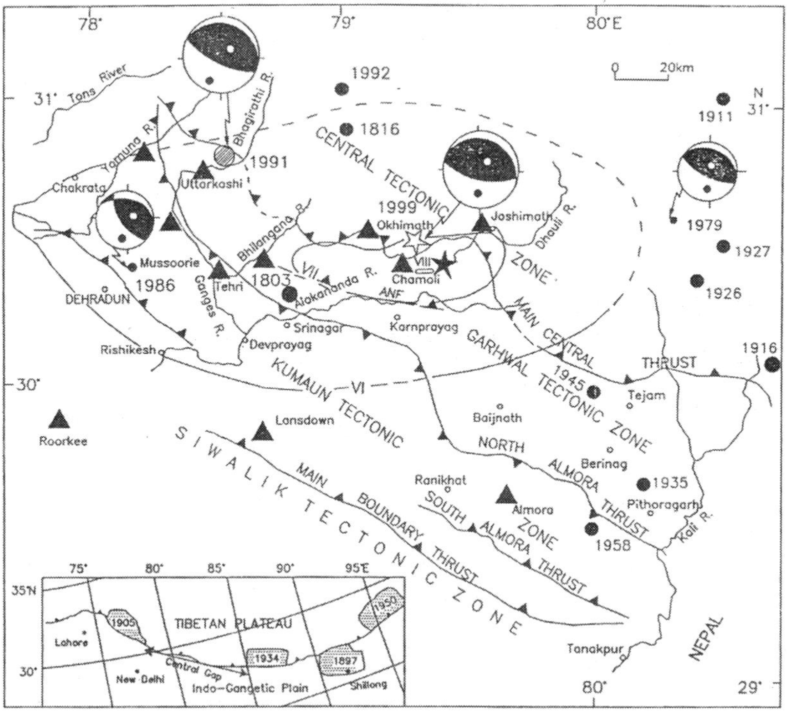

Fig. 5.37 Isoseismals VI and VII of the 1999 Chamoli earthquake and available fault plane solutions in the Garhwal Himalaya; two larger beach balls with usual notations represent solutions of the 1991 and 1999 main shocks and two smaller beach balls represent solutions of the past events (M ~6.0), solid triangles show the permanent seismic network of the WIHG. The USGS location of the 1999 Chamoli earthquake is shown by the open star, and the IMD location by the solid star (Kayal et al., 2003a). Inset: Central gap and rupture zones of the great Himalayan earthquakes (Khattri and Tyagi, 1983).

Macroseismic investigation was carried out by the GSI immediately after the main shock (Ravishankar and Narula, 1999; GSI, 2001). They reported that the maximum intensity VIII (MSK scale) was confined within a small area in the Chamoli town only. The isoseismal VII, however, covers an area of about 90 km × 20 km; it trends E-W (Fig. 5.37). It was further reported that the earthquake caused a loss of 103 human lives, 395 injured, huge damage to about 4500 houses, numerous landslides etc. The major macroseismic effects, like generation of new landslides, reactivation of old landslides, development of cracks in roads, ground fissures, drying up of existing springs and emergence of new springs were reported in the isoseist VII. Within the isoseist VIII, the Chamoli town area suffered heavy damages in the form of collapse of walls of mud houses, and structural damage to brick masonry buildings in the form of extensive shear cracks and partial collapse. Ravishankar and Narula (1999) suggested that the isoseist VII represents the 'epicentral tract' of the main shock. The northern boundary of the tract is delimited by the surface trace of the MCT and the southern boundary by the Alakananda fault (Fig. 5.37). Later on Rastogi (2000) assigned intensity VII (MM scale) to an area 120 km × 70 km, between Tehri and Baijnath (for location see Fig. 5.37), which trends in WNW-ESE direction. Rajendran et al. (2000) reported well developed ground cracks in the E-W direction showing lateral movement up to 12 cm in the Chamoli and adjoining areas. They have also reported that the intensity showed abrupt changes from one location to another implying variation in site amplification (Kayal et al., 2002c; Nath et al., 2002). The earthquake was well felt up to a radial distance of about 500 km; the Delhi area, 280 km southwest of the epicentre, had shown accentuation of intensity VI, where a few buildings sustained non-structural damages. A large number of aftershocks were recorded by a temporary microearthquake network; details of this investigation are given in Section 5.8.2.

5.8.2 MICROEARTHQUAKE NETWORKS

Permanent Network: Northwestern Himalaya

At present northwestern Himalaya is one of the best instrumented regions in the Himalaya. Since 1986, ten permanent observatories with three-component seismometers are running to cater the need of the river valley projects on Beas river. For similar purpose three permanent stations are in operation in Jammu and Kashmir (J&K) state around the Chenab river, Salal dam project. Further, since 1986 five national permanent observatories are established by the IMD; these are at Kalpa (KLP), Dehradun (DDI), Thein Dam (THD), Srinagar (SRN) and at Bhakra (BHK). With the relatively denser network in the northwest Himalaya it is now possible to locate earthquakes of magnitude 2 and above throughout the region and also earthquakes less than magnitude 2 in the area close to the observatories (Chatterjee and Bhattacharya, 1992). A microseismicity map with epicentres located by the permanent network for the period 1986-1989 is shown in Fig. 5.38a. The observed seismic activity is parallel to the trend of the MBT with cluster of events at places. A highest concentration of seismic activity is noted between longitude 76°-77°E and latitude 32°-33°N in the region where the 1905 great Kangra earthquake (M ~8.0) occurred (Richter, 1958). The activity in this small zone is associated with the MBT and with the two N-S tear faults (Fig. 5.38a) (Chatterjee and Bhattacharya, 1992). In a recent study Bhattacharya and Kayal (2005) relocated the earthquakes that were recorded during 1983-1987 and examined a N-S cross section across the Kangra region (Fig. 5.38b). They suggested that the 1905 Kangra earthquake source zone was deeper (30–40 km) at the fault end of the HFF (Fig. 5.38b).

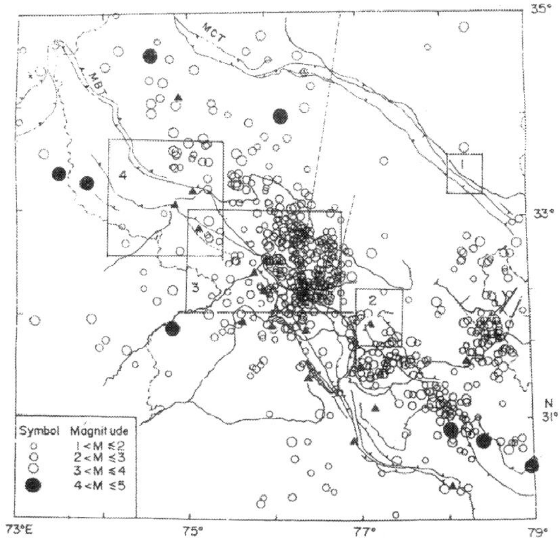

(a)

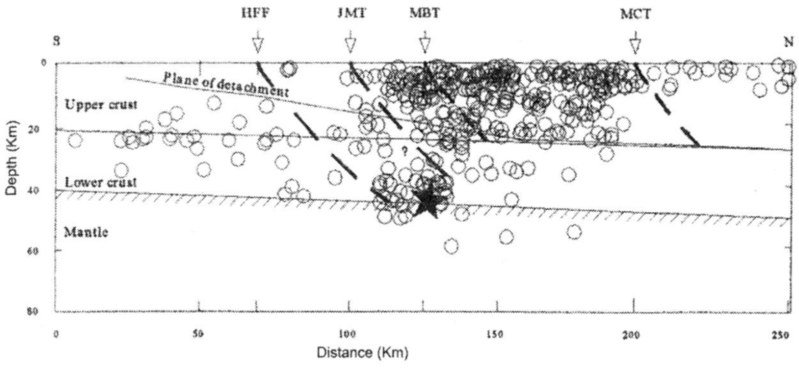

(b)

Fig. 5.38 (a) Microseismicity map of northwest Himalaya for the period 1986-1989, (Chatterjee and Bhattacharya, 1992). Triangles indicate the seismic stations and the broken lines lineaments/tear faults. (b) Seismic cross section of the relocated events in block 3 (Bhattacharya and Kayal, 2005). The star indicates the 1905 great Kangra earthquake. HFF: Himalayan Frontal Fault, JMT: Jwala Mukhi Thrust, MBT: Main Boundary Thrust, MCT: Main Central Thrust.

Temporary Networks: Northwestern Himalaya

During the past two decades much attention was given in obtaining detailed information on seismicity by monitoring microseismicity in selected areas by temporary networks. Monitoring of aftershocks deploying close-spaced temporary microearthquake network has been mandatory, particularly by the GSI and IMD, since early 1990s. The temporary microearthquake network data made a great impact in seismotectonic study in the Himalaya (Kayal, 2001).

Several microearthquake surveys were carried out by the IMD in the Jammu and Kashmir (J&K) state and also in the Himachal Pradesh in the northwestern Himalaya. Five to eight-station temporary networks were in operation in 1978 and 1979 for geothermal exploration in the Puga valley and Parbati valley, Himachal Pradesh (IMD, 1980). Six to thirteen-station microearthquake networks, e.g. Thein dam network and Salal dam network, were in operation for couple of years in 1980s to cater the need of the river valley projects in different parts of the northwestern Himalaya (IMD, 1988). The microearthquakes recorded by these networks are briefly discussed below.

Puga Valley Network, 1978

A microearthquake survey was conducted by the IMD for 54 days during July-August, 1978 by a five-station analog microearthquake network in the geothermal area of Puga valley, Himachal Pradesh (Chatterjee and Bhattacharya, 1992). There are several thermal springs in the area, close to the IST zone. At Puga the hottest spring shows temperature of the order of 85°C, and at Chumathang the maximum temperature is around 87°C. About 150 microearthquakes were located by this survey (Fig. 5.39). Two clusters of events are evident, one to the south of Chumathang, and the other between Puga and south Puga. Focal depths of these microearthquakes are mostly within 10 km (Fig. 5.39); the earthquakes are attributed to the IST zone (Chatterjee and Bhattacharya, 1992).

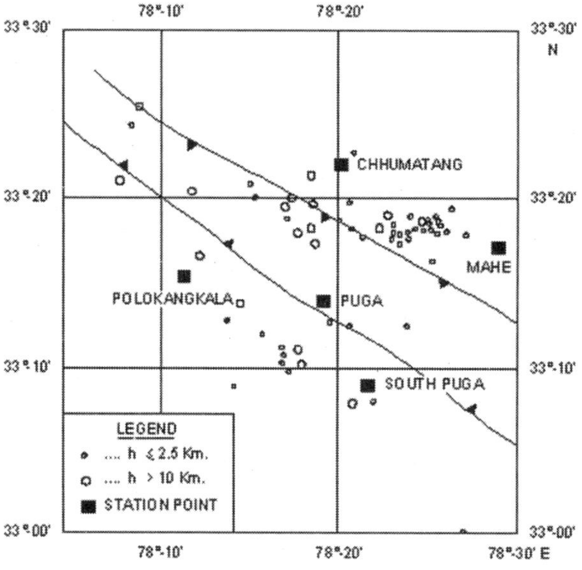

Fig. 5.39 Microearthquakes in the Puga Valley, July-August, 1978 (Chatterjee and Bhattacharya, 1992).

Parbati Valley Network 1979

A eight-station microearthquake network was established by the IMD to monitor seismicity in the Parbati valley, Himachal Pradesh, for geothermal exploration. About 22 geothermal springs are present in the area; the maximum temperature recorded in a spring at Manikaran is of the order of 96°C. The network was run for 67 days during February-April, 1979. 120 microearthquakes were located by this survey (Chatterjee and Bhattacharya, 1992). The events originated at a depth level less than 20 km. The epicentre map shows that the earthquakes are aligned in NW-SE to the south of MCT (Fig. 5.40). Composite fault plane solutions of two clusters show thrust and normal faulting. An isolated NNE-SSW linear trend of activity is observed to the north of MCT. These are shallow events, and may be caused by a tear fault transverse to the MCT trend (Chatterjee and Bhattacharya, 1992).

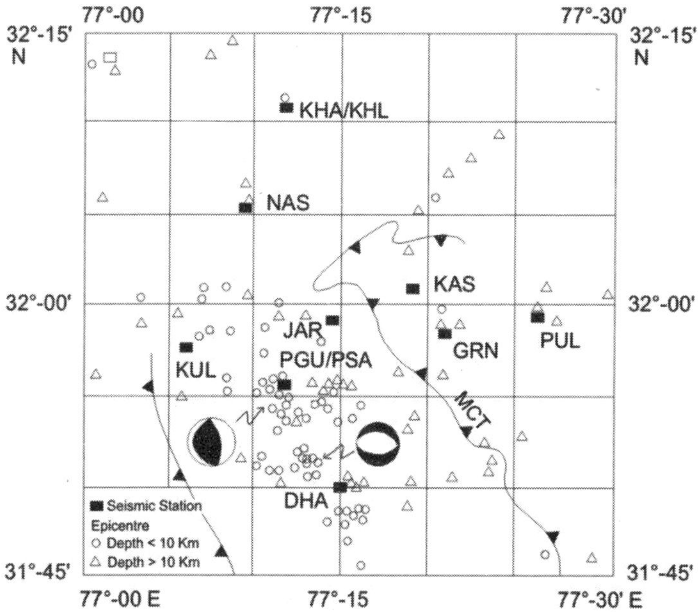

Fig. 5.40 Microseismicity in Parbati valley, February-April, 1979 (Chatterjee and Bhattacharya, 1992).

Thein Dam Network 1983

A microearthquake survey with a seven-station network was conducted around the Thein dam during March-April, 1983, for 34 days (Bhattacharya et al., 1986). 84 microearthquakes were located by this survey; four clusters of events were identified (Fig. 5.41). The earthquakes are of shallow (depth

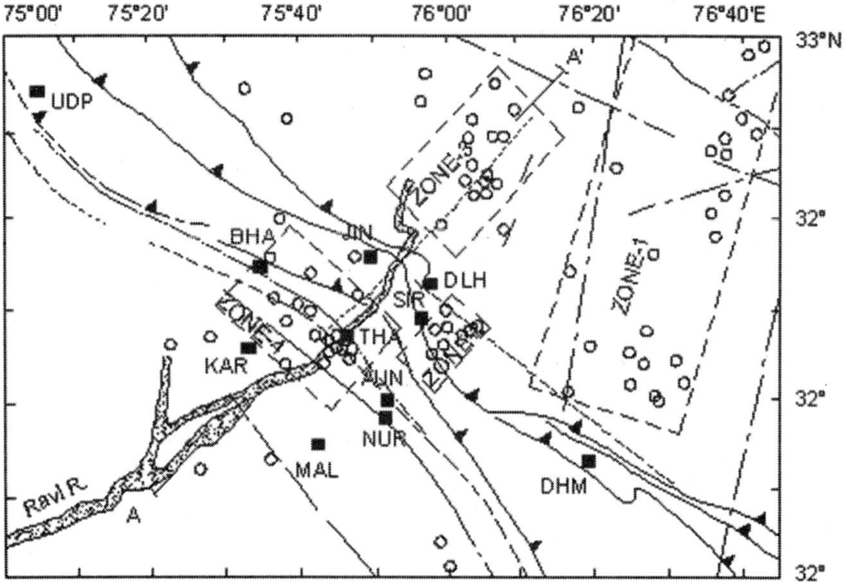

Fig. 5.41 Tectonic features and microseismicity in the Thein dam area, March – April, 1983; open circles represent the epicentres of the earthquakes (M 1.0-3.5), and the solid rectangles the temporary seismograph stations (Chatterjee and Bhattacharya, 1992).

<20 km) origin. The cluster 1 lies up thrust of the MBT and bounded by two lineaments. A N-S tear fault is inferred to be generating the microearthquakes. The cluster 2 lies within the thrust zone, and shallower earthquakes (depth <10 km) are associated with the MBT. The cluster 3 lies up thrust of the MBT and close to the cluster 1. The cluster 4 lies in the Thein dam area. The shallower earthquakes (depth ≤10 km) in the cluster 4 are inferred to the Suruin anticline, whereas deeper (depth ≥ 10 km) events to the Satlitta Thrust at the MBT zone.

Salal Dam Network 1987

A seven-station microearthquake network was operated for 38 days during August-September, 1987 in the Salal dam area, and 19 microearthquakes were located (Fig. 5.42). The microearthquakes between the Raisi Thrust in the MBT zone and the Panjal Thrust in the MCT zone are of shallow origin (depth ≤5 km), and aligned in N-S direction. A composite fault-plane solution shows normal faulting (Fig. 5.42). The NNW dipping nodal plane is the inferred fault plane (Chatterjee and Bhattacharya, 1992).

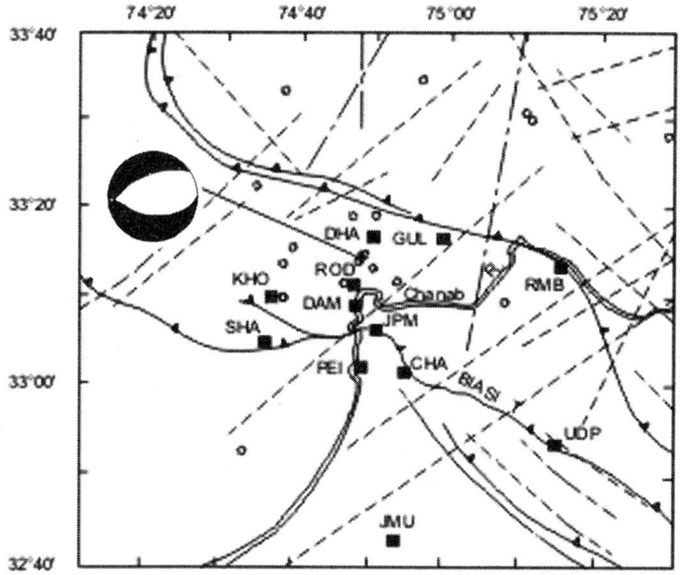

Fig. 5.42 Tectonic features and microseismicity in the Salal dam area, August–September, 1987; other symbols are same as Fig. 5.41 (Chatterjee and Bhattacharya, 1992).

Permanent Network: Western Himalaya

Garhwal Network 1979-92

A local seismological network using five to eight analog stations had been in operation intermittently during 1979-1988 in the Garhwal-Kumaun Himalaya. The network was run by the Department of Earth Sciences, University of Roorkee (presently Indian Institute of Technology, Roorkee, IIT(R)). The seismicity defined by the local network is shown in Fig. 5.43a. The epicentres define a NW-SE trending zone, about 20 km wide (Khattri, 1992). This zone is straddled by the MCT, which is sinusoidal in nature in this part of the region. A NE-SW cross section of the earthquakes indicates that all the earthquakes occurred above the Plane of Detachment, and most of them are of shallower depth <20 km (Fig. 5.43b). Two composite fault-plane solutions are determined; one shows strike-slip and the other shows reverse faulting (Fig. 5.43a). The maximum compressive-stress axis trends in NE direction in both the cases, which is compatible with the regional tectonic stress direction.

The Garhwal network was then redesigned with an eight-station telemetric system by the Department of Earthquake Engineering, IIT(R), and it was operated intermittently from November 1988 to November 1992 (Kumar et al., 1997). The telemetric system did not run smoothly due to various technical problems.

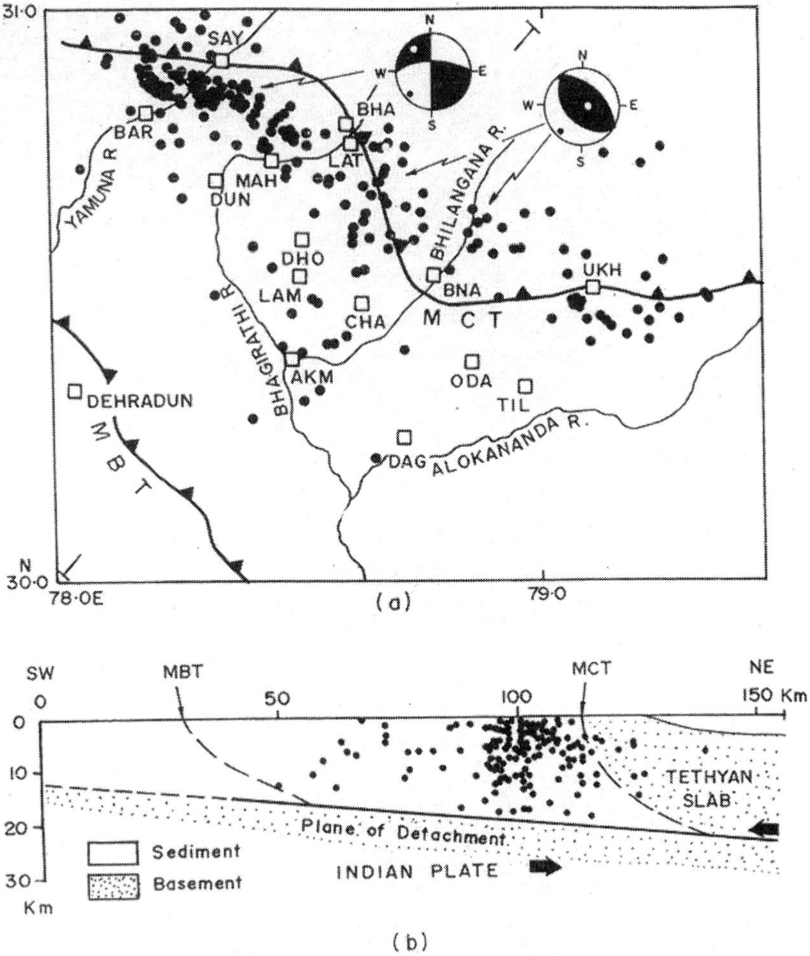

Fig. 5.43 (a) Microseismicity in the western (Garhwal-Kumaun) Himalaya area recorded during 1979-1984, composite fault plane solutions of two clusters are shown by beach ball representations; seismic stations are shown by small rectangles. (b) A NE-SW seismic cross section indicates that the earthquakes occurred above the plane of detachment (Khattri, 1992).

Tehri Dam Network 1993-1999

In recent years a detailed investigation of seismicity in the Garhwal Himalaya has been considered by the Tehri Hydro Development Corporation (THDC) around the Tehri dam area by a permanent network. The THDC has given this job to the Department of Earthquake Engineering, IIT(R). A seven-station analog telemetric network is established with a central recording station at Tehri. The network is in continuous operation since September 1993. Kumar et al. (1997 and 1998) reported that about 500 local events

M≥2.0, S-P≤25s, were recorded in three years, January 1994 to December 1996. 110 events were located; majority of the epicentres follow the trend of MCT.

Two *earthquake swarms* were observed during the reported period (Kumar et al., 1998). The first swarm recorded about 150 events within 45 hours on October 22-24, 1995 with a largest magnitude 3.6, and the second swarm recorded about 290 events on September 13-18, 1996 with a largest magnitude 4.6. Both the swarms were recorded in the Uttarkashi area (Fig. 5.44a). The cross sections of the earthquakes show two distinct zones of activity; one at shallower depth (0-3 km) and the other at deeper depth (6-17 km); all the events occurred above the Plane of Detachment (Fig. 5.44b). No fault plane solution was reported by the authors.

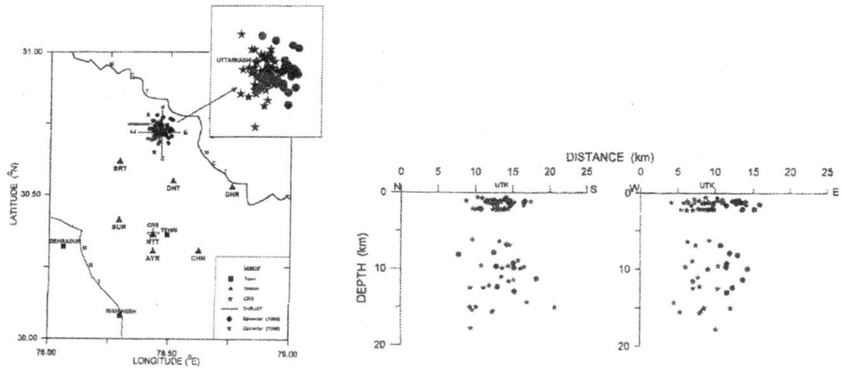

(a) (b)

Fig. 5.44 (a) Swarm activity in Garhwal Himalaya during October 1995 and September 1996 recorded by the permanent network; solid triangles show the seismic stations and the star indicates the central recording station (CRS). (b) N-S and E-W seismic cross sections; circles indicate activity in October 1995 and small stars indicate activity in September 1996 (Kumar et al., 1998).

Kumaun Himalaya Network

A digital telemetric network was established by the Kumaun University in 1999 in Kumaun Himalaya under a DST sponsored project, and the microseismicity of the area was reported by Paul and Pant (2005). The network consisted of five short period digital seismic stations, four remote stations (Almora, Dhaulchhina, Kalakhet and Kausani) and one central recording station at Nainital. About 2000 events were recorded by the network during May, 1999 – April, 2004. About 300 earthquakes were located with S-P<20s. It was reported that the activity is mostly confined between the MBT and MCT in the MHSB. The authors did not present any depth section or fault plane-solution. It is believed that the events are shallow and occurred above the Plane of Detachment.

Temporary Networks: Western Garhwal Himalaya

Uttarkashi Aftershock Network 1991

A temporary five-station microearthquake network was established by the Geological Survey of India (GSI, 1992) to monitor the aftershocks of the Uttarkashi earthquake (m_b 6.6) of October 20, 1991. 124 aftershocks, magnitude 1.0 and above, were located (Fig. 5.45a). A detailed report was given by Kayal et al. (1995). Aftershocks were clustered in two depth ranges. The shallower aftershocks (depth < 5.0 km) showed two elliptical clusters and the deeper aftershocks (depth: 5-15 km) showed one elliptical cluster.

Migration of the aftershock-cluster from east to west with time was observed. Kayal (1996a) suggested that the migration of the aftershocks from east to west or to northwest indicated the rupture direction at depth. The macroseismic investigation also suggested that the source feature was aligned in SE-NW (GSI, 1992). Cotton et al. (1996), by forward modelling of the accelerograms, confirmed the rupture migration to the northwest.

The aftershock characteristics, attenuation and the frequency-magnitude relations, were reported by Kayal et al. (1995). The aftershocks decayed following the modified Omori law: $n(t) \alpha t^{-P}$, with $p = 0.87$, which is slightly low compared to the normal value 1.0 at an active area. The frequency-magnitude relation, the b-value, of the aftershock sequence (M_d 2.0-4.5) was estimated to be 1.23; it is higher than the normal value of 1.0.

A well constrained composite fault plane solution of the aftershocks at a depth range 5-15 km was obtained (Kayal, 1996b). The solution is comparable with the main shock solution; it shows a thrust faulting with strike-slip component (Fig. 5.45a). A near N-S directed compressional stress is dominant. The E–W oriented nodal plane, dipping at 32° to the north, is the inferred fault plane.

A northeast-southwest cross section of the earthquakes shows that all the aftershocks occurred on both sides of the activated fault, hanging wall and foot wall of the Uttarkashi fault, to the south of MCT (Fig. 5.45b). One foreshock F3 and one aftershock A3 (M > 4.0) were, however, located to the north of MCT at shallower depth. It may be mentioned that the foreshocks (F1-F2) and the larger aftershocks (A1-A4), M≥4.0, were located by the IMD (1992) based on large spaced national network data that were recorded before the temporary network was established. It may be mentioned that the 1991 Uttarkashi earthquake occurred much before the upgradation of the national network with broadband seismic stations in 1996. So, locations of the events F_1-F_3 and A_1-A_4 are not much precise. The macroseismic investigation (GSI, 1992) as well as the aftershock investigation by the temporary microearthquake network (Kayal et al., 1995; Kayal, 1996b), however, clearly indicates that the main shock occurred to the south of MCT (Fig. 5.45a and b) as located by Rastogi (1995) using the local seismograph stations. Kayal (1996b) further suggested that the main shock occurred at the

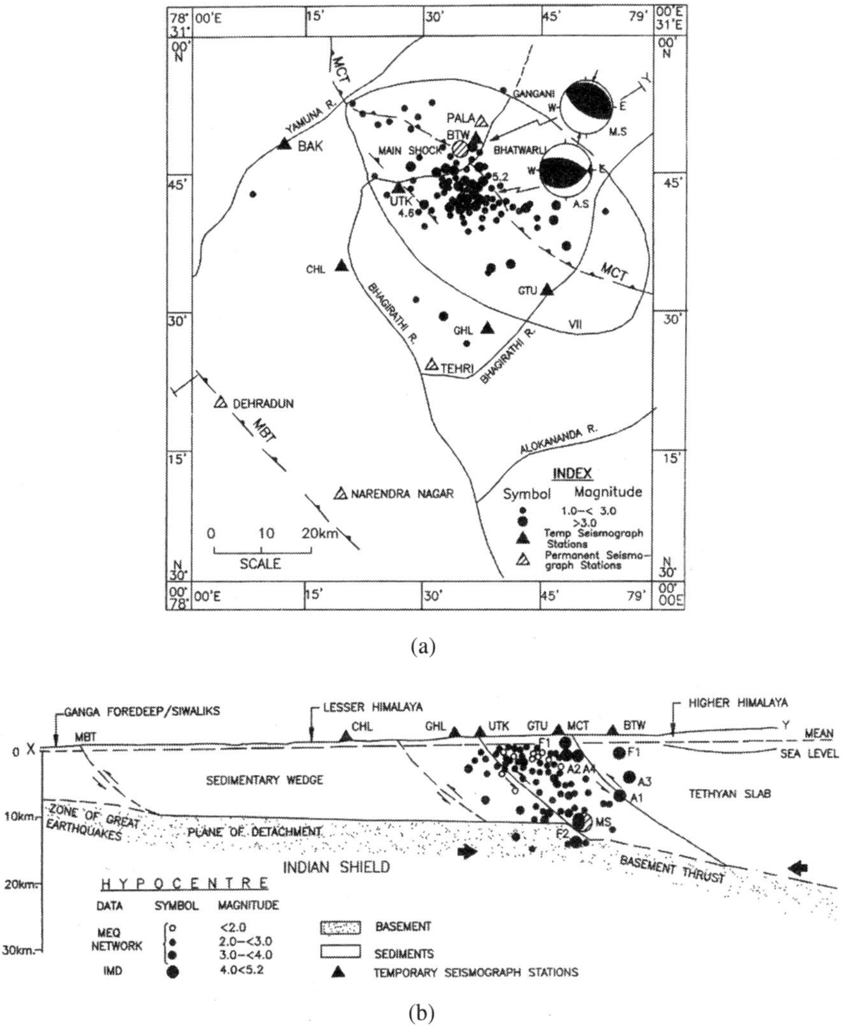

Fig. 5.45 (a) Epicentres of aftershocks of the 1991 Uttarkashi earthquake, isoseismal VII is shown, fault plane solutions of the main shock (MS) and aftershocks (AS) are represented by beach balls. (b) Seismic cross section and seismotectonic model along X-Y is shown (Kayal et al., 1996b).

'fault end' where the Uttarkashi fault meets the Plane of Detachment (Fig. 5.45b), and the aftershocks occurred by shear adjustment at the Uttarkashi fault. It is also observed that all the aftershocks occurred above the Plane of Detachment (Fig. 5.45b).

Chamoli Aftershock Network 1999

The IMD national network, particularly the broadband stations, as well as the Garhwal permanent network of the Wadia Institute of Himalaya Geology

(WIHG) provided continuous monitoring of the aftershocks ($M \geq 2.5$) of the Chamoli earthquake (m_b 6.3) of March 28, 1999 (Fig. 5.46). In addition to this, 26 temporary field stations (analog + digital) were established by various organisations like the GSI, IMD, NGRI and WIHG in the Chamoli area for the aftershock investigation. More than one thousand events, including ultra-microearthquakes ($M<1.0$) were recorded during April 9–June 29, 1999 by the close-spaced temporary seismic stations. These data were collated by the IMD (2000). About 300 aftershocks $M\geq1.0$ were located (Fig. 5.47a). The aftershock attenuation rate was estimated to be p = 0.84 and the frequency-magnitude relation b = 0.61 (Kayal et al., 2003b).

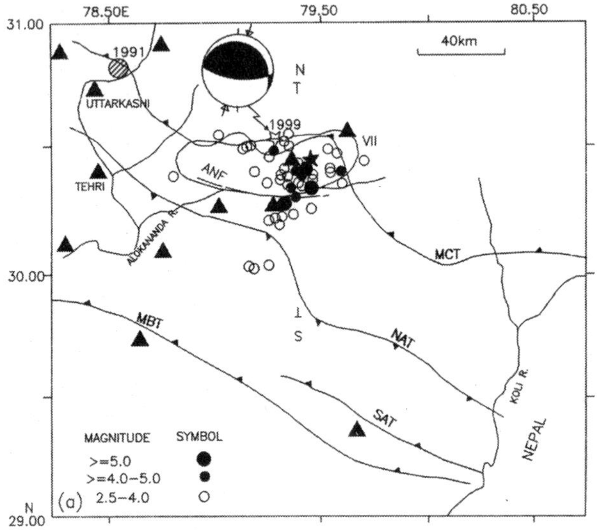

(a)

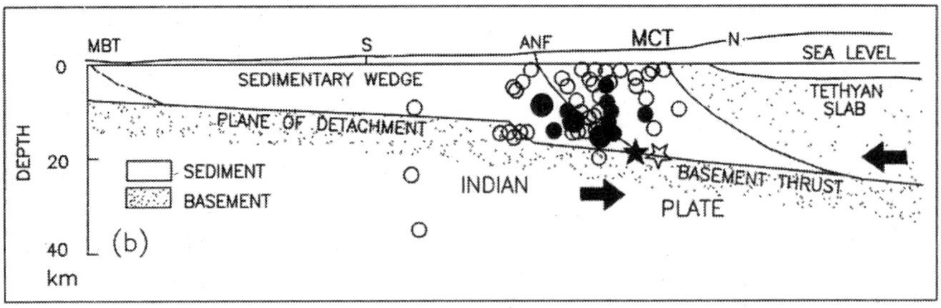

(b)

Fig. 5.46 (a) Epicentres of aftershocks of the 1999 Chamoli earthquake recorded during March 28-April 10, 1999 by the permanent network, fault plane solution of the main shock and the isoseismal VII are shown; permanent seismic stations are shown by solid triangles, main shock epicentre is shown by stars, open located by the IMD and solid by the USGS. (b) N-S seismic cross section and seismotectonic model (Kayal et al., 2003a).

N-S cross-sections of the main shock and the aftershocks are shown in Figs 5.46b and 5.47b. It shows that the main shock was generated at the BTF (Basement Thrust Front) where the Alakananda fault (ANF) ends, and the aftershocks occurred by shear adjustment at the ANF. All the aftershocks occurred above the Plane of Detachment, most of them are at a depth range 11-17 km. Two composite fault-plane solutions of the aftershocks were reported (Kayal et al., 2003a) (Fig. 5.47a). One cluster, that occurred to the

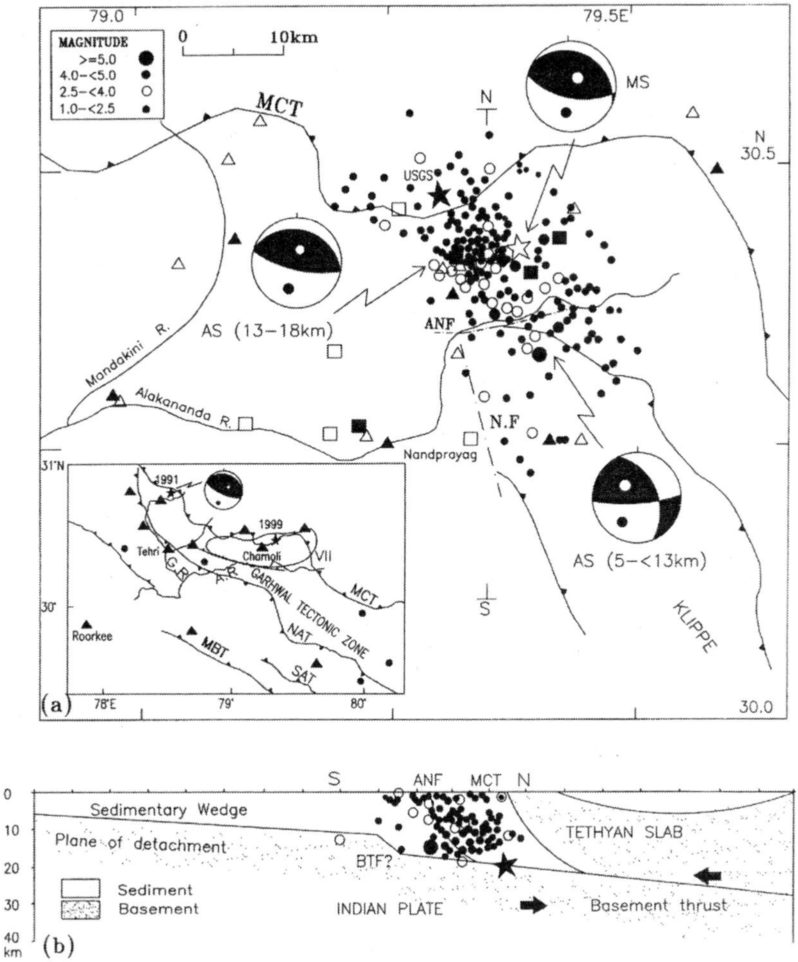

Fig. 5.47 (a) Aftershock epicentres of the 1999 Chamoli earthquake recorded by temporary networks during April 10–June 29, 1999; rectangles indicate the permanent seismic stations and triangles the temporary stations. Fault plane solutions of the main shock (MS) and two clusters of aftershocks (AS) are shown. Inset shows major tectonic features, isoseismal VII of the 1999 Chamoli earthquake and fault plane solution of the 1991 Uttarkashi earthquake, other explanations same as Fig 5.46. (b) N-S seismic cross section and seismotectonic model (Kayal et al., 2003a).

north of ANF, shows thrust faulting and is comparable with the main shock solution. The other cluster shows normal faulting for a group of earthquakes that occurred south of the ANF. The ANF is the inferred fault for the thrust-fault solution, and the N-S trending Nandprayag lineament to the south of ANF is the inferred fault for the normal-fault solution (Kayal et al., 2003a).

Seismic tomography of the main shock source area was studied by Mukhopadhyay and Kayal (2003) using the Local Earthquake Tomography (LET) method. The seismic tomography structures revealed in this study are very interesting, and shed more light to understand the source area. A prominent near-vertical east-west trending LVZ is evident from surface to a depth of 15 km and is comparable with the surface trace of the ANF and the MCT (Fig. 5.48). The LVZ is quite broad at shallower depth (0-10 km), but it narrows down at deeper depth (10-15 km). At shallower depth the LVZ possibly reflects both the MCT and ANF. At deeper depth (15 km) the LVZ reflects the ANF. In the eastern side, this LVZ bifurcates, and a HVZ is seen between the two arms of the LVZ at 10 km depth. The HVZ coincides with an anticlinal structure that is reported on the surface (Valdiya, 1980). The main shock and most of the well-located aftershocks occurred below this anticlinal structure (Fig. 5.48).

In the north-south section, the ANF and the MCT are well reflected in the Vp and Vs images, and are separated by the HVZ (Fig. 5.49). The images indicate that the ANF merges with the Plane of Detachment, whereas the MCT ended at shallower depth. The main shock and most of the well located aftershocks occurred in the high Vp, low Vs and high Vp/Vs zone at the end of the ANF. The MCT seems to have no role for the occurrence of the Chamoli earthquake sequence; it looks to be a dormant feature as suggested by Ni and Barazangi (1984). A similar observation was made for the Uttarkashi earthquake sequence (Fig. 5.45b).

The higher Vp/Vs indicates higher Poisson's ratio, which means the source area may be fluid-filled fractured rock-matrix (Fig. 5.49). Such anomalous Vp, Vs and Vp/Vs structures are prominent in the earthquake source areas (e.g. Zhao and Kanamori, 1995; Kayal et al., 2002b; Zhao et al., 2004). It was suggested that fluid-filled fractured rock-matrix in the source area accentuated the nucleation of the main shock. The heterogeneities in the velocity/geological structure caused stress concentration, and the main shock occurred by thrust faulting on the basement thrust. The main shock triggered the ANF and other conjugate fault(s) to generate the aftershocks in the higher Vp/Vs zone.

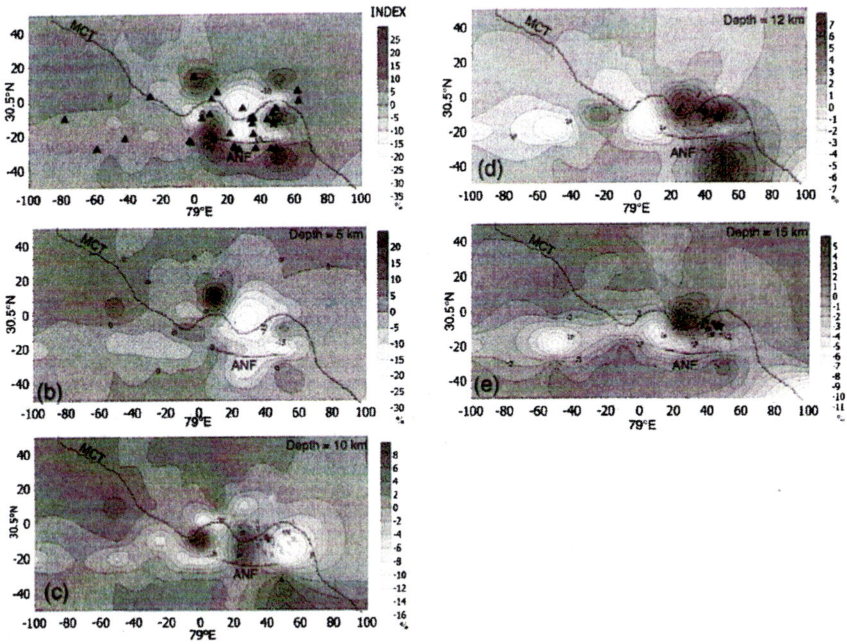

Fig. 5.48 P-wave velocity (Vp) variation at the Chamoli earthquake source area; aftershocks, main shock and seismic stations are shown (Mukhopadhyay and Kayal, 2003).

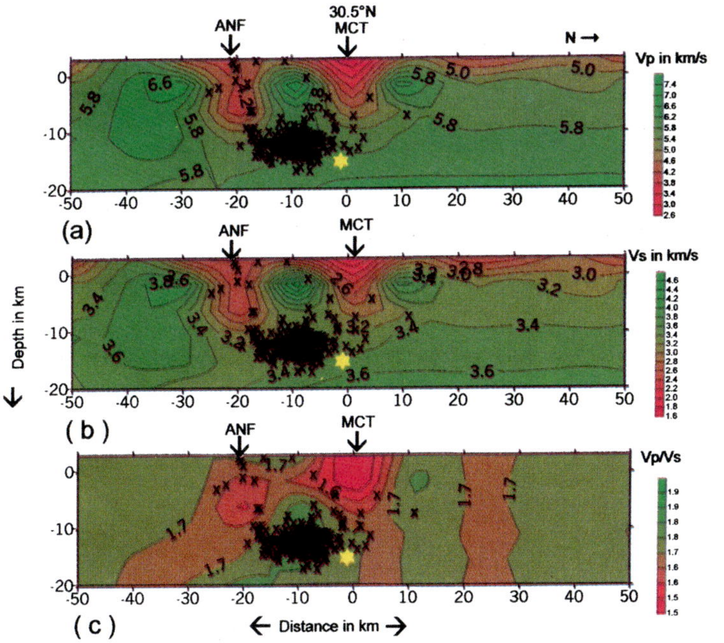

Fig. 5.49 North-south section of the seismic images (Vp, Vs and Vp/Vs) of the 1999 Chamoli earthquake source area (Mukhopadhyaya and Kayal, 2003)

5.9 CENTRAL AND EASTERN HIMALAYAN SEISMICITY

5.9.1 Seismicity Maps

Epicentre map of the earthquakes M ≥ 5.0 in the central Himalaya, longitude 80-88°E, is shown in Fig. 5.50. In central Himalaya (Nepal), most of the events are confined within the MHSB, the area between the MBT and MCT. In the northern Tethyan belt, the activity is much less. The southern foredeep region experienced two devastating earthquakes, the 1934 (M 8.4) and the 1988 (Ms 6.6) events. The available fault plane solutions of the events in the MHSB in the east Nepal Himalaya mostly show dominant strike-slip solutions, and one nodal plane is comparable with the transverse structures/lineaments in the area. To the west Nepal Himalaya, the earthquakes in the MHSB are of low angle thrust faulting (Fig. 5.50).

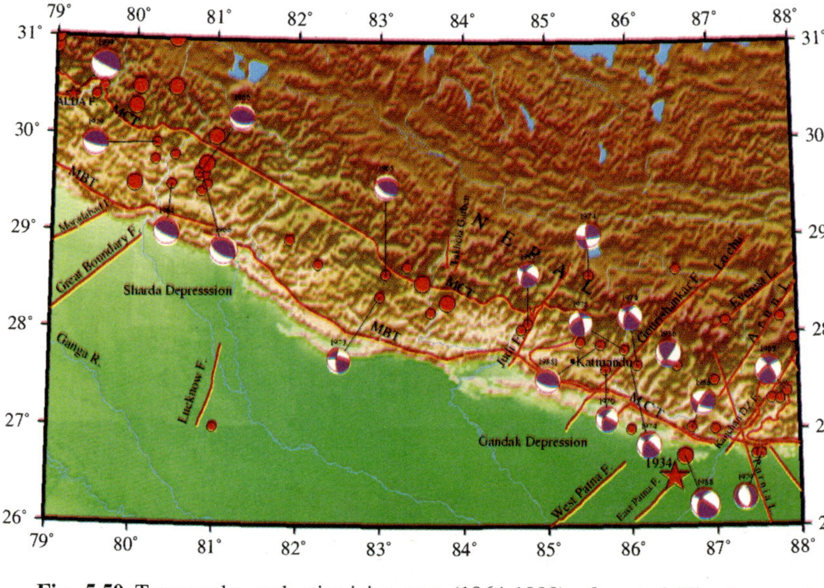

Fig. 5.50 Topography and seismicity map (1964-1998) of central Himalaya and the available CMT solutions; tectonic features are from GSI (2000). Earthquake magnitude (5.0-7.0) is scaled with the size of the circles, the solid star indicates the 1833 and the 1934 earthquakes.

In the eastern Himalaya (east Nepal, Sikkim and Bhutan), on the other hand, activity is more prominent to the north of MCT (Fig. 5.51). In this part of the Himalaya the MCT and MBT run parallel but very close to each other except in the Sikkim Himalaya where the MCT has taken loop-like shape. The available fault plane solutions mostly show dominant strike-slip solutions, and are comparable with the known transverse lineaments/faults(De and Kayal, 2003 and 2004).

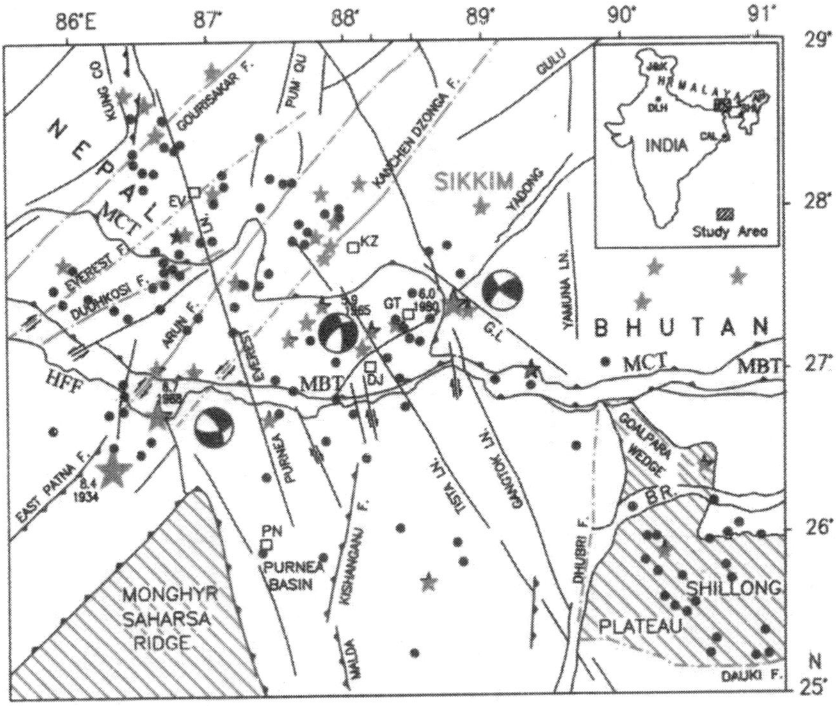

Fig. 5.51 Seismotectonic map of eastern (Darjeeling, Sikkim and Bhutan) Himalaya and its foredeep; stars indicate earthquakes of M ~5.0 and above (De and Kayal, 2003).

Large Earthquakes ≥ 7.0

Nepal Earthquake M> 7.5, 1833

Before the WWSSN was established in 1964, two large (Io $\simeq$ X) earthquakes occurred in this part of the Himalaya. The earliest of these is the Nepal earthquake of August 26, 1833. The epicentre was near Kathmandu (Fig. 5.50). The estimated magnitude, based on the reported intensity, was 7.5<M<7 (GSI, 2000a). Isoseismal of this event is shown in Fig. 5.52. The maximum intensity reached ~X (Rossi Forel scale). The earthquake caused landslides, rock falls and huge damage of about 4600 houses. It took about 500 human lives and affected around one million sq. km (GSI, 2000a). In a recent study a 200-300 km long rupture along eastern Nepal was evaluated (Bilham et al., 2001).

Dharchula Earthquake M_s 7.5, 1916

The August 28, 1916 earthquake, M_s 7.5 (Table 2.3) is one of the largest earthquakes in the west Nepal. This earthquake caused heavy damage to the civil structures in Dharchula area.

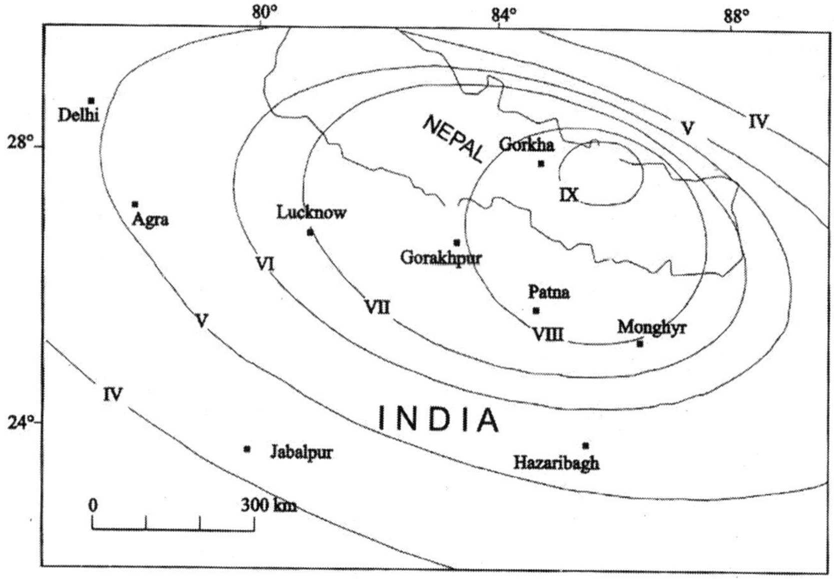

Fig. 5.52 Isoseismal map of the 1833 Nepal earthquake (GSI, 2000a).

Bihar Earthquake M 8.4, 1934

The January 12, 1934 great Bihar earthquake M 8.4 (Richter, 1958) occurred in the foothills of the eastern Himalaya (Fig. 5.51). Based on the national and global seismic station data, the epicentre was estimated at 26.50°N and 86.50°E and depth at 20-30 km. Details of the 1934 great earthquake are given in the Section 2.7.

Recent Strong Earthquakes (M≥6.5)

During the last two decades two strong or damaging earthquakes M≥6.5 occurred in this region. On July 29, 1980 an earthquake M_s 6.5 occurred in the Nepal Himalaya, and the other earthquake M_s 6.6 occurred on August 20, 1988 in the Bihar-Nepal border region, foothills of the eastern Himalaya (Fig. 5.51). The location of the 1988 earthquake is close to the epicentre of the 1934 great earthquake. This earthquake was studied in detail (GSI, 1993); a brief description is given below.

Bihar-Nepal Earthquake (M_s 6.6), 1988

The epicentre of the August 20, 1988 earthquake (M_s 6.6) is located at latitude 26.72°N and longitude 86.63°E, Bihar-Nepal border (Fig. 5.51). Various organisations (USGS, ISC, ERI) estimated focal depths at 51-78 km. This indicates that the earthquake is deep focus below the Gangetic foredeep, within the continental lithosphere. Isoseismal of this event is given in

Fig. 5.53. It took about 1000 human lives and destroyed or badly damaged about 50,000 houses (GSI, 1993).

No immediate *foreshock* was recorded. However, two events, M 4.0-4.5, occurred on April 20 and 25, 1988 in the area about four months before the main shock. 44 aftershocks (M>4.0) were reported during August-September 1988 (GSI, 1993). Six aftershocks reported by the ISC are shown in Fig. 5.51. No temporary microearthquake network was made to monitor the lower magnitude (M<4.0) aftershocks.

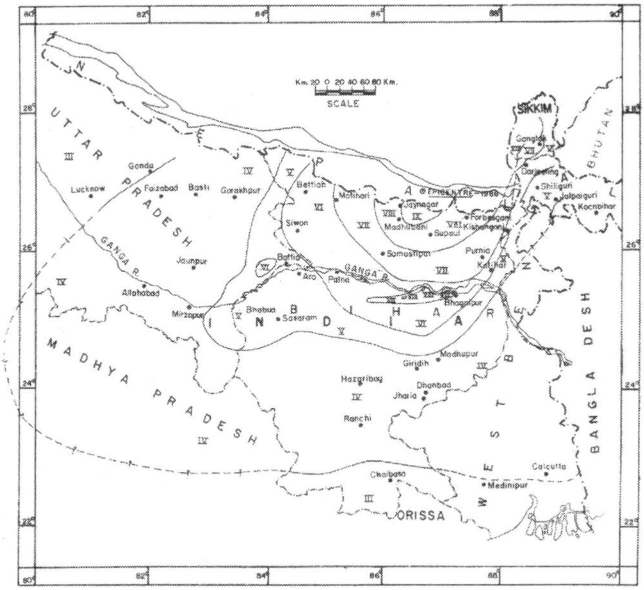

(a)

(b)

Fig. 5.53 (a) Isoseismal map of the August, 1988 Bihar/Nepal earthquake, and (b) damages observed in Munger area, Bihar (GSI, 1993).

Formation of *cracks* and *fissures* in ground was widespread to the north of Ganges; such features are not reported to the south (GSI, 1993). The most spectacular natural phenomenon, like emission of sand and water through fissures and individual point sources or vents were reported at numerous places. The sand vents occurred in random as well as in preferred linear alignment; these were usually aligned in NNE-SSW to NE-SW. Many observers reported that the ground was broken into innumerable fountains of water and sand. The water or sand fountains attained a height of 1.5-2.0 m above the ground. *Earthquake sounds* were reported, which were described variously as of thunder, lightening, an approaching plane, passing by train etc. Damage to buildings varied from minor cracks to almost total collapse. Landslides occurred in the hill terrain of Sikkim and Darjeeling. Based on these observations an intensity *isoseismal map* was prepared by the GSI (1993), (Fig. 5.53). Maximum intensity IX (MM scale) encompasses the epicentral tract, trending roughly NE-SW, extends over an area of 1500 sq km approximately within the Indian territory. Isoseismal VIII covers an area of 10,000 sq km within the Indian territory to the north of river Ganges. There was sudden increase of intensity immediately to the south of the Ganges. A closed contour of isoseismal VIII covering approximately 80 sq km was mapped in Munger and Bhagalpur districts of Bihar (Fig. 5.53). The intensity may have reached IX in Munger town, but these are just spot values and included within the isoseismal VIII.

5.9.2 Microearthquake Networks

Permanent Network in Central Himalaya

Nepal Seismological Network

The territory of Nepal, in the central Himalaya, spans about one third length of the Himalaya between the two syntaxes. Originally a five-station national seismological network was established in the Kathmandu valley in 1982 to monitor seismicity in the valley region. The network was upgraded to a 17-station telemetric network in 1994 in collaboration with France, which was designed to improve the monitoring system of the whole Nepal Himalaya (Pandey et al., 1995). The microseismicity map provides an exceptional view of microearthquake activity. Two seismicity maps are presented; one for the period 1982-85 (Fig. 5.54a) when the five-station network was in operation, and the other for the period 1994-1998 (Fig. 5.54b) when the upgraded 17-station network was in operation. More than 10,000 epicentres are shown in the two maps, which show more or less same pattern. The seismicity belt follows approximately the front of Higher Himalaya. The activity is confined at a depth range 10-20 km. The activity is intense in far western Nepal, west of 82°E, and in eastern Nepal between 86° and 88°E. The central Nepal, between 82°E and 86°E, is comparatively quiet. These lateral variations of seismicity are related to segmentation of the Himalayan thrust faults, at 87°E, 85°E and at 82°E, which coincide with some active normal/tear faults in southern Tibet, the long-lined transverse features in the region (Pandey et al., 1999). Some patches of activity are also detected in southern Tibet. Two depth sections of the Nepal Himalayan earthquakes are shown in Fig. 5.54 c and d. It is observed that the hypocentres are clustered in and around the 'ramp', the Basement Thrust Front.

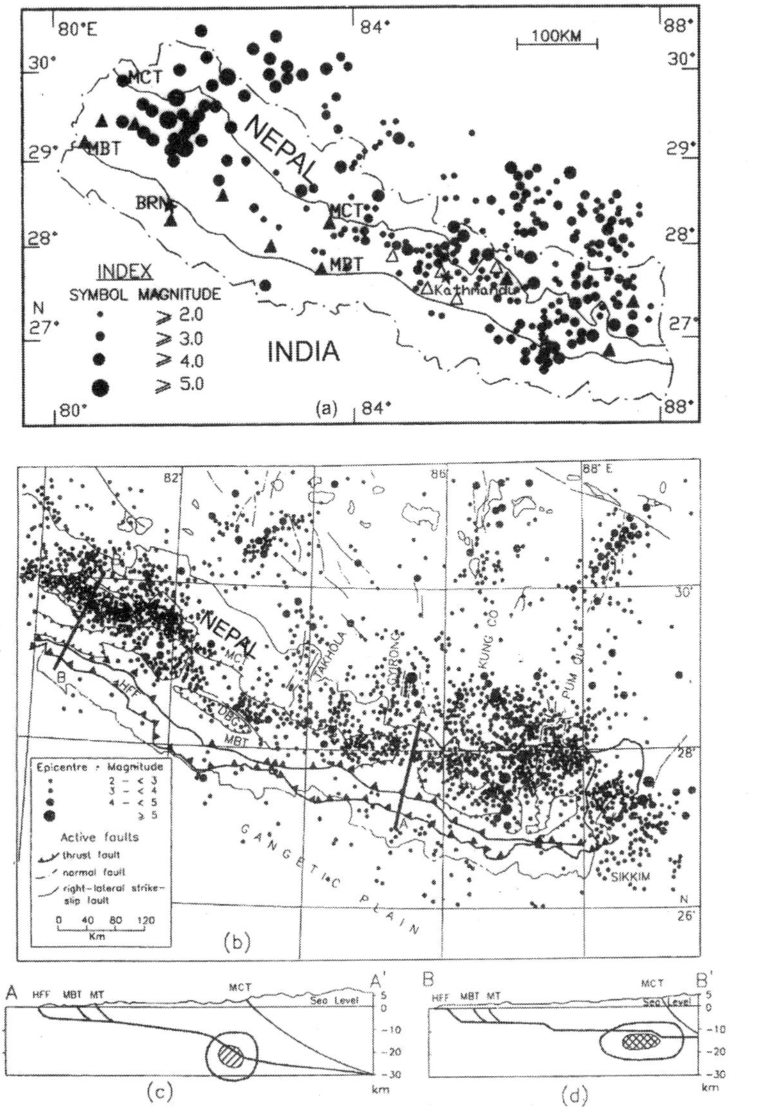

Fig. 5.54 (a) Microseismicity map (1982-1985) of Nepal (Chandra, 1992). (b) Microseismicity map (1994-1998) of Nepal, (c) and (d) seismic cross sections along A-A′ and B-B′ (Pandey et al., 1995).

Temporary Networks in Eastern Himalaya

Darjeeling and Sikkim Himalaya Networks

The Darjeeling-Sikkim Himalaya, in the eastern part of Himalaya, may be regarded as a separate tectonic block. The foredeep is distinctly bounded by the Munger Saharsa Ridge (MSR) to the west and by the Shillong Plateau–Goalpara Ridge to the east (Fig. 5.51). In this tectonic block, the MCT and MBT are not parallel; the MCT arches in the form of a culmination, the core region exposes a vast expanse of pre-Tertiary rocks arranged in a pile of thrust sheets (Ray, 2000). Two significant earthquakes (M $\simeq$ 6.0) occurred in this area, one in 1965 (M 5.9) and the other in 1980 (M 6.0), (Fig. 5.51).

The Geological Survey of India carried out four microearthquake surveys in this area during 1993-2000. About 700 earthquakes including local, regional and teleseismic events were recorded during the four surveys, and about 300 earthquakes (M $\geq$ 2.0) were located within 50 km of the networks (De and Kayal, 2003 and 2004). The epicentre maps show that the activity is mostly clustered to the north of the MBT (Fig. 5.55 a and b). A depth section (T1-T1) of the earthquakes along the strike direction of the Himalayan seismic belt in the area is illustrated in Fig. 5.55c. Critical examination of the epicentre map and the depth section (T1-T1) suggests two tectonic domains, one below the Sikkim/Darjeeling Himalaya and the other below the Goalpara lineament. The earthquakes along the Goalpara lineament are shallower (0-25 km), but the earthquakes below the Sikkim/Darjeeling Himalaya are deeper, down to 45-50 km.

A north-south section across Sikkim Himalaya, along T2-T2, shows that the seismic activity is continuous from surface down to the lower crust (Fig. 5.55d), and the earthquakes also occur below the so-called Plane of Detachment, that was proposed in the tectonic model of the Himalaya (Seeber et al., 1981).

Two composite fault-plane solutions (Fig. 5.55a) using P-wave first motions were obtained for the two clusters of events in the two tectonic zones (De and Kayal, 2003). The solution for the cluster in the Sikkim zone shows thrust faulting; the north-dipping nodal plane is the inferred fault plane. The fault-plane solution and the N-S depth section (T2-T2) suggest that the MBT is seismogenic, and it is possibly a deep-rooted mantle-reaching fault. The fault-plane solution of the cluster along the Goalpara lineament, on the other hand, shows a strike-slip solution (Fig. 5.55a); the near-vertical/southwest dipping NW-SE trending nodal plane, comparable with the Goalpara lineament, is the inferred fault plane (De and Kayal, 2003).

A microearthquake survey was carried in 2000 at the MCT in Sikkim Himalaya making a digital seismic network. The seismic stations were installed on both sides of the N-S segment of the MCT (Fig. 5.55b). Based on the fault-plane solution and geological evidences, De and Kayal (2004) suggested that these events are generated by right-lateral strike-slip motion

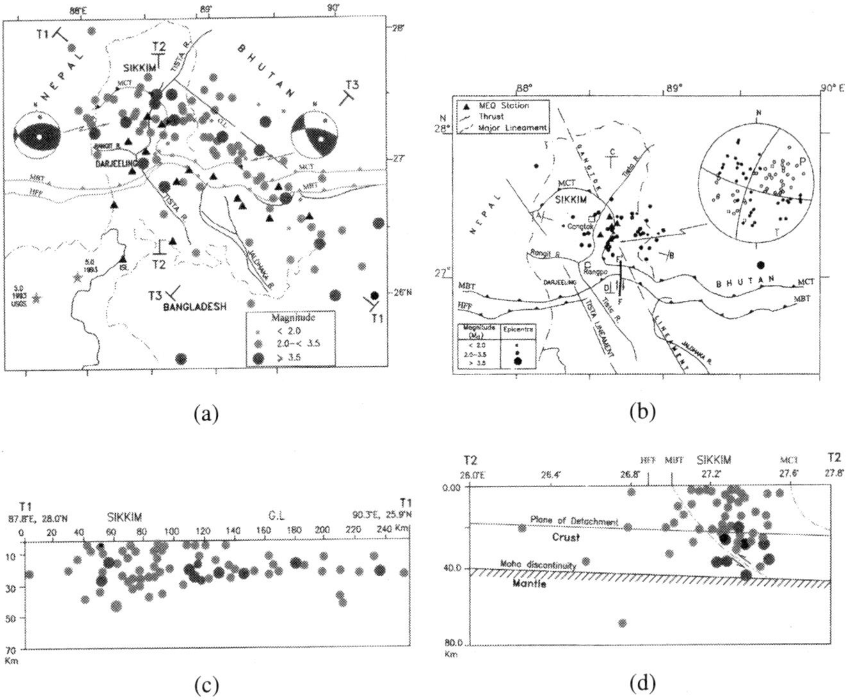

Fig. 5.55 (a) Microseismicity and fault plane solutions in the Sikkim Himalaya (De and Kayal 2003). MBT: Main Boundary Thrust, MCT: Main Central Thrust, GL: Goalpara Lineament. (b) Microseismicity and fault plane solution on the N-S segment of the MCT in the Sikkim Himalaya, the first motion data are also shown (De and Kayal, 2004). (c) Seismic cross section of the events along T1-T1, and (d) seismic cross section of the events along T2-T2 (Fig. 5.55 a), (De and Kayal, 2003).

on a NNE-SSW active fault/lineament. A NNE-SSW fault/lineament F-F′ (Fig. 5.55b) is geologically traced at the MBT zone (Malay Mukul, pers. com, 2004), and it may be deep rooted. This fault, transverse to the Himalayan trend, may be extending below the MCT, that caused right lateral movement on the MCT to produce the curvilinear shape, and the fault is seismically active at depth.

5.10 SEISMOTECTONIC PERSPECTIVE

In the steady state model (Seeber et al., 1981) and/or in the evolutionary model (Ni and Barazangi, 1984), it is suggested that the Himalayan *thrust earthquakes* are shallow and occur on the detachment plane between the MBT and MCT, whereas Tibetan *normal fault* earthquakes occur in the shallow crust as well as in the uppermost mantle (Figs 5.9 and 5.10). Kayal (2001), based on microearthquake network data, argued that this generalized model may fit fairly well in some parts of the western Himalaya, but it differs in other tectonic blocks of the Himalaya as discussed below.

The great Himalayan earthquakes (M≥8.0) which occurred in the Main Himalayan Seismic Belt (MHSB) during the past 100+ years, since the first instrumentally recorded great Shillong earthquake in 1897, provide crucial information on the seismotectonic processes currently happening in the Himalaya region. The available data of the great earthquakes of 1897, 1905, 1934 and 1950, which occurred along the Himalayan arc, indicate that these earthquakes are not the result of slip on the MBT or MCT. The first two earthquakes (1897 and 1905) ruptured a quasi-horizontal surface that extends much south of the MBT, and the 1950 great earthquake occurred at the eastern syntaxis (Assam syntaxis) zone by a more complex tectonics. Krishnaswamy et al. (1970) suggested that tectonically the great earthquakes are similar to those of the MHSB. Seeber et al. (1981), however, argued that the great earthquakes occurred along a seismic belt, which is distinct from the MHSB. The different style of seismic energy-release between these two belts has a physical basis in the tectonic structures associated with each belt. The intensity isoseismals for the four great earthquakes in the Himalaya are shown in Figs 2.13-2.16, and the rupture zones are shown in Fig. 5.56. Rupture zone and isoseismal of the most recent devastating 2005 Kashmir earthquake (M_w 7.8) that caused a loss of ~80,000 human lives, is shown in Fig. 5.34. The data from these great/large earthquakes play a major role in understanding seismotectonics of the Himalaya.

The 2005 Kashmir earthquake (M_w 7.8) occurred at the western extremity of the Himalaya, at the western Hazara syntaxis zone, where the Himalayan arc joins the Karakuram, Pamir and Hindu Kush ranges (Fig. 5.34). The most active thrust fault under the Himalaya is thought to be the Main Frontal Fault (MFT), which marks the emergence at the surface of the Plane of Detachment, the interplate thrust, also named Main Himalyan Thrust (MHT). The zone between the Hazara syntaxis and about 76° E is mostly blind (Seeber et al., 1981). Local seismic network around the Hazara syntaxis revealed a seismicity trend that is called Indus-Kohistan Seismic Zone (IKSZ). The IKSZ strikes parallel to the northwestern Himalaya, but extends beyond the syntaxis. The aftershock trend of the 2005 Kashmir earthquke clearly indicates that the blind MHT extends beyond the Hazara syntaxis (Fig. 5.34). The average slip on the fault ruptured by the 2005 event is ~4.2 m, and if geodetically determined ~14 mm shortening rate across the range were accommodated by repetition of such event, their return period would be about 300 years (Avouac et al., 2006). Despite the complex geological setting associated with the Hazara syntaxis, the slip pattern and kinematics of the 2005 Kashmir earthquake are explainable by the thrust mechanism on the MHT.

Seismotectonics of the two great earthquakes, 1905 and 1934, that occurred in the western and eastern Himalaya respectively, are discussed here. The 1897 and the 1950 great earthquakes occurred in the northeastern Himalaya

Region; seismotectonics of these two great earthquakes are discussed in Chapter 6.

Further, a few strong earthquakes (M≥6.0) occurred in the MHSB during the last two decades; these earthquakes are much well studied by macroseismic as well as by aftershock investigations by temporary networks. Results of these studies and several permanent and temporary microearthquake network data shed more light on understanding of seismotectonics of the Himalaya.

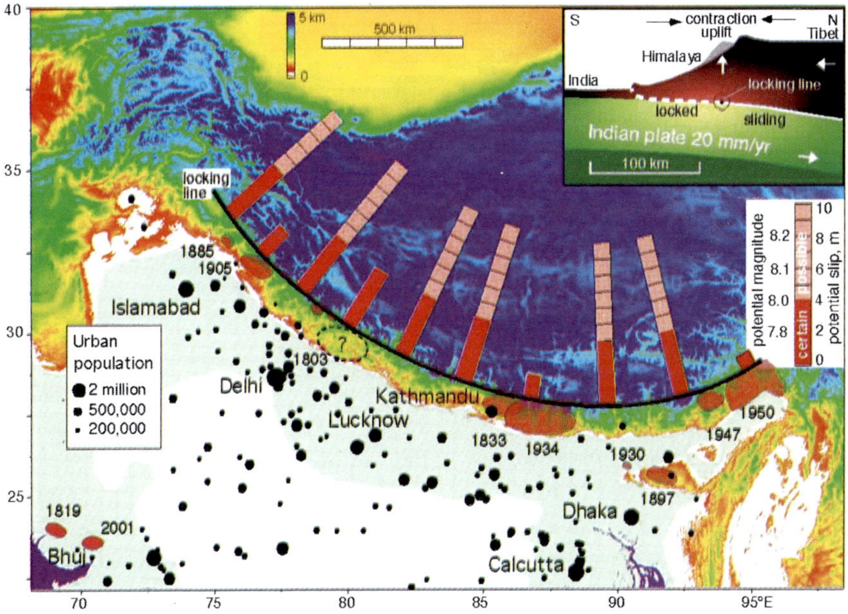

Fig 5.56 Rupture zones of large earthquakes and estimated slip potential along the Himalaya (Bilham et al., 2001).

5.10.1 Northwest Himalaya

In addition to the permanent national (IMD) network data and the ISC data, a large quantity of temporary microearthquake network data are available for this part of the Himalaya, between 75° and 79°E (Fig. 5.38). It is observed that the seismic activity is aligned almost parallel to the MBT, and mostly south of the MCT. A dense cluster of events is observed between latitude 32-33°N and longitude 76-77°E (Fig. 5.38a), which is the epicentre zone of the 1905 great Kangra earthquake. It indicates high stress concentration in the region. In the recent past, two strong earthquakes, the 1991 Uttarkashi earthquake (m_b 6.6) and the 1999 Chamoli earthquake (m_b 6.3) caused severe damages in the western part of the Himalaya, and two significant medium size earthquakes (M>5.0), the 1986 Dharmasala earthquake and the

1997 Surendranagar earthquake, severely rocked the northwestern Himalaya. These earthquakes are fairly well studied. Seismotectonic perspective of these earthquakes along with the 1905 great earthquake are discussed to understand the tectonic model in this part of the Himalaya.

The 1905 Kangra Earthquake M 8.6

Although Richter (1958) assigned magnitude (M_L) 8.6, the magnitude of the Kangra earthquake is now estimated to be M_s 7.8 (Ambraseys, 2000; Ambraseys and Douglas, 2004). Middlemiss's (1910) report contained virtually all the details concerning the macroseismic effects of this earthquake. The highest intensity X (Rossi-Forel scale) was reported near the towns of Kangra and Dharmasala. The detailed isoseismal map is given in Fig. 2.16. The intensity decreased sharply to the west. To the east, the decrease in intensity was gradual. The intensity VIII and IX isoseismals are parallel to the trend of the MBT. An isolated contour of intensity VIII was drawn surrounding Dehradun and Mussoori about 250 km southeast of Kangra (Middlemiss, 1910).

Based on the isoseismals, rupture areas of the great Himalayan earthquakes are given in Fig. 5.56. The extent of rupture length of the 1905 earthquake is, however, poorly defined. It could be that the rupture occurred only beneath the isoseist VIII surrounding Kangra, a zone of about 100 km in length (Wallace et al., 2005). A second possibility is that the two separate areas, I = VIII isoseismals, ruptured with a total length of about 200 km. Third, perhaps the entire zone from Kangra to Dehradun, 280 km in length, ruptured as inferred by Seeber and Armbruster (1981). The rupture area for the 1905 earthquake is, however, much less than for the 1934 earthquake. This difference is inferred to the relatively shallow depth-to-basement associated with the 1905 event compared to that for the 1934 event which occurred in the foredeep region of the central Himalaya (Mathur and Kohli, 1964). Recently, Hough et al. (2005) re-evaluated intensity distributions and re-checked the original seismograms. They interpreted that a second large earthquake (M>7.0) was triggered in Dehradun region within minutes of the 1905 main shock. They argued that two meizoseismals are the product of two main shocks.

Although the maximum intensity in the 1905 earthquake follows approximately the surface trace of the MBT, there was no evidence of coseismic displacement along the trace of the MBT or along any of the other imbricate thrust of the sub-Himalaya. There is a strong argument against a major subsurface rupture on the MBT for the 1905 event. Levelling data, however, show that there was a coseismic uplift of about 14 cm along the Saharanpur-Dehradun-Mussoorie highway, over a period of 130 years since 1861 (Gahalaut and Chander, 1997). Most bench marks again started subsiding in the decades following the earthquake. This is seen as an evidence that the upper crust along and adjoining the leveling line experience a cycling process of strain accumulation and release in great earthquakes. After re-evaluation of raw leveling data, Bilham (2001), however, argued that there was probably little or no uplift in the Dehradun region. This is consistent with the revised magnitude M_w 7.8 and rupture length of 110 km (Ambraseys and Douglas, 2004).

Fault-plane solution of the 1905 earthquake was not obtained. Fault-plane solutions of the moderate magnitude earthquakes, near the epicentre of the 1905 earthquake (Fig. 5.32), indicate under thrusting on planes dipping gently northeast (Molnar and Lyon-Caen, 1989). Molnar and Pandey (1989) assumed such fault plane for this great earthquake.

In a recent study, Bhattacharya and Kayal (2005) analysed a large quantity of microearthquake data that were recorded during the last three decades by semi-permanent and permanent microearthquake networks in the western Himalaya. They reported that the 1905 Kangra earthquake source zone lies beneath the Himalayan Frontal Fault (HFF), south of MBT, and the source zone could be deeper, 30-40 km deep (Fig. 5.38b). This source zone is much below the Plane of Detachment. The HFF is the possible seismogenic fault; at the 'fault end' tectonic stress is accumulated in the lower crust of the Indian shield at a depth 30-40 km and generated the 1905 event (Fig. 5.38b). The conceptual model envisaged by Seeber et al. (1981) for the great earthquakes in the Himalaya possibly does not fit, and the source zone is quite distinct from the events that occur in the MHSB (Fig. 5.38b).

Dharmasala Earthquake (m_b 5.7), 1986

Epicentre of the Dharmasala earthquake of April, 1986 (m_b 5.7) is very near to the epicentre of the 1905 great Kangra earthquake. The epicentre of the 1986 earthquake is well located by the IMD using the local network of WIHG and the national network (IMD) data (Fig. 5.32). The estimated earthquake parameters are: epicentre 32.19°N and 76.29°E; depth 9.0 km and magnitude m_b 5.7. The epicentre lies in the MHSB in the Himachal Pradesh, very close to the thickly populated city of Dharmasala, in the rupture region of the 1905 great Kangra earthquake. The fault plane solution shows a thrust faulting with a small strike-slip component (Fig. 5.32); the low angle north dipping nodal plane is the preferred fault plane, which is comparable with the Plane of Detachment or the Basement Thrust. In the northwest and western Himalaya, the general observation is that the moderate magnitude earthquakes in the MHSB occur on the Plane of Detachment/ Basement Thrust, and rupture propagates towards south. Such observation is also made for the 1991 Uttarkashi and 1999 Chamoli earthquake sequences (Figs 5.45 and 5.46), and discussed in Section 5.11.2 below.

The 1997 Sundernagar Earthquake (M_L 5.0)

This earthquake (M_L 5.0) was well recorded by the permanent network of the WIHG on July 29, 1997 in the northwestern Himalaya. The epicentre was located at 31.63°N and 76.81°E and depth at 10 km in Sundernagar area of Himachal Pradesh (Fig. 5.32). This is the strongest earthquake that was ever felt by the local residents in the recent past. Fault-plane solution of the earthquake was not available, but from the observed NNE-SSW isoseismal trend it was inferred that the earthquake was generated by a strike-slip fault. The NNE-SSW fault/lineament in the area, transverse to the trend of MBT, is the inferred fault for the earthquake (GSI, 2000a). Although earthquakes are mostly generated by thrust faulting on the Basement Thrust, earthquake with strike slip mechanism along transverse lineaments are also observed.

5.10.2 Western Himalaya

Uttarkashi Earthquake m_b 6.6, 1991

The October 20, 1991 Uttarkashi earthquake sequence in the Garhwal tectonic block is studied in more detail to understand the earthquake processes with respect to the present-day conceptual tectonic model in the western Himalaya (Kayal et al., 1995; Kayal, 1996b; Cotton et al., 1996). The focus of the main shock (m_b 6.6) was fairly well determined by the IMD and USGS at a depth of 12 km. The centroid depth was given at 14 km (Cotton et al., 1996). The hypocentre parameters of the lower magnitude (M<4.0) aftershocks were well determined by the GSI temporary microearthquake network (Kayal et al., 1995). The depth section of the main shock and the aftershocks revealed the seismotectonic structure beneath the Garhwal Himalaya (Fig. 5.45). The conceptual tectonic model proposed by Seeber et al. (1981) fits fairly well with the observations. The main shock was generated at the Basement Thrust or Basement Thrust Front (BTF), and the aftershocks were generated by shear adjustment at the MHSB thrust systems to the south of MCT. It was interesting to note that all the earthquakes occurred above the Plane of Detachment as proposed by Seeber et al. (1981) in their steady state tectonic model (Figs 5.9 and 5.45b).

The fault-plane solutions revealed that the earthquakes were generated by thrust faulting (Fig. 5.45a). The main shock shows nearly a pure thrust-faulting with an E-W oriented north dipping low angle thrust plane that is comparable with the Basement Thrust (BT). The aftershocks show thrust-faulting with a small strike-slip component with an E-W oriented north dipping thrust plane comparable with the surface trace of the Uttarkashi thrust to the south of MCT in the MHSB. Both the solutions indicate NNE-SSW directed compressional stress. Kayal (1996b) suggested that the main shock was originated at the low angle north dipping BT/BTF, and the aftershocks were generated by shear adjustment at the Uttarkashi fault to the south of MCT; the main shock rupture propagated to the south along the detachment plane. The 'fault end' of the Uttarkashi fault at the BT/BTF was the source area for concentration of the tectonic stress. The MCT had no role for the Uttarkashi earthquake sequence (Fig. 5.45b); it is a dormant feature as proposed by Ni and Barazangi (1984) in their evolutionary model (Fig. 5.10). The conceptual models fit fairly well in the MHSB in this western part of the Himalaya.

Chamoli Earthquake m_b 6.3, 1999

The Chamoli earthquake (m_b 6.3) of March 28, 1999 (local time 00h, 35m, March 29) occurred in the Chamoli area, which also falls in the Garhwal Himalaya tectonic zone of the western Himalaya (Figs 5.46 and 5.47). The main shock as well as the aftershock sequence was well studied by Kayal et al. (2003 a and b) using the permanent and the temporary network data. The results further enlighten about the seismotectonic structure and the earthquake generating processes in the MHSB of western Himalaya.

The moment tensor solution of the main stock revealed a thrust mechanism similar to that of the 1991 Uttarkashi earthquake (m_b 6.6) (Figs 5.46a and 5.47a). The IMD also reported analogous fault-plane solution of the 1999 main shock using the global and the national network first-motion data (Fig. 5.47a). The north dipping nodal plane (strike 282°, dip 9°N, slip 95°) is the inferred fault plane. The depth-section of the main shock and aftershocks illustrates that the tectonic stress due to the northward subducting Indian and southward overriding Tethyan slab is accumulated at the 'ramp' i.e. at the BTF and/or at the Basement Thrust (BT), and generated the main shock at a depth of 21 km on the Plane of Detachment or BT (Fig. 5.46b). The aftershocks, on the other hand, are produced by shear adjustment of the active segment(s) of the fault(s) to the south of MCT (Figs 5.46b and 5.47b) (Kayal et al., 2003a and b). The MCT is not the causative fault either for the main shock or for the aftershocks (Fig. 5.46b and 5.47b), as observed for the 1991 Uttarkashi earthquake sequence (Fig. 5.45b).

Kayal et al. (2003a and b) suggested that the 'fault end' of the Alakananda fault (ANF) on the Plane of Detachment was the source area for the main shock (Fig. 5.46b). The main shock then triggered the ANF to generate the aftershocks. The aftershock cluster to the north of ANF show similar fault–plane solution as the main shock. The aftershock cluster to the south of ANF, on the other hand, show strike slip solution (Fig. 5.47a) and the N-S trending Nandprayag fault was the inferred fault (Kayal et al., 2003b). All the solutions show NNE–SSW oriented compressional stress, compatible with the regional tectonic stress.

Further, seismic tomography clearly revealed the MCT and the ANF and a high velocity zone on the Plane of Detachment, where the ANF meets the detachment plane at depth (Fig. 5.49). It is interesting to note that the MCT is not extended down to the Plane of Detachment. The E-W trending ANF to its south, on the other hand, is extended down to the Plane of Detachment, and is identified as the seismogenic fault.

These observations, the 1991 Uttarkashi earthquake sequence and the 1999 Chamoli earthquake sequence, support the evolutionary model of Ni and Barazangi (1984), which states that the MCT is the older thrust zone and is less active or a dormant feature now. The main shocks or medium size earthquakes occur on the Plane of Detachment (or BT), and rupture propagates to the south. The fault systems to the south of the MCT get activated to generate the aftershocks. The 'fault ends' of these fault systems where they meet the gentle dipping Plane of Detachment (or BT) are the source areas of the main shocks. These shocks on the BT occur by thrust mechanism. It may also be noted that depth of the detachment plane or BT is more (~20 km) in the Chamoli area compared to that (~15 km) in the Uttarkashi area to the west.

5.10.3 Central Himalaya

Fault plane solutions of several earthquakes (M>5.0) in central Himalaya and the foredeep region are available (Fig. 5.50). Seismotectonics of these events and the recent digital telemetric network data of the Nepal seismological network shed light on understanding the earthquake processes in central Himalaya. Two significant foredeep earthquakes, 1934 (M 8.4) and 1988 (M_s 6.6), are discussed in section 5.11.4. Here we discuss the recent significant events that occurred in the MHSB.

Earthquake of April 20, 1988 (m_b 5.4)

The ISC location of this earthquake gives epicentre at 27.02°N and 86.72°E and depth at 55 km; it occurred to the north of MBT (Fig. 5.50). The CMT solution indicates right-lateral shear movement along the northeast trending nodal plane. Dasgupta et al. (1987) suggested that this could be a foreshock of the August 1988 earthquake (m_s 6.6); deeper focal depths are comparable. It may be noted that the fault-plane solutions for the August 1988 earthquake and this April 1988 event, though identical, display opposite shear movement along the preferred northeast trending fault planes (Fig. 5.50).

Earthquake of October 29, 1988 (m_b 5.5)

This earthquake (epicentre 27.87°N and 85.64°E and depth 15 km) was located northeast of Kathmandu, close to the MCT in the central Himalaya (Fig. 5.50). The CMT solution shows a thrust-faulting solution. The northwest trending nodal plane is the possible fault plane that dips at shallow angle to the northeast.

Earthquake of January 31, 1997 (M 5.6)

Recently an earthquake M 5.6 occurred on January 31, 1997 with epicentre at 28.06°N and 85.34°E and depth at 17 km. The main shock and 36 aftershocks were recorded by the Nepal seismological network (Kumar and Gautam, 1998). The aftershocks define a well constrained NE striking fault. The fault-plane solution of the main shock shows a reverse faulting with the preferred fault plane striking N 60°E dipping 80°SE. The NW directed P-axis is parallel to the regional strike of the geological structure (Kumar and Gautam, 1998). It was suggested that the event occurred on a major transversal structure correlating with the proposed segmentation of the Himalayan arc at 85°E longitude (Pandey et al., 1998).

Seismotectonics

The recent upgraded Nepal network data in the central Himalaya suggest that the earthquakes in the MHSB are shallow (depth >30 km), and occur on and above the BT or BTF. Intense activity is observed at the 'ramp' (BTF) (Fig. 5.54c and d). The 'ramp' zone, just below the MCT, is identified as a conductive zone in the Project INDEPTH investigation (Section 5.3); partially molten and/or fluid filled rock matrix zone is inferred (Fig. 5.8). The fluid filled rock matrix accentuates triggering of earthquakes, and intense activity is observed in the ramp zone in central Himalaya. Fault plane solutions and seismic tomography in the 'ramp' zone would shed more light on the seismotectonics. The available fault plane solutions of the earthquakes to the south of the ramp mostly indicate thrust mechanism in the west Nepal and thrust as well as strike-slip mechanism in the east Nepal Himalayas (Fig. 5.50). The inferred fault planes are comparable with the MHT for the thrust mechanisms, and with the surface faults/lineaments that transect the E-W Himalayan trend for the strike slip mechanisms. Transverse tectonics in the foothills and in the MHSB in the central Himalaya is suggested by many authors (e.g. Mukhopadhya, 1984; Pandey et al., 1995, 1998; Kayal, 2001).

Two microseismicity maps of the central Himalaya clearly illustrate that the seismic belt follows approximately the front of higher Himalaya, and there is lateral variation in seismicity (Fig. 5.54). In this part of the Himalaya, the earthquakes are also confined at depth 10-30 km, and mostly clustered at the 'ramp' on the Plane of Detachment (Fig. 5.54). However, several active transverse faults are identified in the foredeep region, which extends into the central Himalaya (GSI, 1993). These transverse structures may be correlated with the segmentation of the Himalaya arc (Pandey et al., 1998).

5.10.4 Eastern Himalaya and Foredeep

Fault-plane solutions of several significant earthquakes (M>5.0) and a good amount of microearthquake data are studied to understand seismotectonics of the eastern Himalaya (De and Kayal, 2003 and 2004). The foredeep region of the central eastern Himalaya has been the source area for the 1934 great earthquake (M 8.4) as well as for the 1988 strong earthquake (M_s 6.6). These two earthquakes are very different from those in the MHSB.

The 1934 Bihar Earthquake M 8.4

The January 15, 1934 Bihar earthquake is, in many ways, one of the best known Himalayan great earthquakes because it occurred when instrumental seismology was fairly well developed. The epicentre (26.77°N and 86.69°E) was situated in the Himalayan foredeep, about 100 km south of the MHSB (Seeber et al., 1981). Richter (1958) assigned a magnitude (M_L) of 8.4 to this earthquake. Even allowing for reasonable errors in epicentre, it does not seem possible that the rupture started within the MHSB. The meizoseismal zone or the rupture zone primarily lies south of the MBT and extends about 300 km along the strike of the Himalaya (Fig. 5.56). The detailed isoseismal map is given in Fig. 2.15. The northern and southern boundaries of the isoseismal VIII correspond closely to the MBT and the boundary between the foredeep and the Indian shield respectively. The intensity X was assigned to two spots, separated by almost 150 km: one in Munger (Patna) in Bihar to the south, and the other in the Nepal valley to the north (Fig. 2.15).

Although the 1934 earthquake was recorded by more than 100 seismograph stations worldwide, there were not enough consistent P-wave first motions to determine a fault-plane solution. Singh and Gupta (1980) inferred strike-slip faulting from the surface wave data. Chen and Molnar (1977), on the other hand, assumed thrust faulting, and used amplitude spectra of surface waves to infer seismic moment. Dasgupta et al. (1987) argued that the Himalayan foredeep is dominated by conjugate strike-slip faults that face the wedges of the Indian shield. They argued that the 1934 earthquake, located some 100 km south of the MBT, was related to the East Patna fault, and occurred by strike-slip mechanism (Fig. 5.51). It was observed that the slump belt and the Munger high intensity regions were both elongated tracts with their larger axes broadly parallel to the local trend of the Indo-Gangetic basin axis. The width of the slump belt undergoes a notable change near longitude 86.5°E. The East Patna fault appears to be close enough to be linked with the changes in the local site conditions, which led change in the width of the slump belt (Fig. 2.15).

In contrast to the above, Seeber et al. (1981) proposed a model that shows that the 1934 rupture occurred along the Plane of Detachment. The northern limit of the rupture was north of the MBT, and the southern limit of the rupture was suggested to be along the axis of the slump belt about 100 km south of the MBT.

Foredeep Earthquake of June 19, 1979 (m_b 5.2)

The June 19, 1979 earthquake (m_b 5.2) occurred in the foredeep region of the central Himalaya. The CMT location of the epicentre is at 26.29°N and 87.57°E and centroid depth at 24 km. The epicentre of this earthquake falls at the contact of the Siwalik and alluvium, i.e. at the HFT (Fig. 5.50). The fault-plane solution indicates the preferred fault plane trends NNW and dips 60°NE and at the surface the fault correlates with the Purnea-Everest lineament (Dasgupta et al., 1987). Active transverse lineaments in the foredeep region were reported by many authors (Mukhopadhyay, 1984; Kayal, 2001).

Bihar-Nepal Earthquake M_s 6.6, 1988

The August 20, 1988 Bihar-Nepal earthquake (M_s 6.6, depth 50-71 km) occurred in the foredeep region. The epicentre of this event is close to the 1934 great earthquake (Fig. 5.51). Both the earthquakes, 1934 and 1988, produced two meizoseismal zones: one to the north, in the MBT zone, and the other to the south, in the Munger-Bhagalpur (Patna) area (Figs 2.15 and 5.53). The first-motion fault-plane solutions (USGS and Banghar, 1991) and the CMT solution (Dziewonski et al., 1989) are nearly identical. The NE trending nodal plane is the preferred fault plane (Fig. 5.51), which shows left-lateral strike-slip motion (GSI, 1993). The frontal earthquakes in this part of the foothills region, if triggered through activity along the basement detachment, should not have a focal depth more than 10 km, and should be predominantly compressive thrust type event. Neither the focal depth, which is in all probability greater than 50 km as fairly well estimated by various organisations, nor the source mechanism conforms to the 'steady state model' or to the 'evolutionary model' (Figs 5.9 and 5.10). The 1988 earthquake has no equivalent along the entire foredeep region (Chen and Kao, 1996), and cannot be correlated with the MHSB earthquakes or with the Plane of Detachment (Kayal, 2001). Further, in a recent study, relocation of earthquakes in the foothills region shows that deeper mantle earthquakes are common in the region (Monsalve et al., 2006).

It may be mentioned that neither the focal depth nor the fault-plane solution is known for the 1934 great earthquake (M 8.4) to critically evaluate the great detachment-earthquake model in the eastern or in the central Himalaya. One major problem with the medium or large magnitude Himalayan foredeep earthquakes is that the rupture may be associated with the continental type earthquakes. The conceptual model is utilised as a strong evidence for a quasi-horizontal causative fault where dimension of rupture is much larger than the source depth. There are, however, evidences that even shallow foci large events associated with sub-vertical fault have not produced rupture, e.g. the Loma Prieta 1989 event located on the San Andreas fault (Plafker and Galloway, 1989). Absence of rupture can also be reconciled if earthquakes are originated from deeper crustal level associated with deep seated fault, and source dimension of such events are relatively small compared to their focal depth. So, with the Himalayan frontal earthquakes, it is worth scanning the subsurface faults hidden below the foredeep which possibly were formed during the Proterozoic-Mesozoic in response to the then stress system, but capable for reactivation during the present day ambient stress field (Cattin and Avouac, 2000).

East-west trending subsurface normal faults below the Gangetic foredeep are reported by Deep Seismic Soundings (DSS) (Harinarain and Kaila, 1982). Reactivation of these normal faults is possible in response to induced extension due to flexural bending of the Indian lithosphere below the foredeep. These faults seem to be vulnerable for generating the foothill earthquakes, particularly those originating from the upper crustal level. For the deeper events, like the August 1988 earthquake, these faults, however, appear to be less attractive. Some deep seated seismogenic structures in the upper mantle may be causing such deeper continental earthquakes (Monsalve et al., 2006).

It is believed that the deep seated faults oriented highly oblique to the Himalayan trend are potential for the large frontal earthquakes (Dasgupta et al., 1987; Kayal, 2001). The East Patna fault transverse to the Himalaya is inferred as the responsible fault for the August 1988 earthquake. Its epicentre and fault-plane solution are correlated with this fault (GSI, 1993). The strike-slip component of the CMT solution is comparable with the NNE orientation of the fault, and the NNE horizontal compressive stress is compatible with the northward movement of the Indian plate.

The Sikkim Himalaya Earthquakes

The November 19, 1980 Gangtok earthquake (m_b 6.0) occurred within the MHSB, in the Sikkim Himalaya. The epicentre is given at latitude 27.40°N and longitude 88.80°E (Fig. 5.51) and depth at 47 km (ISC). The fault-plane solutions of Ni and Barazangi (1984) and Dziewonski et al. (1988) gave similar results. It shows a strike-slip faulting; either of the nodal planes may be a fault plane. De and Kayal (2003) observed cluster of microearthquakes below the Goalpara lineament (Fig. 5.55). Fault-plane solution of the microearthquakes below the Goalpara lineament shows similar strike-slip solution. They preferred the NW trending nodal plane as the fault plane, which is comparable with the Goalpara lineament (GL). The preferred fault plane indicates right-lateral strike-slip motion. Similarly, the NNW-SSE trending Tista lineament, transverse to the Himalaya trend, is the inferred fault for the 1965 earthquake (M 5.9) (Fig. 5.51); it shows a left-lateral strike-slip motion.

The fault-plane solution of the microearthquakes that were recorded for about three months right on the north-south segment of the MCT near the epicentre of the 1980 Gangtok earthquake shows right-lateral strike-slip motion (De and Kayal, 2004). A N-S trending transverse fault/lineament, extending from the MBT to MCT, is identified to be seismically active (Fig. 5.55).

Microearthquake data in the Sikkim Himalaya also revealed that the MBT possibly is a mantle reaching fault, and the earthquakes occur at a depth range 0–45 km in the Sikkim/Darjeeling Himalaya (Fig. 5.55). Focal depth of the 1980 Gangtok earthquake (M 6.0) was also estimated at 47 km by the ISC. Thus it is observed that seismotectonics of the Sikkim/Darjeeling Himalaya or eastern Himalaya earthquakes are different from those of the western or central Himalaya where earthquakes are much shallow ($\leq$30 km), and occur above the Plane of Detachment. The Plane of Detachment in the Sikkim Himalaya, on the other hand, is not clear. Earthquakes occur mostly by strike-slip faulting by deep rooted or by deep seated transverse faults, rather than on the shallow detachment plane as proposed by Seeber et al. (1981). The area is transected by several gigantic lineaments that are transverse to the Himalayan trend (Fig. 5.51), and these lineaments play a dominant role in seismic activity in the eastern Himalaya and its foredeep region.

5.11 SEISMIC HAZARD POTENTIAL

In a recent work, Bilham et al. (2001) argued that one or more great earthquakes are overdue in large fraction of the Himalayan arc threatening millions of people in the region. Significant ruptures, ~200 to 300 km in length, were produced by the 1803, 1833, 1897, 1905, 1934 and 1950 earthquakes (Fig. 5.56); surface ruptures, however, were not found for any of these events. Further, it may be noted that no historical record is available about the Himalayan earthquakes in the 18th century and before.

GPS measurements are a state of art technique to measure plate movements, that show India and southern Tibet converge at 20.5 $\pm$ 2 mm/year (Bilham et al., 1998; Larson et al., 1999). About 80% of the convergence is observed at the 50 km wide region in the southern edge of the Tibetan Plateau, the rest 20% is accommodated in the surrounding Himalaya region. Thus two metres of potential slip in earthquakes are accumulated in each century. Knowledge of the lengths of ruptures and sensible estimate of seismic moment of the 1934 and 1950 earthquakes yield ~4 m of slip in 1934 and ~8 m of slip in 1950 earthquakes (Molnar and Deng, 1989). Bilham et al. (2001) divided the entire Himalayan arc into 10 regions with lengths (~220 km), roughly corresponding to those of great Himalayan ruptures. With a convergence rate of 20 mm/year along the arc, six of these regions currently have a slip potential of at least 4 m, equivalent to the slip inferred for the 1934 earthquake (Bilham et al., 2001). This implies that each of these six regions now stores the strains necessary for such an earthquake (Fig. 5.56).

Seismologists understand the long-term forces that build the Himalaya, and the reservoir of energy that is stored at the plate boundary, but do not understand the short-term processes that control timing of individual events (Bilham et al., 2001). However, to estimate potential size of future earthquakes we need to know the amount of stored energy and the size of the fault, which would eventually fail to release the energy. For example, the October 8, 2005 Kashmir earthquake that shook Muzafferabad to Kangra exceeded magnitude Mw 7.8, and such event could have struck in any one of the dozen locations from Afghanistan to Assam (Fig. 5.56). Recently, Bilham and Hough (2006) argued that the 1934 great Bihar-Nepal earthquake may have reduced the odds of another large or larger earthquake in eastern Nepal, but no great earthquake has struck in the Sikkim Himalaya in historic times, and it would not be surprising if one occurs there soon. Historical data are silent about seismic segments in Sikkim, Bhutan and in parts of Indo-Burma ranges, and these could be the segments for future large earthquakes like that of the silent segment of the October 8, 2005 Kashmir earthquake (Mw 7.8). Microearthquake data in the Sikkim Himalaya revealed active faults in the region (De and Kayal, 2003 and 2004), and any such active fault can trigger a large earthquake.

Repeat microgravity measurements over last five years across the Sikkim-Darjeeling Himalaya have indicated significant crustal deformation, and recorded a gravity changes of the order of 200 μGal in Gangtok area (Lahiri and Chakraborty, 2004; Lahiri et al., 2007). Recently a medium size felt earthquake M 5.7 occurred on February 12, 2006 about 60 km north of Gangtok. The 200 μGal change in gravity value could be a precursor of the February 12, 2006 felt earthquake. These data indicate vulnerability of a larger earthquake in the Sikkim Himalaya (Lahiri et al., 2007).

Geological and seismological evidences suggest several regional active faults and transverse lineaments in the region. The oblique motion of the Indian plate toward the southern margin of the Eurasian plate is decoupled between interplate under thrusting fault (Plane of Detachment) and intraplate lateral-slip along transcurrent faults like the Chaman, Karakuram and Sagaing faults (Fig. 5.57). Because of interaction of two components of relative plate motion due to oblique convergence, the region bounded by the transcurrent faults and frontal thrusts behaves as a tectonic sliver (Tapponnier and Molnar, 1976; Nakata, 1989). Active faulting of various types and various directions indicate that stress field is not consistent everywhere. Yeats and Lillie (1991) and Yeats and Thakur (1998) argued on contemporary tectonics and reassessment on earthquake hazards based on blind thrust and also on a fault-bend fold model. The microearthquake seismology made a significant contribution in identifying the active seismogenic faults/structure in different tectonic blocks of the Himalaya, but much remains to be done. Future work may fill up the gap areas.

Although our understanding about the earthquake processes or active faults in the Himalaya is enhanced, the general awareness of potential seismic hazard is yet to be developed. Increasing populations, increasing urbanization and construction of non-resistant houses and buildings are the fatal combination to earthquake hazards in the Himalaya. Arya (2000) studied the expected degree of seismic hazard in the western Himalaya due to a hypothetical recurrence of earthquake of M 8.0 in Kangra area, Himachal Pradesh in the seismic zone V (Fig. 5.58a). If all the houses (about 19×10^5 homes) in the Kangra area in zone V are without earthquake safety provisions, the direct loss will amount to about Rs. 51 billion and about 65,000 lives may be lost; the trauma will be too much and cost of emergency relief will be exorbitant. A recent estimate by Wyss (2005) suggests that as many as 150,000 people may die, 300,000 may be injured and 3000 settlements will be effected in a single event in the Himalaya. These results are based on the average rupture dimension of the great historic earthquakes, rupture length ~200 km and width 70 km, magnitude 8.1 and hypothetical depth 25 km. The M_W 7.8 Kashmir earthquake of October 8, 2005 caused approximately 80,000 fatalities, and Wyss (2005) subsequently in the revised estimate suggested that the number of fatalities in future Himalayan earthquakes may even range from 100,000 to 500,000.

Khattri et al. (1984) produced a probabilistic seismic hazard map showing peak ground acceleration in rocks for India and adjoining regions having a 10% probability of being exceeded in 50 years. It was reported that intensity attenuation for the Indian region and the eastern United States was similar. The largest probabilistic accelerations are obtained in the seismotectonic belts of Pamir, Hindukush, Himalaya, Arakan Yoma and the Shillong massif where values of over 70% have been estimated.

In a recent study, a new seismic hazard map for the Indian plate region comprising the Himalaya, northeast India, Andaman regions, south China, Nepal, Burma was prepared under the Global Seismic Hazard Assessment Program (GSHAP) (Fig. 5.58b). The map reveals that the zones of highest risk are the Burmese arc, northeast India and the MHSB and the Hindukush region with PGA values of the order of 0.35-0.40 g. The north Indian plate boundary and the Tibetan Plateau region have hazard level of the order of 0.25 g. In the Indian shield region, it is of the order of 0.5–0.1 g, whereas some locales like Konya depict a hazard level of about 0.2 g (Fig. 5.58a).

Paleoseismology has contributed significantly to earthquake hazard assessment globally. As the instrument records are not available beyond 100-150 years, it is difficult to assess the recurrence period of a large/great earthquake. Paleoseismic studies provide an authentic estimate of past seismicity by identifying, characterizing and dating important geological evidences of *paleoliquefaction*. In India such studies have now started, and recently some significant contributions are made (e.g. Sukhija et al., 1999a and b; Rajendran and Rajendran, 1999, 2002, 2003 and 2004). The meizoseismal area of the 1934 great Bihar earthquake, a 12,000 sq km slump belt in Bihar, was undertaken for paleoseismological investigation (Sukhija et al., 2005). In this investigation, the 1833 or older earthquake was discovered

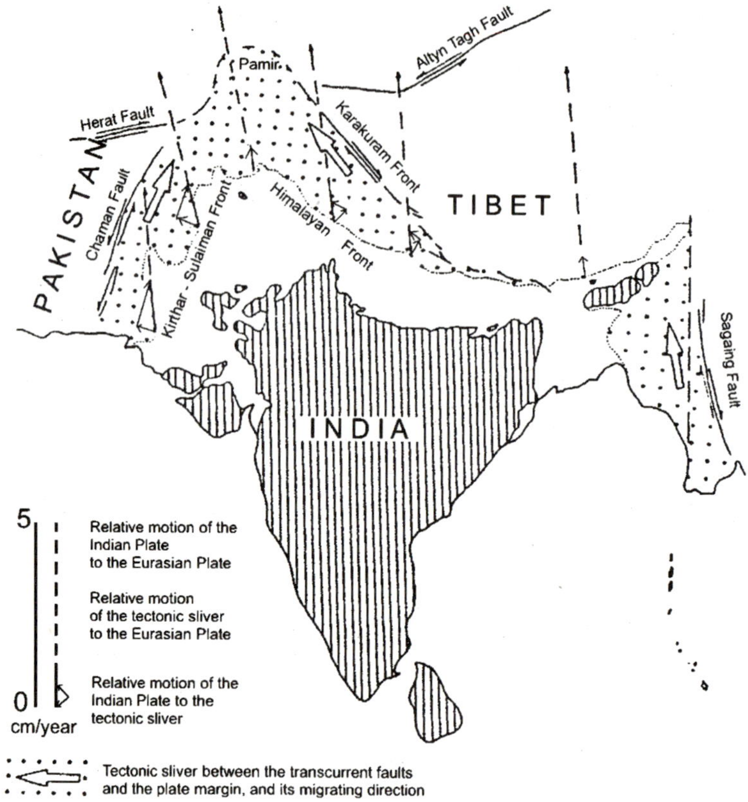

Fig. 5.57 Tectonic model for the relative plate motions along the Indo-Eurasian plate boundary (Nakata, 1989).

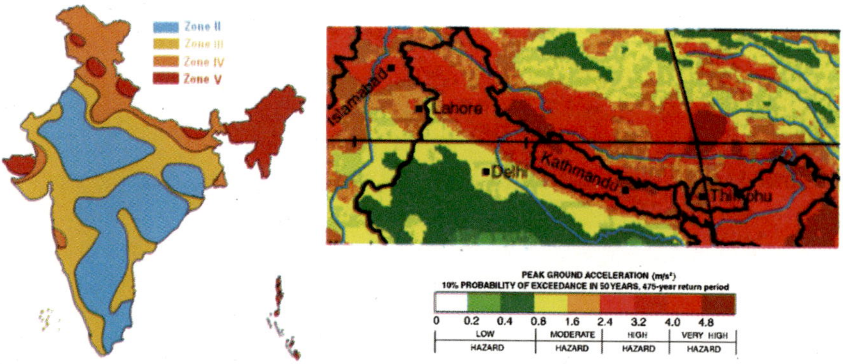

(a) (b)

Fig. 5.58 (a) Sesimic zones (V) in the Himalaya are shown in dark patches (BIS, 2002). (b) Probabilistic seismic hazard map of the Himalaya with peak acceleration having 10% probability of exceeding in 50 years (Bhatia et al., 1999).

as evidenced by sand dykes and sill *seismites* that provide upper bond of 180 ± 80 and 280 ± 80 yr BP. Earthquake occurrences between 1800 and 25000 yr BP were also evidenced; no recurrence period, however, could be established; the river floods seem to have washed away liquefaction features (Shukija et al., 2005). Further, an attempt was made to study paleoseismic evidences in the central Himalaya by the NGRI in collaboration with Tribhuvan University, Nepal. Several sites were identified where paleoliquefaction and soft sediments deformation could be clearly established. Radio carbon dating led to identification of recent earthquakes 1934/1988 and older events between 25,000 to 45,000 yr B.P. (Sukhija et al., 2005). However, much more work is needed to reconstruct the past history of earthquakes in the Himalaya.

Based on the recent studies on seismicity, GPS, paleoseismology and the GSHAP map, Khattri (2006) has suggested that there is a need of further reviewing the seismic zoning map of India (BSI 1983–2002). The isolated zones V in the seismic zoning map (Fig. 5.58a) and the severely threatened zones in the Himalaya are not comparable with the GSHAP map (Fig. 5.58b). These isolated zones V (Fig. 5.58a) are defined based on the high intensity records of the past great earthquakes. The GSHAP map, on the other hand, shows that the entire Himalayan plate boundary is expected to generate higher accelerations uniformly, and indicate a high level of seismic hazard for the entire Himalayan belt (Fig. 5.58b).

6

Northeast India, Myanmar, Bangladesh and Andaman-Sumatra Region

6.1 INTRODUCTION

Northeastern part of India and its adjoining areas of southeast Asia comprise an interesting seismotectonic region of the world. The region includes eastern and northeastern Himalayas to the north, Indo-Burma ranges/Myanmar to the east, Bangladesh to the south and the Andaman-Sumatra region to the southeast (Fig. 6.1). The Shillong Plateau, Mikir hills, Assam valley, Tripura fold belt and the Bengal basin (Bangladesh) in the study region jawed between the two tectonic arcs, the Himalayan arc to the north and the Burmese-Andaman arc to the east (Fig. 6.2a). The present-day tectonics of the northeast India region is complicated because of interaction between the active north-south convergence along the Himalaya (e.g. Molnar et al. 1977; Ni and Barazangi, 1984) and the east-west convergence and folding within the Indo-Burma ranges, a style of deformation that apparently evolved from subduction (e.g. Mitchell, 1981; Curray et al., 1979 and 1982). Several contrasting views are put forward concerning the seismotectonics of northeast India. Uyeda and Kanamori (1979) postulated that the Alpine-Himalayan orogenic belt and the Andaman-Indonesian mobile belt are linked by a transform fault, and the ophiolites of the Indo-Burma ranges are product of a leaky transform fault. Nandy (1983), on the other hand, postulated that the eastern sector of the Indian plate collided with the Burmese block, and the ophiolites were emplaced along the Benioff Zone (Fig. 6.2b). Le Dain et al. (1984) suggested that the Indian plate was subducting beneath Burma in the early Cenozoic, and now the hanging lithoshpere is being dragged northward through the asthenosphere. Mukhopadhyay and Dasgupta (1988), Kayal (1989, 1996a) and Satyabala (2003), however, argued that the Indian plate actively subducts below the Burmese arc.

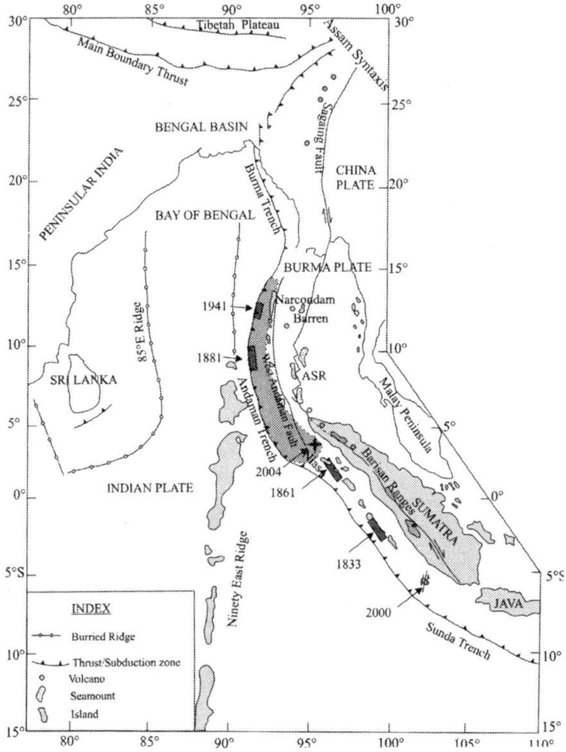

Fig. 6.1 Tectonic setting of NE India, Indo-Burma ranges and Andaman-Sumatra Island arc (modified from Kayal et al., 2004).

The northeast India region is highly populated, and seismically most active. The region falls in the zone V in the seismic zoning map of India (Fig. 2.20). It is, therefore, of great interest to the geoscientists, seismologists and earthquake engineers to understand the mechanism of the seismic events and their effects for designing earthquake resistant civil structures for planning of hazard-free socio-economic projects. The region is fairly well equipped with weak motion as well as with strong motion instruments, compared to other regions in the Himalaya. The Indo-Burma ranges and the Andaman Islands are, however, not so well instrumented except for a few special investigations. We shall discuss the seismotectonics of northeast India and Indo-Burma ranges here, and separately discuss the Andaman-Sunda arc tectonics. Very recently, the mega thrust earthquake (M_w 9.3) of December 26, 2004 occurred in the Andaman-Sunda arc. Another great earthquake (M~ 8.7) occurred on March 28, 2005 about 200 km southeast of the 2004 mega thrust event in the same subduction zone. A brief description of these events are given here. More details about these events are published. A separate volume of book on the seismicity and seismotectonics of southeast Asia needs to be dealt with.

6.2 GEOLOGICAL STRUCTURE

The geological structure and the tectonic units of northeast India region and Indo-Burma ranges are briefed as follows:

6.2.1 Northeastern Himalaya: Himalayan Arc

As discussed in Chapter 5, the major thrust/fault systems are spanning the entire length of the Himalayan arc including the northeastern Himalaya. From north to south these are: the *Indus Suture Thrust* (IST), the *Main Central Thrust* (MCT), the *Main Boundary Thrust* (MBT) and the *Himalayan Frontal Thrust* (HFT) (Fig. 6.2a). In addition to these, the gigantic geomorphological feature, the Brahmaputra river (fault) runs almost parallel to the MBT/HFT along the Assam valley, and takes a turn at 90°E longitude to run parallel to the Dhubri/Jamuna fault (Fig. 6.2a). The northeastern Himalaya may also be subdivided into four major units by the above thrust systems. These from north to south are: the *Trans-Axial Himalaya*, the *Central Himalaya,* the *Lesser Himalaya* and the *Outer Sub-Himalaya*.

The *Outer Sub-Himalaya*, to the south of MBT, i.e. the north Brahmaputra sub-Himalayan basin, mainly consists of Neogene and Quaternary molasse sediments which are terminated to the north by the MBT against the pre-Tertiary rocks of the Lesser Himalaya. The homoclinal northerly dipping formations of the Outer Himalaya are correlated to the Middle and the Upper Siwalik formations of the western Himalaya. The sub-Himalayan sedimentary basin is separated from the Upper Assam Tertiary shelf. The shelf sediments indicate gentle northeasterly dip. The Paleogene sediments thicken southeastward in the Upper Assam shelf (Acharyya et al., 1986; Rao, 1983).

The *Lesser Himalaya*, the well defined Himalayan belt between the MBT and MCT, tectonically underlie the central Himalaya crystallines. While the MBT is clearly defined all along the southern margin of the Himalaya, the position of MCT is debated particularly in the northeastern part (Krishnaswamy, 1981). The MCT have sinuous outcrop pattern in the eastern and northeastern Himalaya. The Lesser Himalayan rocks, specially in the northeastern sector, are associated with basic and acid volcanics and volcanoclastics. The occurrence of the Gondwana (Permian) rocks below the tectonic cover of the Proterozoic metasediments and the outcrops of thick Eocene sequence below the nappes of the Lesser Himalayan rocks corroborate the low dipping nature of the Himalayan nappes viz, MBT, MCT (Acharyya et al., 1983). Subsurface data other than seismicity are not available in the Lesser Himalaya.

The *Central Himalaya* comprises metamorphic and granitoid rocks and higher structural units to the north of MCT. The rocks are affected by Proterozoic metamorphism, granitoid activity with several reactivation and finally with an overprint of Tertiary deformation and neometamorphism

(a)

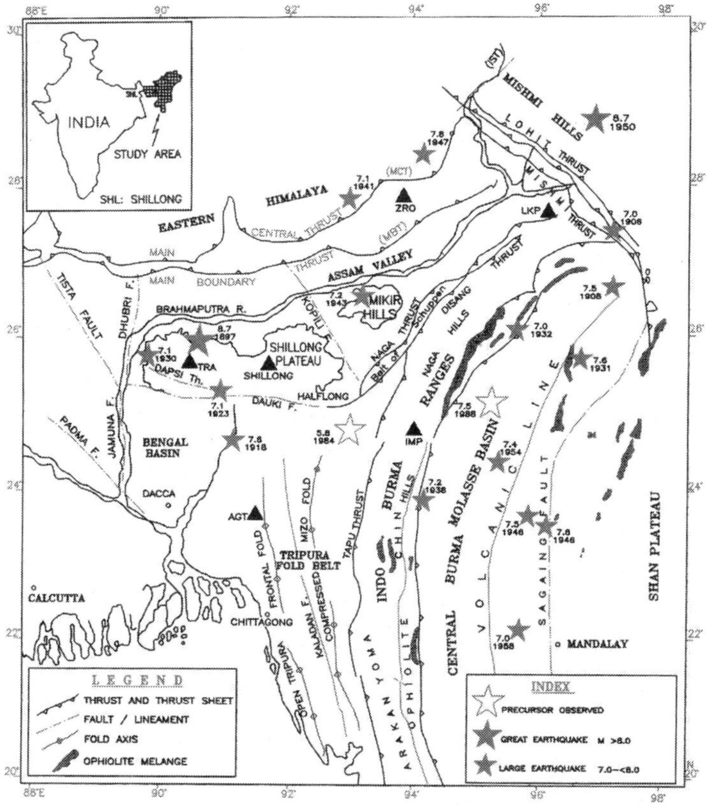

(b)

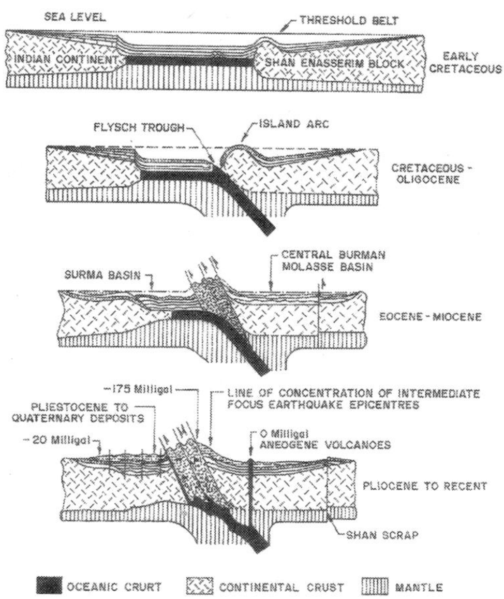

Fig. 6.2 (a) Tectonic map of NE India region showing large earthquakes (M> 7.0) including two great (M ~ 8.7) earthquakes (solid stars), two recent damaging earthquakes are shown by open stars, the permanent stations of the national network (IMD) are shown by solid triangles (Kayal, 1998). Inset: Study area in NE India is shown. (b) Tectonic evolution of the Burmese arc (Nandy, 2001).

(Acharyya et al., 1986). The Himalayan metamorphic rocks are delimited to the north by an E-W lineament, which takes a NE swing in the east. To the north of this line occur folded Palaeozoic-Mesozoic sediments and Mesozoic granite of the Tibetan platform and the Tang-ku-la mountain range with a cover of Tertiary sediments in the platform areas.

In the *Trans-Axial Himalayas*, Phanerozoic metasediments/sediments either tectonically or gradationally overlie the crystallines, the tectonic contact being defined by the Tethyan/Trans-Axial Thrust or Indus Suture Thrust (IST). Cretaceous flysh-ophiolites occur along the Tsangpo suture. The Trans-Axial sedimentary sequence represents nearly complete Phanerozoic epicontinental fossiliferous sediments of Tethyan affinity with a few minor breaks, and is known as the Tethyan Himalaya. The sediments are characterised by intervening flows of basic volcanics (Acharyya et al., 1986). Though definite ages of various fold movement in the northeastern Himalaya are not known, the intensive fold movements in the Trans Himalaya region took place during Cretaceous (Chang Ta, 1963). The Creataceous flysh ophiolites or ophiolite melange occur along the Indus-Tsangpo suture.

6.2.2 Indo-Burma Ranges: Burmese Arc

The Indo-Burma ranges is situated in a complex tectonic zone with oblique subduction at its western boundary, dextral transform fault (Sagaing fault) on the eastern boundary, Mishmi thrust in the north as a buttress, and the Andaman Spreading Ridge (ASR) to the south (Figs 6.1 and 6.2). The plate boundary, the Burmese arc, is nearly 1100 km long in north-south direction, convex westward. The arc comprises a western fold belt, called Indo-Burma ranges, and the Burmese low lands, Chindwin basin, to the east. The Sagaing fault that runs parallel to the Indo-Burma ranges, separates the Shan Plateau to farther east (Fig. 6.2). The Indo-Burma ranges constitute Arakan-Yoma, Chin and Naga hills, which pass northeastward to link with the Himalayas. Southward, the ranges continue into the Andaman and Nicobar islands. The fold-belt width exceeds 150 km in its central part. The north-south fold axes are results of east-west compression acting regionally against the Burmese arc. The average elevation of the Indo-Burma ranges is 1000 m. The Burmese low lands are divided into eastern and western troughs by Burmese volcanic arc (Mitchell and McKerrow, 1975). The prominent transform fault, the Sagaing fault (Fig. 6.2), defines the eastern limit of the Burmese low lands against the Shan Plateau which is part of the southeast Asian plate. Curray et al. (1979) and Le Dain et al. (1984) suggested that the northward translation of India past southeast Asian plate is accommodated, at least by part, by slip on the northerly trending Sagaing fault. The Sagaing fault connects southward with the Andaman back arc spreading ridge (Fig. 6.1).

Several contrasting views are advanced in the literature about the evolution of the Burmese arc. Mitchell and McKerrow (1975) attribute the evolution of the arc to eastward subduction of the Indian lithosphere at the Asian continental margin, which was initiated in late Cretaceous; the subduction is continuing to the present affecting the subduction zone located west of the Indo-Burma ranges. It is generally accepted that the sea-floor spreading along the Carlsberg Ridge pushed the Indian continent northeast and east towards the Shan-Tenasserim block in Burma during the Cenozoic, and the Indo-Burma ranges were formed as a result of subduction of the Indian plate under the *Burma platelet* (Curray and Moore, 1974; Curray et al., 1979; Mitchell, 1981). During subduction, the thick turbidite prism with ophiolite melanges was progressively thrust into the Indian plate. Shortening apparently has taken place more or less continuously from the late Cretaceous to the present. The flysh is tightly folded all along the ranges but more intensely in the Naga hills than further south (Fig. 6.2b). The flysh is overlain by Miocene deltaic or freshwater molasses along the western edge of the ranges (Curray and Moore, 1974). The tectonic evolution is depicted in Fig. 6.2b. In some areas on the eastern side of Arakan Yoma, metamorphic schists and ultrabasics have been overthrust westward onto the flysh. During the close of Oligocene the flysh trough and the threshold belt got uplifted and formed an elevated landmass. Neogene Surma basin was formed to the west of the uplifted landmass, and to the east the central Burma molasse basin continued from Eocene to Recent. The Neogene volcanics along the axis of the central Burma molasse basin has been formed due to the subducting oceanic plate below the *Burma platelet* (Curray et al., 1979; Mitchell, 1981). The Indian oceanic lithosphere dips at an angle 30-60° and seismically active down to a depth of about 200 km (Ni et al., 1989; Kayal, 1996a; Kumar et al., 1996; Nandy, 2001; Satyabala, 2003).

Curray et al. (1979) suggested that the Burma platelet, which includes the Burmese low lands, forms a structural province between the Indo-Burma ranges and the Indo-China block (southeast Asian plate). The Burma platelet has been created since middle Miocene consequent to opening of the Andaman sea by at least 460 km (Curray et al., 1982). Tapponnier et al. (1982) suggested that active spreading under the Andaman Sea and lateral motion along the Sagaing transform fault in Burma are consequences of propagating extrusion tectonics in response to rigid indentation by India into Asia at the northeastern Himalayan syntaxis, called Assam syntaxis (Fig. 6.2a).

The Naga hills, the northeast part of the Indo-Burma ranges, consist of an entire Tertiary succession of Eocene to Pliocene rocks and called a *Belt of Schuppen* (Evans, 1964). It consists of more than eight overthrusts along which the rocks moved northwestwards relative to the *Foreland Spur* (Section 6.2.4). These thrusts form a complex pattern, one thrust overriding the other. This elongated belt, ranging from 13 to 14 km in width, presents a uniform pattern of overthrust masses. The northwestern edge of this belt closely follows the alluvial boundary of upper Assam for over 350 km, and finally ends near Halflong to the south (Fig. 6.2). The northwestern margin of the Belt of Schuppen is called the Naga Thrust. It is not a simple plane throughout, rather it is made up of succession of different thrusts. A single thrust may have a stratigraphic throw of anything upto 9 km and total horizontal movement of all the thrusts may well exceed 200 km (Evans, 1964).

Nagative free air and isostatic anomalies over the Indo-Burma ranges (e.g. Verma et al., 1976a; Warsi and Molnar, 1977) imply a mass deficit that could result from a deeper basement with thicker sediments (see Section 6.3). A downward flexure of the Indian lithosphere, as at deep sea trenches, beneath the Indo-Burma ranges could be responsible for such a pile of sediments. The distribution of intermediate-depth events shows a single inclined zone dipping east beneath the Indo-Burma ranges, which suggests the existence of a subducting slab of ocean lithosphere under the eastern part of ranges (see Section 6.4).

6.2.3 Eastern Himalaya Syntaxis

The Eastern Himalaya Syntaxis (also called Assam syntaxis) joins the MBT in the Himalaya with the Burma arc (Fig. 6.1). At the MBT, the Indian plate is under thrusting the Lesser Himalaya at shallow angles (Gansser, 1964; Fitch, 1970; Molnar et al., 1973; Ni and Barazangi, 1984; Baranawaski et al., 1984); at the Burma arc the Indian oceanic lithosphere dips at angles of 30-60° and is seismically active down to depths of 150-200 km (Ni et al., 1989; Kayal, 1996a). Convergence at the Burma arc is highly oblique. The Burma platelet, which lies between the Burma arc and Sagaing fault, is compressed northward by the oblique subduction (Mitchell and McKerrow, 1975; Curray et al., 1979; Ni et al., 1989; Curray, 1989). Essentially, the Eastern Syntaxis can be viewed as a complex triple junction that joins the Indian and Eurasian plates with the northern end of the Burma platelet (Curray, 1989). The crustal deformation to relative plate motions is complex. In the southeast and east of the Eastern Syntaxis within east Burma and Yunnan (China), the relative motion between India and southeast Asia is dextral; in the west and north of the Eastern Syntaxis, within the Eastern Himalaya and southeast Tibet, the motion between India and Asia is convergent (Holt et al., 1991).

The crustal thickness estimated in this region range from 50 to 60 km (Holt and Wallace, 1990). The great 1950 Assam earthquake occurred within this syntaxis region. Armijo et al. (1989) suggested that the 1950 earthquake involved right-lateral slip, which is consistent with their hypothesis that right-lateral slip on the Jiali-Po-Qu fault zone in southeast Tibet around the Eastern Syntaxis is connected with the right-lateral strike slip motion on the Sagaing fault (Fig. 6.2a). It has been suggested that absence of ophiolite in the Eastern Syntaxis is evidence of right-lateral slip on the faults that wrap around the syntaxis (Mitchell, 1981; Arnijo et al., 1989).

The Mishmi hills occur at the junction of the northeastern Himalaya and the Indo-Burma ranges. The Himalayan arc takes a sharp turn of about 90° and meets the Indo-Burma ranges. The rocks of the eastern Lesser Himalaya and the Central crystallines appear to be largely attenuated and truncated in the Mishmi hills by Lohit thrust, Tidding suture and the Mishmi thrust (Figs 6.2a and 6.3). The frontal Mishmi thrust overrides the Quaternary sediments of the Mishmi foot hills and finally truncates the Indo-Burma fold belt and associated thrust system (Acharyya et al., 1986).

6.2.4 Shillong-Mikir Massif

The Shillong-Mikir massif, forming a northeastern prolongation of the Indian shield, occurs as a pivot to the west of the Naga hills and to the south of the Eastern Himalayas. The Shillong Plateau has an average elevation of about 1000 m (Fig. 6.3), and consists of high grade metamorphic basement rocks of Archaean age which are unconformably overlain by the mildly deformed Proterozoic shelf sediments. The shelf sediments comprise the Shillong Group with the metavolcanic Khasi greenstone and late Precambrian granite plutons (Nandy, 2001). To the south the Plateau is limited by the ~320 km long E-W Dauki fault system, and to the west by the Jamuna and Dhubri faults (Fig. 6.2 a). The Dauki fault separates the Precambrian basement of the Shillong Plateau to the north and the thick Tertiary sediments of the Bengal basin to the south. Within the Plateau region, the contact between the Tertiary sediments to the southwest and the Archaean gneiss to the northeast is demarcated by the 90~100 km long NW-SE trending Dapsi thrust, a northwest extension of the Dauki fault (Kayal and De, 1991; Nandy, 2001). Oldham (1899) reported that a 19~20 km long NW trending fault, called Chedrang fault, was developed by the great Shillong earthquake of 1897 within the Plateau (Fig. 6.3); about 10 m vertical displacement was observed during the earthquake. Besides, there are several NE-SW and NW-SE criss-cross photo-imagery lineaments in the whole region (Nandy, 2001). The Mikir hills are detached from the Plateau by the NW-SE trending Kopili fracture zone, called the Kopili lineament (Fig. 6.3a). This is about a 300-400 km long lineament that crosses the Assam valley and reaches to the MBT; this lineament is transverse to the Himalayan trend and parallel to the Bomdila lineament in the Arunachal Himalaya. This is an important active tectonic feature lying between the Shillong Plateau and the Mikir hills; the area between the two masiffs is called the Kopili gap (Nandy, 2001).

There are controversies on the movements along the Dauki fault. Results of drilling show that the sedimentary rocks in the Bengal basin are underlain by a basement of basaltic lava flows of early Cretaceous/late Jurassic (Bakshi, 1965; Sengupta, 1966). These flows are correlated with the Rajmahal volcanic group of eastern India (Sengupta, 1966; Paul and Lian, 1975). The large contrast in the depth of basement, from 1 km above sea level in the Shillong

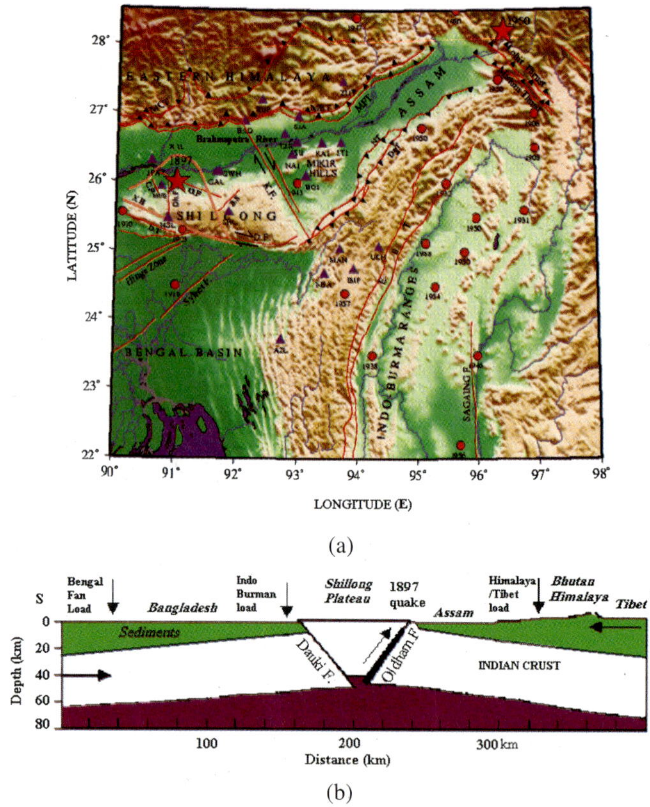

Fig. 6.3 **(a)** Topographic map with tectonic features of NE India region, permanent digital seismic stations are shown by triangles. Epicentres of the large earthquakes are annotated, the two great earthquakes are shown by star symbols. The maximum intensity (XII, MM Scale) of the 1897 great earthquake is shown, MCT: Main Central Thrust, MBT: Main Boundary Thrust, NT: Naga Thrust, DsT: Disang Thrust, EBT: East Boundary Thrust, KF: Kopili Fault, DF: Dauki Fault, DT: Dapsi Thrust, DhF: Dudhnoi Fault, ChF: Chedrang Fault, OF: Oldham Fault, BS: Barapani Shear zone (Kayal et al., 2006). **(b)** Pop-up tectonic model of the Shillong Plateau (Bilham and England, 2001).

Plateau to 13 km or more below sea level in the Bengal basin (e.g. Mathur and Evans, 1964; Sengupta, 1966; Paul and Lian, 1975; Hiller and Elahi, 1984), indicates that the Dauki fault is a major fault. Evans (1964) inferred right-lateral strike-slip movement on this fault, and suggested that 200 km or more of right-lateral displacement occurred on this fault since Miocene to detach the Plateau from the Indian shield. He further inferred the Dauki fault to be a tear fault, therefore extending only through the upper crust. Many authors (e.g. Murthy et al., 1969; Desikachar, 1974; Dasgupta, 1979; Rao, 1983; Hiller and Elahi, 1984; Khan et al., 1988), on the other hand, emphasized vertical displacement across the fault. Molnar (1987) suggested, from Oldham's (1899) interpretation of the rupture zone of 1897 great earthquake, that probably the rupture involved largely thrust faulting on a plane dipping gently northward. The apparent right-lateral offsets inferred by Evans (1964) can be explained by thrust faulting on the Dauki fault, if the strikes of the stratigraphic markers are northeast, oblique to the E-W trend of the fault (Chen and Molnar, 1990). Nandy (2001), however, argued that the large contrast in the depth of basement (~18 km) between the Shillong Plateau and Bengal basin is difficult to explain by any kind of thrust tectonics at whatever angle it might have moved southward.

Geologically the entire area has evolved during the Mesozoic to Tertiary. Along with the mountain building in Burma and in Himalaya, large scale vertical movements have taken place resulting in uplift of the Shillong Plateau in northeast India and the Shan Plateau in Burma during the Tertiary (Krishnan, 1960; Murthy, 1970). Kailasam (1979) reported a relative uplift of the Shillong Plateau of the order of 2.5 cm in 70 years. On the southern side of the Plateau, Sylhet traps (Jurassic?) are present. During Jurassic to Cretaceous the Plateau broke up along its southern margin by the E-W trending Dauki fault; the vertical uplift of the Plateau caused subsidence to the north as well as to the south of the Plateau. Subsidence to the north led to the deposition of sediments (3-4 km) in the lower Assam (Brahmaputra) valley up to the MCT, while subsidence to the south led to the deposition of thicker (~15 km) sediments in the Bengal basin. The Shillong Plateau, Mikir massif and the Brahmaputra valley together is called *Foreland Spur* (Evans, 1964).

Recently, Bilham and England (2001) suggested 'pop-up tectonics' for the Shillong Plateau, which is bounded by two reverse faults, the north dipping Dauki fault to the south and WNW-ESE trending south dipping deep seated 'Oldham fault' at the boundary of the Plateau and the Brahmaputra valley (Fig. 6.3b). From the geodetic observations they argued that the Shillong Plateau rose violently by at least 11 m during the 1897 great earthquake and this was due to a burried reverse fault, named it Oldham fault, approximately 110 km in length and dipping steeply away from the Himalaya (Fig. 6.3b).

6.2.5 Assam Valley

The Assam valley flanks the Shillong-Mikir massif with the basement gently slopping towards northeast as well as towards southeast (Dasgupta, 1977). The Cenozoic sedimentation on the Assam-Shillong shelf begins with the deposition of the Jaintia Group (Paleocene-Eocene), which has a large spread in the upper Assam sub-surface and in the Mikir hills. From bottom to upwards it is mainly made up of basal arkoses, blanket limestone and black shale. The deposition of the Eocene sequence on the platform under shallow marine to lagoonal environments was primarily controlled by vertical movements along the basement faults (Dasgupta, 1977). The Oligocene sequence in Assam shelf is represented by 600-1000 m thick sediments. The upper Assam valley, between the Himalaya and Naga hills, is flat, about 100 km in width and 300 km in length, and extends northeast-southwest (Fig. 6.3). Beneath the alluvium, 4-6 km thick succession of Tertiary rocks overlie the Precambrian basement complex that is exposed in the Mikir hills and in the Shillong Plateau.

6.2.6 Bengal Basin

The Bengal basin together with its off-shore continuation, the *Bengal Fan* in northeast Indian ocean, constitutes the Bengal geosyncline that evolved in the Mesozoic-Cenozoic (Evans, 1964). This is one of the largest geosynclinal basins of the world, almost twice as large as that of Mississippi delta. Its western part is called the Gangetic West Bengal basin, and its eastern part that forms the deeper basin is now the Bangladesh. Geologically this is an interesting region because it contains ~15 km of sediments (Evans, 1964; Talukdar, 1982). The area of the geosyncline shrunk with time in consequence of the eastward subduction of the Indian plate below the Andaman-Burmese arc (Curray et al., 1979). Several stages are considered significant in the geological history. These are: collision of India with Eurasia, uplift and erosion of the Himalaya in the Tertiary-Quaternary, transport of the eroded materials by confluent Ganges and Brahmaputra rivers and deposition of the sediments in the newly formed Bengal geosyncline (Curray and Moore, 1971; Johnson et al., 1976; Curray et al., 1982). It is interesting to note that due to continued spreading of the Indian ocean floor, the above stages are still in operation. The most interesting tectonic feature of the Bengal basin is perhaps the *Eocene Hinge Zone,* which separates the continental shelf in the west and the geosynclinal facies in the east (Evans, 1964), (Fig. 6.3a). The Hinge Zone is atleast 500 km long. Its width varies from 25 km in the north to 110 km in the central part and 35 km in the south. Curray et al. (1982) suggested that the Hinge line represents the boundary between the continental crust and the young oceanic crust that extends southwards into the Bay of Bengal. They further inferred that the boundary of continent-ocean transition beneath the Bengal basin intersects the Dauki fault near 92°E, and at least part of the Bengal basin might be underlain by the oldest part of the Indian ocean floor. The E-W Dauki fault would separate ancient continental crust of the Indian shield to the north from the Cretaceous ocean floor to the south. Paul and Lian (1975) suggested that part of the Bengal basin indicates oceanic crust beneath the area (now Bangladesh). From surface wave analysis of the teleseismic events Brune and Singh (1986), on the other hand, advocated continent like crust beneath the Bay of Bengal.

6.3 GRAVITY ANOMALY

Evans and Crompton (1946) first published a composite Bouguer gravity map of northeast India and adjoining areas. Since 1946 various geophysical surveys including gravity, magnetic, seismic and drilling have been carried out by many organisations like the Assam Oil, Oil India, Standard Vacuum Oil Company, Indo Stanvac Petroleum Project (ISPP), Oil and Natural Gas Commission (ONGC) and Geological Survey of India (GSI) (e.g. Mathur and Evans, 1964; Ray et al., 1964; Murti, 1969; Chowdhury and Dutta, 1973). The investigations made by the Burma Oil Company (Evans, 1964) and Pakistan Petroleum Ltd (Khan and Azad, 1963) provide considerable information on the subsurface geology of deeper basin areas. Ten deep boreholes were drilled between 1957 and 1960 in the Gangetic West Bengal basin under the ISPP, which penetrated sediments ranging in age from Cretaceous to Recent overlying basaltic lava flows of early Cretaceous. Under the regional survey programme of the National Geophysical Research Institute (NGRI), a gravity survey was conducted in northeast India during 1972-73. About 400 gravity stations were made, specially in areas where gravity data so far had been insufficient. Verma and Mukhopadhyay (1977) prepared Bouguer and isostatic gravity maps of the region (Figs 6.4 and 6.5) using the data of Survey of India (Gulatee, 1956), Burma Oil Company (Evans and Crompton, 1946), ONGC and NGRI. The maps show several distinctive features; these are briefly discussed below.

Two major anomaly trends prevail in the Bouguer gravity map. One is east-west trend, almost parallel to the structural trend of the eastern Himalaya, and the other is the north-south trend parallel to the structural trend of the Indo-Burma ranges. The east-west trend predominates over the Brahmaputra valley, Shillong Plateau and Surma valley. The E-W trend changes to northeast-southwest over upper Assam valley conforming to the structural trend of the geologic formations in the area. A north-south trend predominates over the Bengal basin and the fold belt of Tripura. The gravity field shows large variations, from +44 mGal over Shillong Plateau to –255 mGal over upper Assam valley and eastern Himalaya. In the Bengal basin, the coastal anomalies are close to zero. Steep gravity gradient is observed over the outer edge of the Mikir massif and southern edge of the Shillong Plateau, along the Dauki fault (Fig. 6.4).

The isostatic gravity map shows a variation from +100 mGal over the Shillong Plateau to -130 mGal over the upper Assam valley and eastern Himalaya (Fig. 6.5). The isostatic anomaly is close to zero in the Bengal basin. Relating isostatic anomaly with the tectonic evolution, Verma and Mukhopadhyay (1977) suggested that along with the root formation in the uplifted blocks of the Himalayas and the Naga hills, the sinking foreland, the Assam valley, was possibly forced to sink into the mantle creating isostatic

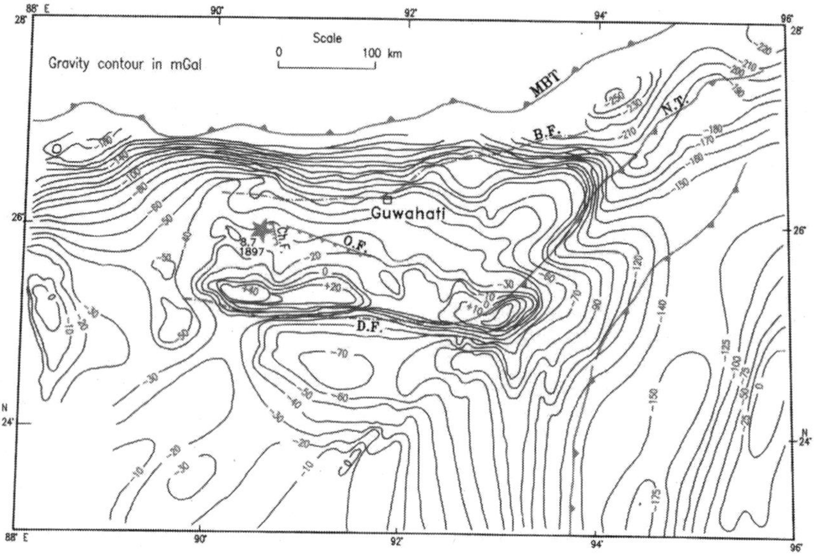

Fig. 6.4 Bouguer gravity map of NE India (Verma and Mukhopadhyay, 1977) and major tectonic features are shown (see Fig. 6.3a); OF: Oldham Fault (Bilham and England, 2001), and BF: Brahmaputra Fault (Nandy, 2001).

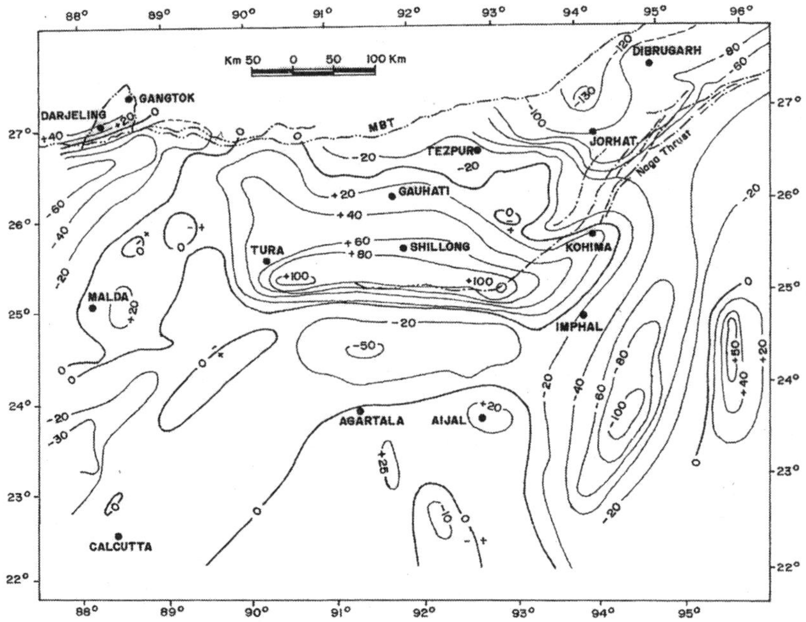

Fig. 6.5 Isostatic anomaly map of NE India (Verma and Mukhopadhyaya, 1977).

imbalance. It is, therefore, possible that the upper Assam valley has reached a state of isostatic imbalance since the Plio-Pleistocene, the period of the main uplift of the Himalayas. The vertical movements played a major role in tectonic evolution of the Shillong Plateau, which could have resulted in densification of the crust and the positive isostatic anomaly. In the Bay of Bengal, on the other hand, the crust may be of intermediate type, between continent and ocean. The crust possibly subsided in the late Cretaceous to late Tertiary, and is at present in isostatic equilibrium.

Chen and Molnar (1990), however, argued that if local isostatic equilibrium prevailed and if the elevation of the Shillong Plateau was compensated by thick root, a negative anomaly would have resulted instead of the prevailing positive anomaly. They have explained the high elevation of the Plateau by the support of a strong lithosphere; the oceanic crust below the Bengal basin might have under thrust beneath the Plateau. They also suggested that the Dauki fault could be a thrust fault, more in keeping with the concept of Oldham (1899) than the idea of right-lateral strike-slip proposed by Evans (1964).

Increasingly negative Bouguer and isostatic anomalies, as low as -150 to ~175 mGal, toward the eastern part of the Indo-Burma ranges suggest that the Indian plate has flexed down considerably during the active subduction beneath the ranges. A mass deficit because of thicker sediement (~70 km) due to downward flexure of the Indian lithosphere is responsible for this large negative anomalies (e.g. Mukhopadhyay and Dasgupta, 1988; Kayal, 1989 and 1996b) (Section 6.4.3). The distribution of intermediate depth earthquakes, which shows an inclined seismic zone (Benioff zone) dipping east at about 30°–45° beneath the Indo-Burma ranges, suggests the existence of the subducting slab (e.g. Mukhopadhyay and Dasgupta, 1988; Kayal, 1989 and 1996a). The weight of the lithospheric slab attached to the Indian plate and hanging beneath the eastern edge of the Indo-Burma ranges would provide the force that holds the region to the west out of isostatic equilibrium (Le Dian et al., 1984).

6.4 CRUSTAL STRUCTURE

Based on available geological information and nature of isostatic anomalies, possible crustal thicknesses beneath the Shillong Plateau, Assam valley and Bengal basin were estimated by Verma and Mukhopadhyay (1977). These are briefly discussed below.

6.4.1 Crustal Thickness

The gravity field over Shillong Plateau is distinctly positive (+40 mGal) as compared to its surroundings. This is a distinct feature of the Plateau, which makes it one of the most anomalous parts of the Indian shield. The Shillong Plateau has an average elevation of 1 km, and according to Airy's theory for perfect compensation, it should have a crustal thickness of 40.6 km with mean density 2.84 gm/cc (Verma and Mukhopadhyay, 1977). Such a crustal column would produce zero isostatic anomaly, but the Plateau is characterised by positive Bouguer as well as positive isostatic anomaly indicating mass excess in the crustal and subcrustal levels (Figs 6.4 and 6.5). Tandon (1954), Qureshy (1971) and Mukhopadhyay (1974), based on seismological and gravity data, suggested thicker and denser crust beneath the Plateau. Verma and Mukhopadhyay (1977) modelled the crustal thickness of the order of 46 km with density 2.96 gm/cc beneath the Plateau to interpret the gravity anomaly over the Plateau.

The Bouguer anomaly varies from –30 mGal at the outer edge of the Shillong-Mikir massif to –220 mGal near the upper Assam depression (Fig. 6.4). The Assam valley has an average sedimentary thickness of about 4 km with average surface elevation of about 0.1 km. Assuming that the sediments of mean density 2.40 gm/cc are underlain by a crust of normal density 2.84 gm/cc, the crustal thickness, excluding sediments, was estimated to be 24.74 km for isostatic equilibrium. It is, however, observed that the isostatic anomaly over the Assam valley is negative (Fig. 6.5). Verma and Mukhopadhyay (1977) have shown that a crust of 29.65 km underlying sediments of 4.1 km could explain the isostatic anomaly of the order of –120 mGal. Hence for a crust of mean density 2.84 gm/cc, the thickness of crust under the Assam valley should be of the order of 31-34 km inclusive of sediments.

The Bengal basin is characterised by nearly zero isostatic anomalies over major parts except in Surma valley (Fig. 6.5). This is inspite of large thickness of sediments of the order of 13-15 km. Applying geological correction for sediments it is seen that the isostatic anomalies would become positive over most parts of the Bengal basin. This suggests that the overall density of the crust underlying the sediments should be higher than normal. Based on this observation, Verma and Mukhopadhyay (1977) estimated thickness of the crust underlying the sediments would be about 20 km. Thus the total crustal thickness including sediments would be about 33-35 km in the Bengal basin.

6.4.2 Mohorovičić Discontinuity

The depths to the Mohorovičić (Moho) discontinuity have been computed by Gaur and Bhattacharji (1983) from the gravity data of northeast India region lying between latitude 22°-29°N and longitude 88°-98°E (Fig. 6.6). The computed depths from the Bouguer and isostatic gravity maps are compatible within half a kilometre. The Moho surface map is presented in Fig. 6.6. The Moho depth beneath the Shillong Plateau is of the order of 46 km but deepen to the north below the eastern Himalaya as well as to the east below the

Indo-Burma ranges. The estimated Moho depth beneath the Shillong Plateau is comparable with the estimated crustal thickness as reported by Verma and Mukhopadhya (1977). The northward slope of the Moho is estimated to be 4°; it is comparable with the estimated slope of the basement below the Indo-Gangetic Plain further west (Hayden, 1918).

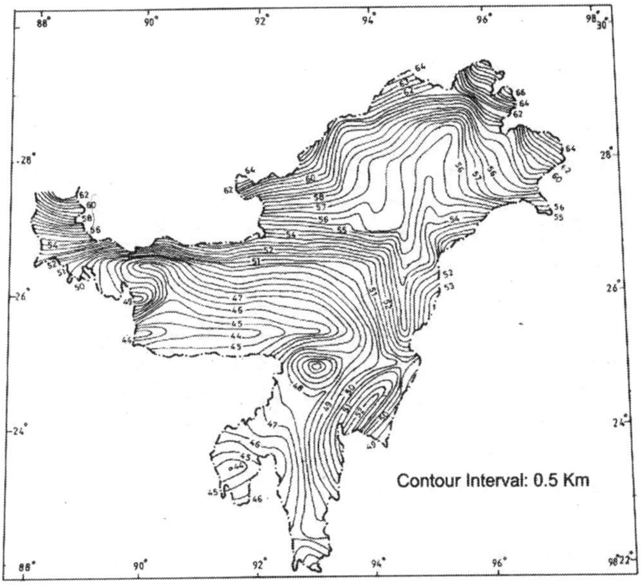

Fig. 6.6 Moho level map of NE India, depth contours are in km (Gaur and Bhattacharji, 1983).

Receiver function analysis of the recent broadband teleseismic data (Kumar et al., 2004; Mitra et al., 2005; Ramesh et al., 2005), however, shows that the crust is thinner (~35-38 km) beneath the Shillong Plateau. The Moho beneath the northern part of Bengal basin to the south of the Plateau is at a depth of ~44 km and beneath the Brahmaputra valley to the north, the Moho depth is 40-42 km, deeper by ~5–7 km than below the Shillong Plateau (Fig. 6.7). Further north of the foredeep, the Moho dips gently northward reaching a depth of ~48 km beneath the Arunachal Himalaya, and 88 km below Lasha in Tibet (Mitra et al., 2005).

Kaila et al. (1992) conducted deep seismic soundings (DSS) in the West Bengal basin. They reported that the Moho occurs at a depth 26-36 km. The configuration of Moho indicates that the crust is more akin to continental in nature in the western and middle parts of the West Bengal basin.

The early crustal models across the Himalaya derived from modelling gravity data (Kono, 1974) in this area showed that Bouguer anomaly increased from –150 mGal at the MBT to –300 mGal beneath the Great Himalayas and

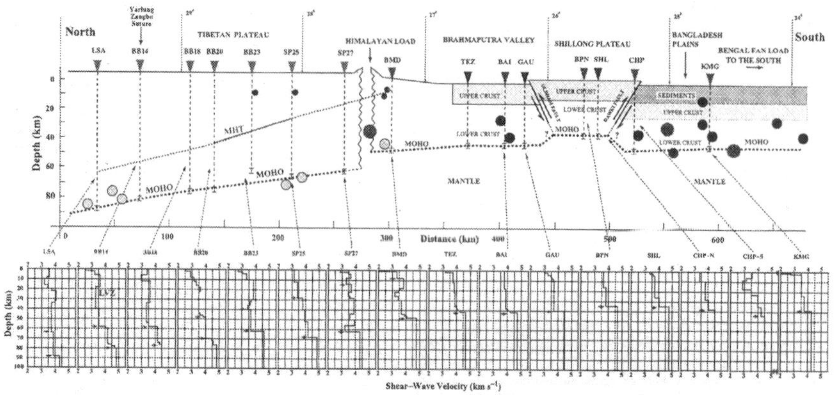

Fig. 6.7 Moho depth from receiver function across the Bengal basin-Shillong Plateau-Brahmaputra valley-Himalaya-Tibetan plateau; earthquakes are shown by solid and open circles (Mitra et al., 2005, copy right @ AGU).

–500 mGal beneath southern Tibet. This gravity variation was interpreted as due to moderately dipping Moho from Himalayan foothills into Tibet and approximate doubling of crustal thickness from ~35 km in shield area to ~70 km in southern Tibet. The DSS across the southern Tibet, great Himalaya and lesser Himalaya revealed crustal thickness ~75 km below southern Tibet and ~55 km beneath the great Himalaya. These results are consistent with the Receiver function analysis (Mitra et al., 2005).

6.5 SEISMOLOGICAL DATA

A major constraint in preparing a detailed seismological catalog for northeast India is the sparse seismic network in the northeastern Himalaya and in the Indo-Burma ranges, particularly before 1964. Since the inception of the WWSSN in 1964 and with the increase of local networks in the region since 1980s, location quality of the epicentres is much improved over the years. The catalogs of the earlier data are prepared by several authors (e.g. Dutta, 1964; Gupta et al., 1986; Bapat et al., 1983). The existing seismological data can broadly be classed under the following heads:

6.5.1 Historical Earthquakes Prior to 1897

The June 12, 1897 great earthquake (M 8.7) appears to be the first Indian earthquake for which instrumental records, recorded outside India, are available (Oldham, 1899). Gupta et al. (1986) have treated the earthquakes prior to 1897 as historical earthquakes. Based on a detailed literature survey, a descriptive list of the historical earthquakes was given by Oldham (1883). The earthquake of 1548 is the earliest one for which some description is available (Table 2.3). Epicentres of the reported historical earthquakes, magnitude 7.0 and above or intensity more than equal to VIII (MM scale), are shown in Fig. 6.2. A brief description of a few significant historical earthquakes in northeast India and adjoining region is given below.

Cachar Earthquake (M 7.5), 1869

The January 10, 1869 Cachar earthquake (M 7.5) is one of the largest historical earthquakes in northeast India region (Fig. 6.2.). Some descriptions of this earthquake were given by Oldham and Oldham (1882). The felt area was about 6,65,600 sq km. Several damages were reported in Guwahati, Cachar, Nagaland, Manipur and Sylhet (Bangladesh). Oldham and Oldham (1882) deduced the epicentral tract 5-6 km wide and 32-48 km long, trending nearly in the E-W direction around the intersection of latitude 26°0′N and longitude 92°40′E, at the northern border of the Jaintia hills, Meghalaya. The epicentral tract is located in the Kopili gap area (Nandy, 2001), the area between the Shillong Plateau and Mikir hills; the focal depth was estimated at 50 km.

Other Large Earthquakes in Adjoining Area

Several large historical earthquakes in the adjoining areas of northeast India region along the Burmese-Andaman-Sumatra arc, seem to have been associated with continued seismic activity in the region. An extensive change of elevation on the Burma coast apparently was caused by the great *Arakan earthquake of April 2, 1762* (Richter, 1958) (Fig. 6.2). Then following an earthquake in 1878, uplift of 6 m occurred on the west coast of Ramree island (at 19°10′N) while another island further north, False island, seems to have disappeared (Richter, 1958).

A few large earthquakes are reported along the Andaman-Sumatra arc. In 1881 a large event (M 7.9), west of the Andaman Islands, generated a small tsunami. The rupture zones of 1833, 1861, 1881 and 1941 large earthquakes (M 7.7~8.4) are shown (Fig. 6.1). The 1881 earthquake is known to have been accompanied by several months of aftershocks and to have caused landslides. Tipper (1906) remarked that the stability of the islands is such that not much change was observed. The recent December 26, 2004 mega thrust event, however, overshot the historic rupture zones, and it propagated from north Sumatra to north Andaman, about 1300 km long and 200 km wide rupture caused devastating Tsunami and about 3,00,000 casualties. A brief description of this mega thrust event is given in Section 6.10.

6.5.2 Recent Large and Damaging Earthquakes Since 1897

There are many shortcomings, like inaccurate timing, location, magnitude etc. in the earlier catalogs. Gupta et al. (1986) prepared a corrected list of earthquakes, which has a total of 504 entries for the period 1897 to 1964. It may be noted that 22 large earthquakes (magnitude ≥7.0), including the two great earthquakes (M >8.0), occurred during the last 100+ years, since 1897 in northeast India and its surroundings (Fig. 6.3). The two great earthquakes of 1897 and 1950 respectively are discussed in Chapter 2. Brief descriptions of the four recent large and damaging earthquakes (1918, 1930, 1984 and 1988) that occurred in the study area are given below. A reassessment of the 1897 earthquake is also briefly given here.

Shillong Earthquake (M 8.7, 1897)

The 12 June, 1897 Shillong earthquake (M 8.7) is the largest intraplate event in the last two centuries; details of this great earthquake is given in Section 2.7. Oldham (1899) devised a scale for intensity assessment that consists of six grades. He assigned maximum intensity on grade 6, which is equivalent to X in the Rosi Forel (RF) scale; it covered an area of 160 km radius. Ambraseys and Bilham (2003) reevaluated the maximum intensity to VIII on the MSK-1981 scale, which covers an area of ~90 km radius. They also estimated the magnitude ~7.0 only, which is not comparable with the reported ground acceleration by Oldham (1899).

Oldham (1899) postulated that the event occurred on a north-dipping thrust fault. For many years a gentle north-dipping thrust beneath the Shillong Plateau, linked structurally to the style of Himalayan thrust faulting, was considered to be the source area (Seeber and Armbuster, 1981; Molnar and Pandey, 1989; Kayal and De, 1991). This interpretation is recently shown to be untenable. Bilham and England (2001), based on geodetic and GPS data, inferred that the 1897 earthquake was generated by a rupture on a south-dipping 110 km long reverse fault at the northern boundary of the Plateau; they named it the 'Oldham fault' that extends from a 9 km depth to more than 40 km depth. A detailed discussion on the seismotectonic model of the Shillong Plateau is given in Section 6.7.

Srimangal Earthquake (m_b 7.6), 1918

The July 8, 1918 earthquake (m_b 7.6) in the Bengal basin (Fig. 6.2), also known as the Srimangal earthquake, is one of the largest earthquakes that was recorded by a few local seismographs, viz. Bombay, Colaba and Calcutta. The epicentre was given at 24.5°N and 91.0°E, and depth at 15 km (Table 2.3). The meizoseismal zone covered an area of 332 sq km and was felt over 122,000 sq km. The maximum intensity X (Rossi-Forel scale) was reported around Srimangal (24°15′, 91°42′) in Sylhet district, Bangladesh. Many tea estates were ruined but no casualty was reported. Post earthquake investigation was carried out by Stuart (1926), and he reported that major axes of the isoseismals were aligned in the NW-SE (Fig. 6.8a). Nandy (2001) had, however, redrawn the major axes in the ENE-WSW direction, and suggested that the earthquake was originated by the high angle reverse Sylhet fault (Fig. 6.3). Stuart (1926) estimated the focal depth at about 14 km from the isoseismals.

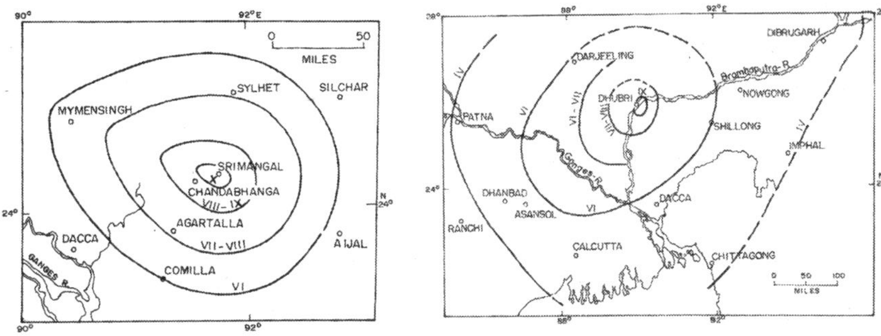

(a) (b)

Fig. 6.8 (a) Isosismal map of the 1918 Srimangal earthquake m_b 7.6. (b) Isoseismal map of the 1930 Dhubri earthquake M 7.1 (GSI, 2000a).

Dhubri Earthquake (M 7.1), 1930

The July 2, 1930 earthquake (M 7.1) occurred in Dhubri area of Assam. Epicentre of this earthquake was estimated at latitude 25°.5N and longitude 90°.0E. (Fig. 6.3). The maximum intensity was reported to be IX on the Rossi-Forel scale around Dhubri town and adjoining area (Fig. 6.8b). A considerable depth (~60 km) was suggested from the isoseismal map (Table 2.3). It was perhaps remarkable that no loss of life was reported; only a few instances of injuries were reported. Effected area was of the order of 322,000 square miles extending from Dhubri to Manipur valley in the east, to Patna in the west, to Calcutta and Chittagong (Bangladesh) in the south, and to Bhutan and Sikkim in the north. A detailed description of the damages was documented by Gee (1934). The shock was estimated to have lasted for three to five minutes as reported from different places. It was preceded, for a period of several seconds, by a rumbling sound, which was described as an approaching train, thunder, passing motor car etc. by the observers at different places. A number of clocks in the town were reported to have stopped at 2-35 to 2-37 A.M., July 3, 1930, Indian Standard Time (IST). From the local geology and damage pattern, the N-S Dhubri or Jamuna fault (Fig. 6.2) was assigned to be the source of this event (Nandy, 2001).

No evidence of *foreshock* was reported preceding the main shock. Numerous *aftershocks* followed the earthquake, which, however, diminished considerably with time. During the first 24 hours, 54 aftershocks were recorded. On 4th and 5th July, 37 and 34 aftershocks respectively were recorded, whereas after 5th July on average about five shocks per day were recorded during the first week (Gee, 1934).

Cachar Earthquake (M 5.8), 1984

The Cachar earthquake, M_L 5.8, m_b 5.6, origin time: 1984, December 30, 23h 33m 39.1s, latitude: 24°.598N, longitude: 92°.939E, depth: 33 km restricted (USGS report), although not a large event magnitude-wise (Fig. 6.3), but it rocked the entire area in the early hours of December 31, and was responsible for a loss of about 20 human lives and considerable damage to properties (GSI, 2000).

The maximum intensity of the earthquake reached VIII on the Modified Mercalli (MM) scale (GSI, 2000). The isoseismal map indicated the intensity VIII with a linear and narrow extent (Fig. 6.9). The VIII intensity was assigned in view of the formation of fissures in flat ground through which sand and water were ejected. In the meizoseismal area of about 250 sq km, a considerable damage was caused to a well designed concrete structure (bridge) and total collapse of the weak structures (Fig. 6.9b).

(a) (b)

Fig. 6.9 (a) Isoseismal (sketch) map of the 1984 Cachar earthquake, m_b 5.8, and (b) an example of a damaged mud-house in the maximum intensity zone (GSI News Letter, 1985).

Based on the rapid attenuation of the isoseismals, it was suggested that the earthquake was probably focussed at a shallower depth (≤10 km) (GSI, 2000). From the instrumental records the depth of this event is poorly constrained. Dziewonski et al. (1985) computed a depth of about 100 km, and the USGS gave a restricted depth of 33 km. Unfortunately the long period P and S phases for this event at most of the WWSSN stations were masked by the coda of a large earthquake near New Zealand; no clear depth phases could be seen on the available WWSSN seismograms (Chen and Molnar, 1990). Thus the focal depth could not be estimated precisely.

The earthquake was followed by seven felt aftershocks in the subsequent 11 days (Dube et al., 1986). Focal mechanism of the main shock indicated a thrust-fault mechanism with a compressional stress in the NW-SE direction (Dube et al., 1986). Dziewonski et al. (1985) reported the CMT solution which also shows a pure thrust faulting (Section 6.5.4).

Indo-Burma Earthquake (M 7.5), 1988

The Indo-Burma earthquake of August 6, 1988, M_L 7.5, M_s 7.2, m_b 6.8, origin time: 00h 36m 29.6s (GMT), latitude 25°.116N, longitude 95°.171E, depth 115 km (USGS report) (Fig. 6.3), rocked the whole northeastern region of the country in the early morning of August 6 at 6.10 AM (local time). The tremor was felt all over northeast India, Bangladesh and parts of Burma. It lasted for about two minutes, and caused a loss of four human lives, considerable damage to buildings, railway tracts, roads etc. Field survey showed that the maximum intensity reached VIII on the MM scale (GSI, 1988, unpublished report). The survey revealed three areas (viz. Jorhat, Silchar and Diphu) where maximum damages were recorded. Landslides, formation of fissures, ejection of sand, mud and water were observed at several places.

The focal mechanism study based on P-wave first-motions indicated thrust-faulting solution with a compressional stress parallel to the trend of tectonic trend of the Indo-Burma ranges (Banghar, 1990). The result of inversion, however, shows a mixture of strike-slip and reverse-fault mechanism (Chen and Molnar, 1990). They have well estimated the focal depth at 90 $\pm$ 10 km from the unambiguous P_p and S_p phases on both the long- and short-period seismograms. Considering the depth and the fault-plane solution, there is little doubt that this large earthquake occurred within the subducted Indian plate. The deeper focal depth provided a plausible explanation for small number of casualties and less damage (Person, 1989). The centroid moment tensor solution is similar to that obtained by Chen and Molnar (1990) (Section 6.5.4).

6.5.3 Regional Seismicity Since 1964

Regional seismicity maps are prepared by several workers considering the data from various sources (e.g. Santo, 1969; Fitch, 1970; Chandra, 1975; Le Dain et al., 1984; Mukhopadhyay and Dasgupta, 1988). The earthquake listings published by the International Seismological Centre (ISC) and the United States Geological Survey (USGS) since 1964 were used by Kayal (1996a, 1998) to prepare a seismicity map of the northeast region (Fig. 6.10). It is observed that during this period, almost all the earthquakes of body-wave magnitude (m_b) ≥4.5 have been uniformly recorded and located. In addition to this, the great earthquakes (M>8.0) and large earthquakes (M>7.0), that occurred during the last 100+ years, since 1897 are also shown in this map. The seismic activities in different tectonic zones are characteristically different as discussed below.

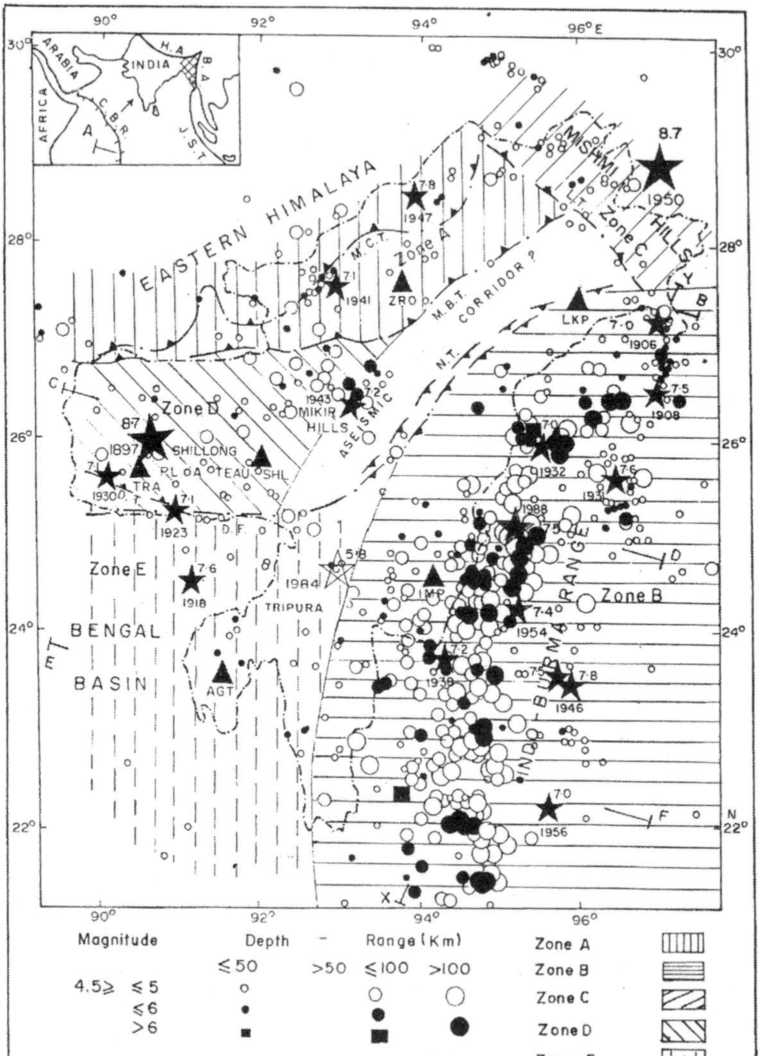

Fig. 6.10 Seismicity map (1964-1996) of NE India (data source: USGS and ISC), stars indicate the large earthquakes (M> 7.0), two larger stars are the epicentres of the two great earthquakes (M~ 8.7) in the region, the open star indicates the 1984 Cachar earthquake m_b 5.8, triangles are the permanent stations of the national network (Kayal, 1998). Inset: Regional tectonic setting, CBR: Carlsberg Ridge, HA: Himalayan Arc, BA: Burmese Arc, JST: Java-Sumatra Trench

Northeastern Himalaya

In the northeastern Himalaya the seismic activity though appears to be sparse, two large earthquakes (M>7.0) occurred in this zone: one on January 21, 1941 (M 7.0) and the other on July 29, 1947 (M 7.8) respectively. The seismic activity is, however, high in the syntaxis zone (Fig. 6.10). The syntaxis zone was also the source area of the great 1950 Assam earthquake (M 8.7). A vertical section of the earthquakes along A-B (Fig. 6.11a), almost in perpendicular direction of the Himalayan trend, indicates that the earthquakes are confined within a depth of about 70 km, and the activity is more or less uniform along the entire section except a cluster of activity beneath the Mishmi hills in the syntaxis zone (Fig. 6.11a).

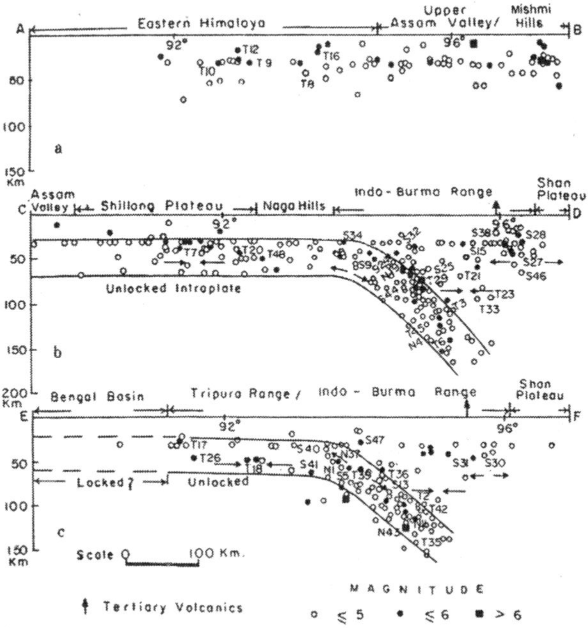

Fig. 6.11 Vertical cross sections of the earthquakes along the selected profiles, NE India (Fig. 6.10), (a) cross section along A-B across the eastern Himalaya, (b) cross section along C-D across the Shillong Plateau-Indo-Burma ranges, (c) cross section along E-F across the Bengal basin-Tripura fold-Indo-Burma ranges, T: thrust, N: normal and S: strike-slip faulting earthquakes (Kayal, 1998).

Indo-Burma Ranges

In the Indo-Burma ranges, the earthquake epicentres are highly concentrated. As many as 10 large earthquakes M>7.0 occurred during the last 100+ years since 1897. The years of occurrence and magnitudes of the large earthquakes are annotated (Figs 6.3 and 6.10). A depth section of the earthquakes in the NNE-SSW strike direction of the ranges (along X-Y, Fig. 6.10), shows that the activity is intense and more or less uniform down to a depth of about 200 km, but beyond the latitude 26° N there is a marked shallowing of the lower limit of activity where collision process has taken over the subduction process (Fig. 6.12) (Mitchell and McKerrow, 1975; Kayal, 1989 and 1996a). Two vertical cross sections of the earthquakes in perpendicular direction of the strike of the Indo-Burma ranges (along C-D and E-F, Fig. 6.10), which fall to the south of 26°N latitude, evidently show a (40-45°) dipping seismic zone or the *Benioff zone* (Fig. 6.11b and c). An envelope of the seismic zone beneath the Indo-Burma ranges is clearly indicated. The upper boundary of the envelope demarcates normal faulting in the subducting plate, and thrust-faulting as well as strike-slip faulting in the overriding plate (Kayal, 1996a and 1998). The lower boundary completes the envelope of the seismic zone or the Benioff zone. The depth of the top-boundary is about 25 km below the Shillong Plateau/Tripura fold belt, and the Benioff zone flexes beneath the Indo-Burma ranges. The thickness of the Benioff zone varies from 40 to 45 km. The Benioff zone structure is conformable with the observed Bouguer gravity anomaly (Kayal, 1989 and 1996a) (Fig. 6.13). A similar observation was made by Mukhopadhyaya and Dasgupta (1988).

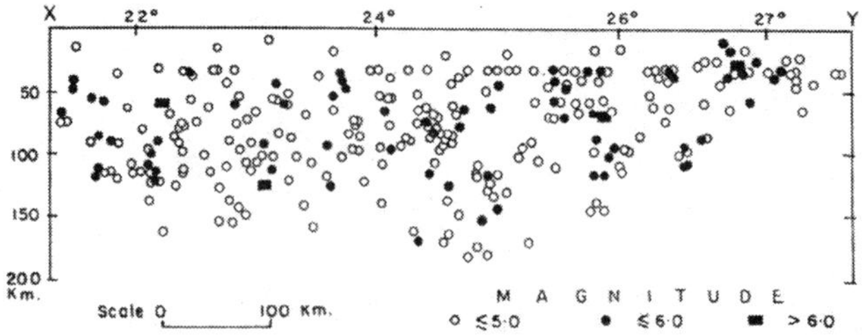

Fig. 6.12 Vertical cross section of the earthquakes along X-Y, strike direction of the Indo-Burma ranges (Fig. 6.10) (Kayal, 1998).

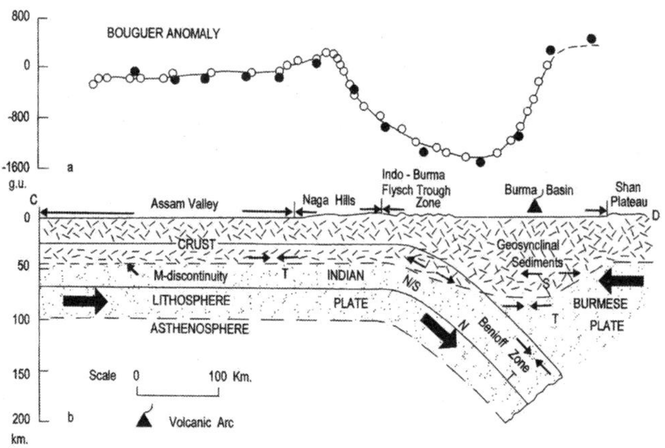

Fig. 6.13 (a) Bouguer gravity profile along C-D (Fig. 6.10), open circles indicate the observed gravity values and the solid circles computed. (b) Model of the dipping Indian lithosphere below the Indo-Burma ranges, the Benioff zone, is adopted from Fig. 6.11b. T: Thrust, N: Normal and S: Strike slip fault regime (Kayal, 1996a).

The foci distribution in the inclined seismic zone suggests that depth penetration of the subducting lithosphere is about 200 km in the central part of the Indo-Burma ranges (Fig. 6.11b), and about 150 km to the south (Fig. 6.11c). The volcanic arc position corresponds to the deepest part of the Benioff zone. The shallower seismicity in the Burmese platelet to the east of the Benioff zone is related to the overriding-plate seismicity and the shallower seismicity to the west of the Benioff zone is related to the plate-boundary/intraplate seismicity in the Indian plate (Kayal, 1996a). Bevis and Isacks (1984) studied in the ISC data, 1964-1986, and presented the hypocentral trend on a surface contour map (Fig. 6.14). The eastern edge of the Indo-Burma ranges roughly coincides with the map projection of the 60 km depth contour of the Benioff zone (Fig. 6.14). The map projection of 60 km depth contour approximately separates the accretionary complex from the southeast Asian continental crust. They further estimated the dip of the inclined Benioff zone which varies from about 50^0 southeast in the north near the syntaxis to about 30° north-northeast in the Bay of Bengal area.

The absence of intermediate-depth earthquakes beyond about 26°N marks the northern end of the Burma arc (Fig. 6.12). In the southern Burma arc (south of 20°N and north of 14°N), it is reported that there are no well constrained hypocentres with depths greater than 50 km, and it is not possible to characterise the shape of the Benioff zone in this part of the arc (Ni et al., 1989; Kumar and Rao, 1995). This is possibly a transition zone between the Burma arc and the Andaman-Sunda arc (Kayal et al., 2004).

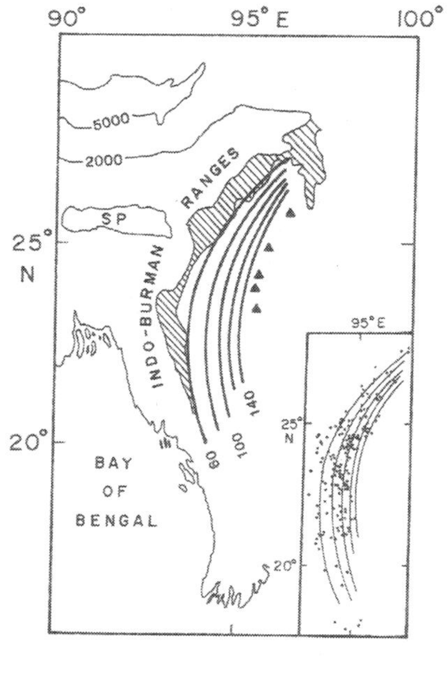

Fig. 6.14 Surface contours of the hypocentral trends in the Indo-Burma ranges (Bevis and Isacks, 1984).

Shillong Plateau and Assam Valley

The seismic activity is quite high in the Shillong Plateau as well as in the Mikir hills area (Fig. 6.10). Three large earthquakes M>7.0 occurred in this zone, and as we discussed earlier the Shillong Plateau was the source area of the 1897 great Shillong earthquake M 8.7 (Fig. 6.3). The hypocentres of earthquakes in the Plateau region are mostly within 50 km as seen in the depth section (Fig. 6.11b). The large/damaging earthquake M 7.1 of July 2, 1930 occurred on the Dhubri fault, at the western margin of the Shillong Plateau, and the September 9, 1923 earthquake M 7.1 occurred at the southern edge of the Plateau, on the Dauki fault (Fig. 6.10). The focal depths for both the events were estimated at about 50 km (Table 2.3).

The activity in the Assam (Brahmaputra) valley, to the north of Plateau, is much less (Fig. 6.10). It is interesting to note that the upper Assam valley, the area between the syntaxis and the Shillong Plateau, is quiescent. Khattri and Wyss (1978) demarcated this area as *Assam Gap* for an impending large earthquake. Mukhopadhyay (1984) and Kayal (1996a), however, marked this zone as *Aseismic*; Kayal (1996a) named it *Aseismic corridor* (Fig. 6.10). The trend of this aseismic zone is conformable with the NE-SW trend of the upper Assam valley. The zone is about 100 km wide and 300 km in length.

Bengal Basin and Tripura Fold Belt

The Bengal basin is characterised by low seismic activity, whereas a moderate activity is observed in the Tripura fold belt. The activity in the Bengal basin may be related to intraplate seismicity. One large earthquake m_b 7.6 occurred beneath the Bengal basin on July 8, 1918 (Fig. 6.10). The epicentre of this event falls in the Srimangal area, Bangladesh, and is known as the Srimangal earthquake of 1918 (Section 6.5.2). The felt area was about 8,00,000 square miles. The activity in the Tripura fold belt, on the other hand, may be related to the plate-boundary activity. Recently a damaging earthquake, M_L 5.8, known as the 1984 Cachar earthquake, occurred in the Tripura fold belt (Fig. 6.10); it is discussed in Section 6.5.2.

6.5.4 Fault-plane Solution

P-wave First-motion Solutions

Fault-plane solutions of about 50 events were published by several authors (e.g. Fitch, 1970; Rastogi et al., 1973; Tandon and Srivastava, 1975; Verma et al., 1976a and b; Mukhopadhyay, 1984, Dube et al., 1986; Mukhopadhayay and Dasgupta, 1988) using the polarities of P-waves reported in the published bulletins. These solutions were compiled by Kayal (1996a). These solutions are annotated in the depth sections as N (Normal faulting), T (Thrust faulting) and as S (Strike slip faulting) (Fig. 6.11), which fairly explain the Benioff zone earthquakes and the overriding plate earthquakes. Although some of these solutions are fairly consistent with the CMT solutions, it is difficult to assess the quality of these solutions because of inconsistent readings and often unreliable bulletin data (Chen and Molnar, 1990).

Chen and Molnar (1990) presented 17 reliable fault-plane solutions for the region after reanalysing the source parameters using the original seismograms and taking care of necessary corrections. These solutions are shown in Fig. 6.15. Four events (1, 2, 3 and 6) occurred close to the Dauki fault on the southern edge of the Shillong Plateau. The event 6 (February 6, 1988, m_b 5.8, depth 31 $\pm$ 3 km) is the largest one known in the immediate vicinity of the Dauki fault. They argued that although a nearly pure strike-slip fault-plane solution is obtained, it cannot be associated with the Dauki fault as the solution indicates compression in N-S direction, normal to the fault. A N-S compression is dominant in all the fault-plane solutions occurred in or near the Shillong Plateau (events 1-6). All the solutions show a component of reverse faulting, but slip does not seem to have occurred on a gently dipping plane except possibly for the event 4 (August 18, 1968; M 5.1, 29 $\pm$ 3 km), north of the Plateau.

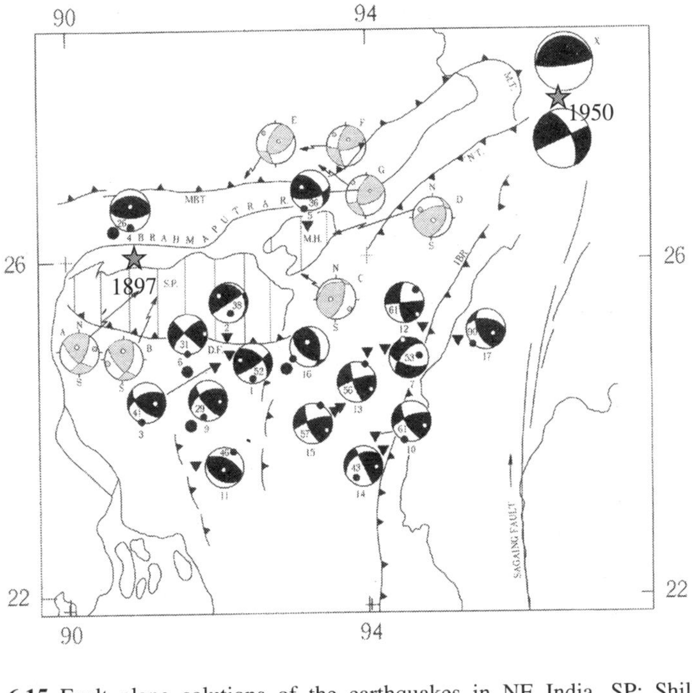

Fig. 6.15 Fault plane solutions of the earthquakes in NE India, SP: Shillong Plateau, MH: Mikir Hills, IBR: Indo-Burma Ranges. Tectonic features are same as Fig. 6.3. The CMT solutions are represented by the conventional black beach-balls (Chen and Molnar, 1990) and the composite solutions (A-G) of microearthquakes by grey beach balls (Kayal, 2001), X and Y represent two solutions of the 1950 great earthquake, the two great earthquakes (M~8.7) are shown by star symbols (see text).

The event 17 (August 6, 1988, m_b 6.8, M_S 7.2, depth 90 ± 10 km) is the largest earthquake that occurred beneath the Indo-Burma ranges in the recent years (Section 6.5.2). The fault-plane solution obtained by Chen and Molnar (1990) is almost similar to the CMT solution. The solution is a mixture of reverse and strike-slip faulting showing a NNE-SSW compressional stress (Fig. 6.15). The events 13 (May 31, 1973) and 15 (May 6, 1984) occurred close together, both having similar magnitude ~5.7 and depth 57 ± 5 km, and show almost similar solution with a large strike-slip component. Similar fault-plane solutions are also obtained for the events 10 (May 29, 1970), 12 (December 29, 1991) and 14 (May 21, 1975) magnitude ~5.1 –5.6, focal-depths 40-60 km in the Indo-Burma ranges (Fig. 6.15). The solution for the event 7 (February 18, 1965, M 5.4, depth 53 ± 6 km), which shows a large component of normal faulting, is the only solution that differs from the oblique thrust faulting observed for the other events beneath the Indo-Burma ranges.

The two events 9 (December 27, 1968, M 5.1, depth 29 ± 4 km) and 1! (February 02, 1971, m_b 5.4, depth 46 ± 5 km) that occurred beneath the Bengal basin show large component of thrust faulting with NNE-SSW compressional stress. These two events occurred at a depth of 30-40 km, below the folded sedimentary rocks, possibly in the lower crust. None of the nodal planes of the solutions are parallel to the trend of the local NNE thrust/ folds in the area (Fig. 6.15). The event 16 (Cachar earthquake of December 30, 1984, m_b 5.8) at shallower depth, the damaging earthquake that occurred in the Tripura fold belt in the recent years (Section 6.5.2), shows a near thrust faulting on either of the two moderately dipping nodal planes that strike northwesterly. Again the strikes of the nodal planes are oblique to the local NNE trend of the Indo-Burma ranges (Fig. 6.15).

Two fault plane solutions were given for the 1950 great Assam earthquake (Fig. 6.15), the solution X by Chen and Molnar (1977) and the solution Y by Ben-Menahem et al. (1974). Chen and Molnar (1977) used first-motion data, and they inferred a shallow dipping thrust fault much like other thrusts in the Himalaya. Ben-Menahem et al. (1974), on the other hand, reported a right lateral pure strike-slip fault from amplitude observation.

Seven composite solutions (A-G), reverse faulting with strike slip motion, were obtained from microearthquake data (Fig. 6.15), four in the Shillong Plateau-Assam Valley area and three in the Arunachal Himalaya. These solutions are discussed in Section 6.6.

Ni et al. (1989) reported 15 reliable fault-plane solutions in the Indo-Burma ranges (Fig. 6.16). Solutions of five earthquakes (6, 7, 11, 13 and 14) that occurred within the Indian slab below the central and northern Burma ranges, show strike slip faulting with P-axes that plunge towards north-northeast, parallel to the oblique convergence. Three events (2, 3 and 10) that occurred in the deeper part (depth ≥ 90 km) of the Indian plate indicate thrust faulting; the P axes are almost parallel to the strike of the subducting plate. Two normal fault solutions (1, 4) are obtained at intermediate depth range (55-75 km) within the Indian slab. Two shallow (depth ≤ 10 km) earthquakes (5 and 12) on the Sagaing fault in the Burma plate show strike slip faulting, and three shallow (depth > 12 km) earthquakes (9, 8, 15) in the Indian plate at the northern most end of the arc show dominant thrust faulting.

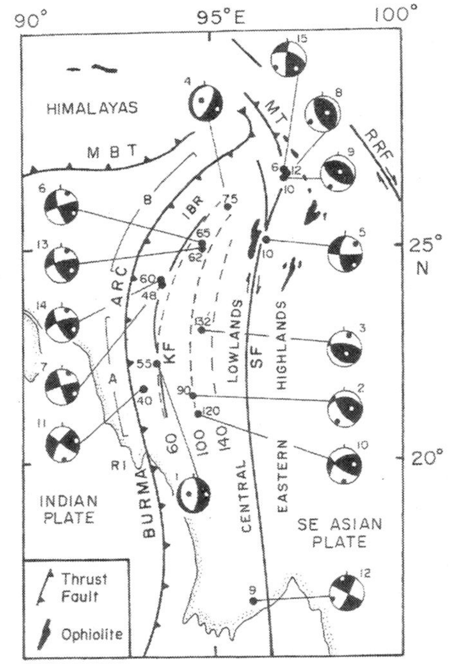

Fig. 6.16 Fault plane solutions of the earthquakes in the Indo-Burma ranges (Ni et al., 1989).

Holt et al. (1991) have also reported 15 reliable fault plane solutions in the region; six to the north of MBT zone in the northeast Himalaya, six in the eastern syntaxis zone, and three on the Sagaing fault (Fig. 6.17). All the solutions (1-5) to the north of MBT show northeast dipping low angle thrust fault comparable with the MBT that accommodate underthrusting of the Indian plate and the overlying Himalaya. Although at the surface, the MBT is steeply dipping, it rapidly shallows north of the surface trace (Valdiya, 1980). One well constrained solution (event 6) that occurred at a deeper depth (35-40 km) shows south dipping low angle nodal plane. The crustal thickness estimated in this region ranges from 50 to 65 km (Tan, 1987; Holt and Wallace, 1990); this indicates that the event 6 occurred in the lower middle crust within the underthrusting Indian crust, much below the envisaged Plane of Detachment. To the south of MBT, deeper earthquakes (depth 50-80 km) were reported in Arunachal Pradesh by microearthquake surveys; a transverse seismogenic structure across the Himalayan trend was inferred (Kayal et al., 1993) (Section 6.6). Six fault plane solutions (event 7-12) are reported in the southeast corner of the syntaxis zone on the southeast extension of the Mishmi thrust. All show dominantly thrust faulting with a northeast dipping nodal plane, except one (event 9) with a strike-slip component. Three solutions of the events (12-15) that occurred on the northern end of the Sagaing fault show strike-slip motion on this fault. It may be noted that the solutions 9, 10 and 11 (Holt et al., 1991) in Fig. 6.17 are identical to 15, 8 and 9 respectively (Ni et al., 1989) in Fig. 6.16 for the same events.

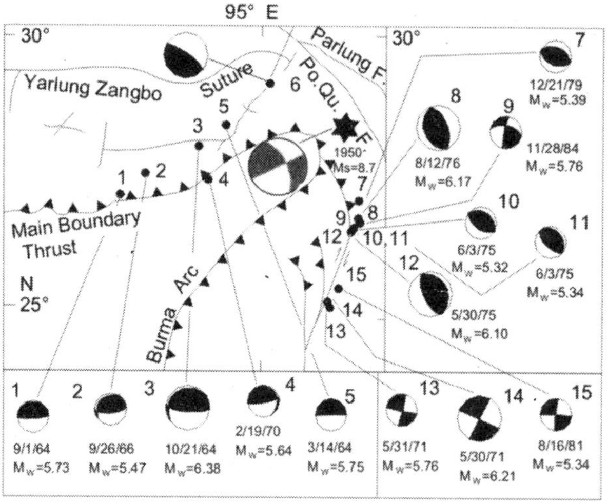

Fig. 6.17 Focal mechanisms of the earthquakes in eastern syntaxis zone (Holt et al., 1991).

CMT Solutions

Rao and Kalpana (2005) studied CMT solutions of 60 earthquakes (M 4.5-6.5) within the subducted Indian plate in the Indo-Burma region (Fig. 6.18a) for stress inversion. It may be noted that the solutions show three types of faulting (Fig. 6.18a). There is a clear segregation of normal, strike-slip and thrust type solutions along the dipping slab; the thrust/reverse solutions are observed in the lower portion of the down going slab below 90 km, above that depth normal and strike-slip solutions are observed (Fig. 6.18b). The stress inversion indicates distinct stress fields above and below 90 km along the subducted Indian lithosphere. In the upper part of the slab the P and T axes trend NNE and ESE respectively, in conjunction with the ambient stress field of the Indian plate. In the lower part of the slab, however, there is no preferred orientation of the P and B axes but orientation of the T axes is well defined in the down dip direction of the slab. These observations suggest that the upper part of the slab is governed by the NNE oriented horizontal plate tectonic forces, whereas the lower part is governed entirely by tensile forces due to gravitational loading on the subducted slab.

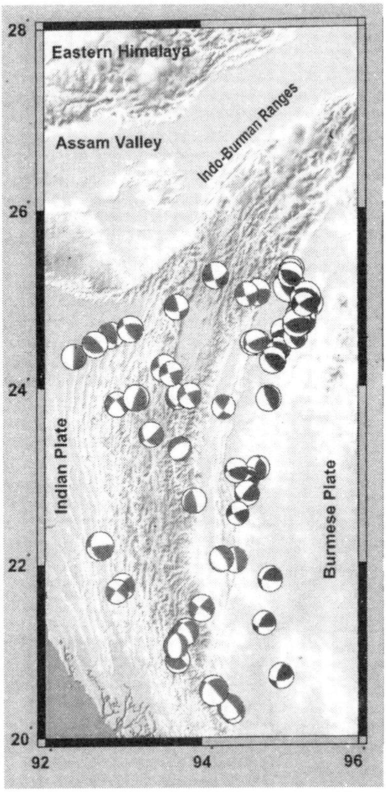

(a)

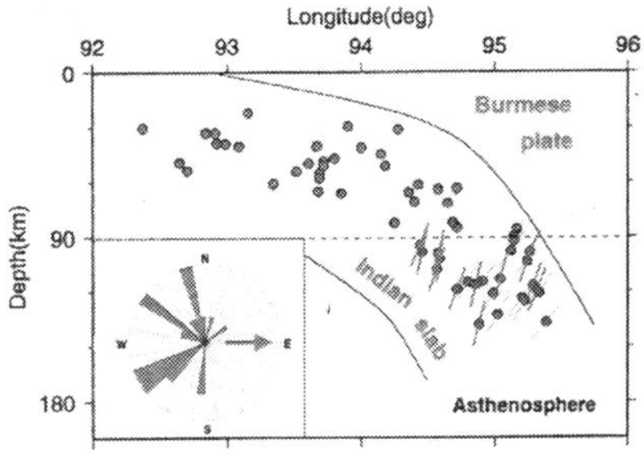

(b)

Fig. 6.18 (a) CMT solutions of the earthquakes (1977-2003) in the Indo-Burma region; grey beach balls represent the events at 0-90 km and the black beach balls at 90-140 km depths. (b) E-W depth section showing shallow (0-90 km) strike-slip and normal faulting events and deeper (depth > 90 km) reverse faulting events. The rose diagram shows azimuthal distribution of the dip directions, the small arrow indicates down dip direction of the Indian plate (Rao and Kalpana, 2005, copy right AGU).

6.6 MICROEARTHQUAKE DATA

6.6.1 Temporary Microearthquake Networks

Shillong Plateau Networks (1980-1986)

A few microearthquake surveys were carried out in the Shillong Plateau and Assam valley area by the Geological Survey of India (GSI) in collaboration with the University of Roorkee (now IIT Roorkee) during 1980-1986. Temporary networks were established in selected areas deploying analog Teledyne Geotech portacorders (Model RV-30) and short period vertical seismometers (Model 13). In each year the temporary networks were run for about four-five months in different selected areas of the region. Preliminary reports were published by Saha et al. (1981), Gaur et al. (1982), Khattri et al. (1983), Kharshiing et al. (1986) and Khattri et al. (1986).

The temporary networks were not much close-spaced during the surveys in 1980-82. Well designed close-spaced networks were made during 1983-86. Kayal (1987) and Kayal and De (1991) reanalysed the data of the close-spaced networks and relocated the microearthquakes. About 5000 earthquakes were recorded by the temporary networks during 1983-86 in the Shillong Plateau-Mikir hills-Assam valley area. Out of these 5000 recorded events, about 400 earthquakes, magnitude 2.0 and above, which occurred within 50 km of the networks were relocated by the *homogeneous station method* (Kayal, 1987; Kayal and De, 1991) using the Hypo71 computer program of Lee and Lahr (1975). These locations were further improved by 3D ray-tracing technique by Kayal and Zhao (1998).

A microseismicity map of the Shillong Plateau-Mikir hills-Assam valley area is presented in Fig. 6.19a, the epicentres are relocated by 3D ray-tracing (Kayal and Zhao, 1998). The map shows a high seismic activity in the Tura area, western part of the Shillong Plateau. It is interesting to note that the highest seismic activity is delimited within the maximum intensity (XII) zone of the 1897 great earthquake, and the activity is confined to the north of the Dapsi thrust. The high intensity zone is also delimited by the Dapsi thrust to the south. This northeast dipping thrust is demarcating the Archaean gneiss to the north and the Tertiary sediments to the south within the Shillong Plateau (Figs 6.2 and 6.3). Among the other faults within this zone, the Chedrang fault was reported to be coseismic, developed by the 1897 great earthquake (Oldham, 1899). A criss-cross pattern of photo-imagery lineaments is reported in the area (Nandy, 2001). Correlation of epicentres with the individual surface fault/lineament is difficult (Figs 6.2 and 6.19). The depth-sections and the composite fault-plane solutions (A&B), however, suggest that a northeast dipping fault is seismically active in the area, which is comparable with the Dapsi thrust on the surface (Kayal, 1987; Kayal and De, 1991) (Fig. 6.19). It may be mentioned that (Oldham, 1899) first inferred a northward dipping fault-plane for the 1897 great Shillong earthquake which was later supported by Seeber and Armbruster, (1981), Molnar (1987) and Chen and Molnar (1990). It is further observed that the maximum number of earthquakes, recorded by the temporary microearthquake networks, occurred at a depth range 20-30 km. A few sub-crustal earthquakes (depth > 40 km) are located at the eastern segment of the Dauki fault, but no fault-plane solution of these earthquakes could be made due to meagre first-motion data. In the eastern part of the Plateau, in Shillong area, the seismic activity may be correlated with the Barapani thrust/shear zone (Fig. 6.3). The depth-section and the composite fault-plane solution (C) of the earthquakes suggest that the Barapani shear zone and its parallel lineaments are seismically active (Kayal and De, 1991) (Fig. 6.19).

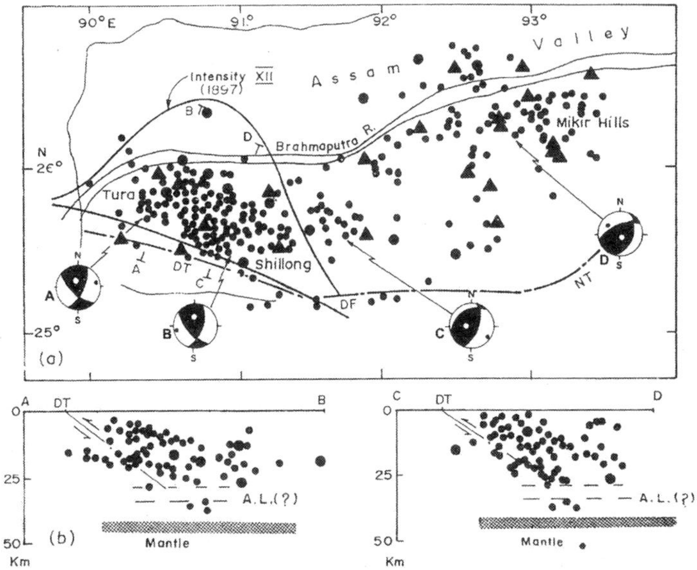

Fig. 6.19 (a) Microseismicity (1983-86) map of the Shillong Plateau, solid triangles indicate the temporary seismic stations, four composite fault plane solutions (beach balls) are shown for the selected four clusters, tectonic features are same as Fig 6.3. (b) Cross sections of the events along A-B and C-D , A-L: Aseismic layer (?) (Kayal, 2001).

In the Mikir hills-Nowgong area (Assam valley), a high seismic activity is also observed (Fig. 6.19). No surface fault is mapped in the area. Among the major lineaments, the NE-SW trending Guwahati-Golaghat lineament, parallel to the Brahmaputra river (fault), is prominent. The trend of activity as well as the composite fault-plane solution (D) is correlatable with the major fault/lineament (Fig. 6.19). Depth section of the earthquakes showed southeast dipping plane (Kayal and De, 1991). A few crustal (depth < 40 km) and a few sub-crustal (depth ≥ 40 km) earthquakes are located along the Kopili river/lineament.

Arunachal Pradesh Network

Two temporary microearthquake networks were established in the Arunachal Himalaya. One was run for about three months during 1986 in the Tipi area under the GSI and University of Roorkee collaboration. The other was run in the Ranga valley for about 10 months during 1990 by the GSI using the Kinemetrics PS-2 analog recorders and vertical seismometers SS-1. A total of about 1000 earthquakes were recorded by the two surveys, and about 150 earthquakes (M ≥ 2.0) were located within 50 km of the networks. The results of these surveys were reported by Kayal et al. (1993). The epicentral map is shown in Fig. 6.20a. The rate of seismic activity is much low compared to that in the Shillong Plateau-Mikir hills area (Kayal, 1996a). Temporal as well as spatial variation of seismic activity was reported by Kayal et al. (1993). The trend of activity is not correlatable with the trend of MBT or with the known photo-imagery lineaments; a N-S trend of activity is apparent (Fig. 6.20). The depth-section of the earthquakes indicates that, unlike in the western Himalaya, the activity in the northeastern Himalaya is not confined above the Plane of Detachment at a shallower depth <20 km as was proposed by Seeber et al. (1981) (Fig. 6.20b). A good number of earthquakes occurred below 50 km depth at the MBT and to its south, i.e. much below the Plane of Detachment (Kayal et al., 1993). A similar observation is made in the Sikkim Himalaya by De and Kayal (2003) and in the central Himalaya (Monsalve et al., 2006) (see Chapter 5).

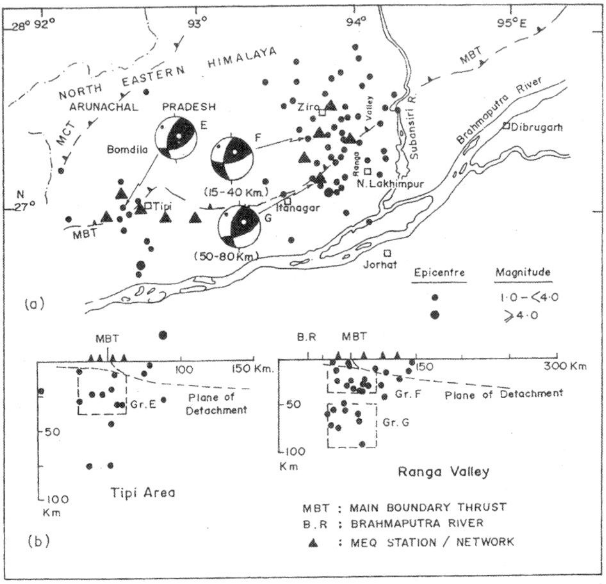

Fig. 6.20 (a) Microseismicity map of the Tipi and Ranga valley area, Arunachal Himalaya; fault plane solutions of the two clusters are shown. (b) N-S cross sections of the events in the selected two blocks (Kayal et al., 1993).

Fault-plane Solutions of Microearthquakes

In the Shillong Plateau-Mikir hills-Assam valley area, four well constrained composite fault-plane solutions of four selected clusters/groups of earthquakes (depth 11-25 km) were obtained (Kayal, 1987 and Kayal and De, 1991) (Fig. 6.19a). Two solutions for groups A and B earthquakes in Tura area, within the Shillong Plateau, show reverse faulting with strike-slip component. The northeast dipping NW-SE trending nodal plane is the inferred fault plane (Kayal and De, 1991). It is correlatable with the Dapsi thrust. The fault-plane solution of the group C earthquakes in the eastern part of the Plateau indicates a thrust faulting with a strike-slip component. The southeast dipping

NE-SW nodal plane is the inferred fault plane, which is comparable with the Barapani thrust and its parallel lineaments in the Shillong area. The group D earthquakes also show a thrust faulting with a strike-slip component. The southeast dipping NE-SW trending nodal plane is the inferred fault plane. This fault plane is correlatable with the Brahmaputra fault and its parallel lineaments.

In Arunachal Himalaya, three composite fault-plane solutions were obtained (Kayal et al., 1993) (Fig. 6.20a): two for the groups E and F earthquakes (depth 15-40 km), and one for the group G earthquakes (depth 50-80 km). These solutions show strike-slip cum reverse faulting. The NE-SW trending nodal plane in all the three solutions are compatible with the trend of MBT, whereas N-S trending nodal plane is comparable with the seismicity trend. Kayal et al. (1993) argued that the MBT cannot represent the fault plane because the depth sections show that these earthquakes not only occurred to the south of MBT but also occurred at a depth range 15-80 km (Fig. 6.20b). The N-S nodal plane is the preferred fault plane, and is comparable with the trend of observed seismicity, which suggests a transverse tectonics in the area (Kayal et al., 1993).

6.6.2 Permanent Microearthquake Networks

In addition to four to five permanent observatories of the national network (IMD), there are about 20 permanent microearthquake seismic stations that are in operation in the region since 1980s. In 1986, the NGRI (Hyderabad) and the RRL (Jorhat) jointly started two six- to eight-station analog microearthquake telemetric networks in the region. Since 2000 the RRL(J) intensified the microearthquake monitoring and upgraded its network to 13 stations with stand alone 24-bit digital instruments with short period sensors, and a few with broadband sensors. Similarly, the NGRI network is also upgraded to a six-station 24-bit digital network, all with broadband sensors. Further, the University of Gauhati, University of Tezpur, University of Aizwal (Mizoram) and University of Manipur established four to six permanent stand alone 24-bit digital broadband seismic stations in the region. These permanent seismic stations in the region are shown in Fig. 6.3. All the phase data recorded by the permanent networks/stations are compiled and the seismological bulletins are jointly published by the NGRI/RRL(J).

Relocation by Hypo71

A good data set, recorded by the permanent and temporary networks during the period 1993-99, was re-analysed by Bhattacharya et al. (2002); about 2000 events were relocated by the Hypo71 program with a fair degree of precision; average residual method was used for relocation; the average rms was brought down to ≤0.56 s from 0.75 s. The microseismicity map is shown in Fig. 6.21.

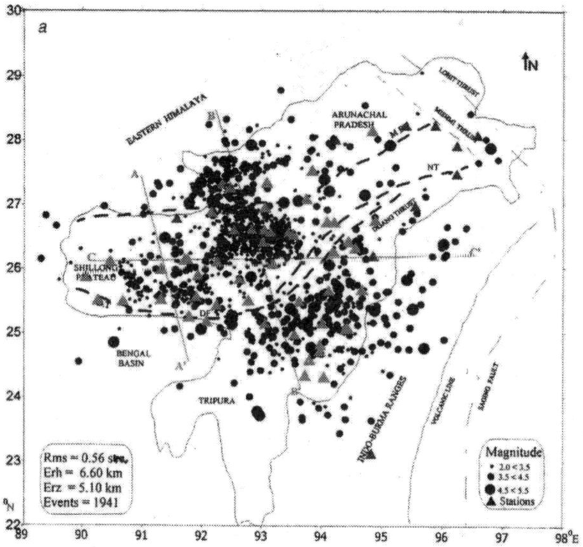

Fig. 6.21 Microseismicity (1993-1999) map of NE India based on permanent and temporary network data (Bhattacharya et al., 2002).

The microseismicity map reveals a few high active zones; among these the NNW-SSE trending active zone along the Kopili fault is very prominent in this map (Fig. 6.21). The active zone/fault is transverse to the MBT/MCT trend. Such intense activity was not recorded by the temporary networks in 1980s as seen above in Fig. 6.19. Among the other active zones, the Shillong Plateau and the Manipur fold belt in the Indo-Burma ranges are also prominent in this map (Fig. 6.21). Fault-plane solutions of these earthquakes could not be made as the first-motion data were meagre in the published bulletins.

Relocation by JHD

Bhattacharya et al. (2005) made an attempt to relocate these events by JHD method (see Section 3.12). About 300 events were relocated with average rms 0.26 compared to that of 0.56s by Hypo71 (Fig. 6.22a and b). This method uses 1-D model, but simultaneously determines station corrections, which indirectly indicates lateral variations in geologic/velocity structure at shallower depth (Fig. 6.23). The positive station correction implies low velocity and negative correction higher velocity structure beneath the seismic stations. It may be noted that station corrections are positive in the Shillong Plateau, Assam valley and Manipur fold belt, and negative in the Mikir hills area (Fig. 6.23).

A visual examination of the two maps (Fig. 6.22a and b), prepared by using the same events located by the Hypo71 and by the JHD analysis, reveals almost similar pattern of seismicity. The true pattern is, however, depicted in the depth sections of the events. In the Hypo71 locations, the focal depths of many events were fixed at the trial depth (20 km) and the true pattern was masked. Focal depths are well estimated by the JHD method and true pattern of the seismicity at depth is well reflected.

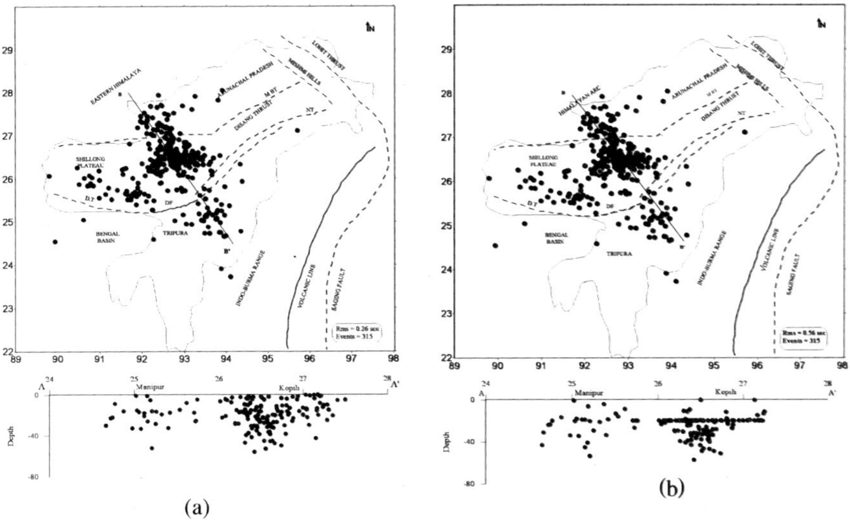

Fig. 6.22 (a) Relocated 300 events by the JHD method and depth section along A-B. (b) Relocated the same 300 events by Hypo71 and depth section along A-B (Bhattacharya et al., 2005).

Analysis of Digital Network Data

An Indo-Russian collaborative project was taken under ILTP (Integrated Long Term Program) to re-analyse the digital network data recorded by about 20 permanent digital seismic stations during the period 2001-2003 in the Shillong Plateau area (Kayal et al., 2006). About 200 best located events that were relocated with rms ≤ 0.003 s and with horizontal and vertical error < 1 km were studied. The epicentre map is shown in Fig. 6.24. They reported 27 fault plane solutions, 17 by using first motion data and 10 by waveform inversion (Fig. 6.24). A north-south cross section of the events that occurred in the Shillong Plateau was examined across the Dapsi thrust/Oldham/ Brahmaputra fault zone (Fig. 6.25); the considered events are shown by a shaded zone A-A′ in Fig. 6.24. Nine fault-plane solutions were obtained in the western Plateau region within the considered shaded zone A-A'; the region is dominated by thrust/reverse faulting (Fig. 6.24). Two more solutions were obtained (events 9 and 10) in the shaded zone; one in the Bengal basin that shows a strike slip faulting (event 9), and one in the Himalaya that shows a normal faulting (event 10). The strike-slip solution 9 was correlated with the NE-SW trending Sylhet fault in the Bengal basin (Fig. 6.3), and the normal solution 10 was correlated to the Kopili fault (Fig. 6.3), that transversely extends into the Himalaya (Kayal, 2001; Bhattacharya et al.,

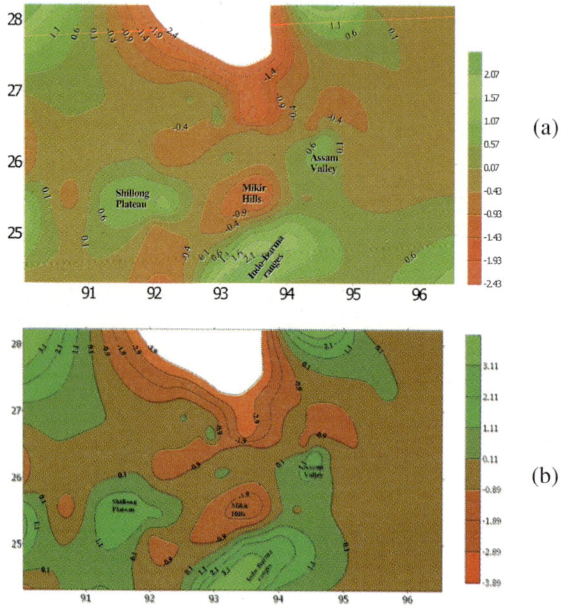

Fig. 6.23 Contour map of (a) P-wave station correction and (b) S-wave station correction (in seconds) obtained by the JHD method (Bhattacharya et al., 2005).

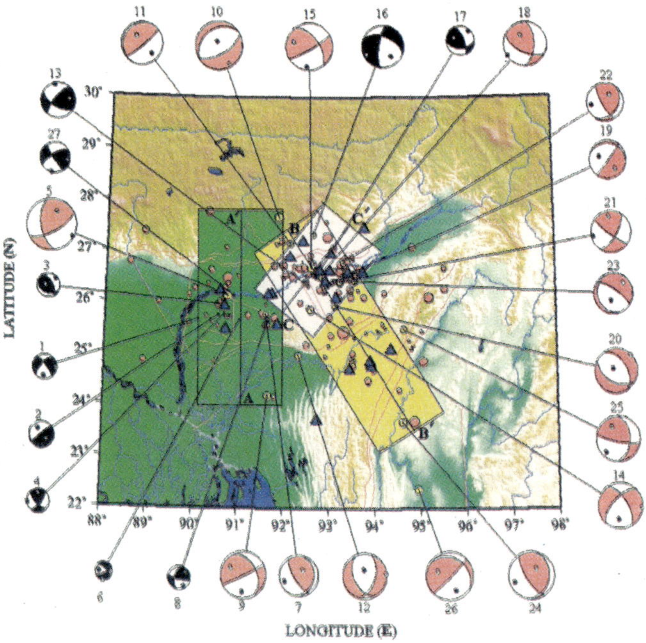

Fig. 6.24 Seismicity map of the Shillong Plateau region using the digital data (2001 and 2003) and fault plane solutions by first motion and waveform inversion (Kayal et al., 2006).

2002). The cross section of the earthquakes and the fault planes were examined with the proposed pop-up tectonic model (Fig. 6.25a). The program RAKE (Louvari and Kiratzi, 1997) was used to prepare the cross section of the fault planes. The dipping inferred fault planes of the nine solutions (events 1-8 and 27) that were obtained in the Plateau region and the fault planes of the events 9 and 10 were plotted at the respective depths in this section (Fig. 6.25c). The section shows that the earthquakes in the Shillong Plateau are bounded by two boundary faults: the north dipping Dapsi thrust (northwestern extension of the Dauki fault) and the south dipping Brahmaputra fault. Kayal et al. (2006) proposed that the pop-up tectonics of the Shillong Plateau is possibly happening in between these two boundary faults.

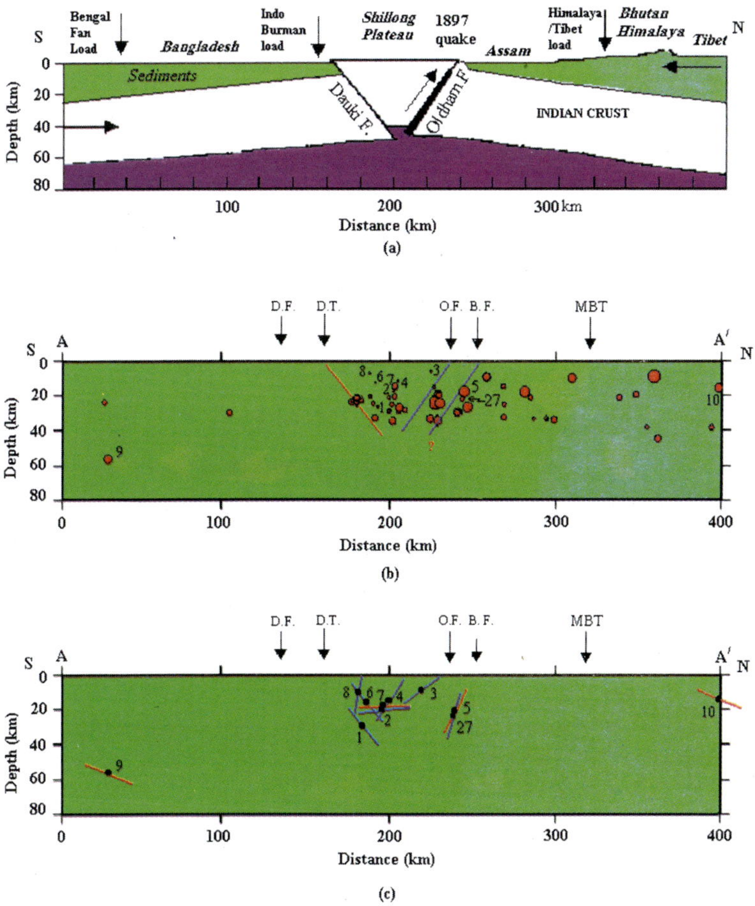

Fig. 6.25 (a) Pop-up tectonic model of the Shillong Plateau (Bilham and England, 2001). (b) Depth section of the events those fall in the shaded zone A-A′ in Shillong Plateau (Fig. 6.24). (c) Inferred fault planes of the events plotted at depth. DF: Dauki Fault, DS: Dapsi Thrust, OF: Oldham Fault, BF: Brahmaputra Fault, MBT: Main Boundary Thrust (Kayal et al., 2006).

Two cross-sections of the events were examined for the Kopili cluster; one NW-SE section along the Kopili fault (Fig. 6.26a), and one NE-SW section across the fault zone (Fig. 6.26b); the corresponding considered events are shown in the shaded zones B-B′ and C-C′ in Fig. 6.24. A cross section of fault planes of the 12 fault plane solutions, mostly normal and strike slip faulting, that were obtained for the considered cluster was also examined (Fig. 6.26c). All these sections reveal that intense activity along the Kopili fault is dominated by a transverse tectonics (Kayal et al., 2006). In addition to the 12 fault plane solutions (11, 13-23), three fault plane solutions were obtained for the events 24-26 in the Indo-Burma ranges and one for the event 12 near the Dauki fault that occurred outside the shaded zone C-C′ (Fig. 6.24). The solution (event 12) near the Dauki fault shows a normal faulting. The Indo-Burma range earthquakes (events 24-26) show dominantly thrust faulting with easterly/northeasterly dipping fault planes; these solutions may be attributed to the subduction tectonics as reported by many authors (e.g. Chen and Molnar, 1990; Kayal, 1996a; Kumar and Rao, 1995).

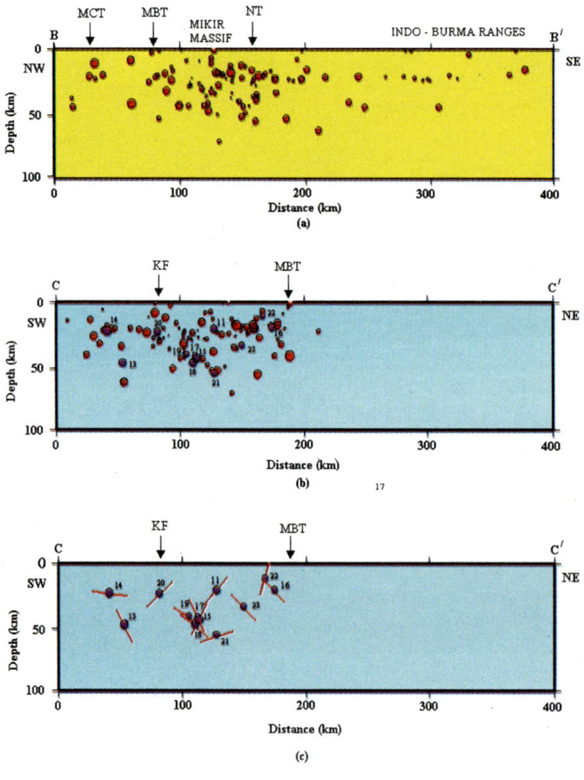

Fig. 6.26 (a) Vertical cross section of the events along B-B′ and (b) along C-C′ (shaded zones in Fig. 6.24). (c) Vertical cross section of the inferred fault planes. MCT: Main Central Thrust, MBT: Main Boundary Thrust, NT: Naga Thrust, KF: Kopili Fault (Kayal et al., 2006).

6.6.3 b-value Estimation

In frequency-magnitude relation of earthquakes (Gutenberg and Richter, 1944), the b-value is an important seismological parameter to understand seismotectonics of a region (Sec. 3.15.1). Verma et al. (1993) made an attempt to study b-value of the region using the teleseismic data of earthquakes $M \geq 4.0$. They, however, found that the teleseismic data were meagre for $1° \times 1°$ area for a period of 11 years (1979-1990). So, they chose $2° \times 2°$ area, and estimated the b-values. They estimated highest b-value (1.26) for the Shillong Plateau, and the second highest value (b = 1.20) for the Indo-Burma ranges. The Arunachal Himalaya depicted a normal value (b = 1.00), and the Bengal basin the lowest (b = 0.80).

A temporal change of b-value, before the August, 1988 Indo-Burma border earthquake (M_L 7.5), was studied by Sahu and Saikia (1994). They analysed the data for the period 1976-1987, with a threshold magnitude of earthquake m_b 4.5, within the latitude 21°-25°N and longitude 93°-96°E in the Indo-Burma ranges. It was reported that for the deeper earthquakes (depth > 70 km), the b-value gradually increased from 0.6 in 1976 to a maximum value of 1.33 in July, 1987, followed by a short-term drop before the occurrence of the August, 1988 earthquake at a depth 90 $\pm$ 10 km. The temporal variation of the b-value for the shallower earthquakes (depth $\leq$ 70) in the same area, on the other hand, indicated no significant change, i.e. the frequency of deeper seismic activity changed before the large earthquake.

Kayal (1998) estimated b-values in the Shillong Plateau and Arunachal Himalaya using the temporary microearthquake network data for the earthquakes in the magnitude range 2.0-4.0. The estimated b-value was 1.29 in the Shillong Plateau for the period 1984-86, and 0.86 in the Arunachal Himalaya for a period of about one year in 1990.

A comprehensive study on b-value was then made by Bhattacharya and Kayal (2003) for the whole region using the data recorded by the permanent networks/stations during the period 1993-99. About 2000 earthquakes in the magnitude range 2.0-5.5 were used for this study (Fig. 6.21). The b-value, Gutenberg-Richter log-linear relation, was estimated using both the methods; least square fit method and the maximum likelihood method (Section 3.15.1). The estimated b-values with standard deviation (δb) and the constant 'a' for each tectonic block of the region are given in Table 6.1. It shows that the b-value not only varies in the different tectonic blocks, but it also varies with depth. At greater depth (> 36 km), for the upper mantle earthquakes, the b-value decreases in all the tectonic blocks. The average b-value for the crustal earthquakes (depth $\leq$ 36) in the Shillong Plateau is about 0.68 which is much lower than the earlier reports (e.g. Verma et al., 1993 and Kayal, 1998). One implication of the different values may be that the mean magnitude of events is different in all the three cases; the data sets are different for three different type of networks. Inspite of the differences in data set and in mean magnitudes, the b-value estimate for the period 1979-1990 and that for the period 1983-86 are compatible; the value is in the range 1.12-1.29 in the Shillong Plateau. The microearthquake network data for the period 1993-99 show lower b-value (0.68) in the Plateau area. Bhattacharya and Kayal (2003) suggested that it could be a temporary change in b-value in the area, or it could be due to the fact that the permanent network is far away from the west Garo hills area of the Plateau to record the events M < 3.0. Further, the data set was limited in a narrow range (3.0-4.5) of magnitude for the Plateau earthquakes (Fig. 6.21). In the west Arunachal Himalaya, the estimated b-value ranges between 0.63 and 0.73 at different depth ranges, and it is not very much different from the earlier report (Kayal, 1998). The b-value is, however, much lower (0.51-0.61) in the east Arunachal Himalaya. In the Indo-Burma ranges, the b-value ranges between 0.85 and 0.88 for the crustal earthquakes (depth <36 km), but it is much lower 0.67 for the deeper earthquakes (depth >36 km) within the subducting slab. Lower b-values for deeper earthquakes are normally attributed to increased tectonic stress at depth; this is quite evident for the subducting slab earthquakes in the Indo-Burma ranges.

Table 6.1: Details of b-value estimate and fractal dimension

Block	*Depth range km*	*Maximum likelihood*	*Least sq. ft.*		*Standard error*	*Fractal dimension*
		b	*b*	*a*	*δb*	*D*
1 Shillong Plateau	0-<15	0.70	0.65	3.70	0.001	1.60
	15-35	0.71	0.66	3.40	0.015	1.47
	>35	0.48	0.50	2.60	0.013	1.19
2 Mikir Hills	0-<15	0.78	0.68	4.03	0.001	1.66
	15-35	0.79	0.81	4.48	0.001	1.74
	>35	0.77	0.71	4.50	0.001	1.65
3 Lower Assam and	0-<15	0.60	0.66	3.40	0.007	1.50
Western Arunachal	15-35	0.71	0.75	3.20	0.006	1.45
Pradesh	>35	0.66	0.70	2.28	0.006	1.30
4 Upper Assam and	0-<15	0.50	0.52	2.90	0.001	1.30
Eastern Arunachal	15-35	0.61	0.62	3.50	0.001	1.57
Pradesh	>35	0.44	0.48	1.80	0.001	1.40
5 Indo-Burma Range	0-<15	0.86	0.84	4.20	0.006	1.56
	15-35	0.89	0.87	4.90	0.005	1.80
	>35	0.64	0.66	4.20	0.005	1.40

(Bhattacharya and Kayal, 2003)

To visualize the frequency-magnitude distribution as a function of space, a b-value map is prepared (Fig. 6.27). In doing this map, 1° × 1° sample volume was considered with an overlapping of 0.5° for the crustal (depth < 36 km) earthquakes (Bhattacharya et al., 2002). The higher b-value contours represent the higher frequency of earthquake zones. The Kopili fault, the Indo-Burma

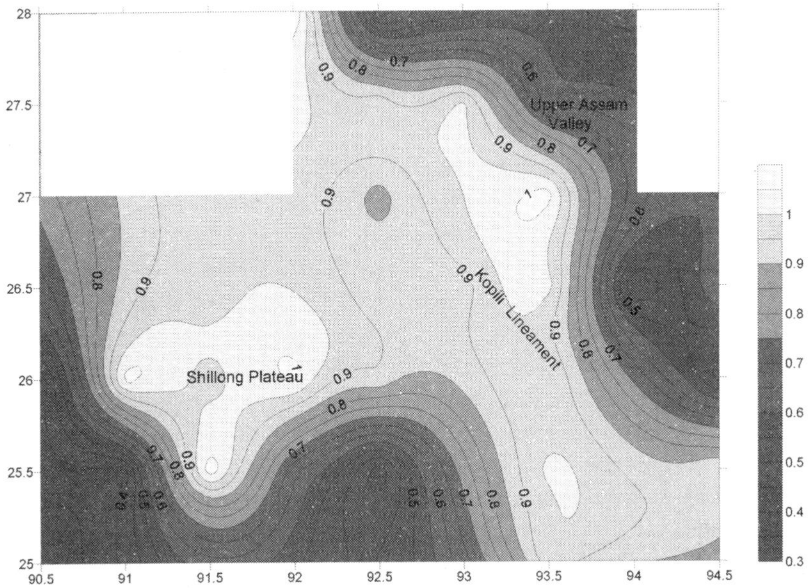

Fig. 6.27 Contour map of b-value (microearthquake network data: 1993-99) of the Shillong Plateau-Mikir hills area (Bhattacharya et al., 2002).

ranges and the Shillong Plateau are well reflected as higher b-value zones in the map.

6.6.4 Fractal Dimension

Bhattacharya and Kayal (2003) estimated fractal dimension, more strictly correlation dimension (D_2) of the epicentre distribution in the selected five tectonic blocks in this region (Fig. 6.21, Table 6.1). Methodology of estimating fractal dimension is given in Section 3.15.1. The estimate was made for the earthquakes at the three different depth ranges; a total of 15 observations were obtained (Table 6.1). For all the fifteen observations it is found that the estimated fractal dimensions are greater than 1.0; it ranges between 1.19 and 1.80. This indicates that the distribution of the epicentres reflects the fractal nature of the causative faults at different depth ranges.

The observed fractal dimensions (D_2), we call it *D* hereafter, are comparable to some recent analysis in other active regions of the world. Hirata (1989) estimated *D* equal to 1.6 for an active fault system at the central part of Japan, which corresponds to the theoretical value 1.65 derived for percolating crack network (Korvin, 1992). Angulo-Brown et al. (1998) found *D* equals 1.64 for an active fault system linked to the Middle American trench in southern Mexico. They correlated *D* values with depth of the seismogenic zone; the larger the seismogenic depth the smaller the *D*. Idziak and Teper (1996) estimated D = 1.60 for the Silesian coal basin in Poland. Hirata (1989) argued that 1.60 is an upper limit to the fractal dimension of the fracture geometry that can be explained by the Griffith energy balance concept. Tosi (1998) illustrated that possible values of fractal dimension are bound to range between 0 and 2, which is dependent on the dimension of the embedding space.

The observed *D* values 1.60-1.68 in the upper crust (0≤15 km) beneath the Shillong Plateau (block I) as well as beneath the Mikir hills area (block II) (Table 6.1) are within the normal range for seismogenic fault systems. The *D* value decreases with depth in the Shillong Plateau; the value goes down to 1.19 below the lower crust (depth >35 km). Such decrease of the fractal dimension indicates certain degree of clusterisation and presence of non-random structures (Tosi, 1998). In the Mikir hills, on the other hand, the *D* value is higher (1.74) in the lower crust (15-35 km) and again decreases to 1.65 below the lower crust (depth >35 km). The higher value 1.74 in the lower crust indicates that the event distribution is more random and unpredictable in a two dimensional space below the Mikir hills.

The estimated *D* values in the block V, Indo-Burma ranges, are also interesting (Table 6.1). In the upper crust (0-15 km) it is 1.56, and in the Benioff zone at depth 35-70 km it is 1.40, which indicate the seismogenic structure/fault dimensions are equally dominant in the upper crust and in the depth range 35-70 km within the subducting plate. The *D* value in the depth range 15-35 km is much higher (1.80) than the normal limit of 1.60 for a seismogenic fault system. This observation indicates that the event distribution is more random and unpredictable in this depth range, which may be caused by maximum external forces due to bending of the subducting Indian plate as well as by forces due to the overriding Burmese platelet at this depth range (Bhattacharya and Kayal, 2003). The maximum interaction between the two plates causes more external forces for the more random seismic activity at shallower depth beneath the Indo-Burma ranges (Kayal, 1996a).

The epicentre distribution and the estimated fractal dimension suggest that the faults are spatially distributed in the whole region, and the whole region is seismically active. Higher fractal dimension (1.65-1.74), particularly in the Mikir hills, is worth noting; this observation is also reflected in the b-value map (Fig. 6.27). Since the data set was constrained within 150 km distance of the network, the Bengal basin and Tripura fold belt areas could not be examined (Figs 6.21 and 6.28). The networks have also the limitations to record the distant lower-magnitude earthquakes. Sunmonu and Dimri (1999), however, studied fractal dimension in these two tectonic blocks using the teleseismic data (M≥4.5) from the ISC catalog. They estimated *D* values 1.66 in the Tripura fold belt and 1.41 in the Bengal basin for the crustal earthquakes. Their results are comparable with the observations made in the other five tectonic blocks of the region (Table 6.1).

Bhattacharya et al. (2002) presented a fractal-dimension map using the microearthquake network data of 1993-99 (Fig. 6.28). The data were grided at 1° spacing with an overlapping of 0.5° as was done for the b-value mapping (Section 6.6.3). The fractal dimension map is shown in Fig. 6.28. The higher contours (1.65-1.85) along the Kopili fault reflect a prominent seismogenic structure. The higher values indicate more heterogeneity. A deep rooted fault, transverse to the Himalayan trend, from Manipur fold belt to Arunachal Himalaya is identified, and they named it a Kopili fault. A similar observation is made in the b-value map (Fig. 6.27). These observations are pertinent in seismotectonic studies of the region.

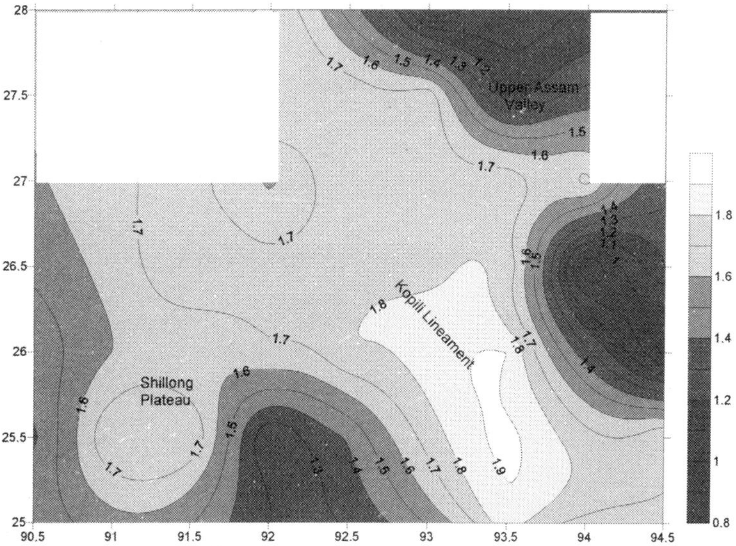

Fig. 6.28 Fractal dimension map using the epicentres (data: 1993-99) of the Shillong Plateau-Mikir hills area (Fig. 6.21) (Bhattacharya et al., 2002).

6.6.5 Seismic Tomography

Determination of the crust and upper mantle velocity structure, particularly in such a tectonically complex region, plays an important role for better understanding of seismotectonics, identification of source areas of large and great earthquakes and evaluation of earthquake risk/hazards. Only a few limited studies have been made to determine the crust and upper mantle velocity structure in this region. Tandon (1954) studied travel-time curves of P and S phases of the great Assam earthquake of 1950 and its aftershocks, and estimated P_n velocity 7.9 $\pm$ 0.02 km/s. Sitaram et al. (1986) estimated the upper mantle P_n velocity 7.85-8.05 km/s in some parts of the region by using teleseismic P-wave arrivals. Kayal and De (1987) studied the seismic velocity of the uppermost mantle by using P_n arrivals of local earthquakes recorded by the temporary microearthquake networks. They used the time-term method and reported high mantle velocities (8.1~8.5 km/s) beneath the Shillong Plateau (Fig. 6.29). Gupta et al. (1990) studied crust and mantle velocities in the Indo-Burma ranges using local earthquakes (ISC data), and reported Vp_g 6.5 km/s and Vp_n 8.1 km/s. De and Kayal (1990) and Mukhopadhyay et al. (1995) studied the crustal velocity beneath the Shillong Plateau by the time-distance and Wadati-plot (Wadati, 1933) methods using the P-wave arrivals of microearthquakes, and reported high velocity upper and lower crustal layers, 5.6 km/s and 6.5 km/s respectively. Normal velocity ratios (Vp/Vs), 1.71-1.73, were reported for the crustal layers beneath the Plateau (De and Kayal, 1990) (Fig. 6.30). All these preliminary studies, however, shed little light on the detailed 3D velocity structure in the region, and lateral heterogeneities in the structure were not understood well.

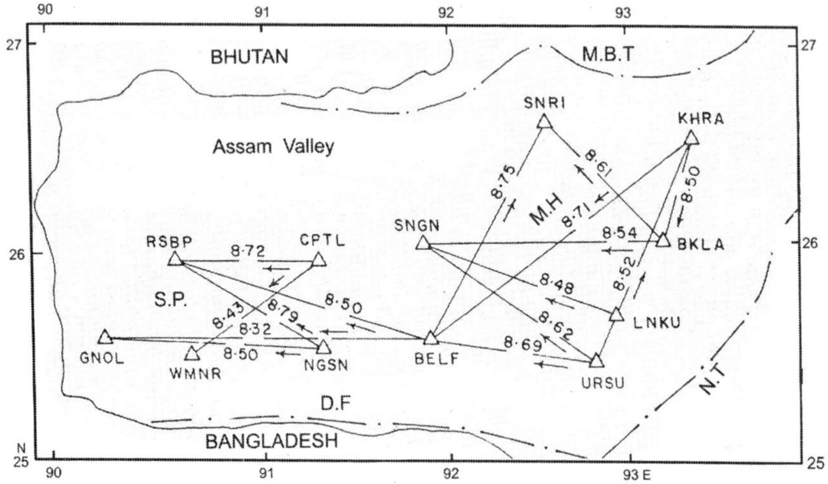

Fig. 6.29 Pn velocity (km/s) in the Shillong Plateau area by time-term method. SP: Shillong Plateau, MH: Mikir Hills, MBT: Main Boundary Thrust, NT: Naga Thrust, DF: Dauki fault (Kayal and De, 1987).

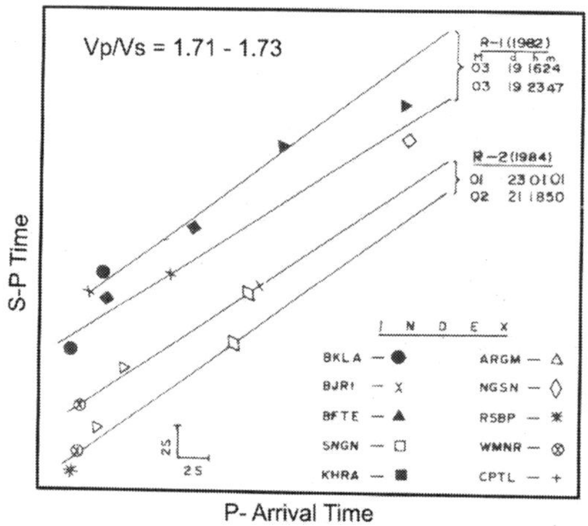

Fig. 6.30 Vp/Vs in the Shillong Plateau (R1) and Mikir Hills (R2) areas by Wadati plot method (De and Kayal, 1990).

Kayal and Zhao (1998) first attempted to study 3D velocity structure in this region. They used the program of Zhao et al. (1992) to evaluate the P- and S-wave *tomographic images* of the crust and upper mantle beneath the Shillong Plateau and Assam valley region. 364 earthquakes with about 3000 high quality P and S arrivals, which were well recorded within 50 km of the close-spaced temporary networks, were used for the seismic tomography study. The relocated events are shown in Fig. 6.19. The P-wave tomographic images at five depth slices (2, 10, 22, 35 and 47 km) are shown in Fig. 6.31. The observed S-wave velocity images are compatible with the P-wave images (Kayal and Zhao, 1998). The V_p images include the epicentres of the earthquakes at the respective depth ranges, which are relocated using the estimated 3D velocity structure.

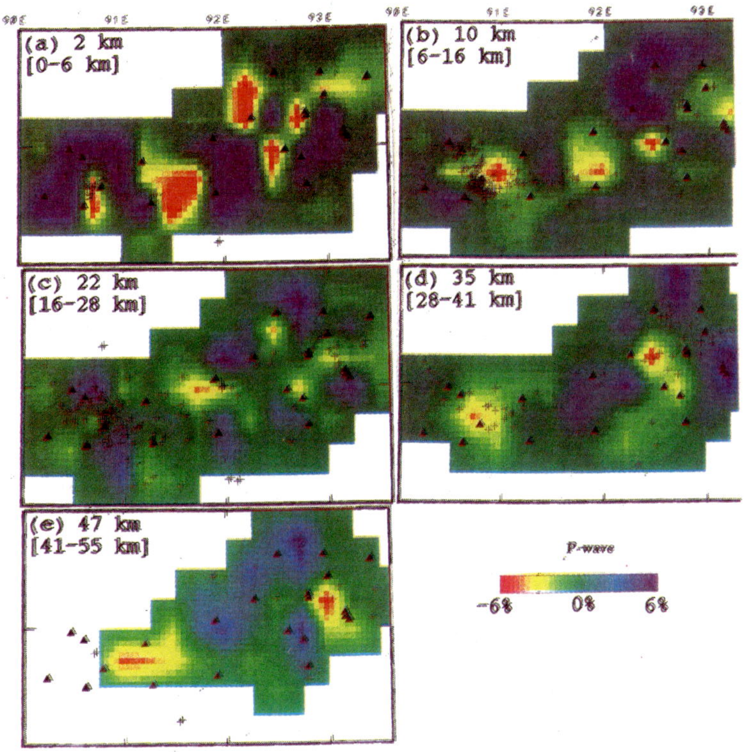

Fig. 6.31 Seismic images of the crust beneath the Shillong Plateau region (Kayal and Zhao, 1998).

The tomographic images reveal the crustal heterogeneities beneath the Plateau region (Fig. 6.31). One of the most interesting results is the compatibility of the low velocity zones (LVZ) with the surface geological features at shallower depth (2-10 km). The Dudhnoi fault, Barapani shear, Kopili fault zone and other shallow geological fractures are imaged as LVZ. The parallel trend of high velocity zones (HVZ) on both sides of the LVZs are possibly the zones of denser rocks (Archaean gneiss) under compressional stress in the region. At 22 km depth slice a HVZ is prominent beneath the Shillong Plateau. The HVZ in the lower crust is conformable with the high seismic activity in the western part of the Shillong Plateau. The HVZ may be a stress concentrator at the lower crust, the earthquake source zone.

Recently, Bhattacharya et al. (2006) imaged 3D P- and S-wave velocity structures in the region using a large data set recorded during the period 1993-1999 by the permanent/temporary networks in the region. About 80 seismic stations, analog and digital, were in operation in different times during the said period, and about 3000 events were recorded. Out of 3000 events about 2000 events were initially located by 1D inverted velocity model using the Hypo71 program of Lee and Lahr (1975) (Fig. 6.21). The 1D inversion of the velocity was carried out by the Seisan program (Havskov and Ottemoller, 2000). Then the Local Earthquake Tomography (LET) program of Thurber (1983), modified by Eberhart-Phillips (1993), was used to relocate the events and to image the 3-D velocity structure beneath the region. The seismic images along with the seismic stations and the relocated events are shown in Fig. 6.32; the location quality was much improved by this method. The average rms 0.49s obtained by the Hypo71 using 1D inverted velocity model was reduced to 0.11s by the LET method.

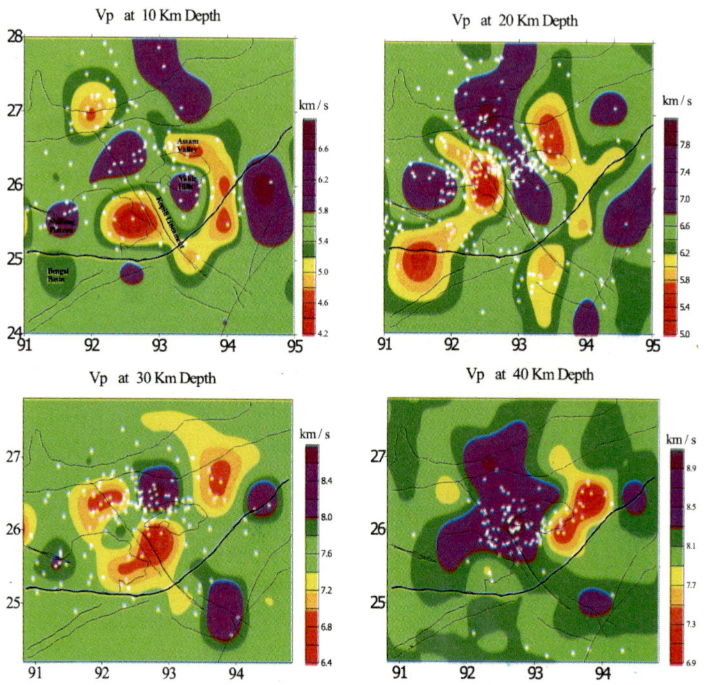

Fig. 6.32 P-wave velocity images of the crust beneath the northeast region of India (Bhattacharya et al., 2005).

The LET method, in addition to high precision locations of the earthquakes, provided 3D seismic images. The P- and S-wave velocity structures and Vp/Vs were imaged at different depth levels. The P-wave images at 10, 20, 30 and 40 km depth levels are shown in Fig. 6.32. The S-wave images are similar to P-wave images (Bhattacharya et al., 2006). The upper most crust, the depth slice at 10 km, is characterized by low Vp (5.3 km/s) as well as by high Vp (6.1 km/s); there is lateral heterogeneity. The high velocity lower crust beneath the Shillong Plateau-Mikir massif is well distinguished from the low velocity thicker sediments below the Bengal basin at the 20 km depth slice (Fig. 6.32) The lower crust at 20-30 km was also identified as a high velocity zone beneath the Shillong Plateau by Kayal and Zhao (1998). Very prominent HVZs are, however, observed at depth slices 30 and 40 km beneath the Kopili fault zone (Fig. 6.32). The high and deeper seismic activity below the Kopili fault zone may be correlated with this HVZ. The Vp/Vs images also showed lateral heterogeneity in the upper as well as in the lower crust; it, however, varies within a normal range of 1.65 to 1.75 (Bhattacharya et al., 2006).

The 3D seismic velocity structures beneath the region shed new light on the seismogenic structures at depth. The high velocity structures in the mid and lower crust (20-30 km) beneath the Shillong Plateau are identified as the source areas for earthquakes. The intense activity beneath the Plateau is confined within 30 km depth. Beneath the NW-SE trending Kopili fault lies a high velocity structure at 30-40 km depth, and intense seismic activity is observed in this depth range along this fault. The seismic images beneath the Indo-Burma ranges were, however, not much resolved (Fig. 6.32) as the area falls away from the permanent microearthquake networks (Fig. 6.21).

6.7 SEISMOTECTONIC PERSPECTIVE

Based on the distribution of epicentres and geological/tectonic setting, the whole region is divided into five seismotectonic blocks or zones (Kayal, 1996a); Zone A: northeast Himalayan collision zone, Zone B: Indo-Burma subduction zone, Zone C: the Eastern Himalaya syntaxis zone, Zone D: the Shillong Plateau-Mikir hills-Assam valley, and Zone E: the Bengal basin and the Tripura fold belt (Fig. 6.10). Seismotectonics in each zone is characteristically different as discussed below.

Northeast Himalaya: Collision Zone

The seismicity of the eastern/northeastern Himalaya is commonly accepted as due to *collision tectonics* between the Indian plate and the Eurasian plate (e.g. Dewey and Bird, 1970; Tandon and Srivastava, 1975; Molnar and Tapponnier, 1977; Verma et al., 1977; Chandra, 1978; Seeber et al., 1981;

Ni and Barazangi, 1984; Molnar, 1990; Holt et al., 1991). The seismicity is broadly correlated with the MBT and MCT in the Himalayan Seismic Belt (HSB).

A detailed microearthquake survey during 1990-91 by a temporary network in Arunachal Pradesh, however, revealed that the earthquakes are not correlatable with the MBT or MCT in this part of the Himalaya (Kayal et al., 1993). The earthquake trend as well as the focal-mechanism solutions (Fig. 6.20) suggest a NNW-SSE trending deep seated seismogenic fault in the Arunachal Himalaya. Thc best located epicentres by the LET method, based on the permanent network data recorded during 1993-1999, clearly show a NNW-SSE trend of intense seismicity along the Kopili lineament which is transverse to the MBT/MCT trend (Fig. 6.21). The recent digital network data (2001–2003) support this observation (Fig. 6.24) (Kayal et al., 2006). The b- value and fractal dimension maps also revealed the seismogenic Kopili fault (Figs 6.27 and 6.28), a seismogenic fault transverse to the Himalayan trend. Further, the deeper shocks (depth 50-80 km) at the MBT and to its south occurred much below the Plane of Detachment, the interplate thrust (Fig. 6.20); which was envisaged in the proposed conceptual tectonic model of the Himalaya (Seeber et al., 1981; Ni and Barazangi, 1984). This model was based on the teleseismic data, that fits fairly well in the western Himalaya (see Chapter 5), but does not fit well in the eastern (Sikkim) or in the northeastern (Arunachal) Himalaya (Kayal, 2001; De and Kayal, 2003).

Three composite fault plane solutions were obtained for the two clusters of microearthquakes in the Arunachal Himalaya (Kayal et al., 1993) (Fig. 6.20). The NW-SE trending nodal planes of the solutions were interpreted as the fault plane, which is transverse to the Himalayan trend. This inference was drawn based on the seismicity trend and depth section of the events (Fig. 6.20) as mentioned above. Holt et al. (1991) reported five fault plane solutions for the teleseismic events to the north of MBT in northeastern Himalaya (Fig. 6.17). Two events (1 and 5) show pure thrust faulting, and three events (2-4) show oblique slip. All the five events are shallow (depth $\geq$ 15 km). P-axes are NNW directed (Fig. 6.17), and are comparable with that obtained by the composite solutions of the microearthquakes (Fig. 6.20). The NNW directed compression in the northeast Himalaya is supported by the NNE directed relative motion of the Indian plate as it under thrusts at the boundary that strikes oblique to the direction of motion (Holt et al., 1991). One well constrained thrust event (6) at greater depth (36 km) shows a south dipping low angle thrust plane (Fig. 6.17). The crustal thickness estimates in this region range from 50 to 65 km (Holt and Wallace, 1990) indicating that the event 6 occurred in the lower-middle crust within the dipping Indian plate. This event occurred much below the Plane of Detachment. Deeper events are not produced on the north dipping shallow Plane of Detachment; it was evidenced by the microearthquake survey also (Fig. 6.20).

Four great earthquakes (1897, 1905, 1934 and 1950) occurred in the Himalaya region between the longitude 75° and 98°E, during the last 110 years since 1897. It was reported that the 1905 Kangra, the 1934 Bihar and the 1950 Assam earthquakes have many uniformities in the rupture geometries (Seeber et al., 1981; Chander, 1988; Molnar, 1990). These caused ruptures along the Himalaya up to 300 km in length on the Detachment Plane with the southern limits close to the surface trace of the HFT (Chapter 5). Bhattacharya and Kayal (2005), however, argued that these events were not generated on the Plane of Detachment. Molnar (1987) and Chander (1989) considered that the rupture geometry of the 1897 Shillong earthquake was different; there is a N-S offset of about 200 km between the southern limit of the 1934 and 1897 ruptures. Chander (1989) suggested that the process of subduction or under thrusting of the Indian subcontinent beneath the Himalaya is perturbed in the region of Shillong Plateau. The Shillong Plateau, a fragment of the Indian shield, is bounded by the Dauki fault to the south and by the Brahmaputra river to the north, which lie respectively 200 and 100 km south of the MBT (Fig. 6.2). The western boundary of the Plateau merges with the 90°E (East Ridge) where the Brahmaputra river after flowing a long distance parallel to the Himalaya takes an abrupt N-S turn along the Dhubri fault at 90°E. Kayal (2001) and Bilham and England (2001) argued that the Plateau earthquakes are characterized as shield seismicity rather than Himalayan seismicity. Bilham and England (2001) further argued that the Plateau earthquakes are generated by pop-up tectonics. Recent analysis of the digital network data in NE India region supports the pop-up tectonics of the Plateau between the Dapsi thrust and the Brahmaputra fault (Fig. 6.25) (Kayal et al., 2006).

The August, 1988 strong Bihar/Nepal earthquake (M_S 6.6) occurred to the south of MBT in the foredeep region; the epicentre is near to that of the 1934 great earthquake (Chapter 5). The depth of the 1988 Bihar/Nepal earthquake was well estimated to be between 50-77 km (USGS, ERI, ISC), and the focal mechanism solution indicates a left-lateral strike-slip movement along the NE trending plane, which is correlated with the East Patna fault (GSI, 1993) (Chapter 5). These observations do not support the *steady state tectonic model* of the Himalaya for generating deeper earthquakes by rupture on the Plane of Detachment to the south of MBT (Kayal, 2001; Bhattacharya and Kayal, 2005).

A similar observation was made by the microearthquake survey in the Arunachal Himalaya (Fig. 6.20), where the earthquakes are deeper (50-80 km) and the conceptual tectonic model does not hold good. It was suggested that transverse tectonics are the possible explanations for the deeper earthquakes to the south of MBT in the eastern and northeastern Himalayan foredeep region (Mukhopadhyay, 1984; Dasgupta et al., 1987; Kayal et al., 1993; Kayal, 2001). In a more recent study, Monsalve et al. (2006) analysed the Himalaya Nepal Tibet Seismic Experiement (HIMNT) data of the 29 broadband-station network in central Himalaya, and argued that most of the earthquakes are clustered in the lower crust and upper mantle, between near surface and around 100 km below sea level. The well defined seismicity at depths between 50 and 100 km confirm the presence of earthquakes in the upper mantle in the region of continental collision. Cluster of earthquakes at depths 40-70 km located to the vicinity of the August 1988 earthquake (Ms 6.6) to the south of MBT clearly indicate the source zone of the 1988 large earthquake. The deeper earthquakes (50-100 km) further north of the MBT and MCT in the High Himalaya and south Tibet were explained by different tectonics within the Indian shield (Holt et al., 1991). These deeper clusters of events to the south of MBT and to the north of MBT/MCT cannot be explained by the interplate thrust plane, the Plane of Detachment.

Indo-Burma Ranges: Subduction Zone

The earthquakes in the Indo-Burma ranges and its adjoining areas can be classified as (a) plate-boundary earthquakes, (b) Benioff-zone earthquakes and (c) overriding-plate earthquakes. The Benioff zone earthquakes are clearly indicated in Fig. 6.11. Focal mechanism solutions of the earthquakes in the Indo-Burma ranges are shown in Figs 6.15-6.18. The events within the Benioff zone (depth ≤ 90 km) show normal faulting as well as strike-slip faulting due to tension in the zone where the plate flexes or bends. The deeper (> 90 km) events in the Benioff zone, on the other hand, show thrust/reverse faulting due to a compressional force for increasing strength of the rocks/gravity loading of the subducted India plate at deeper depth. The shallower earthquakes in the overriding Burmese plate show thrust faulting as well as strike-slip faulting. The thrust faulting may be explained by the compressional force due to interaction of the overriding plate with the subducting plate, and the strike-slip faulting by the regional compression and the internal tension within the overriding plate; the strike-slip faulting events occur mostly on the Sagaing transform fault. The depth below 90 km appears to be significant; it separates the regime of normal and strike-slip faulting at the shallower depth (< 90 km) from that of thrust/reverse faulting within the slab at deeper depth (>90 km) (Fig. 6.18) (Rao and Kalpana, 2005). Kumar and Rao (1995) propose this depth level as the zone of contact between the Indian and Burmese lithospheres.

The Indo-Burma ranges is one of the few intercontinental regions in the world where intermediate focus earthquakes occur. The other regions are the Hindu Kush, the Betic Mountain Belt in Spain and the Carpathian Arc in Romania. The Indo-Burma ranges is also characterised by a negative Bouguer gravity anomaly (Fig. 6.4). Le Dian et al. (1984), on the basis of landsat-imagery data and focal-mechanism solutions, suggested that the subducted

Indian lithosphere is now being dragged northward. Based on CMT solutions of the earthquakes a similar view is given by Kumar and Rao (1995). The NNE trend of the P-axes in the lower part of the plate implies that the maximum compressional stress in the Indian plate is oblique to that responsible for N-S trending folds in the Indo-Burma ranges. Thus either in the recent geologic time (since 1 Ma) the orientation of P-axis within the Indian plate changed dramatically, or more likely the deformation in the Indo-Burma ranges is decoupled from that in the subducting Indian plate (Chen and Molnar, 1990). The T-axes of the normal-fault or strike-slip solutions in the upper part of the slab are oriented predominantly in the E-W direction, possibly indicating down-dip extension due to bending of the slab (Ni et al., 1989; Kumar and Rao, 1995; Kayal, 1996a, Rao and Kalpana, 2005). The down going Indian slab is well imaged as a high velocity zone like a subducting cold oceanic lithosphere beneath the Indo-Burma ranges (Mishra et al., 2006). Satyabala (2003) strongly argued active subduction of the Indian lithosphere beneath the Indo-Burma ranges.

In a recent stress inversion study, Rao and Kalpana (2005) suggested that distinct stress fields exist above and below 90 km along the subducted Indian plate (Fig. 6.18b). In the upper part, the compression (P) and tension (T) axes trend NNE and ESE respectively, in conjunction with the ambient stress field of the Indian plate. In the lower part of the slab there is no preferred orientation of the P or B axes, but a very well defined T axis is observed that trends steeply in the down dip direction. It has been suggested that while the upper part is controlled by the NNE directed plate motion, the lower part is governed by the gravitational loading on the subducted plate. Rao and Kalpana (2005) have presented a model that illustrate various stages of subduction of the Indian plate (Fig. 6.33). In this model they have shown an overturned Indian plate at the 410 km discontinuity (410 D). In reconstruction of the Cenozoic history of the Indian-Eurasian plate kinematics, Karig et al. (1979) suggest a fast (~ 18 cm/y) subduction of the Indian plate during early Eocene (~ 51Ma), and relatively slow subduction during early Oligocene. Rao and Kalpana (2005) suggested that rapid sinking coupled with slow descent of the plate resulted in a resistance at 410D along with an overturn (Fig. 6.33). This resulted in an attempted detachment beneath the lithospheric contact zone of the two plates, manifesting as reverse faulting on steep planes. This scenario looks like normal faulting or gravitational sliding with a difference that the footwall and hanging wall slide towards rather than away from each other (Rao and Kalpana, 2005) (Fig. 6.33d). One dip direction of these fault planes indicate slab detachment mostly in southwestern and northwestern directions (Fig. 6.18b), but clearly away from the eastward slip direction in agreement with detachment model (Fig. 6.33).

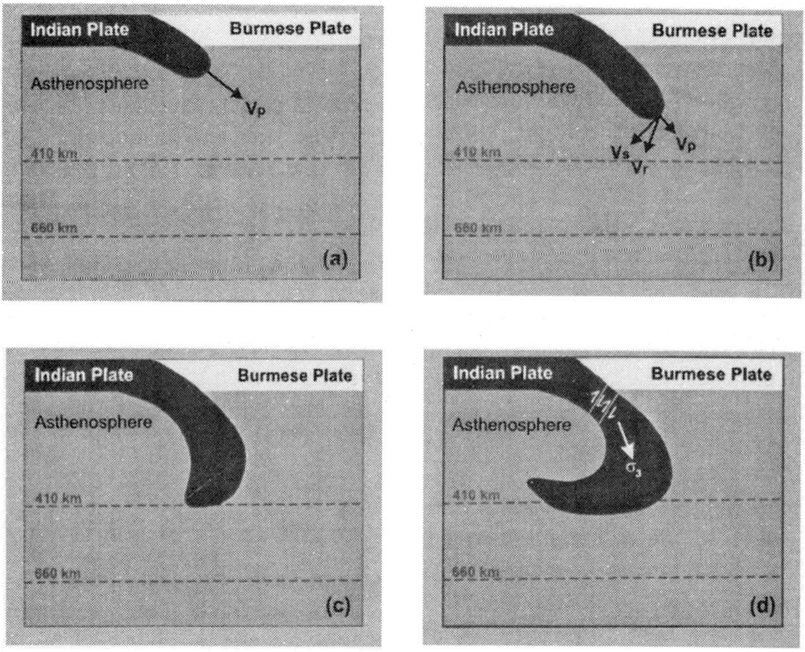

Fig. 6.33 Subducting Indian plate beneath the Indo-Burma ranges (Rao & Kalpana, 2005) (Copyright AGU).

Eastern Syntaxis Zone

The great 1950 Assam earthquake occurred within this zone. Ben-Menahem et al. (1974), from amplitude observations, obtained a right lateral strike-slip solution, and Chen and Molnar (1977), from first-motion data, determined north dipping thrust faulting for this event (Fig. 6.15). The aftershock events show both strike-slip and shallow under thrusting (Chen and Molnar, 1977). Molnar and Pandey (1989), however, reported that all the relocated aftershocks recorded by 50 or more stations, lie east of the main shock in a 250 km long 100 km wide zone, suggesting that the 1950 Assam earthquake occurred on a gently north-northeast dipping thrust plane. Armijo et al. (1989), however, preferred the right-lateral strike-slip solution of Ben-Menahem et al. (1974); this is consistent with their hypothesis that right-lateral slip on the Jiali-Po-Qu fault in southeast Tibet wrap around Eastern Syntaxis and connect with the right-lateral strike-slip on Sagaing fault (Fig. 6.17). It has been suggested that the absence of the ophiolite in the region of Eastern Syntaxis is an evidence of ~450 km right lateral displacement on the fault system that wrap around the syntaxis (Mitchell, 1981; Armijo et al., 1986 and 1989).

Five thrust-fault mechanisms were obtained for the events located on the southeast extension of the Mishmi thrust in the syntaxis zone (Fig. 6.17), and if we assume slip occurs on the Mishmi thrust, the northeast dipping nodal planes would represent the fault. Valdiya (1976) reported that the Mishmi metamorphics thrust over the Tertiary sediments eroded from the northern Indo-Burma ranges. Three of the events have source depth greater than 8 km, only one event (8) of August 12, 1976, M_W 6.2 has a shallow source (depth 3 km). One event (9), the event of November 28, 1984 (depth 4 km), shows right-lateral strike-slip, which may indicate shallow crustal deformation. Three events (13-15) on the Sagaing fault, south of 26°N, clearly show right lateral motion (Fig. 6.17). The region north of 26°N appears to mark the transition from strike-slip to thrust as predominant mode of deformation in this area of Eastern Syntaxis (Holt et al., 1991; Kayal, 1996a). In the depth section along the Indo-Burma ranges it is evident that the earthquakes are much shallower to the north of 26°N; the subduction is taken over by collision tectonics (Fig. 6.12) (Kayal, 1996a).

Shillong Plateau-Mikir hills-Assam Valley

The earthquakes in the Shillong Plateau-Mikir hills-Assam valley area may be referred to the category of plate-boundary earthquakes due to its proximity to the Himalayan arc to the north and Burmese arc to the east. Although the Shillong Plateau-Mikir hills is a fragmented portion of the Indian shield, the seismic activity is much high, unlike less activity in the shield area of other part of peninsular India (see Chapter 7). Evans (1964) suggested that the Dauki fault acted as a tear fault for separation of the Shillong Plateau from the peninsular shield. Fault-plane solutions and tectonic implications of six moderate-size earthquakes (events 1-6) in this zone are illustrated in Fig. 6.15. Chen and Molnar (1990) argued that the four events (1-3 and 6) (depth 30-50 km) in the Plateau region are too deep to be associated with this tear fault in the upper crust as suggested by Evans (1964). Similarly, the events 4 and 5 in the Assam valley are also too deep (30-35 km) to be associated with a major Himalayan thrust or interplate thrust, while moderate-size events along the Himalaya typically occur at depths 10-20 km (Baranowski et al., 1984; Ni and Barazangi, 1984). Further, these six events took place not only at much greater depths (30-50 km), but 100-200 km south of the MBT in the Himalaya. The deepest event (No. 1) with a depth of about 50 km probably occurred in the mantle, suggesting that the mantle beneath the Bengal basin might be deforming in response to continental convergence. Based on consistent orientation of the P-axes, and lack of consistency in the slip vectors, Chen and Molnar (1990) suggested that the present seismicity occurs within a "diffuse zone of convergence" extending from the foothills of the Himalaya to south of the Shillong Plateau. Kayal (2001) referred the Plateau earthquakes to intraplate seismicity, which is about 100 km wide and seems to extend down to a maximum depth of 50 km into the uppermost mantle; the activity is, however, maximum in the lower crust (20-30 km) beneath the Shillong Plateau (Fig. 6.19). Occurrence of earthquakes in the uppermost mantle has been observed in several regions of recent continental convergence where an aseismic lower crust seems to be straddled between the seismically active lower crust and the uppermost mantle (Chen and Molnar, 1983). These observations are also made in the Plateau region from the microearthquake data (Kayal, 2001) (Fig. 6.19). Kayal (1987) and Kayal and De (1991) reported microearthquake activity at a depth range 10~35 km with maximum activity at 20~30 km, and a few as deep as 50-55 km, but almost no activity between 35-45 km (Fig. 6.19). These observations suggest that two layers of seismicity exist beneath the Plateau. The lower crust is most active and some activity is found in the uppermost mantle.

Fault-plane solutions of microearthquakes in the Plateau area are shown in Figs 6.19 and 6.24. Fault plane solutions of the seismic activity in the Shillong Plateau were critically examined using the digital data in Fig. 6.24(a), (Kayal et al., 2006). The N-S depth section of the earthquakes and the inferred fault planes in the western Plateau reveal that the earthquakes are generated by thrust faulting at the two boundary faults, the Dapsi thrust and the Brahmaputra fault, and the Shillong Plateau is popping up between these two boundary faults (Fig. 6.25). This observation supports the pop-up tectonics of the Shillong Plateau proposed by Bilham and England (2001). They, however, proposed that north dipping Dauki fault and south dipping Oldham fault are the boundary faults for pop-up tectonics. On the basis of digital seismic data, Kayal et al. (2006), on the other hand, argued that Dauki fault is neither north dipping nor much seismogenic, the north dipping Dapsi thrust and the south dipping Brahmaputra fault are the likely boundary faults for the Shillong Plateau earthquakes. The geological and geophysical data support this observation (Rajendran et al., 2004). Bilham and England (2001), however, suggested that a south dipping fault at the boundary of the Shillong Plateau and Brahmaputra valley was the causative fault for the 1897 great earthquake (Fig. 6.3). They named it 'Oldham fault', and suggested 'pop-up' tectonics of the Shillong Plateau between the north dipping Dauki and south dipping Oldham faults (Fig. 6.3). Geologically it is believed that the Dauki fault is a near vertical or southerly dipping fault (Evans, 1964). Nandy (2001) argued that the large (~ 20 km) difference in basement depth between the Shillong Plateau and the Bengal basin cannot be in any way explained by a thrust movement. The microearthquake data, however, suggest that the Dapsi thrust, western extension of the Dauki fault, is north dipping and seismogenic (Kayal, 1987; Kayal and De, 1991; Kayal, 2001). About 100 km long Dapsi thrust is separating the Archaean gneiss to the north and the Tertiary metasediments to the south within the Plateau (Fig. 6.3). Kayal et al. (2006) suggested that the 'pop-up' tectonics may occur between the Dapsi thrust and the Brahmaputra fault, not between the Dauki fault and Oldham fault. The intense seismic activity and the dipping fault planes clearly reveal the south dipping Brahmaputra fault (Fig. 6.25).

Seismic tomography study revealed LVZs at shallow depth 0-10 km; these are comparable with the surface faults in the Plateau region (Kayal and Zhao, 1998; Bhattacharya et al., 2006) (Figs 6.31 and 6.32). HVZs are prominent in the lower crust and in the upper mantle. The high velocity structures in the lower crust are identified as the source areas for stress accumulation for generating earthquakes beneath the Plateau. High isostatic anomaly (Verma and Mukhopadhyay, 1977), high P_n velocity (Kayal and De, 1987; Rai et al., 1999b) are reported in the area, which require support of a large excess mass beneath the Plateau either by dynamic process underlying the mantle or by strong lithosphere (e.g. Chen and Molnar, 1990; Rai et al., 1999b).

It is, however, surprising to note that in the microearthquake surveys practically no seismic activity is observed to the south of the Dapsi thrust; only a few earthquakes occurred along the eastern segment of the E-W trending Dauki fault (Fig. 6.19). The microearthquake activity is delimited to the north of Dapsi thrust. It is also interesting to note that the high intensity zone of the 1897 great Shillong earthquake was also delimited by the Dapsi thrust (Fig. 6.19). At the fault end of the Dapsi thrust lies a HVZ at 20-30 km depth (Figs 6.31 and 6.32), which may be the stress concentrator, the source area, in the lower crust for the Plateau earthquakes.

A large data set was analysed using the permanent/temporary network data recorded during 1993-1999 (Bhattacharay et al., 2002; Bhattachaya and Kayal, 2003). This study showed an intense seismic activity beneath the 300-400 km long Kopili fault/lineament from south of the Mikir hills to Arunachal Himalaya (Fig. 6.21). The seismic tomography revealed a prominent HVZ in the lower crust at the 'fault end'. This HVZ is identified as the source area, stress concentrator, for the earthquakes along the Kopili lineament (Fig. 6.32). High b-value and high fractal dimensions are also estimated in this zone as mentioned above (Figs 6.27 and 6.28). The seismogenic zone beneath the Kopili fault is well reflected in the b-value and fractal dimension maps, and may be identified as a high risk zone for an impending large earthquake in the area (Bhattacharya and Kayal, 2003; Kayal et al., 2006).

Kayal et al. (2006) studied digital seismic data of the permanent network, that were recorded during 2001 and 2003 as mentioned above (Fig. 6.24). In the Kopili fault zone the seismic activity is more intense, and base of the seismic zone extends down to 45 km (Fig. 6.26). The cross section of the events across the fault zone shows two possible northeasterly dipping faults (Fig. 6.26). The cross-section of the fault planes at depth is fairly conformable with the two dipping faults, one being very prominent for the solutions 15, 18 and 19, and the other is evident with the solutions 23 and 16, though other complex fault-plane solutions exist in the zone. The entire Kopili fault zone is, however, dominated by normal/strike-slip faulting, unlike the thrust/

strike faulting earthquakes at the Dapsi/Brahmaputra fault zone in the western Plateau where 'pop-up' tectonics is dominated. The Kopili fault zone is believed to extend transversely below the Himalaya (Kayal, 2001; Bhattacharya et al., 2002). The solutions 10 and 16, that occurred to the north of Main Boundary Thrust (MBT) in the Himalaya (Fig. 6.24), may be explained by this transverse seismogenic structure.

Bengal Basin and Tripura Fold Belt

The portion of the Indian plate which lies beneath the Bengal basin, is seismically less active; only few mantle earthquakes are reported (Chen and Molnar, 1990). Some seismic activity is observed in the Tripura fold belt (Fig. 6.10). The earthquakes in the Tripura fold belt are interpreted as the plate-boundary activity; the area is in proximity to the Burmese arc (Kayal, 1996a). The low activity in the Bengal basin may be attributed to intraplate activity (Kayal, 2001). The low activity is also interpreted due to thick sediments. Kayal (1989 and 1996a) suggested a *locked* portion of the Indian plate below this basin (Fig. 6.11c). The gigantic E-W Dauki fault is probably the surface expression of the lateral segmentation of the Indian plate which is *locked* below the Bengal basin and *unlocked* below the Shillong Plateau and Tripura fold belt. About ~20 km thick sediments of the Bengal basin is well reflected as LVZ (Fig. 6.32). Verma and Mukhopadhyay (1977) using gravity observations estimated the sediment thickness of about 13 km. Metivier et al. (1999), on the other hand, reported the Cenozoic sediments which may go down to 22 km; this observation supports the tomography results. Due to the thicker sediments, the crustal activity is much less. A few deeper earthquakes M>4.5 in the lower crust/mantle are, however, recorded below the Bengal basin (Fig. 6.10). One event (9) was located in the Bengal basin by the digital seismic network in Northeast India. It shows a strike slip faulting, and is correlated with the Sylhet fault (Fig. 6.24); the compressional stress is nearly in the NNE-SSW direction, compatible with the direction of plate motion.

Teleseismic data analysis (Chen and Molnar, 1990) of the three medium magnitude (4.0~5.0) earthquakes (events 3, 9 and 11) in this zone show broadly an NNE-SSW compressional stress (Fig. 6.15), which is also comparable with the motion of the Indian plate in the NNE direction. The event 3 occurred to the south of the Dauki Fault in the Bengal basin, and it cannot be associated with the Dauki fault as discussed above. Similarly, the events 9 and 11 in the Tripura fold belt cannot be associated with the local trend of folds or faults (Chen and Molnar, 1990). The northeast dipping nodal planes are the likely fault planes, and these are oblique to the local fold trends. The NNE-SSW compressional stress is dominant in this zone.

Aseismic Zone

The quiescent zone or the *aseismicity* in the upper Assam valley, between the Shillong Plateau and Eastern Syntaxis, is worth noting (Fig. 6.10). Khattri and Wyss (1978) and Khattri et al. (1983) demarcated the upper Assam valley *aseismicity* as the *Assam Gap* for an impending large earthquake. Mukhopadhyay (1984), on the other hand, explained the *aseismicity* of the upper Assam valley by the reactivation model of Jackson (1980), which suggests that the tectonic deformation is taken up by the boundary thrusts, the MBT to the north and the Naga thrust to the east, and the valley sediments take up the shortening by folding, the basement remaining *aseismic* itself. The historical earthquake data since 1548 (Bapat et al., 1983), though limited with documentation in earlier times, show no record of any medium or large earthquake in this elongated zone. The *aseismicity* or very low activity is also observed in this zone by the microearthquake surveys (Khattri et al., 1983; Kayal, 1987; Kayal and De, 1991). Kayal (1989 and 1996a) marked this elongated zone as an *Aseismic Corridor* (Fig. 6.10), which runs parallel to the Naga thrust to the east, and ends at the Dauki fault to the south. The Mikir massif within this corridor is seismically active. However, consequent upon the deadly Killari earthquake M 6.3 of September 30, 1993 in the so-called stable land area of the peninsular India, it is now dangerous to say that any part of the continent is aseismic (Kayal, 1998). The recent digital network data show activity along the Kopili fault/Mikir hills region which extends to the southeast within the corridor (Fig. 6.24).

Seismic Anisotropy

In a recent study, Singh et al. (2006) analysed SKS/SKKS phases in northeast India using the digital seismic data. They reported distinct shear wave anisotropy in E-W direction in the northeast Himalayas and its foredeep region, in N-S direction in the Indo-Burma ranges and in NE-SW direction in the Shillong Plateau (Fig. 6.34). Anisotropy of the Indian lithosphere in this region is contrary to earlier findings in southern most Tibet by INDEPTH study (McNamara et al., 1994; Hirn et al., 1995)

The E-W anisotropy in the northeast Arunachal Himalayas (Fig. 6.34) is in agreement with the tectonic stress obtained by strike-slip fault plane solutions (E, F and G) of the microearthquakes (Fig. 6.15). In the Shillong Plateau the NE-SW direction of anisotropy is also parallel to the NE-SW tectonic stress of the fault plane solutions (A and B) (Fig. 6.15). The anisotropy in N-S direction in the Indo-Burma ranges is parallel to the northward drag of the Indian lithosphere. It seems local tectonic stress plays an important role in the S wave anisotropy.

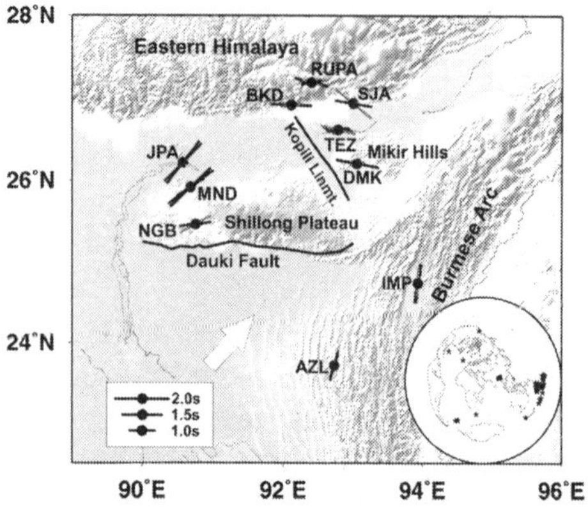

Fig. 6.34 Shear wave fast axis direction is indicated by thicker black bar with results from different back-azimuth in grey. Inset shows distribution of events used (Singh et al., 2006; Copy right @ AGU).

6.8 STRONG MOTION DATA

In an International Workshop on Strong Motion Instrumental Array held in 1978 at Hawaii, northeast India was considered as one of the six most potential areas in the world for deployment of Strong Motion Array (SMA). As a follow up action, the Department of Earthquake Engineering, University of Roorkee (now IIT Roorkee) installed an array of 45 analog accelerographs in the Shillong Plateau and Assam valley area in 1985 under a DST project. In 1990s, 17 digital accelerographs have been installed covering the Arunachal Himalaya, Shillong Plateau and Assam valley. The reports were published by Chandrasekaran and Das (1990, 1992 and 1994). At present the region is well instrumented with some 50-60 digital strong motion accelerographs.

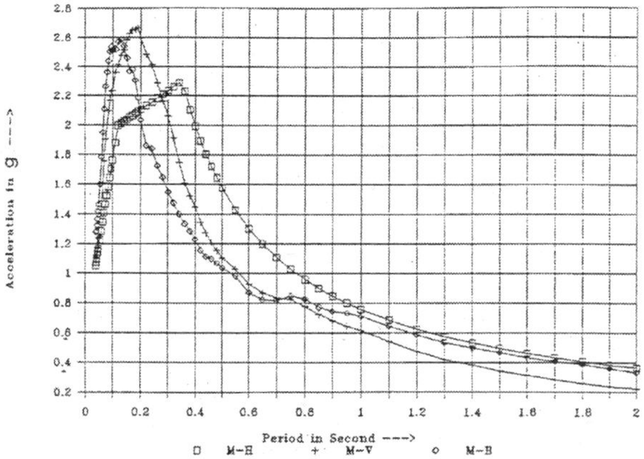

Fig. 6.35 Strong motion spectra normalized to peak ground acceleration in NE India (Chandrasekaran and Das, 1992).

The analog instruments (model SSA-1) provides triaxial photographic recording when triggered by strong local earthquakes. The SMA-1 records 1 g (full scale) with a triggering level of 0.01 g. The digital instruments (SSA-2), on the other hand, can record a full scale of 1 g as well as perturbations as small as 0.0005 g in the same record.

Analysis of strong ground-motion records provides the knowledge in understanding of factors which affect the character of strong motion, such as source mechanism, influence of wave propagation, effects of local topography, geology and soil conditions and evaluation of design parameters (Chandrasekaran, 1994). Using the arrival times, reliable hypocentre parameters are also determined (Chandrasekaran and Das, 1992). Valuable records for five earthquakes (M>5.0) are obtained till 2000 by the SMA in northeast India since its installation in 1985. These are summarised in Table 6.2.

Table 6.2: Details of Strong Motion Records

Year	*Month*	*Day*	*Magnitude* (m_b)	*Record Nos.*	*Max. Duration (sec)*	*Peak Acceleration (g)*
1987	5	18	5.7	14	42	0.09
1988	2	6	5.8	18	45	0.11
1988	8	6	7.2	33	120	0.34
1990	1	10	5.5	14	60	0.14
1996	19	10	5.2	12	29	0.13

The range of acceleration is 0.09 to 0.34 g in horizontal direction during the five events in northeast India. Figure 6.35 shows the mean shape of spectra normalised to peak ground acceleration of unit gravity.

The records are extremely useful for the dam sites, thermal and hydro-power project sites for the purpose of design parameters and earthquake risk analysis. The shape of the spectra clearly points out that long period structure like earth and rock-fill dams and long bridges would get a relief in design, if the shape of the spectra as obtained from the actual events are used. These data are also useful for microzonation in the region.

6.9 SEISMIC HAZARD POTENTIAL

The seismological data recorded over the last 100+ years since 1897, particularly the seismological data of the ISC and USGS since 1964, the microearthquake analog data of the temporary networks for the period 1980~91 and that of permanent network for the period 1993-1999 and the digital network data for 2001 and 2003, have been useful to a great extent to understand the earthquake source processes and the active faults in some parts of the northeast India region.

The Shillong Plateau region, which was the source area of the 1897 great earthquake, is identified as one of the most potential area for seismic hazard/ impeding large earthquake (Bilham and England, 2001). Any large/great earthquake in the Plateau region would also cause devastation in Bangladesh (Bengal basin) adjoining areas. The pop-up tectonics of the Plateau between the two identified active boundary faults may cause this devastation. These active boundary faults are well identified by the microearthquake network data. The early triangulation data and the recent GPS data support the observation of the Shillong Plateau uplift (Kailasam, 1979; Bilham and England, 2001). An abnormal uplift is reported in the recent past (Nagarajan, 2006; pers. comm.) after the 2004 mega thrust Sumatra earthquake (M_w 9.3). These observations suggest a potential earthquake hazard in the Shillong Plateau area in the future.

Analysis of recent (1993-2003) microearthquake network data has also established the long Kopili lineament as the most active fault in the region. It is a 300-400 km long and ~50-100 km wide fault zone, a deep rooted seismogenic feature, transverse to the Himalayan trend. A large earthquake M ~7.0 occurred in 1941 in this zone (Fig. 6.3). Kayal et al. (2006) identified this zone as the most vulnerable for a large/great earthquake in the region. Although earthquake prediction/precursor is yet to be understood well, they argued that in addition to monitoring seismicity, the zone requires a permanent multi-parameter geophysical observatory for precursor studies and close monitoring of crustal deformation across the fault zone by GPS measurement. Like the assessment of large casualty and huge property loss in the western Himalaya (Arya, 2000), a large/great earthquake in the Plateau or in the Kopili fault zone would also cause similar devastation with the increased population growth and non-resistant dwellings in the region. In this respect some investigations like paleoseismology, precusrsor and microzonation studies may be useful to mitigate seismic hazards. A few Research and Development (R&D) efforts were made to mitigate earthquake hazards in the region (GSI and DST, 2006; unpub. reports), but much remains to be done.

Paleoseismic Study

Paleoseismic investigations were carried out at the Chedrang fault that was developed by the 1897 great Shillong earthquake (Rastogi et al., 1993; Sukhija et al. 1999). Field investigations revealed that liquefaction features were wide spread along the Krishna river, close to the Chedrag fault (Fig. 6.36). Sukhija et al. (1999) reported four paleoseismic events, and characterized as (i) 1897 great earthquake, (ii) 300-500 yr. BP event, (iii) 900-1250 yr BP event and (iv) an event earlier than 1500 yr BP, thus providing a recurrence of about 500 $\pm$ 100 years. A similar investigation is recently conducted in the meizoseismal area of the 1950 great Assam earthquake in the Eastern Syntaxis. Field investigations along the Buri-Dihang river, a tributary of the

Brahmaputra river, resulted in identification of a dozen of liquefaction features; dating of these features is being attempted (Shukija, 2006; pers. com.) . Thus instrumental and paleoseismological data in the Plateau region call for a reassessment of seismic hazard. The GSHAP map (see Fig. 5.58), however, identifies northeast India as a zone of highest risk with PGA values of the order of 0.35-0.40 g. A great earthquake in the Shillong Plateau like that of the 1897 that caused PGA of the order of 1 g may cause more devastation as normally assessed for other regions.

(a) (b)

Fig. 6.36 Paleoseismic evidences, sand dykes in the epicentre zone of the 1897 earthquake (Sukhija et al., 1999).

Precursor Study

Precursor studies were started in a big way in early 1970s around the world to forecast earthquakes. Several precursor studies were undertaken in northeast India (Guha and Bhattacharya, 1984; Gupta and Singh, 1986; Kayal and Banerjee, 1988; Kayal, 1991). Khattri and Wyss (1978) had reported decrease in seismic activity and identified a seismic gap between the Shillong Plateau and Mishmi thrust zone for an impending large earthquake in the region. Guha and Bhattacharya (1984) reported a significant precursory decrease in b-value and indicated a probability of occurrence of a great earthquake (M>8.0) in near future. Kayal and Banerjee (1988) made deep resistivity measurements in the Shillong region and reported correlation between

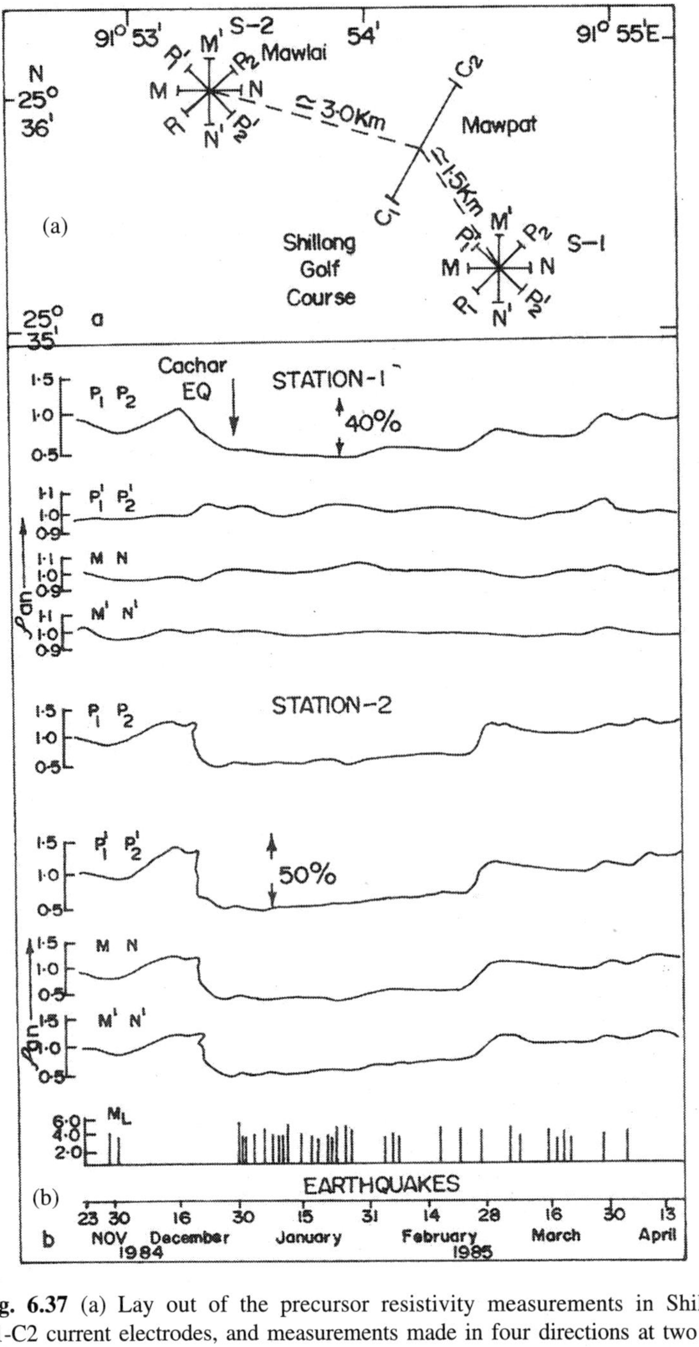

Fig. 6.37 (a) Lay out of the precursor resistivity measurements in Shillong, C1-C2 current electrodes, and measurements made in four directions at two fixed stations. (b) Temporal changes of normalized apparent (ρ_{an}) resistivity of the Earth and occurrence of the 1984 Cachar earthquake (M_w 5.8) and Shillong earthquake sequence (Kayal and Banerjee, 1988).

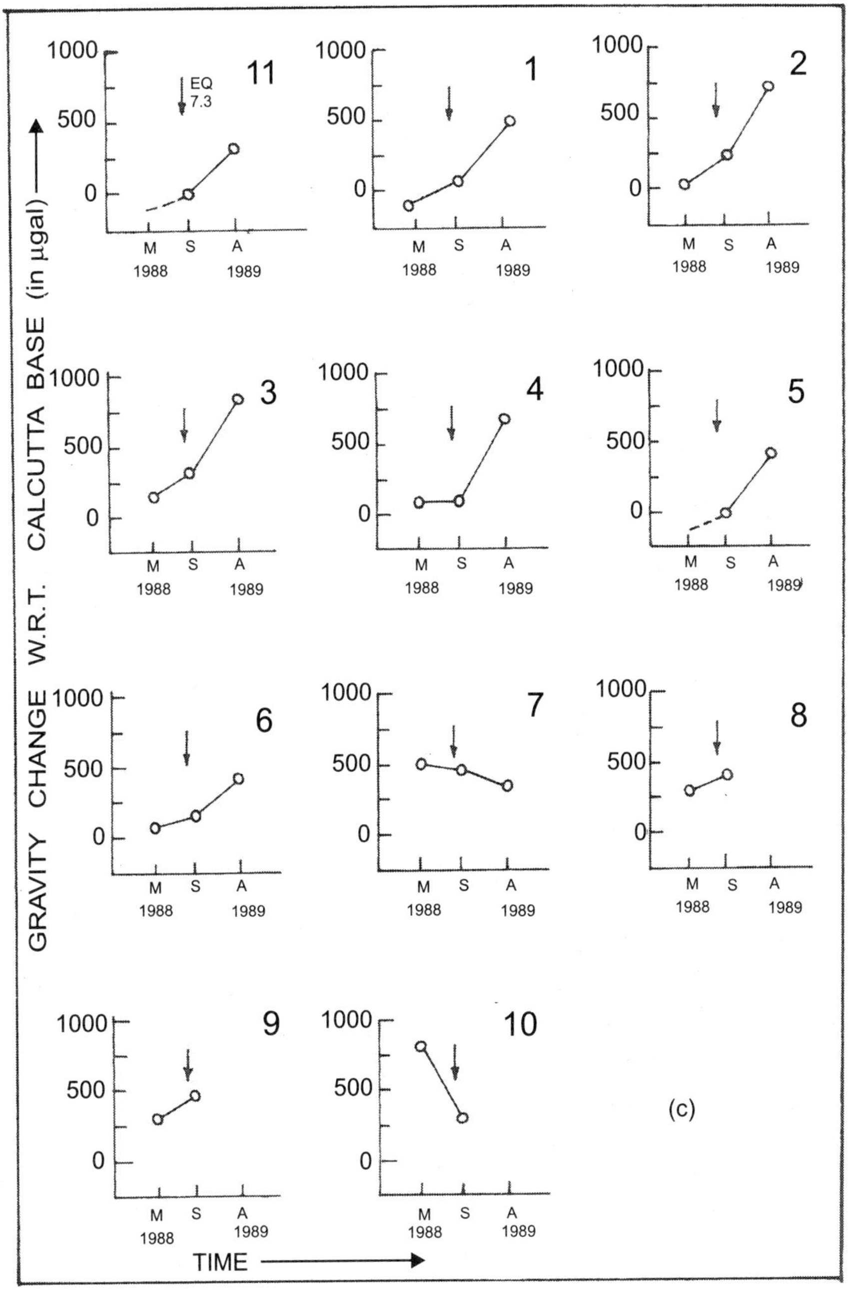

Fig. 6.37 (c) Changes of microgravity values at fixed stations along a N-S transect across the Shillong Plateau and Brahmaputra valley, the stations in the Plateau show increasing and that in the valley show decreasing trend before the 1988 Manipur-Burma border earthquake (M 7.2), (Kayal, 1991).

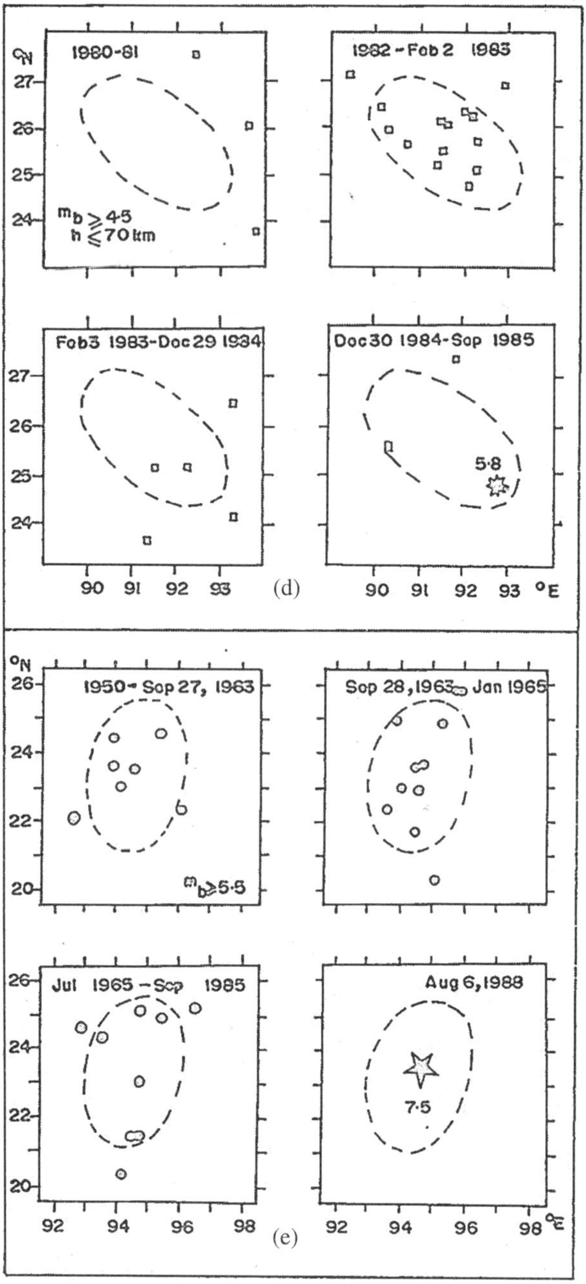

Fig. 6.37 (d) Swarm hypothesis tested for the 1984 Cachar earthquake, and (e) the 1988 earthquake forecasted by Gupta and Singh (1986 and 1989).

precursor resistivity anomaly and occurrence of the December 1984 Cachar earthquake (Mw 5.8) (Figs 6.3 and 6.37a and b). They suggested that negative results or insignificant changes in the resistivity values are possible, and the changes are not only dependent on rock types but also on the direction of measurements. Some anomalous changes in seismological parameters like decrease in b-value, V_p/V_s, and in seismicity rate were also reported before the Cachar earthquake (Kayal, 1991). In his review work, Kayal (1991) reported precursor gravity anomalies, though limited with a few repeat measurements, before the August 1988 Manipur-Burma border earthquake (M 7.2) (Figs 6.10 and 6.37c).

Based on the earthquake *swarm hypothesis* of Evison (1982), Gupta and Singh (1986) tested the precursory phenomena of the 1984 Cachar earthquake (Fig. 6.37d), and made a *forecast* of the August 6, 1988 Manipur-Burma border earthquake (M 7.2). The 1988 earthquake occurred within the stipulated time and space window (Fig. 6.37e), and they claimed it to be a successful forecast (Gupta and Singh, 1989). Continuous monitoring of geophysical parameters like gravity, magnetic, deep resistivity, radon content in soil or well water, and monitoring of seismic activity may prove useful for forecasting a large earthquake in the region.

Although there are debates on the topic of *earthquake prediction,* great researchers like Evison, Wyss and many others (see *Tectonophysics* sp vol. 338, 2001) believe that *synoptic earthquake forecasting* may be possible with proper funding and leadership on this research. A book review of this volume is given by Kayal (2002), where the problem is candidly discussed. Globally, superior earthquake catalog started in early 1980s, hardly three decades ago, whereas return period of large earthquake extend over 100 years. We have not done enough research on earthquake prediction. With the advent of modern technology like Global Positioning System (GPS), Synthetic Aperture Rader Interferometry (InSAR), Ultra low frequency Electromagnetic Emission (ULF) etc., our future researchers may reach to the level of predicting earthquakes. The DST is promoting establishment of permanent geophysical observatories in the western and northeastern Himalaya region with modern instruments. Precursor studies are, however, yet to be developed to forecast a large earthquake with space and time. Even if it is successfully predicted the property loss, however, cannot be avoided unless earthquake resistant constructions are made, building codes etc are followed. This is most important to save life as well as to mitigate earthquake hazards.

Microzonation Study

Microzonation studies took over the precursor studies in early 1990s. Numerous studies on devastating earthquakes world over amply demonstrated that concentration of damage is site dependent, related to geologic conditions and local site effect or site response (e.g. Aki, 1988; Field et al., 1992; Field and Jacob, 1995). Site response study is a major input for microzonation, that plays an important role to retrofitting old buildings or to construct new ones. The DST and the GSI made a considerable effort to prepare microzonation maps for the city areas, like Guwahati (Assam) and Agartala (Tripura), in this region. Higher site amplification zones and the corresponding resonance frequency are mapped. It was observed that the river valley sediments as well as depth of water table plays an important role for higher site amplifications (GSI and DST, 2006; unpub. reports). This topic needs to be dealt separately; detailed discussion is beyond the scope of this book.

6.10 TECTONICS AND SEISMICITY OF ANDAMAN-SUMATRA ARC

6.10.1 Introduction

The Burmese-Andaman-Sumatra-Sunda arc defines a ~5500 km long boundary between the Indo-Australian and Eurasian plates, from Myanmar to Sumatra and Java to Australia (Fitch, 1970; Curray et al., 1979). The plate boundary separates the NE moving Indian plate from the southeast Asian plate that includes Burma, Andaman and Sunda *microplates*. Global reconstructions suggest that the Indian plate converges obliquely toward the Asian plate at the rate of 54 mm/year (De Metes et al., 1994). The result of the oblique convergence has been the formation of a *sliver plate* between the subduction zone and a right lateral fault system, which has evolved as the Sumatra Fault System in the southern part of the subduction zone and the Sagaing Fault in the northern part in Myanmar, as well as the opening of the Andaman sea, the Andaman Sea Ridge (ASR) (Fig. 6.1). The nature of convergence varies from continental type in the Burmese arc to oceanic type in the Andaman-Sunda arc. Like the Burmese arc, the Andaman-Sunda arc is also seismically very active, and falls in the category of highest seismic hazard zone at par with the Himalayan arc (Fig. 5.1).

Varying degrees of tectonism and volcanic activity occur along this subducting margin (Curray, 2005). The Andaman-Sumatra section of the subduction zone had produced many large earthquakes in the past, some of which generated destructive tsunamis. Largest among the historical earthquakes occurred in 1833 (M~8.7), 1861 (M~8.5), 1881 (M_w 7.9) and 1941 (M_w 7.7) (Fig. 6.1). While these large earthquakes ruptured only a few hundreds of kilometres (200-300 km) of the plate boundary, the 2004 earthquake has ruptured more than 1200 km length of the arc, stripping the regions that were ruptured in the past as well as the intervening unbroken patches (Bilham et al., 2005). Historical seimicity of the Sumatra region is dominated by two great earthquakes, the 1833 (M~8.7) and 1861 (M ~8.5) events. This region has also witnessed two of the recent large earthquakes in the central Indian ocean; these occurred on 4 and 18 of June 2000 (M_w

~ 7.8), located south of the 1833 rupture (Abercrombie et al., 2003). The Indonesian part of the trench that produces deeper events down to 600 km in the subduction zone, has been extensively studied in the recent years using a variety of techniques including GPS based deformation studies as well as coral microatolls (Natawidjaja et al., 2004). Similar studies are, however, only in its early phases in the Andaman part of the arc. In this section the seismicity and tectonics of the Andaman-Sumatra region is briefly discussed. A great detailed study on seismotectonics of southeast Asia needs to be dealt separately.

6.10.2 Tectonic Setting

The N-S trending Sagaing transform fault demarcates the eastern margin of the Burma microplate. It separates the central low lands from the eastern high lands of Burma, and the fault continues to the Andaman Sea Rift system, known as the Andaman Spreading Ridge (ASR) (Figs 6.1 and 6.38). The Andaman sea basin is considered to be a complex back arc spreading centre; it is categorized as a 'pull apart' or 'rip off' basin rather than a typical back-arc-extensional basin. Seismic reflection studies across the trench slope indicate folding and thrusting in the accretionary prism, where a major component of convergence occur normal to the trench axis (Curray, 2005). Among the series of thrusts/faults in the Andaman subduction zone, the West Andaman Fault (WAF) is most prominent (Fig. 6.1). This thrust/fault appears to be continuous from the west of Sumatra in the south to Irrawady basin in the north, where it is buried beneath the sediment cover. The region between the Andaman-Nicobar Islands and the volcanic arc is the foredeep sedimentary trough (Dickinson and Seely, 1979).

Subduction along the Andaman-Sumatra trench system has given rise to a discontinuous belt of submarine ridges and volcanic seamounts. The andesite volcanoes of the Barren and Narcondam Islands are prominent among them; the Narcondam is now extinct, but the Barren is still marked by an active volcano; it erupted last in March, 1991 after lying dormant for about two centuries (Halder et al., 1992). Further south, this volcanic belt is represented by the Barisan range in Sumatra, and the trend is correlated with the central molasse basin of Burma in the north (Curray et al., 1982).

Along the Sumatra subduction zone, plate convergence is partitioned into dip-slip and right lateral strike-slip components, the former being accommodated by slip on the subduction interface and the latter by the Sumatran Fault System (Fitch, 1972). The present-day tectonic processes are controlled by three major fault systems, the most prominent being the subduction thrust, which outcrops in the Sunda trench. Inland, the trench-parallel Sumatra Fault that runs through the entire length of the island from Banda Aceh to Sunda Strait accommodates oblique convergence through strike-slip faulting. The Mentawai fault at the outer margin of the forearc basin is another important fault system in the Sumatra region (Sieh and Natawidjaja, 2000).

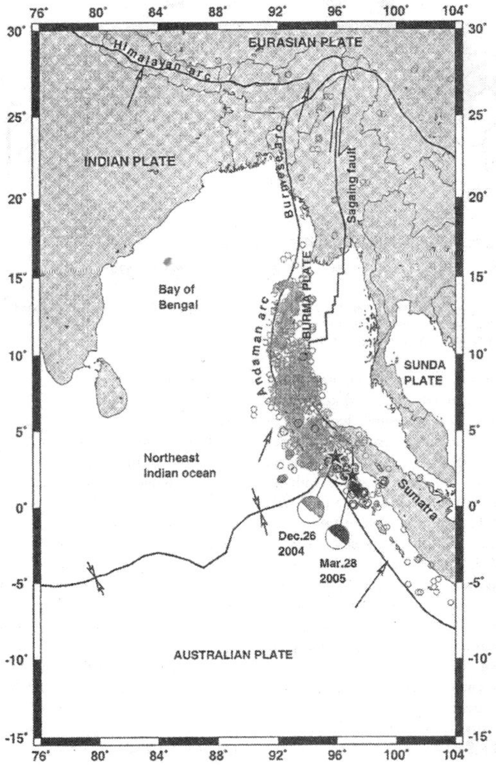

Fig. 6.38 Tectonic setting, plate boundaries and epicentres of the two great earthquakes in Andaman-Sumatra region, the 2004 mega thrust event (M_w 9.3) and the 2005 great (M_w 8.7) earthquake and their aftershocks (Rao and Chary, 2005).

6.10.3 Significant Earthquakes in the Past

The Andaman-Sumatra section of the subduction zone has produced many large and destructive earthquakes in the past, some of which have also generated destructive tsunamis (Bilham et al., 2005). The M 8.7 earthquake of 1833 is reported to have ruptured about 550 km segment of this arc; it also generated a tsunami (Natawidiaja et al., 2004). Another great earthquake of 1861 (M 8.5) broke a segment north of the equator, and triggered a tsunami. The 1833 and 1861 earthquakes and the attendant tsunamis occurred before the introduction of harbour tide gauges in most parts of the world and no tidal gauge data exist for these events. However, better documentation exists for the 31 December 1881 and 26 June 1941 tsunamis, both of which caused run-up in India's eastern coast. The 1881 tsunami is reported to have reached an amplitude of 1.04 m at Chennai (Ortiz and Bilham, 2003). This earthquake (M_w 7.9) is the earliest for which slip geometry has been inferred. The 1941 earthquake (M_w 7.7) also appears to have ruptured the region (~200 km) near the Andaman Island (Ortiz and Bilham, 2003).

A detailed study of past seismicity for the period of 1916-1975 in the Andaman Sea was illustrated by Verma et al. (1978). Le Dain et al. (1984) and Kumar et al. (1996) have reported the distinct tectonic patterns between the northern and southern parts of the Andaman using the focal mechanisms of shallow and deeper earthquakes from the regional seismicity. Dasgupta and Mukhopadhyay (1993) and Dasgupta et al. (2003) studied subduction tectonics at the Andaman arc and reported an east-dipping (40-55°) Benioff zone down to about 200 km depth in the forearc. Further south, the Benioff zone below the Sumatra-Java goes down to 600 km (Rao and Chary, 2005).

6.10.4 Regional Seismicity

A regional seismicity map of the Andaman-Sumatra is shown in Fig. 6.39. The entire Burmese-Andaman-Sunda arc is seismically very active, but for a small patch north of 15°N latitude. Interestingly, this patch represents the region where the sediment thickness due to the Bengal fan exceeds nearly 6 km and it is quite possible that most of the deformation in this part of the region is being accommodated by aseismic slip. It is also noted that most of the large earthquakes occur within the upper part of the lithosphere, giving due consideration to the fact that depths are not well constrained for a large number of earthquakes. The shallow seismicity east of 93.5°E is representative of the Andaman Spreading Ridge (ASR). Based on the analysis of focal mechanisms of 148 earthquakes recorded during 1970-1993, Dasgupta et al. (2003) have suggested predominance of shallow thrust earthquakes as well as down-dip tensional events within the Benioff zones, indicating slab-pull extensional tectonics as a dominant factor. Figure 6.39 shows Benioff zone configurations from north to south in the Andaman-Sumatra region based on the selected E-W cross-section of the earthquakes. Figure 6.40 illustrates an overall view of the Benioff zones below the Burma arc, Andaman arc and the Sumatra-Java arc.

The earthquakes in the Andaman trench zone (7°-15° N latitude) show distinct characteristics of the subduction and back arc spreading. This part of the arc exhibits more complex pattern and segregation into clusters of thrust, normal and strike-slip faulting earthquakes (Kumar et al., 1996; Dasgupta et al., 2003). Earthquakes mostly follow a NNE trend, the western arm representing the Andaman trench, the Narcondam and Barren Volcanic Islands forming part of the fore arc setting. The eastern arm that branches off from about 11°N is characterized by shallow seismicity associated with the back arc extension in the ASR (Fig. 6.39). Cross section of hypocentres within the selected block shows that the Benioff zone is about 120 km deep along this part of the trench compared to ~170-200 km in the Burmese arc (Fig. 6.40). The last major earthquake in this region occurred on June 26, 1941; this one affected the middle and south Andaman regions, including the town of Port Blair. The other significant earthquake in this region occurred on December 31, 1881, whose source area is thought to be in the open ocean (Ortiz and Bilham, 2003). The latest earthquake in this region, prior to the 2004 sequence, occurred on September 2002 (M_s 6.8), offshore, near Diglipur, north Andaman (Rajendran et al., 2003 and Kayal et al., 2004). Aftershocks of this event were recorded by making a temporary network in the Andaman Islands (Kayal et al., 2004), and discussed later. It may be noted that to the north of the Andaman basin, between latitude 15° and 20°N, not much seismic activity is observed (Fig. 6.40). The Indo-Burma Range activity, further north, latitude 20°-26°N, is well explained by the dipping Indian lithosphere (Fig. 6.11). The low seismic activity in the central portion, latitude 15°-20°N, may be explained due to presence of thick pile of Irrawady sediments that takes up the stress by creeping.

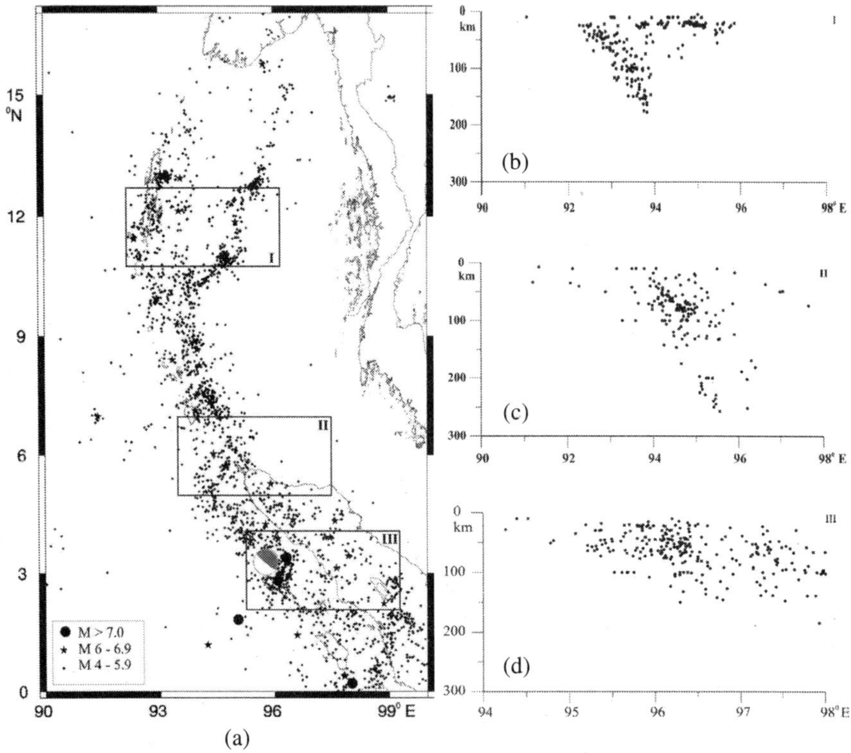

Fig. 6.39 (a) Andaman-Sumatra seismicity (USGS data: 1964-2004), and (*b-d*) E-W seismic sections of the events of the selected blocks I, II and III (modified from Kayal et al., 2004).

To the south of the Andaman arc, the orientation of the fore arc gradually changes from NNE to NW-SE to E-W in the South Nicobar-Sumatra Islands and to Java Islands (Fig. 6.40). The subduction zone can be seen as belonging to two sectors, northern sector and southern sector. The northern sector corresponds to the great Nicobar and the Sumatra islands, where the Indian plate and the southeast Asian plate (Burma platelet) are interacting. The southern sector, on the other hand, corresponds to the Java trench where the Australian plate is subducting beneath Sunda plate. The Benioff zone is about ~300 km deep in the northern sector (Fig. 6.40d); the plate convergence is portioned as dip-slip and a right lateral strike-slip components respectively along the Andaman-Sumatra trench and the Sumatra Fault system. In the southern sector, the Benioff zone is much deeper, down to 600 km (Fig. 6.40e). Various workers have analysed the focal mechanisms from this region as well as the pattern of stress orientation (Dasgupta et al., 2003; Rao and Chary, 2005). While some minor along-the-arc variations have been noted, the general pattern is that of active subduction dominated by shallow thrust events.

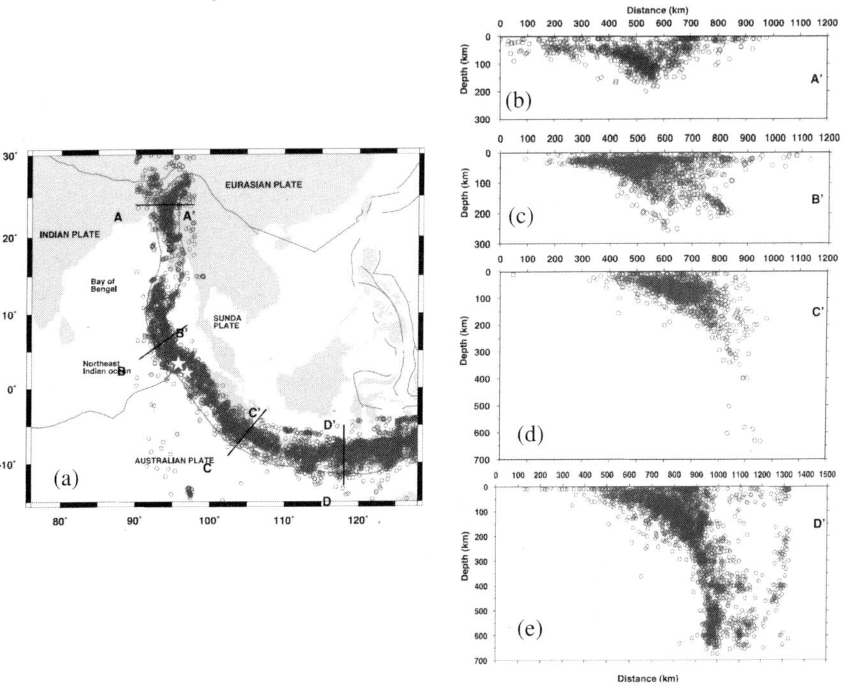

Fig. 6.40 (a) Seismicity along the Burma-Andaman-Sumatra-Java arc, and (*b-e*) seismic cross sections along A-A′, B-B′, C-C′ and D-D′ (Rao and Chary, 2005).

The M_W 9.3, December 26, 2004 earthquake is the largest to have occurred in this zone; it ruptured ~1300 km length of the arc in the north direction, which included segments that have ruptured in the past large earthquakes (Figs 6.1 and 6.41).

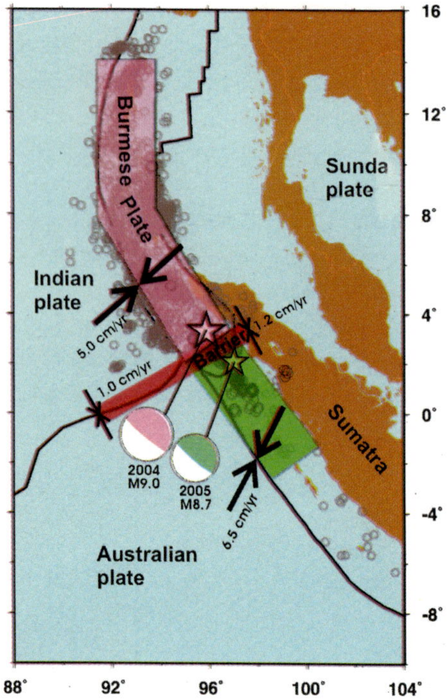

Fig. 6.41 Two rupture zones, the 2004 mega thrust rupture (~1300 km) propagated to the north, and the 2005 rupture (~300 km) to the south, epicentres are shown by star symbols, a barrier zone is suggested between the epicentre regions of the two events, fault plane solutions of the main shocks and epicentres of the aftershock sequences are shown (Rao and Chary, 2005).

The March 28, 2005 great earthquake M_w 8.7 occurred further south, and it ruptured ~300 km toward south, within three months of the December 2004 mega thrust event (Fig. 6.41). It is interesting to note that the December 2004 rupture (~1300 km) propagated northward from the source, and the March 2005 rupture, though much less (~300 km) compared to that of the 2004, propagated southward. Rao and Chary (2005) suggested a barrier zone that separates the two source zones (Fig. 6.41). Recent global seismicity map suggests a plate boundary zone between the Indian and Australian plates (Fig. 2.1), this boundary zone may act as a barrier zone between the two segments of the Andaman-Sunda trench. Relative plate movements are reported to be different in the two segments (Fig. 6.41). Fault plane solutions for both the main shocks are reported to be similar, both occurred by thrust faulting in the subducting plate (Fig. 6.41). The 2004 event caused a maximum slip of ~20 m, and the 2005 caused a maximum slip ~5 m. The significant difference in vertical slip, rupture length and the bathymetry played a major role in generating huge tsunami for the 2004 mega thrust event, and much less tsunami for the 2005 event.

6.10.5 Aftershock Investigation in Andaman-Nicobar region

Andaman Earthquake M_s 6.8, September 2002

Main Shock

Seismological parameters of the main shock of September 13, 2002 in the Andaman Sea are estimated by the U.S. Geological Survey (USGS) as follows: moment magnitude M_w 6.5; M_s 6.8; epicentre 13.087°N, 93.112°E; origin time 13 September 2002, 22h 28m 31.93s; centroid depth 31 km; seismic moment $M_o = 6 \times 10^{18}$N m. The moment tensor solution shows reverse faulting (Fig. 6.42a). The maximum intensity reached VII on MM scale in Diglipur and Mayabadar area (Kayal et al., 2004).

Larger Aftershocks

Five larger (M≥5.0) aftershocks were recorded within about one month of the main shock. The largest aftershock (M_w 5.8) occurred within 22 hr with a magnitude 0.7 units lower than the main shock; it occurred at a shallower depth (21 km) compared to that of the main shock (31 km). Normally the largest aftershock occurs within 24 hr with 1.0-1.2 magnitude units lower than main shock. In this case, magnitude of the largest aftershock is significantly higher. The USGS also reported its moment tensor solution; it shows a similar reverse-faulting mechanism (Fig. 6.42a).

In addition, the global network recorded eight aftershocks with magnitude 4.3≤5.0 (USGS). All these aftershocks were located using fixed depth 33 km, except one at 83 km depth. No fault plane solutions for these aftershocks were reported.

Temporary Microearthquake Network

Four temporary digital seismic stations were installed by the GSI in the Andaman Islands to monitor the aftershocks (Kayal et al., 2004). The GSI was the only organization to establish a temporary microearthquake network to monitor the aftershocks, and this was the first microearthquake investigation in this Island. Short-period SS-1 seismometers and digital (Reftek and Kinemetrics) recorders were used to establish the four stations. The seismic stations were established at Diglipur, Mayabandar, Havelok and Port Blair. The digital instrument in Port Blair was installed at the IMD permanent observatory. The Port Blair IMD observatory was equipped with an analog instrument. This analog instrument was, however, useful to make comparative study.

Epicentres of all 145 aftershocks located by the temporary network show a NW-SE trending aftershock cluster area of 140×70 km^2 (Fig. 6.42b). The USGS locations of the main shock and the larger aftershocks fall within this aftershock zone. The aftershock zone is, however, well defined by the close-spaced temporary network.

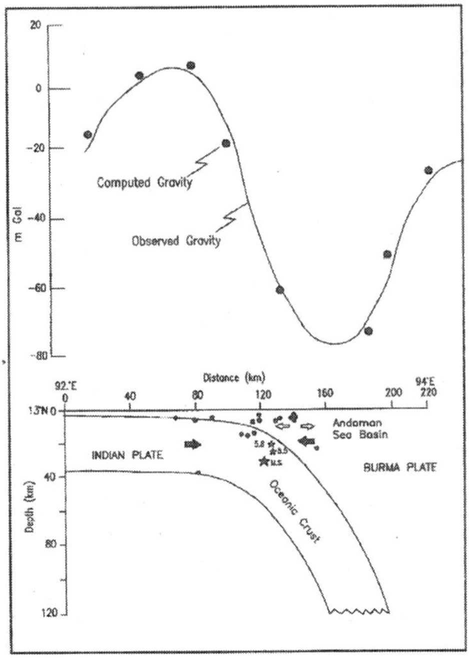

Fig. 6.43 Seismotectonic model for the 2004 main shock and its aftershocks, the Benioff zone obtained from the seismic cross section (Fig. 6.39 b). The aftershocks off the subducted plate show normal faulting (Fig. 6.42c). The gravity anomaly is fitted with the subducted Indian lithosphere (after Kayal et al., 2004).

Sumatra Mega Thrust Earthquake Mw 9.3, 2004

Main Shock

The December 26, 2004 devastating mega thrust earthquake (M_W 9.3) occurred at the interface of the India and Sunda plates in the Indian Ocean (Fig. 6.38), which caused a huge loss of property and person in most of coastal countries of Southeast Asia (e.g. Bilham, 2005; Lay et al., 2005; Rajendran et al., 2005). The main shock was felt in different parts of the world with varying intensities: VIII-X at the Banda Aceh, VI-VIII in Andaman-Nicobar and Sumatra, and II-V in parts of Bangladesh, east coast India, Malaysia, Maldives, Myanmar, Singapore, Sri Lanka and Thailand (Martin, 2005). The earthquake was severely felt in the entire Andaman-Nicobar Islands of India. The damage survey reported that the Andaman-Nicobar Islands is the worst affected state of India due to strong shaking and tsunamis (Mishra et al., 2007). The severe shaking and damages were experienced in the Diglipur (north Andaman) and the Mayabandar (middle Andaman) areas despite their distant locations of about 1200-1300 km from the main shock epicentre. A mud volcano near Baratang in the middle Andaman witnessed a massive eruption of mud and slurry materials following the December, 2004 main shock (Fig. 6.44).

Fig. 6.44 Macroseismic observations of the 2004 mega thrust event in the Andaman Islands: (a) A bridge in Mayabandar, north Andaman, broke and separated by strike slip movement, (b) A tree trunk was split and separated by about 80 cm by strike slip movement in the middle Andaman, (c) The mud volcano activated in Baratang, middle Andaman, and (d) a liquefaction observed in Port Blair area (GSI, 2007).

Earthquake parameters of the December 26, 2004 main shock and four large aftershocks ($6.3 \leq M \leq 7.3$) that occurred within 12 hours of the main shock were estimated by the US Geological Survey (USGS) and the National Earthquake Information Centre (NEIC). The main shock shows 1.48×10^{30} dyne-cm as its seismic moment that is almost equivalent to the 1960 great Chilean earthquake (M_w 9.5) that was associated with a seismic moment of 2.7×10^{30} dyne-cm (Kanamori and Cippar, 1974). The USGS determined the Centroid Moment Tensor (CMT) solution of the main shock that shows a low angle thrust fault (Fig. 6.45).

The world's largest recorded five earthquakes have been the mega thrust events and they occur at the interface of the subducting plate beneath another. The five mega thrust events include the 1960 Chile earthquake (M_w 9.5), the 2004 Sumatra earthquake (M_w 9.3), the 1964 Prince William Sound, Alaska earthquake (M_w 9.2), the 1957 Andreanof Islands, Aleutian earthquake (M_w 9.2) and the 1952 Kamchatka earthquake (M_w 9.0). All these events generated large rupture (800 ~ 1200 km) zones. The occurrence of the largest earthquake (M_w 9.3) in the northern Sumatra and its series of aftershocks along the ~1300 km long rupture zone posed a gigantic task to seismologists/

geoscientists to resolve the complex geodynamical processes of the region in the light of subduction dynamics and seismic imaging. The huge data that are obtained by the temporary network in the Andaman-Nicobar Islands coupled with global earthquake data would be of immense use to better understand the seismotectonics and the heterogeneous seismic structures of the entire rupture zone extending from the northern Sumatra to northern Andaman.

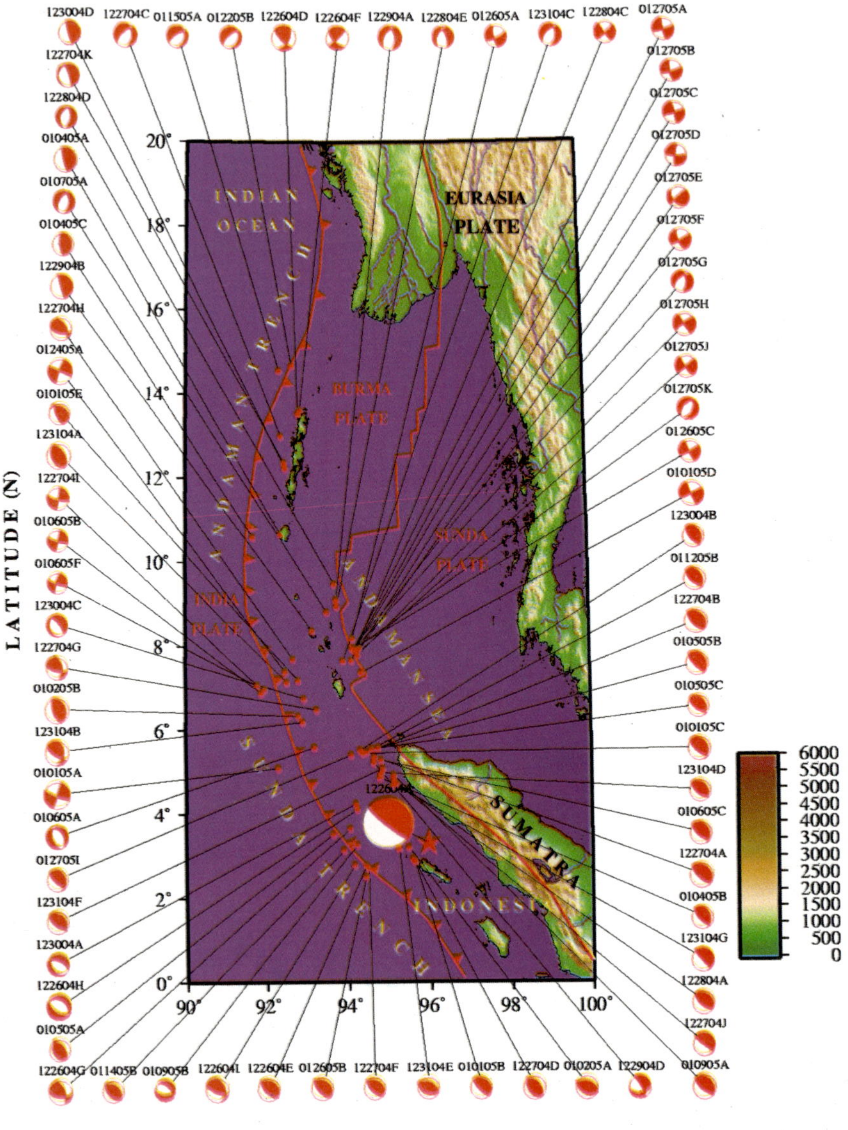

Fig. 6.45 CMT solutions of the main shock and aftershocks that were recorded during December 26, 2004 -January 31, 2005 (USGS data source).

Estimates of speed and size, energy radiation pattern, co-seismic stress, cumulative slip and slip rate and static displacement have been made for the 2004 Sumatra-Andaman earthquake from waveform analysis (Ammon et al., 2005), spectral studies (Lay et al., 2005; McCloskey et al., 2005) and from studies of global positioning system (Banerjee et al., 2005; Vigny et al., 2005; Gahalaut et al., 2006). The Sumatra earthquake is different from other big earthquakes elsewhere in the world because of its very anomalous longest rupture length of ~1300 km. A large quantity of aftershock data are recorded by a temporary six-station seismic network set up by the Geological Survey of India in the Andaman-Nicobar Islands following the December 26, 2004 Sumatra-Andaman main shock (Mishra et al., 2007). The network was run for more than a year, and about 30,000 aftershocks (M> 3.0) are recorded. These data are still in process for detailed tectonic study and seismic imaging of the region.

Aftershocks and rupture zone

The USGS/NEIC locations of the larger aftershocks ($M \geq 5.0$) clearly show the ~1300 km long and 200 km wide rupture zone (Fig. 6.41). The aftershocks ($M \geq 3.0$) recorded by the temporary network reflect a broader rupture area beneath the Andaman-Nicobar Islands, about 800 × 300 km (Mishra et al., 2007). The aftershocks are concentrated around the WAF, the volcanic zones and around the ASR (Fig. 6.41). The entire region is associated with all three classes of aftershocks: (i) initial aftershocks that describe the main shock rupture (Fig. 6.45), (ii) secondary aftershocks that represent the growth of aftershock zone by events (Fig 6.41), and (iii) aftershocks by reactivation of pre-existing fault system.

Fault-plane solutions of the main shock (M_w 9.3) and the larger aftershocks (M_w 5.0-7.0) that were recorded till January 31, 2005 are shown in Fig. 6.45. The main shock occurred by a low angle thrust fault. Fault-plane solutions of the larger aftershocks that fall in the different tectonic zones show different solutions. The near-trench shocks show low angle thrust faulting, the fore arc and the volcanic shocks show normal and strike-slip faulting and the spreading ridge shocks dominantly show normal and strike-slip faulting. The tectonic models that explain different fault plane solutions in a tectonically complex spreading zone and subduction zone are illustrated in Figs 4.23 and 4.24. All these types of solutions are obtained in the aftershock zone. A more detailed study is required for precise locations of the aftershocks, well constrained fault plane solutions and high resolution seismic images to better understand seismotectonics of the geologically so complex region. The high resolution digital data of the large aftershock sequence, that are recorded by the global network and by the local temporary network, provide an excellent opportunity to the seismologists for a comprehensive collaborative research to understand more about the source region of the mega thrust event and its rupture zone. Readers are referred to special volumes of publications on this mega thrust event.

7

Seismotectonics of Peninsular India and Sri Lanka

7.1 Introduction

Peninsular India constitutes one of the prominent and largest Precambrian shield areas of the world. It is exposed to the south of the Indo Gangetic Alluvial Plains (IGAP); in other words the IGAP is separating the Himalayas to the north and the peninsular India to the south. The Shillong Plateau in the northeast constitutes an outpost separated from the main shield by the Bengal basin (Bangladesh Plain), and from the Himalaya by the Brahmaputra river (Fig. 7.1). While the Himalaya is a region of dominant compressional tectonics and the IGAP is a region of relatively less eventful recent sedimentation, the peninsular India, in contrast, is a region marked by early Archaean cratonisation with associated Proterozoic belts. The Indian peninsular is known as a stable cratonic subcontinent. It was reported that the seismicity of peninsular India is low (Chandra, 1977), despite its ongoing collision with central Asia (Tapponnier et al., 1986). During the last decade (1993-2003), however, three strong/large earthquakes (M 6.0~7.7) occurred in peninsular India (Fig. 7.2).

Earthquake occurrence in stable continental interiors is far from a fully explained phenomenon. In many cases, the relationship between intraplate earthquakes and the subsurface seismogenic structures is enigmatic. In order to perform a systematic global study of *intraplate* earthquakes, Johnston (1989) and Johnston and Kanter (1990) introduced the term *Stable Continental Region* (SCR) to eliminate the vagueness of the term 'intraplate' used for both active and stable plate-interior zones. The SCR term also eliminates the sea floors. They found that SCR have strong association with *rift* and *extended crust*. The term extended crust is defined as passive margins and areas close to the rifts. The New Madrid (USA) earthquake of February 7, 1812, $M_w = 8.3$,

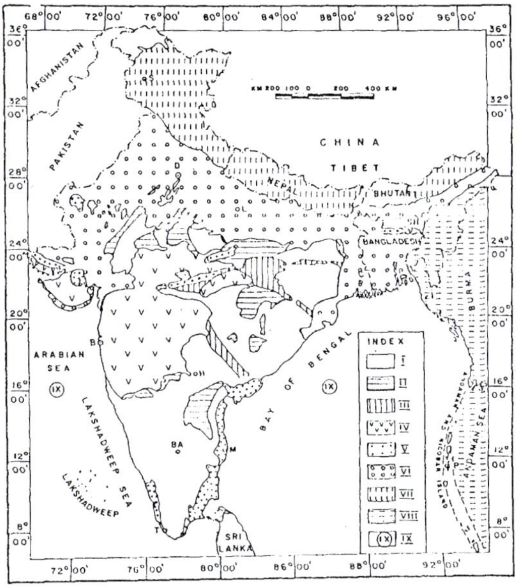

Fig. 7.1 (a) Major tectono-stratigraphy divisions in India and adjoining regions. I: Precambrian shield, II: Proterozoic basins, III: Gondwanas, IV: Deccan and Rajmahal traps, V: Phanerozoic belts and basins, VI: Indo-Gangetic-Brahmaputra alluvial plains and Bengal basin, VII: Himalayan belt, VIII: Andaman-Nicobar Island arc, IX: Submerged continental margins and islands in the Arabian Sea and Bay of Bengal (Mahadevan, 1994).

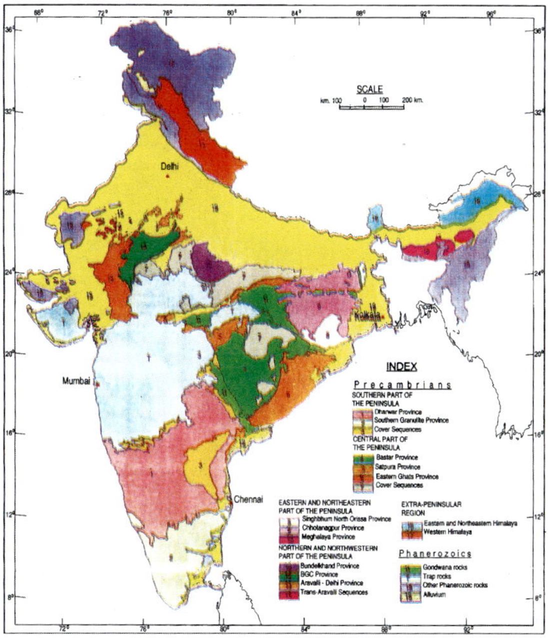

Fig. 7.1 (b) Geological provinces of India (GSI, 2006).

is the largest known SCR earthquake, and is rift associated. In the 'non-extended' (or non-rifted) crustal environment, the largest known SCR earthquake is M_S=6.7 in Australia (Table 7.1). Table 7.1 illustrates some significant historical SCR earthquakes around the world. Peninsular India has produced both rift and non-rift associated SCR earthquakes of magnitudes 6.0~8.0. Investigation of characteristics of these earthquakes played a very important role in understanding the seismotectonics of SCR earthquakes in peninsular India to a great extent (Fig. 7.2). The geological evolution, its geophysical signatures, the past seismicity of the region and detailed study of the recent damaging earthquakes (M>6.0) are discussed in this chapter.

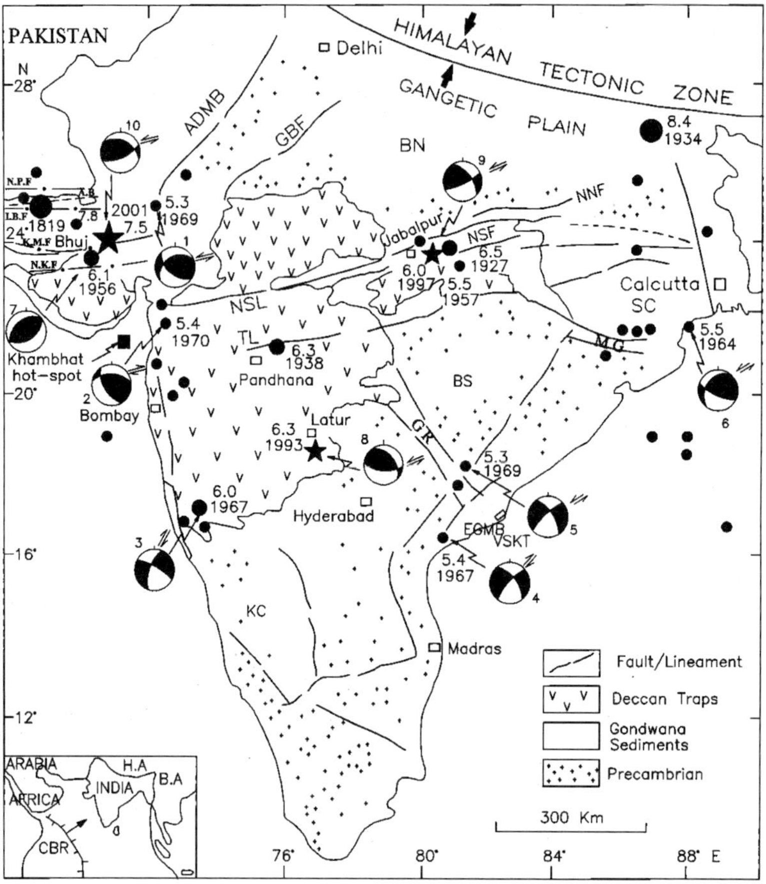

Fig. 7.2 Map showing seismotectonic domains in peninsular India and the significant earthquakes with fault plane solutions, ADMB: Aravalli Delhi Mobile Belt, GBF: Great Boundary fault, NSL: Narmada Son Lineament, NNF: Narmada North Fault, NSF: Narmada South Fault, TL: Tapti Lineament, EGMB: Eastern Ghat Mobile Belt, GR: Godavari Rift, MG: Mahanadi Graben, BN: Bundelkhund, SC: Singhbhum Craton, BS: Bastar Craton, KC: Karnataka Craton (Kayal, 2000).

Table 7.1: World Examples of SCR Events with Surface Rupture (Bhattacharya and Dattatrayam, 2003)

Sl. No.	*Location of Earthquake*	*Date*	*Earthquake magnitudes*	*Dominant faulting*	*Scarp length (km)*	*Max. scarp height (m)*
01	Kachchh, India	June 16, 1819	M 8.4 M_w 7.8 (revised)	Thrust (Surface fold, not a rupture; see Rajendran and Rajendran, 2001)	>90	4.3 (coseismic 5.1 (peak height))
02	Accra, Ghana	June 22, 1939	M 6.5	Sinistral ?	9-17	0.46
03	Centtral Sudan, Africa	October 9, 1966	m_b 5.1	Sinistral	6	0.00
04	Meckering, WA, Australia	October 14, 1968	M_s 6.8; m_b 6.0	Dextral>reverse	37	3.5
05	Calingiri, WA Australia	March 11, 1970	M_s 5.7; m_b 5.7	Sinistral>reverse	3	<0.4
06	Cadoux, WA Australia	June 6, 1979	m_b 6.3 Ms 6.4	Dextral>reverse	28	<1.4
07	Guinea, Africa	December 22, 1983	M_s 6.2; m_b 6.4	Dextral	9.4	0.13
08	Marryat Creek, SA, Australia	March 30, 1986	M_s 5.8; m_b 5.7	Reverse>dextral and sinistral	13	~0.9
09	Tennant Creek, NT, Australia	January 22, 1988	M_s 6.3; 6.4, 6.7; m_b 6.1, 6.1, 6.5	Reverse	32	1.8
10	Ungava, Quebec, Canada	December 25, 1989	M_s 6.3; m_b 6.2	Reverse>sinistral	10	1.8
11	Killari (Latur), India	September 29, 1993	M_s 6.4; m_b 6.3	Reverse	3	0.8

7.2 GEOLOGICAL EVOLUTION

The Indian subcontinent, covering some 3.2 million sq km in area, constitutes three major physiographic provinces: The Himalays, the Indo Gangetic Alluvial Plains (IGAP) and the Peninsular Shield.

The peninsular shield is made up of three main cratonic regions: the Aravalli, the Dharwar and the Singhbhum Protocontinents (Fig. 7.3), which are separated by 'rifts' (Burke et al., 1978). The Singhbhum block is further divided by two cratons separated by the Mahanadi graben (rift). The craton to the southwest of the Mahanadi graben is called the Bhandra craton. The Bhandra craton and the Dharwar craton are separated by the Godavari rift. The major prominent rifts, which separate the northern and southern blocks of the shield, are the Narmada Son Lineament (NSL) and the Tapti Lineament (TL), together called SONATA (Son-Narmada-Tapti Lineament) zone. It is about 1000 km long and 50 km wide (Fig. 7.2). The other prominent rift is the Kutch rift at the northwest margin of the Indian shield. The SONATA zone reactivated several times in the geological past (West, 1962); block faulting as well as horst and graben tectonics is suggested by many workers for its origin (e.g. Choubey, 1971; Roy and Bandopadhyay, 1988). There are

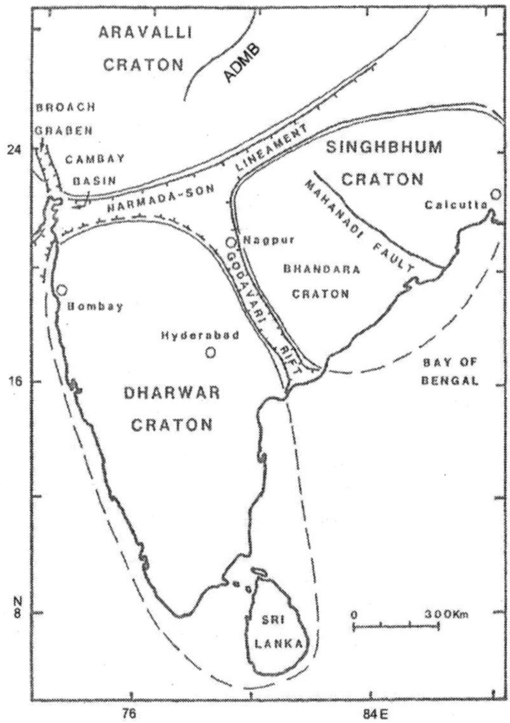

Fig. 7.3 Three main cratonic regions in peninsular India and Sri Lanka (Burke et al., 1978).

two major boundary fault systems with the NSL: a northerly dipping Narmada North Fault (NNF) and a southerly dipping Narmada South Fault (NSF) (Fig. 7.2). These are mantle reaching faults as confirmed by Deep Seismic Soundings (DSS) (Kaila, 1988) and by gravity study (Mishra, 1992; Verma and Banerjee, 1992; Agarwal et al., 1995).

The Central Core Region (CCR) of the shield area comprises granite gneisses and granites; metasedimentary and metavolcanic rocks are also associated. The SCR was cratonised before 2600 m.y. (Radhakrishna and Naqvi, 1986). The Precambrian shield witnessed broadly two orogenies falling in the time bands 2600-2000 m.y. and 2000-1600 m.y.; the first forming the early Proterozoic mobile belt and the second the middle Proterozoic mobile belt. Tectonism and dominantly acid-alkaline magmatism continued in the 1600-540 m.y. time-band, specially in parts of western Rajasthan, Kerala and in parts of eastern India (Mahadevan, 1994). The late phase of 750-540 m.y. time-band seems clearly related to a late anorogenic crustal up warping and rifting. The last tectono-magmatic episode in the peninsular shield is represented by the Deccan traps, the basaltic flows of Cretaceous-Eocene. This pile of volcanic rocks have concealed major tectonic features in the area (Fig. 7.1). Thickness of the traps is estimated to vary between 60 m and 2000 m. The Deccan traps appear to be fissure type of eruptions (West, 1958), and no large crater like structure are observed except one at Lonar lake, for explaining the eruption in terms of large volcanic eruptions. The palaeontological and geochronological data indicate that the basaltic flows become progressively younger from west to east. On the basis of K-Ar dating method, Rama (1968) reported the ages between 60-65 m.y.

The southern peninsular region constitutes three major tectonic domains; these are: Dharwar craton, Eastern Ghat mobile belt and Southern Granulite Terrain (SGT). The Dharwar craton is characterized by the Dharwar schist belt, Kolar schist belt and the N-S trending closepet granulite (Fig. 7.4). The closepet granulite belt separates the eastern Dharwar and western Dharwar blocks. To the south of this block, the Moyar-Bhavani and Palaghat shear systems separate the Dharwar craton from the SGT. The shear zones represent a transition zone marked by gradation in metamorphism from gneisses to charnokites. The gravity anomalies across this zone are smoothly varying. The SGT is further subdivided into three major blocks. The northern granulite block occupies the area between the Dharwar craton and the Palaghat-Cauvery shear zone; a transition zone between the low and high grade terrain (age ~2.5 Ga). The southern granulite block lies between the Palaghat-Cauvery shear and the Achankovil fault; it includes most of the highlands with a maximum elevation of 2600 m. To the south of Achankovil shear zone is the Kerala Khondolite block, a seat of metasedimentary granulites (2100~2830 Ma). A detailed tectonic map with aeromagnetic anomalies are discussed later (Section 7.3.2).

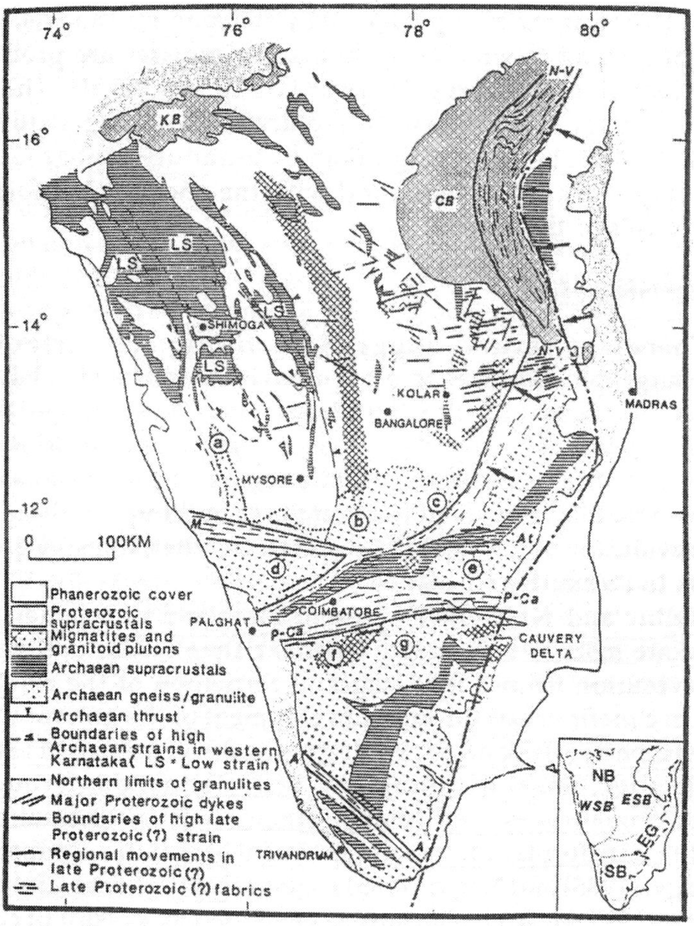

Fig. 7.4 Detailed tectonic features of southern peninsular India, CB: Cuddapah basin, KB: Kaladgi basin, N.V: Nallamalal-Velikonda, M-B: Moyar-Bhavani; At: Attur; P-Ca: Palghat-Cauvery; A. Achankovil; lower case letters a-g are granulite massifs. a-Coorg; b-Biligirirangan (B.R); c-Sheveroi; d-Nilgiri; e-Kollimalai; f-Anaimalai; g-Palni. The inset shows the main Archaean blocks in south India. EG: Eastern Ghats; NB-northern block; WSB-western sub-block; ESB-eastern sub-block; SB-southern block (Mahadevan, 1994).

The other prominent structural units are the NNW-SSE trend of the Western Ghats, NE trend of the Eastern Ghats and the major basins developed in the shield area. The major basins are discussed below.

7.2.1 Proterozoic Basins

A number of Proterozoic basins are developed mainly in the region south of the SONATA zone. These are the (i) Cuddapah, (ii) Godavari, (iii) Bhima, (iv) Kaladgi, (v) Bastar and (vi) Chattisgarh basins (Fig. 7.5). To the north of the SONATA, however, the large Vindhyan basin and smaller Gwalior basin are conspicuous. The Vindhyan basin is known to extend below the Gangetic alluvium and emerges in the Himalaya (Mahadevan, 1994). The Cuddapah, Chattisgarh, Bastar and Godavari basins to the south are regarded as middle Proterozoic. The Vidhyan, Kaladgi and Bhima are considered younger (Raha and Sastry, 1982). Krishnan (1953), however, suggested that Cuddapah, Godavari, Bastar and Chattisgarh basins possibly developed in a single basin of which they are now erosional remnants.

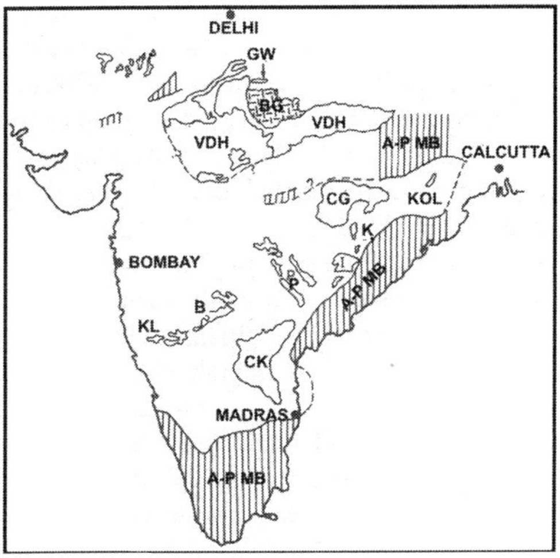

Fig. 7.5 Sketch map showing Proterozoic Basins of India; CK: Cuddapah-Karnool, P: Pakhal, B: Bhima, KL: Kaladgi, K: Khariar, CG: Chattisgarh, KOL: Kolhan, VDH: Vindhyan, BG: Bundelkhund Granite, A-P MB: Archaean-Proterozoic Mobile Belt, GW: Gwalior (Mahadevan, 1994).

All the basins to the south are largely platformal, interior fracture type basins on an ensialic crust. The basin margins have thrust contacts with the Proterozoic mobile belts. To the north, the Vindhyan basin is surrounded by the Delhi mobile belt in the west; the Great Boundary Fault (GBF) possibly separates the two (Fig. 7.2). At the southern margin, the Vindhyans are in contact with the Satpura formations and are folded. The Proterozoic sediments of the Satpura belt are reported to have been thrust over the Vindhyans (Choubey, 1971). Some details of the Cuddapah basin and the Vindhyan basin are given below.

Cuddapah Basin

The Cuddapah basin to the south of the SONATA is one of the major basins which is well studied by aeromagnetic, gravity and DSS. 2D modelling of magnetic anomalies indicate thick sediments that vary from 1.8 km to 6.5 km (Atchuta Rao et al., 1970).

The regional Bouguer anomaly map shows a gravity low of the order of -100 mGal in the basins (Fig. 7.6). There is a sharp rise in the gravity field from -100 mGal in the basin to -10 mGal over the Precambrians in the eastern side. The gravity gradient is very steep. In a detailed gravity map (Kailasam, 1976), the other major features include a rise in gravity field over Srisailam Plateau in the north and a low over Kurnool. A prominent 30 mGal rise is also mapped at Mehboob Nagar, NW of Cuddapah basin. Several attempts are made to explain these anomalies. The steep gradient of gravity anomaly along the eastern margin of the basin is attributed to ~12 km thick sediments (Kailasam, 1976), and to an intracrustal discontinuity. (Subrahmanyam and Verma, 1982). Magnetic and gravity profiles along major traverses across the Cuddapah basin indicate that the basin is made up of a number of up and down thrown blocks (Balakrishnan et al., 1967). The depth of the Mohorovičić (Moho) discontinuity estimated from a mean Bouguer anomaly map is of the order of 38 km (Fig. 7.7). Interestingly, the Moho depths estimated by the DSS profiles range from 36 to 40 km within the basin (Kaila et al., 1979). The Moho depth increases eastwardly towards the coast. The gravity modelling suggested that the Cuddapah basin could be the result of en-echelon faulting of the Moho which has an eastward down-throw leading to deepening of the basin eastward (Balakrishnan et al., 1967). A prominent reflector has been identified at a depth range 18-25 km with P-wave velocities (V_p) ranging between 6.7 and 6.9 km/s. The velocities being typical of a basaltic layer, a *Conrad discontinuity* is inferred to this reflector (Kaila et al., 1979).

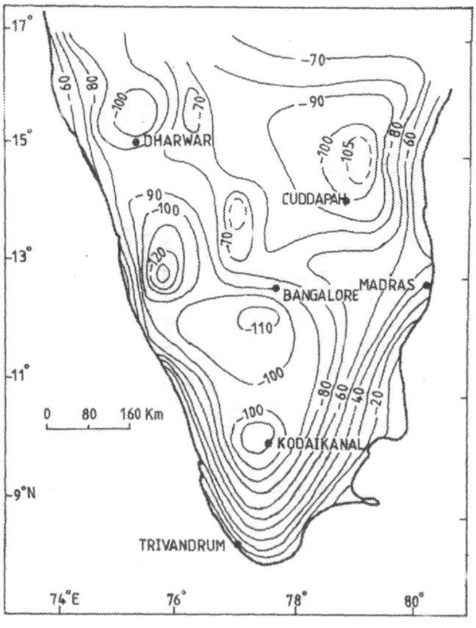

Fig. 7.6 Regional Bouguer gravity anomaly (mGal) map of south India (NGRI, 1975).

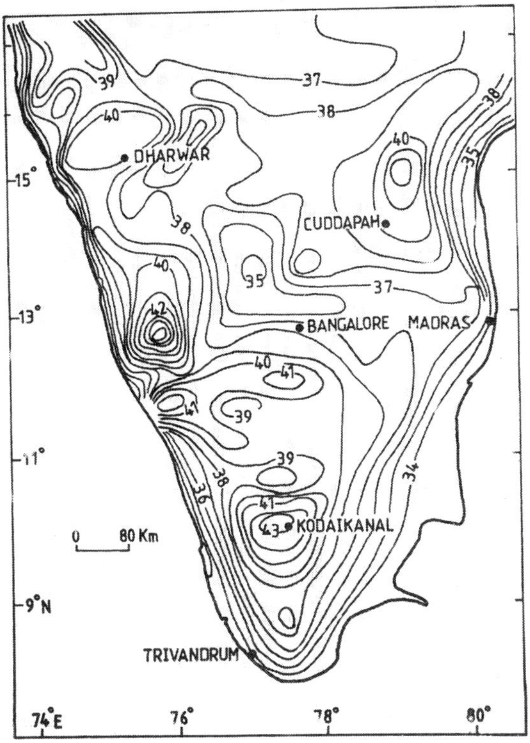

Fig. 7.7 Crustal thickness (km) map of south India, based on gravity (NGRI, 1975).

Two E-W DSS profiles across the Cuddapah basin were taken, one along (i) Kavali-Paranapalli (near 14°N), and the other along Alampur-Ganapeswaram (near south of 16°N) between 78° and 81°E (see Section 7.3.5). The DSS profiles revealed that the crust is dissected into blocks by numerous steep faults, some of which displace the Moho. The DSS results and gravity-magnetic modelling infer that the Cuddapah basin is the result of combination of factors: (i) steady dip of the Moho from west to east, (ii) thickening and thinning of the granitic and basaltic layers and the overlying sediments, and (iii) intrusion of basic magmatic material above the Moho; intrusions are exposed in the western margin but concealed at shallow depths in the eastern margin.

Vindyan Basin in the CCR

The Vindhyan basin to the north of the SONATA forms a characteristic half griddle around the Bundelkhund granite massif, covering an outcrop area of about 100,000 sq km. Some 65,000 sq km of area is under the cover of the Deccan traps (Fig. 7.1). Gravity and borehole data suggest that the Vindhyans also underlie the Indo-Gangetic alluvium (Kailasam, 1976). The real northern limits of the Vindhyan basin must possibly be in the now uplifted Lesser Himalayan region in Garhwal and Nepal. Hari Narain and Kaila (1982) suggested that the Vindhyan basin extending below the Indo-Gangetic alluvium may be divided into western and eastern parts, separated by the NE trending Faizabad Ridge, a tectonically active and growing ridge at the time of Vindhyan sedimentation. Horst and graben tectonics seem to have been prevalent resulting in a number of basement faults which led to the formation of many sub-basins. Many of these faults are active. The gravity map has been interpreted by many authors (Qureshy and Warsi, 1975; Kailasam, 1976; Mishra and Tiwari, 1981). A steep gravity gradient characterizes the western boundary of the Vindhyan basin along the Great Boundary Fault (GBF). The Archaean basement in the west thrust over the Vindhyan sediments along the GBF over a zone of 20 km wide, and a throw of the order of 6 km is estimated (Kailasam, 1976). The NE-SW trending gravity high of about 18 mGal in the centre part of the basin, supported by a positive magnetic anomaly, indicates the possible presence of basic intrusive into the Vindhyans. Spectral analysis of the Bouguer anomaly field reveals an interface, possibly the Conrad discontinuity, at about 16 km depth and the Vindhyan basement at 4.5-6.5 km depth (Mishra and Tiwari, 1981; Agarwal et al., 1995).

A DSS as well as a MT profile (Mandla-Hirapur) across the SONATA (Fig. 7.8a) that transects a part of the Vindhyan between the Bundelkhund craton exposed in the core of the Vindhyan basin in the north and Satpura formation in the south were taken (also see Section 7.3.5). It is seen that the crustal block between Narsinghgarh and Katangi forms a graben in the crystallines with ~5.5 km thick Vindhyan sediments (Fig. 7.8b). The granitic layer with Vp ~5.8 to 6.2 km/s is absent between Damoh and Katangi, the deepest part of the Vindhyan basin. The Vindhyans (Vp ~4.4-5.4 km/s) directly overlie the deeper layer (Vp ~6.0 km/s). Kaila (1988) suggested a pre-Vindhyan uplift along one of the deep faults, erosion, removal of the granitic layer and subsidence followed by Vindhyan sedimentation. Such reversals of uplift and subsidence is a characteristic of block tectonics in the SONATA belt. The MT soundings show approximately 5 km thick conductive sediments with resistivities in the range 6-30 Ωm over a resistive layer of ~200 Ωm down to a depth of 10-12 km corresponding to pre-Vindhyan crust (Fig. 7.8b).

Based on Bouguer gravity (Fig. 7.8c), Verma and Banerjee (1992) interpreted the gravity high between Jabalpur and Mandla in terms of high density in the upper crust with a density contrast of ± 0.1 gcm^{-3}. They further identified the positive Bouguer anomaly as a part of the NE-SW chain of gravity features aligned parallel to the NSL. The negative Bouguer anomaly to the north of Jabalpur is related to the Vindhyan basin sediments.

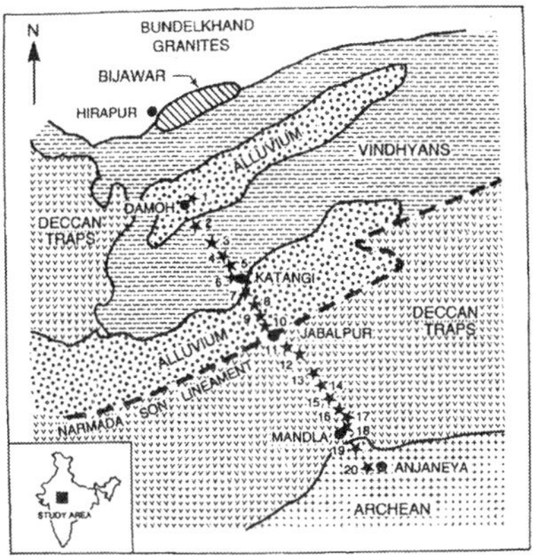

(a)

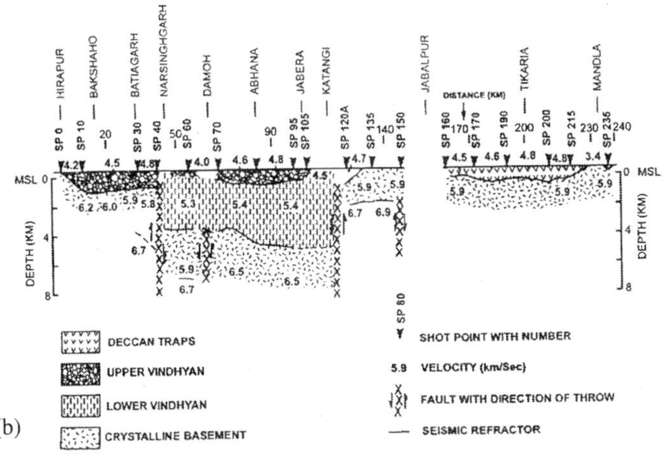

(b)

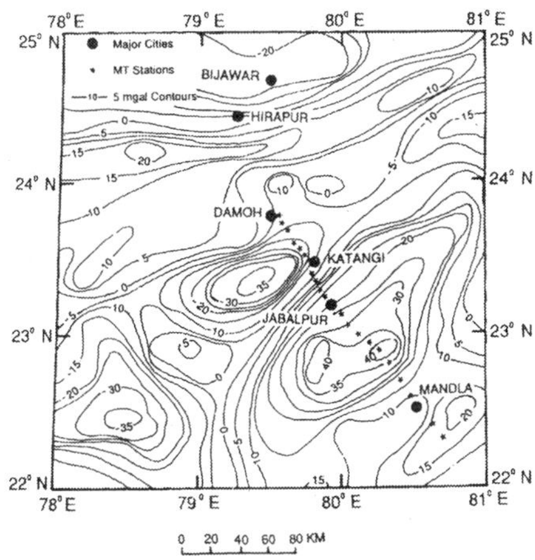

(c)

Fig. 7.8 (a) DSS and MT profiles are shown across the SONATA, from Mandla to Hirapur. (b) DSS crustal (shallow) section (Kaila et al., 1989). (c) Detailed Bouguer gravity map of the Jabalpur area (Verma and Banerjee, 1992).

7.2.2 Gondwana Basins

The Gondwana basins in India are mostly developed in major linear tectonic troughs with faulted boundaries in graben like structures. These basins are shown in Fig. 7.9. The Gondwana formations also occur as thrust sheets overlying the Siwalik and lower Tertiary formations in the Abor hills in northeastern Himalaya, and are well exposed in the Sikkim Himalaya as well.

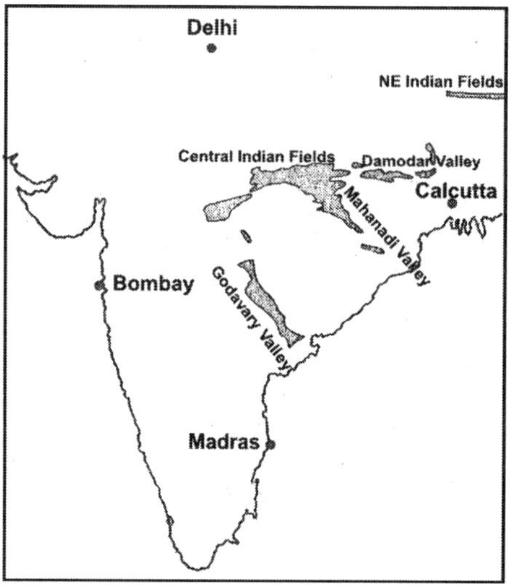

Fig. 7.9 Sketch map showing the Gondwana basins (shaded areas) in peninsular India (Mahadevan, 1994).

The Gondwanas are generally believed to have been deposited in extensive sags, which developed into grabens due to faulting contemporaneous with sedimentation. The faulted long linear belts of grabens and horsts are indicative of passive type of rifts, internal fracture basins. A Precambrian ancestry was assigned to the Godavari and Mahanadi basins (Dar and Viswanathan, 1964), and these may be the areas of merger of the Precambrian protocontinents or zones of intracontinental passive rifts (Mahadevan, 1994). Igneous activity in the Gondwana basins is rather restricted to some specific fields. Lamprophyric magmatic intrusions are observed in some parts of the Gondwana basin in eastern India, which are well recorded by borehole geophysical logs (Kayal, 1979). The dolerite/gabbro intrusions are observed in the western basins in Madhya Pradesh falling in the domains of the Mesozoic Deccan or Rajmahal igneous episodes. These intrusives are, however, not directly related to the tectonic evolution of the Gondwana basins (Mahadevan, 1994). The Godavari and Mahanadi basins are totally devoid of any igneous activity, except at Rajahmundry in the Godavari valley where Deccan traps (called Rajahmundry traps) occur on both banks of the river Godavari (Krishnan, 1960). This volcanic activity does not find expression in the lower Gondwana coal fields in the basin.

Based on gravity data, Verma and Subrahmanyam (1984) suggested that the crust is able to bear the additional load of sediments without any major readjustment. Thus there is a difference in interpretation whether the Gondwana basins are active rifts or massive interior fracture basins. Sedimentation and evolution of the basins have been controlled by block movements. The DSS profiles do corroborate with the block faulted nature of the upper crust. The crustal thickness in the Gondwana basins is not markedly different from the average crustal thickness 35-40 km for the Indian shield. There is, however, some thinning of the crust along the east coast of Mahanadi and southeast flank of Godavari, which may be due to tectonics not totally related to the graben-horst mechanisms of the Gondwana basin formation but to the later evolution of the continental margins. The Mahanadi and Godavari basins seem to end up against the Eastern Ghat mobile belt and do not extend into the coastal basins or into the Bay of Bengal.

7.2.3 Volcanic Provinces

Two major regions of Phanerozoic volcanism in India are the older Rajmahal-Sylhet Volcanic Province (RSVP) and the younger but more extensive Deccan Volcanic Province (DVP) (Fig. 7.10). The RSVP, the major Mesozoic event, is characterized by block faulting, deep crustal faults and dykes. The DVP is, however, many orders of magnitude larger than the RSVP, and it received much attention. The DVP covers some 500,000 sq km of area of peninsular India (Fig. 7.10). It is viewed as tensional tectonics and related to break up of the Gondwana land, opening of the Indian ocean and development of Arabian Sea. De Paolo et al. (1991) proposed *plume tectonics* which could emanate from differing depths, even as deep as the core-mantle thermal boundary. Dietz and Holden (1970) suggested that the Deccan volcanism manifested when the western continental margin of India passed over a thermal centre at 7° S lat and 72° E long. White and Mckenzie (1989) proposed a narrow (~150 km wide) rising plume formed by spreading out below the lithospheric plate. The model requires a steady-state long-lived plume (say ~20 m.y) beneath the slow moving lithosphere so as to generate large volume of continental basalts. These concepts are supported by many authors (e.g. Gupta and Gaur, 1984; Raval, 1989). Addressing a question whether rifting was the driving force for volcanism, Hooper (1990) suggested that primary event was the upwelling of mantle plume that caused subsequent extension, rifting and decompression.

Mahadevan (1994) classified the Deccan volcanic province into four major zones: (i) the western pericontinental Belt of Active Rift Systems (BARS), (ii) the Saurashtra Continental Block (SCB), (iii) the SONATA belt and (iv) the South East Platformal Block (SEPB).

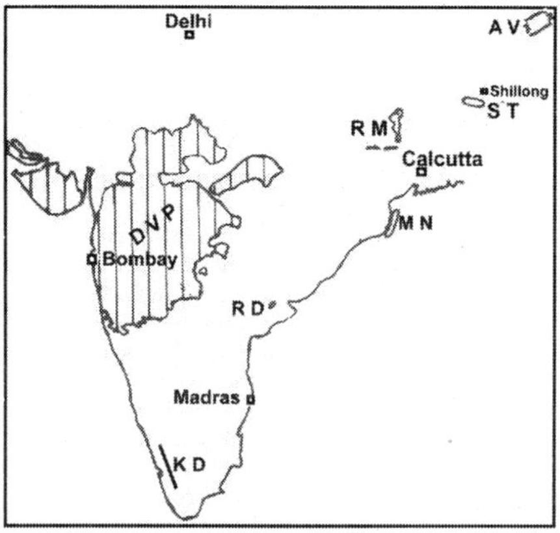

Fig. 7.10 Sketch map showing volcanic provinces of peninsular India. DVP: Deccan Volcanic Province, RM: Rajmahal and ST: Sylhet Traps, AV: Abhor Volcanism, MN: offshore Mahanadi Traps, RD: Rajahmundri Dykes, KD: Kerala Dykes (Mahadevan, 1994).

The BARS is the most dynamic of the four crustal provinces, which witnessed the most tectono-magmatic evolution in the last 60–70 m.y. It comprises the Cambay rift to the north extending southward to Bombay and offshore region. The Cambay rift has a crust thinned down to some 33 km and is characterized by 'rift cushions' of about 10 km thick. P-wave delays reveals a prominent low velocity zone at a depth range 80-200 km in the western coastal area (Iyer et al., 1989; Kennett and Widiyantoro, 1999), and the feature has been related to possible zones of magma generation in the mantle and rift system that provided channel for the Deccan volcanism.

The SCB, on the other hand, has a Precambrian thickened crust 38-43 km. It is intruded by magmatic complexes mantle lineage. The SONATA belt is characterized by ENE-WSW trending horsts and grabens, and also by thickened crust 38-45 km. The SONATA crust is highly reflective; it may be due to magmatic under plating or fluid enriched layers of the lower crust (Mahadevan, 1994). The Deccan volcanism manifests an extensive flow to the western part of the belt bordering BARS. There is lack of well developed 'rift-cushions' in the DSS profiles across the SONATA (Kaila et al., 1990).

The SEPB has normal crust of about 38 km thick. The DSS and geoelectric modelling reveals about 500 m of Deccan traps overlying the granitic basement, and the *Conrad discontinuity* at a depth of 12-18 km (Kaila et al., 1990). Teleseismic tomography reveals the presence of a high velocity (1-4% faster) layer at 100-400 km beneath most of the SEPB (Rai et al., 1992). Gravity field in the vast area of SEPB is negative, whereas it is positive in the SONATA and SCB. The positive trends are inferred to up warp of the Moho, presence of large plugs and intrusives, relative thickening of the upper granitic crust and thickening of the sedimentary supracrustals.

7.2.4 Younger Sedimentary Basins

The younger sedimentary basins of India can be classified into two groups: (i) the pericratonic sedimentary basins of Mahanadi, Godavari and Cauvery deltas along the east coast, and the Cambay basin at the western margin of the peninsular shield, (ii) the intracratonic Punjab-Rajasthan basin, the Ganga basin and the Brahmaputra-Bengal basin fringing the peninsular shield along its northern and eastern borders. The geographical locations of these basins are shown in Fig. 7.11. These basins are different in tectonic setting, but leave some common basement tectonic features, primarily one of block faulting and vertical tectonics.

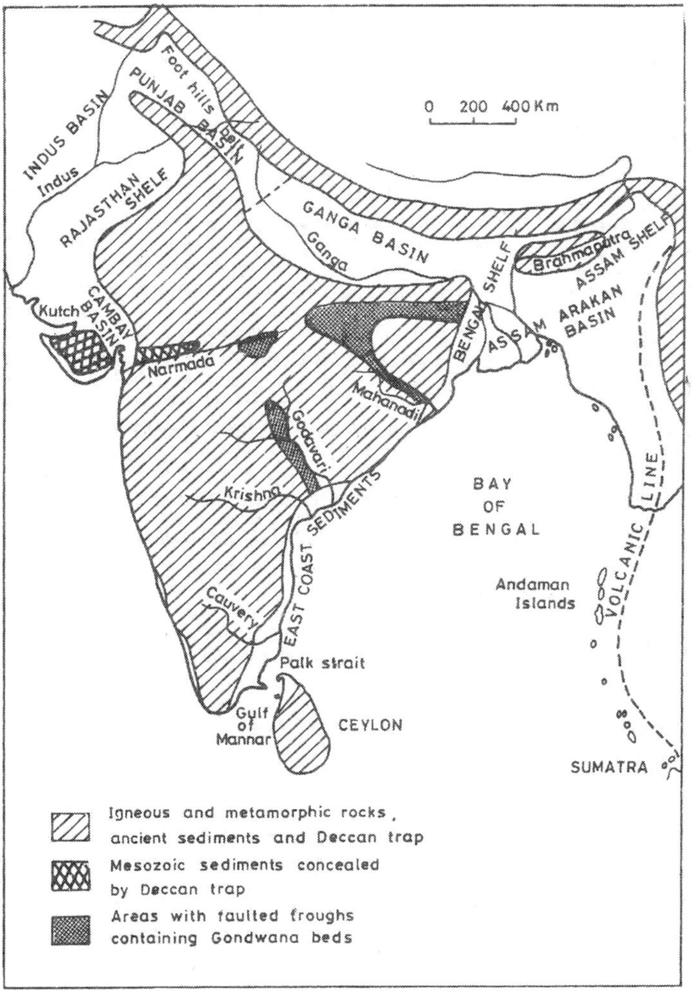

Fig. 7.11 Sketch map showing younger sedimentary basins in India, Bangladesh, Myanmar and Sri Lanka (Mathur and Evans, 1964).

All these basins are made up of horsts and grabens, and arched up basement evolved over a period of time. These are separated by many deep mantle faults with histories of more than one subsidence-sedimentation-uplift cycle. These are products largely of dilatational tectonics, the dominant Precambrian shield have participated and on which are superimposed effects of the Himalayan and coastal neotectonics (Mahadevan, 1994). The Cambay basin has all the features of an active rift and opens into the Arabean sea. The down faulted Bengal basin, on the other hand, is bounded by gravity faults in the west as well as in the north and by the Arakan Yoma anticlinal uplift, fold belts in the east. It represents a deltaic transition opening into the Bay of Bengal in the south, and is underlain by oceanic crust.

Offshore of the Mahanadi basin are basalts, which may be oceanic or representative of Rajmahal volcanics (Mahadevan, 1994). The Gondwana grabens of the Mahanadi and Godavari valleys seem to end up against the Eastern Ghat Mobile Belt, and do not extend into the coastal basins or into the Bay of Bengal. The Cauvery basin, however, occupies a layer, down faulted Precambrian central block, which may extend into the Bay of Bengal.

The Rajasthan shelf, the Indo-Gangetic Alluvial Plains (IGAP) and the upper Assam Brahmaputra valley are characterized by foredeep along the Himalayan fronts. The structure of the IGAP is debated. The structure has been projected as rift or as that of a basin. If the Himalayan major lineaments, the MCT and MBT, had a Precambrian ancestry, the IGAP may have been an early rift. Modern concept of '*skin tectonics*' or the tectonics of low angle under thrusting by intracrustal wedges require that crust below the IGAP to have a dynamic character, but in contrast, the IGAP seem to be passive, adjudged from the undeformed character of a large part of the overlying sediments (Mahadevan, 1994).

The IGAP has been commonly interpreted as a foredeep (see Section 5.6), as a rift filled with alluvium (Burrard, 1915), as a trough of steeply subducting Indian plate (Holmes, 1965) and as a basin in which the Indian shield dips at a low angle below the IGAP and under thrusts the overlying Lesser and Higher Himalayan sedimentary wedge. The basement contour map of the IGAP is presented in Fig. 5.14 (Chapter 5). The Bouguer gravity anomalies run almost parallel to the length of the basin with some prominent northeast inflections (see Fig. 5.16, Chapter 5). These northeast inflections may reflect the extension of the shield elements and subsurface ridges.

Geophysical and drilling investigations revealed many basement ridges and depressions defining the basement geometry of the IGAP in relation to the Siwalik foredeep to the north and peninsular shield to the south (see Fig. 5.15). The major subsurface ridges beneath IGAP include the Delhi-Aravalli ridge, the Faizabad ridge, the Mungher-Saharsa ridge and the Goalpara ridge. The Delhi-Aravalli horst has been redefined and expanded from the original concept of the Delhi-Hardwar ridge (Eremenko and Negi et al., 1968). Based on structural pattern and gravity observations most of these ridges are inferred to be bounded by faults.

7.3 GEOPHYSICAL FEATURES

The major geophysical inputs comprise isolated detailed studies using specific techniques like ground magnetic, aeromagnetic, gravity, DSS, magsat, magnetotelluric etc. These studies revealed the subsurface major tectonic features; a few are illustrated below.

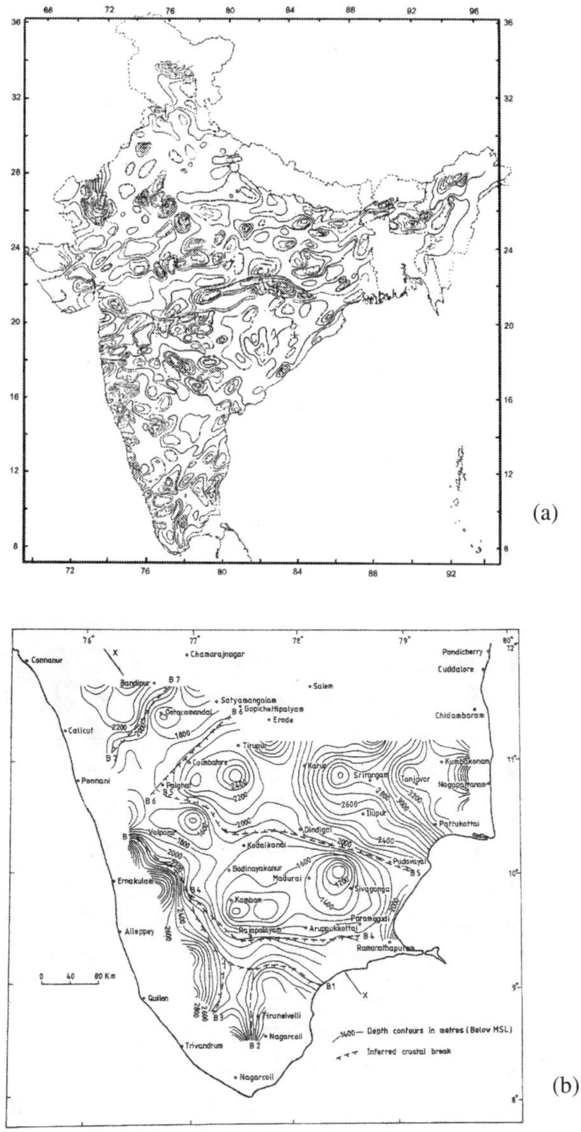

Fig. 7.12 (a) Total intensity magnetic map of India (Qureshy and Midha, 1986). (b) Base of the magnetic crust derived from Magsat data by inversion, south India (Mishra, 1986).

7.3.1 Ground Magnetic Map

Qureshy and Midha (1986) published a total intensity residual magnetic anomaly map of India based on available ground magnetic observations (Fig. 7.12a), which marks many broad features. The SONATA is clearly revealed by a prominent ENE trending magnetic feature. The NE trending Aravalli-Delhi systems also find an expression in the magnetic trends. A prominent low magnetic feature over the Mahanadi Gondwana basin continues further NW over the Deccan trap region. Qureshy and Midha (1986) infer that this magnetic feature indicates the existence of Gondwana beneath the Deccan traps. In the Indo-Gangetic plain, the magnetic contours distinguish a NW-SE trending feature in the west and an E-W trending feature in the east with a central low. There are many other strong magnetic features, which are comparable with the geologic setting of the region.

7.3.2 Aeromagnetic Map

Aeromagnetic survey is the most economical method of undertaking a geophysical reconnaissance. The aeromagnetic map provides data on broad scale structural trend, basement features and faults. A national programme was first taken up by the NGRI and then by the GSI in 1980s. Some results were published in the form of total intensity contour maps but not corrected for variations due to the main field of the Earth. Some regions were surveyed under the projects such as *Operation Hard Rock, Project Vasundhara* for mineral exploration (Reddi et al., 1988). Harikumar et al. (2000) and Rajaram et al. (2000) prepared the aeromagnetic anomaly map up to 16°N and made an attempt to interpret the anomalies in relation to deeper crust and basement features in the southern part of the peninsular shield (Fig. 7.13).

The aeromagnetic anomaly map depicts the major tectonic elements in the region with change in gradient and trend. A striking contrast in gradient is visible across the Moyar-Bhavani shear zone at ~12°N, which indicates change in magnetic sources (Fig. 7.13b). Further, looking from north to south, the predominant trends change from NW-SE to E-W. Amongst two major NW-SE trending lineaments, one cuts across the Cuddapah. These features are comparable with the gravity observations (Fig. 7.6). The Wajrakarur (WK) lineament/fault is possibly responsible for the crescent shape of the Cuddapah basin (Harikumar et al., 2000). The other major NW-SE lineament to the northeastern edge of closepet granite is called Sandur lineament. The various basins such as Cauvery, Krishna-Godavari, Palar and Cuddapah are characterized by magnetic lows. In southernmost part, the NW-SE trending major feature is comparable with the Achankovil (AK) shear zone. Prominent magnetic high is observed over Khondolite to the south of AK shear zone. The Palghat-Cauvery shear zone seems to extend into the offshore as evidenced in the aeromagnetic anomaly map. A larger variation in the aeromagnetic map compared to the gravity potential field is attributed to larger susceptibility variations than density variations of the crustal rocks (Harikumar et al., 2000).

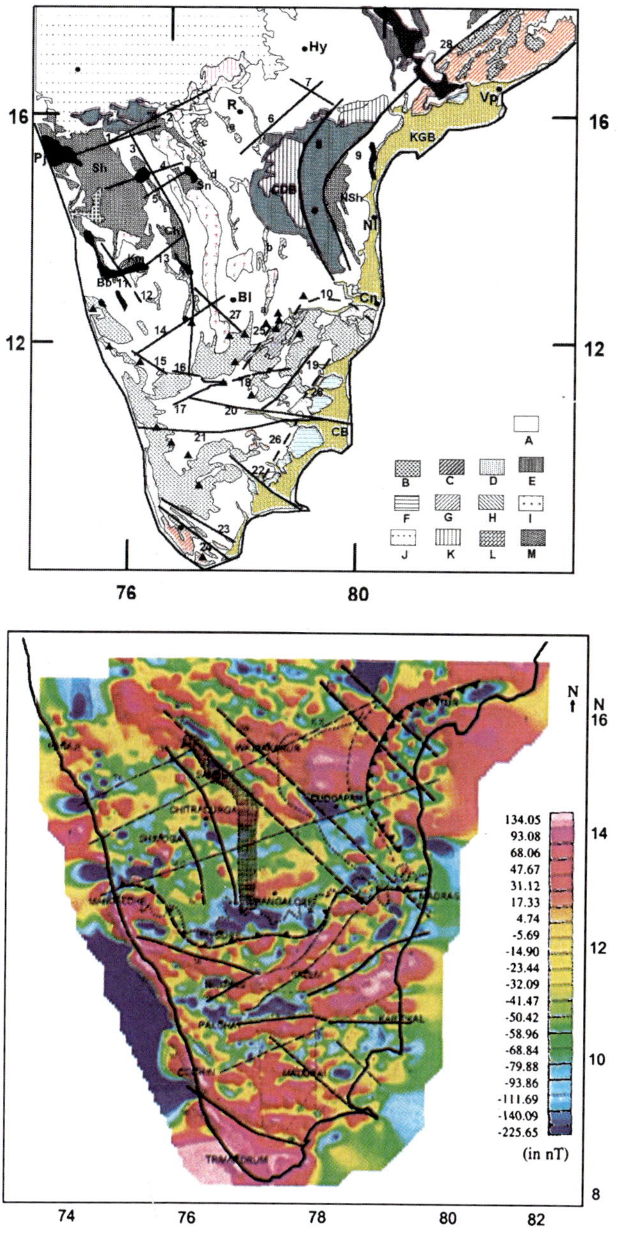

(a)

(b)

Fig. 7.13 (a) Tectonic features and (b) aeromagnetic map of south India (Rajaram et al., 2000).

7.3.3 Magsat Map

Magnetic field satellite (magsat) was put into orbit in October, 1979 by NASA and it was the first satellite dedicated to the measurement of Earth's magnetic field; both vector and scalar data are recorded at low altitudes using Fluxgate and Cesium vapour magnetometers. The crustal magnetic field maps prepared from magsat data have been interpreted in terms of regional geology, tectonics and thermal characteristics of the crust (e.g. Singh and Gokarn, 1989; Rajaram, 2000). Great care has been taken in preparation of the maps taking into account necessary corrections for removal of magnetospheric noise.

Figure 7.14 shows the magnetic moment distribution. The positive magnetisation overlie the continental cratons and stable provinces. Three old fold belts, Dharwar, Aravalli and Singhbhum, show positive which have lower heat flow values than younger fold belts of Delhi and Eastern Ghats. The Delhi and Eastern Ghat fold belts show low magnetization. The IGAP, SGT and the basins of Cambay, Cauvery, Cuddapah and Godavari also depict lows, while the Shillong Plateau shows a high. The Trans Himalayan and the Palk Strait conductors appear to lie on the magnetic lows. Magnetic lows indicate that these are associated with either high heat flow and/or low magnetic susceptibility. The continental crust seems to extend into the Arabean sea across a part of the west coast, which is corroborated by the Bouguer gravity anomaly. A sharp transition is, on the other hand, seen on the eastern coast (Singh and Rajaram, 1990). The magnetic low in the Himalaya-Tibet region is modelled by Purucker et al. (1998), and they have interpreted that the susceptibility of the Asian and Himalayan region is typical of continental

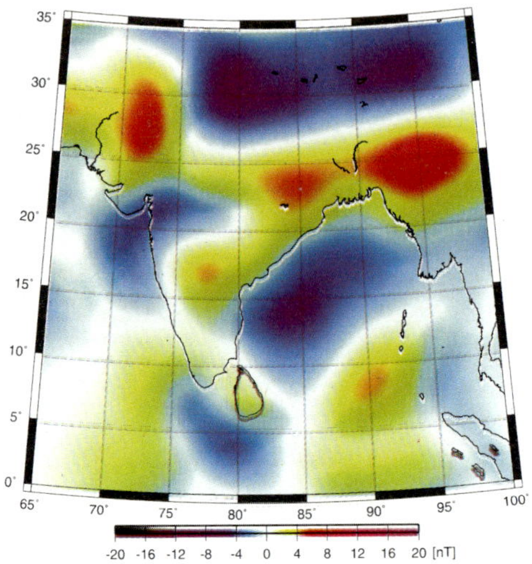

Fig. 7.14 Vertical Field Magnetic anomaly (nT) at 400 km altitude (Rajaram and Anand, 2003).

type, while that of the Tibet is lower or the Curie isotherm in Tibet region is elevated compared to the Asian and Himalayan regions. These observations are comparable with the results of the INDEPTH Project (Nelson et al., 1996).

7.3.4 Gravity Map

Regional gravity anomaly maps provide valuable information on the major tectonic structures, structural lineaments of the crust, lithosphere, geodynamic aspect etc. The regional Bouguer gravity map of India in 1:5 million scale was first published by the NGRI (1974). It brings out striking gravity features, which are comparable with the regional features of peninsular India. The map is, however, revised incorporating the data that were generated over the two decades by the GSI, NGRI, SOI, ONGC and OIL; this revised map is in the press (GSI, 2007). An image of the digitized Bouguer gravity map, using the data published by NGRI (1974), is presented by Rambabu (1999) (Fig. 7.15). These images are maintained in the revised map (GSI, 2007). The images display the subsurface features more clearly than the contour map. A number of interesting features, not reflected in the geological map, can be seen in this map. The Deccan traps, despite being the single largest unit in the geological map of India (Fig. 7.1), do not appear as a single unit in the gravity image (Fig. 7.15). Belts of positive Bouguer anomalies are seen along the west coast, Panvel flexure zone and in the Cambay-Kutch-Rajasthan region. Positive anomaly trends are also seen along the east coast in the Eastern Ghat Mobile Belt (EGMB) and in the Rajmahal-Midnapur zone. The other major anomaly trends are along NNW-SSE Dharwarian trend in South India, NE-SW Aravalli trend in the north and the ENE-WSW SONATA trend in central India. The gravity value associated with the trend located in the south further divide the Dharwar craton into northern and southern parts clearly; the two are separated by the distinct signature of Palghat-Cauvery shear zone. In the northeast, the Shillong Plateau and Tripura fold belt are characterized by positive gravity values. Sediments of Vindhyans and Gondwana basins and sedimentary tracts of east coast are characterised by gravity lows. Reddi et al. (1967) and Ramakrishna and Chayanulu (1988) suggested that the Cuddapahs show a major gravity low due to deeper crustal structure. The Vindhyans, on the other hand, do not show a commensurate low possibly because of up warp of the basement. The Pakhals appear as a low due to overlying thicker Gondwana sediments. Only Chattisgarh basin shows a moderate high. The four basins are also characterised by steep gravity linears, which possibly represent deep crustal faults. The gravity lows over the two other basins, Kaladgi and Bhima, seem to correspond with simple depressions. The gravity low over the Karnool basin is too large to be accounted for sediments alone; this low possibly represents eastern synclinal axis of Dharwar anticlinorium (Qureshy, 1971). The low and high Bouguer anomaly zones in the Proterozoic basins possibly indicate that the basins have horst and graben structures, and not a simple basin structure.

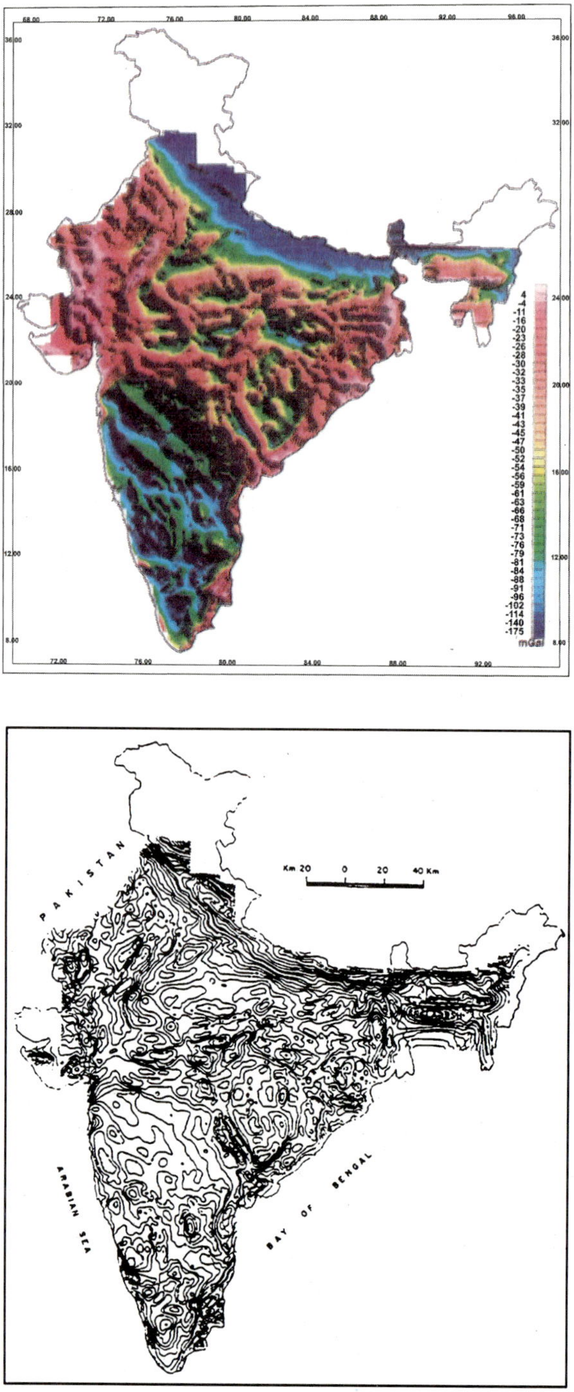

(a)

(b)

Fig. 7.15 (a) Bouguer gravity image of India with inclination and azimuth of the sun at 45° (Rambabu, 1999). (b) Bouguer anomaly map of India (NGRI, 1974).

Regional isostatic anomaly map was prepared by Qureshy (1981); the map is shown in Fig. 7.16. Four types of regions may be identified in peninsular India: (i) regions of zero anomalies covering the major land features which are interpreted to be isostatically compensated, (ii) regions of moderate positive anomalies comprising mainly Aravallis which are over-compensated due to thinner rock formations, (iii) regions of strong negative anomalies encompassing the Caddapah basin and along the Western Ghat, and (iv) regions of moderate negative anomalies.

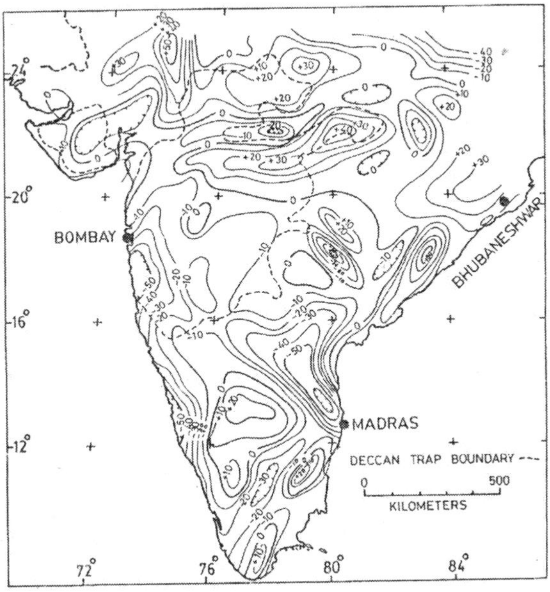

Fig. 7.16 Regional isostatic anomaly map of peninsular India (Qureshy, 1981).

7.3.5 Crustal Thickness Map

Gravity Estimate

Gravity estimate of crustal thickness is mapped by Qureshy (1970) using the relationship $T = 35+5.9\,E$ based on Bouguer anomaly and elevation regression analysis, where T = crustal thickness and E = elevation in km. The crustal thickness map is presented in Fig. 7.17. A $15' \times 15'$ square grid was used in preparing this map, and highest elevation in each grid is taken into account.

The crustal thickness estimates range from 35 to 40 km for a major part of peninsular India. In the uplifted and mountain regions, like Himalaya, Aravallis, Karnataka Plateau, Eastern Ghats and Western Ghats, substantial crustal thickening is inferred. Thicker crust of 40-80 km below Himalaya, 50 km below Karnataka Plateau, 44 km below Eastern Ghats and 40-50 below Western Ghats are inferred.

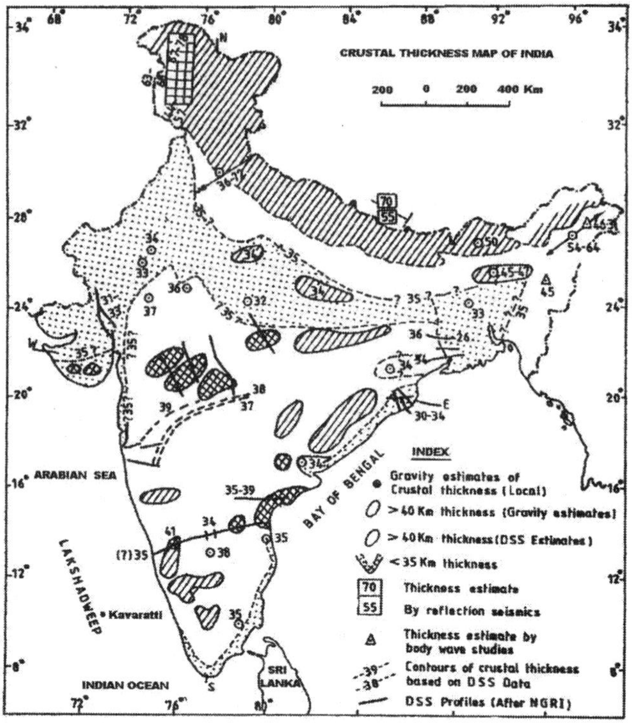

Fig. 7.17 Crustal thickness map of India based on gravity (Qureshy, 1970).

Das et al. (1996) made a 2D spectral analysis of the Bouguer anomaly data in central India, and brought out a Moho relief map. This map reveals two notable up warps in Moho interface, one under Satpura Mountains and other parallel to it but displaced by 70 km to its north. The SONATA occupies central down warp block, where crust is thicker by 5 km. The two up warps are interpreted as the shoulders of the Proterozoic SONATA rift. Similar Moho up warps are also noted on either side of the Gondwana basin near Sahdol. The Moho depth increases (>40 km) under the Gondwana basin, Ganga basin and in some parts of the Vindhyan basin.

Estimate of crustal thickness in the Indian region has been primarily obtained by the Gravity study. A good amount of work, however, has been done by Deep Seismic Sounding (DSS) by the NGRI group of scientists; a review of the work is given by Kaila and Krishna (1992). The crustal and seismic velocity structures of the Indian lithosphere are separately discussed in Section 7.3.6.

Receiver Function Estimate

Receiver Function technique has been used by Kumar et al. (2001) to study the crustal thickness using teleseismic earthquake waveform data from the

10 broadband (BB) stations spread over the Indian shield. These stations are of GSN (Global Seismograph Network) standard, and are in operation since 1996. The station distribution is shown in Fig. 7.18. They reported 33-39 km thick crust below the south Indian shield, a simple continental Archaean crust with a Poisson's ratio close to 0.25. The crust below the stations PUNE and KARD in the Deccan Volcanic Province (DVP) is similar to that below the HYB (Hyderabad) situated on the Archaean basement, the HYB is about 500 km outside the DVP. This indicates that Deccan basalts have not significantly affected the underlying crust. The velocity contrast at the well developed Moho is large, resulting in very clear P to S conversions as well as first order multiples. The predominant Proterozoic crust in the northern and eastern part of the shield, on the other hand, exhibits a complex character. The Moho conversions are considerably weaker compared to that in the Archaean terrains, and crustal thickness is more than 40 km. Rai et al. (2003) carried out joint inversion of the receiver function and surface wave phase velocity, and reported almost uniform seismic characteristics in the southern peninsular region; crustal thickness 35 ± 2 km, average V_s = 3.79 ± 0.09 km/s and V_p/V_s = 1.746 ± 0.014.

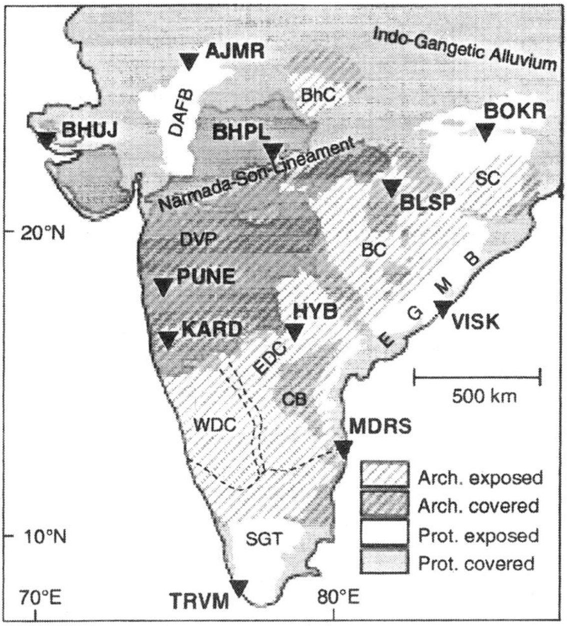

Fig. 7.18 General geology and locations of the 10 BB stations (inverted triangles) in peninsular India shield area established in 1996; SGT: Southern Granulite Terrain, WDC: Western Dharwar Craton, EDC: Eastern Dharwar Craton, EGMB: Eastern Ghat Mobile Belt, DVP: Deccan Volcanic Province, Cuddapah Basin, BC: Bhandra Craton, SC: Singhbhum Craton, BhC: Bundelkhund Complex, DAFB: Delhi Aravalli Fold Belt (Kumar et al., 2001).

Subsequently more BB stations were installed in peninsular India. Gupta et al. (2003) reported crustal thickness and Poisson's ratios from receiver function analyses at 32 BB stations on the Archaean and Proterozoic terrains of south India (Fig. 7.19). The crustal thickness in the eastern Dharwar craton, DVP and Cuddapah basin varies from 34 to 39 km, whereas beneath the western Dharwar it is anomalous, varies from 42 to 50 km. Beneath the exhumed granulite terrain in southern-most India the crustal thickness varies between 42 and 60 km. These observations of the thickened crust suggest significant crustal shortening in south India during the Archaean (Gupta et al., 2003). The Poisson's ratios between 0.24 and 0.28 beneath the Precambrian terrains indicate intermediate rock type in the lower crust.

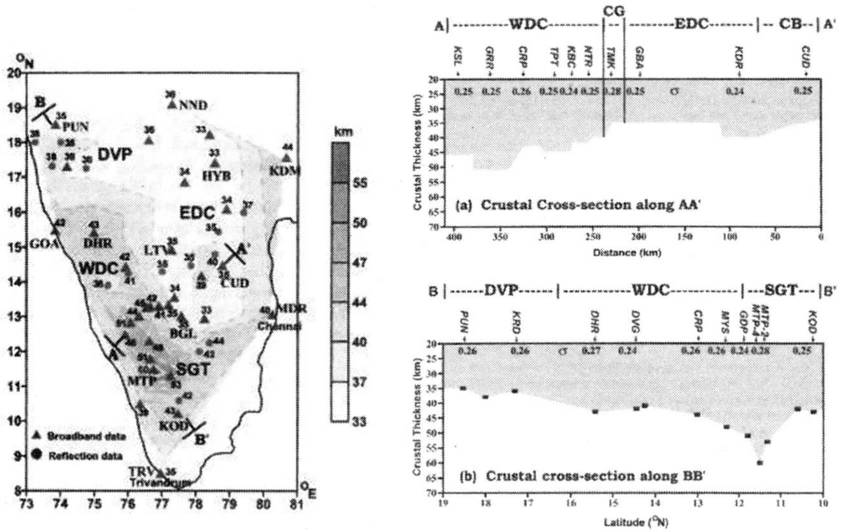

(a) (b)

Fig. 7.19 (a) Crustal thickness map of south India derived from receiver function analysis; Moho depths are indicated at each site. (b) Two cross sections are shown indicating variations in crustal thickness and Poisson's ratio (Gupta et al., 2003).

Rai et al. (2005) further reported receiver function analysis at nine BB stations along a 250 km profile across the Narmada rift valley, SONATA zone; they identified Moho down warp to ~52 km beneath the rift zone in contrast with an average Moho depth 40 km elsewhere along the profile. In addition, the crust beneath the Narmada rift shows higher V_p/V_s of 1.84 compared to 1.73 in the surrounding; this suggests high density mafic mass at depth.

The lithospheric and upper mantle structures of the Indian shield are studied by Saul et al. (2000). They used teleseismic P waveforms of about 300 earthquakes recorded at the BB station HYB (Hyderabad). They reported normal crustal structure of 33 ± 2 km thick crust with a normal Poisson's ratio 0.25 ± 0.01. A seismic discontinuity at 90 km depth is inferred, which possibly indicates a depth-located anisotropy in the lithospheric mantle. The two principal mantle discontinuities are reported at depths of 406 and 659 km, which are close to the global average (410 and 660 km).

Recently Kumar and Mohan (2005) studied about 900 seismograms from a network of six BB stations in the DVP along a 350 km long north-south profile, paralleling the west coast of India. The Moho and the 410 and 660 km discontinuities are well identified by the receiver function analysis. They suggested that the upper mantle beneath the DVP, south of SONATA, does not exhibit any anomalous signatures associated with a mantle plume.

7.3.6 Seismic Velocity Structure and Crustal Thickness

Seismological Data

Kaila et al. (1968) analysed the travel times of P_g, P* and P_n waves for shallow earthquakes occurred in the Himalaya, that were recorded at New Delhi, Chatra, Dehradun, Bokaro and Shillong observatories. Seismic refraction data of the ONGC across the foot-hills enabled correction due to effect of sedimentary layers. The crustal velocity structure observed below the IGAP is given in Table 7.2.

Table 7.2: Crustal Velocity Structure below the IGAP (Kaila et al., 1968)

Layer	*Depth (km)*	*P-wave velocity (km/s)*
Sedimentary	0	3.4
Granitic layer	6	6.2
Basaltic layer	20	6.9
Upper mantle	43	8.2

Seismological studies on the velocity structure beneath Indian subcontinent have led to a few 1-D velocity models (e.g. Tandon and Choudhury, 1967; Kaila et al., 1969; Dube et al., 1973 and Kaila and Krishna, 1983). Based on the teleseismic data and synthetic seismograms, Bhattacharya (1992) presented a crustal velocity model for peninsular India (Table 7.3), which is accepted widely for local earthquake location.

Table 7.3: Crustal Velocity Structure below Peninsular India (Bhattacharya, 1992)

Depth (km)	*P-wave velocity (km/s)*	*S-wave velocity (km/s)*
0.0	5.78	3.53
20.4	6.58	3.92
38.7	8.19	4.60
100.0	8.30	4.60

Huang and Mitchell (1987) estimated shear-wave velocity and its lateral variations in the Indian subcontinent and in the American continent based on Rayleigh wave dispersion data. They reported that the shear-wave velocity models for the Indian subcontinent and South American shield are almost similar, but have lower velocities than those reported for the North American shield.

Deep Seismic Sounding (DSS)

Shallow and deep structural features of the Indian continental crust have been reported by extensive Deep Seismic Sounding (DSS) studies carried out by the DSS group of NGRI during 1972-1993. The study started in India in 1972 as an Indo-Soviet collaboration project for three years. Under this project, a 600 km long Kavali-Udipi DSS profile in south India was first studied that brought out detailed crustal structure down to the Moho discontinuity. Subsequently, extensive DSS studies have been carried out along various profiles across the Deccan trap covered regions and peninsular shield, across the SONATA and Himalaya, and across the Mahanadi, Bengal and Godavari basins (Kaila et al., 1985, 1987, 1988; Kumar et al., 2000). Further, deep seismic reflection profiles were taken across the Aravalli-Delhi fold belt and across the central India under the *deep continental studies* programme of the DST (Kaila and Sain, 1997; Reddy et al., 1999). A status map of DSS profiles and a few 1-D crustal sections are given in Fig. 7.20; a total of 25 DSS profiles covering more than 5000 line km are studied.

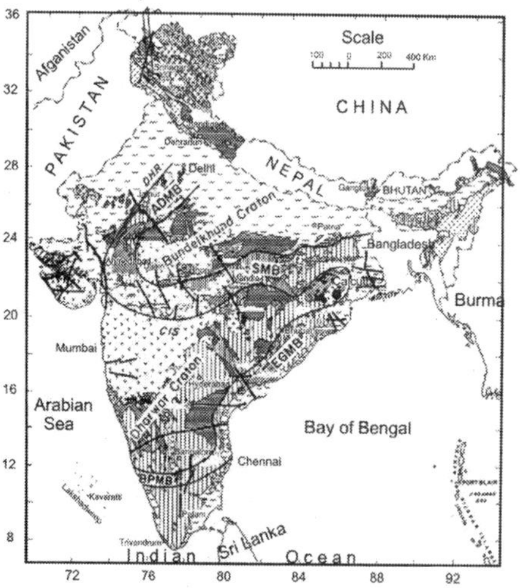

(a)

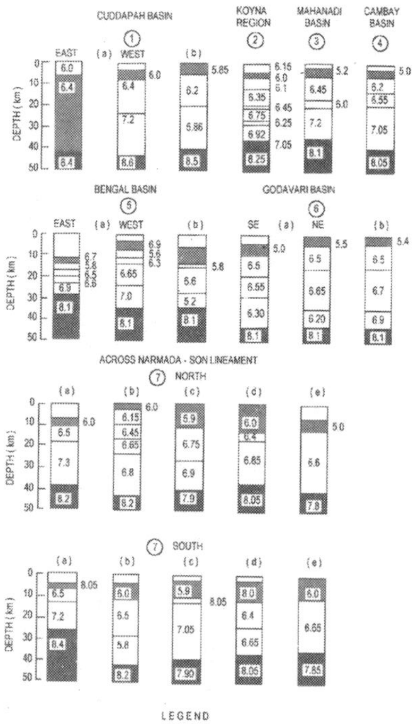

(b)

Fig. 7.20 (a) Map showing DSS profiles (thicker lines), 1972-92. (b) 1-D sections showing P-wave velocity Vp (km/s) of the crust and upper mantle beneath different basins and tectonic zones (Kaila and Sain, 1997).

The results of the DSS from refraction and wide angle reflection studies revealed many subsurface information; these are summarized below:

(i) It is found that thickness of the Deccan traps varies from about 100 m in the northeastern part to about 1500 m in the west coast. Hidden Mesozoic basins are delineated below the Deccan traps.

(ii) Number of faults and their displacement pattern indicate block tectonics in the peninsular India region.

(iii) Thinner crust (28-35 km) is mapped beneath the Bengal, Mahanadi and Cambay regions, and thicker (33-43 km) beneath the shield region (Fig. 7.20 b).

(iv) Several DSS profiles across the SONATA region were taken (Fig. 7.20a), and the results brought out anomalous structure with the presence of high velocity layers (7.0-7.3 km/s) from a depth of 8-12 km down to the Moho discontinuity. These are interpreted in terms of mafic and ultramafic intrusives (Kaila et al., 1985). The DSS profile along the Ujjain-Mahan, in the western part of the SONATA, is exemplified in Fig. 7.21a. The DSS results reveal undulation in the Moho beneath the area, with individual blocks bounded by deep faults extending almost down to the Moho (Fig. 7.21b). Undulating Moho is also mapped along the Hirapur-Mandla profile (Fig. 7.8c), in the eastern part of the SONATA.

(v) Recent reanalysis of the DSS data in the Mahanadi delta in eastern India indicates crustal thinning (34-37 km), presence of mid-crustal low velocity zone, and an approximately 10 km thick high velocity zone at the base of the crust (Behera et al., 2004); basaltic underplating has been suggested, that is synchronous with the Rajmahal volcanism (~117 Ma).

(vi) Sarkar et al. (2001) reinterpreted the wide angle seismic data in the western Dharwar craton. They derived a simplified crustal model with an average Moho depth of 40 km, relatively low velocity (6.8-7.0 km/s) in the lower crust and unusual P_n velocity 8.4 km/s, higher than global average (8-8.1 km/s). This observation suggests that this part is not typical as many other Archaean crust of the world, which is believed to be thinner than Proterozoic crust (Durrheim and Mooney, 1994).

(vii) A seismic section across the Central India Suture (CIS) from Seoni to Kalimati, indicates an average depth of crustal thickness 42 km. Moho offset and reflectivity pattern of two adjacent blocks or protocontinents show opposing reflection segments. The southeast dipping reflector represent the Bundelkhund protocontinent and the northwest dipping segment represents the Deccan protocontinent (Reddy et al., 2000).

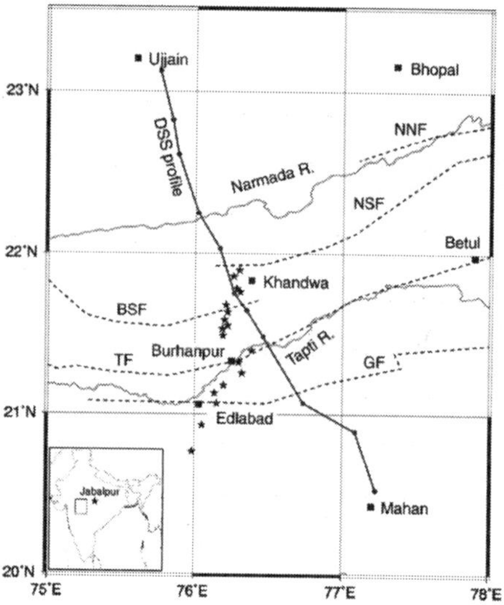

(a)

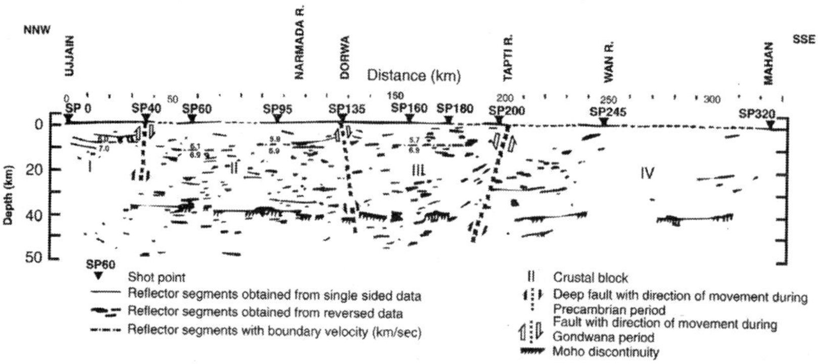

(b)

Fig. 7.21 (a) The Ujjain-Mahan DSS profile across the SONATA. NSF: Narmada South Fault, BSF: Barwani-Sukta fault, TF: Tapti fault, GF: Gavligarh Fault. (b) Crustal section along the profile from the DSS results (Kaila et al., 1985).

(viii) A 400 km long profile, from Nagpur in the Marwar Basin in northwest to Jhalawar in the Vindhyan basin in southeast traversing through the Aravalli Delhi fold belt (Fig. 7.20a), indicated an average depth of Moho 46 km (Tewari et al., 1998; Vijaya Rao et al., 2000). A high reflective thrust fault (named Jahazpur thrust), from Moho upward to a depth of 5 km near Jahazpur, a major crustal boundary, was identified that demarcates two tectonic domains, the green schist of the Hindoli group and the amphibolites of the Mangalwar complex. The oppositely dipping reflection fabric indicates crustal shortening and thickening due to collision process that developed Aravalli suture between the Bundelkhund craton to the east and the Marwar craton to the west (Fig. 7.20a).

(ix) A 300 km long Kuppam-Palani transect in the Southern Granulite Terrain (SGT) also indicated oppositely dipping bands, that implies collision tectonics between Dharwar craton in the north and a crustal block in the south which could be a part of the EGMB. A four-layer thicker crust (40-45 km) with a 7-15 km thick mid crustal low velocity (6.0 km/s) was suggested (Reddy et al., 2003).

(x) Mesozoic sediments (1.16 km thick) masked below 1.4 km thick Deccan trap in Saurashtra peninsula of western India was delineated from the wide angle seismic data. Sain et al. (2002a) and Sarkar et al. (2005) imaged reflective Moho at a depth of 45 km at the Bhuj earthquake epicentre region.

Seismic Imaging

Not much seismic imaging is done using the body wave tomography, except a few detailed study of tomography that used large quantity aftershock data of the recent large/strong earthquakes in peninsular India (Kayal and Mukhopadhyay, 2002; Kayal et al., 2002b). Since the inception of the broadband seismic network in 1996 in peninsular India, large high quality data are now available for seismic imaging.

Mitra et al. (2006) presented seismic images using Rayleigh wave group velocity dispersion results of the Indian crust and uppermost mantle. 1-D path-averaged dispersion measurements for 1001 source-receiver paths are combined to produce seismic images between 15-45 s period. The seismic images demonstrate that the Indian shield is characterized by high crustal and uppermost mantle velocity (Fig. 7.22). Lower crustal velocities exist beneath the Himalaya and Gangetic plain due to molasse sediments and alluvium cover. The Shillong Plateau also displays higher velocities, similar to that beneath the south Indian shield. The northern Bay of Bengal shows very low velocity, which is due to sediment blanket of the Bengal fan. The Katawaz basin in southern Pakistan also shows lower velocity that resembles those of the Bay of Bengal.

Previous studies (Bhattacharya, 1974 and 1981), although limited with number of seismograph stations in peninsular India, showed shield like crustal velocities extended below the Himalayan foredeep sediments, that matches well with the results of Mitra et al. (2006). Mohan et al. (1992) determined Rayleigh group velocities for periods 15-60 s that also showed similar high velocities in the Indian shield, and low in the Bay of Bengal, Gangetic plain and Himalaya. Brune and Singh (1986) inverted surface wave dispersion data, and estimated the crustal thickness ~15 km at the latitude of the southern tip of India to ~35 km at the northern end of Bay of Bengal. Singh (1988) made a similar study for paths crossing the Arabian sea, and found somewhat higher velocities for paths crossing the Indus fan compared to that of the Bengal fan in the Bay of Bengal.

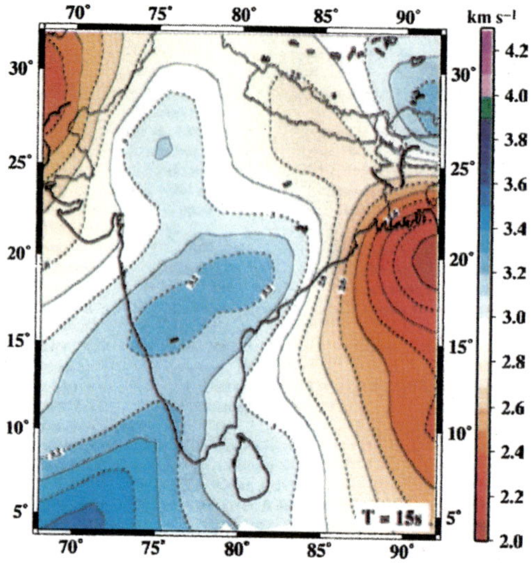

Fig. 7.22 Seismic images of the Indian crust at 15 s period (Mitra et al., 2006).

7.3.7 Electrical Structure

A few magnetotelluric (MT) profiles are studied in peninsular India (Gokarn et al., 1992; Sarma et al., 1994; Gokarn et al., 2001 and 2004; Patro et al., 2004). The MT profile along Damoh-Mandla (Fig. 7.8), across the SONATA, identifies detailed electrical structure beneath this profile. The Vindhyan sediment is found to be about 5 km thick, below which lies the lower crust with characteristic lower resistivity of about 200 Ωm and high P-wave velocity 6.5 km/s. It is conjectured that the upper crust is eroded in the uplift and erosion process, and the Vindhyan sedimentation occurred directly over the lower crust (Gokarn et al., 2001). An anomalous conductivity (resistivity less than 1 Ωm) is observed at depth 10-12 km. This conductivity is attributed to the fluids in the upper crust. It is reported that the thickness of Deccan trap is about 100 m below Jabalpur, and it further decreases near Mandla (Fig. 7.8). The Archaean is exposed on the south of Mandla. Two crustal conductors are delineated below the Deccan volcanics with resistivity of about 30 Ωm. The first conductor to the immediate south of Jabalpur may be indicating the southern boundary of the Jabalpur horst block. The second conductor, about 40 km southeast of Jabalpur, is coincident with positive Bouguer anomaly of 30 mGal (Fig. 7.8). Gokarn et al. (2001) attributed these observations in this Proterozoic mobile belt to the recent (~65 m.y.) plume volcanic activity.

Patro et al. (2004) presented a MT traverse from Edlabad to Khandwa, sub-parallel to the Ujjain-Mahan DSS profile in the western SONATA zone (Fig. 7.21a). This traverse cut across major structural features like Narmada

South Fault (NSF), Barwani-Sukta Fault (BSF), Tapti Fault (TF) and Gavligarh Fault (GF). It also passes through major gravity anomalies in the region (Fig. 7.8). The geoelectric structures imaged from 2D inversion of the MT data are shown in Fig. 7.23. The modelling results reveal four high conductive structural features (A, B, C and D) that extend from mid to deeper crustal levels; these high conductive features are comparable with the major tectonic faults, GF, TF, BSF and NSF respectively in the area.

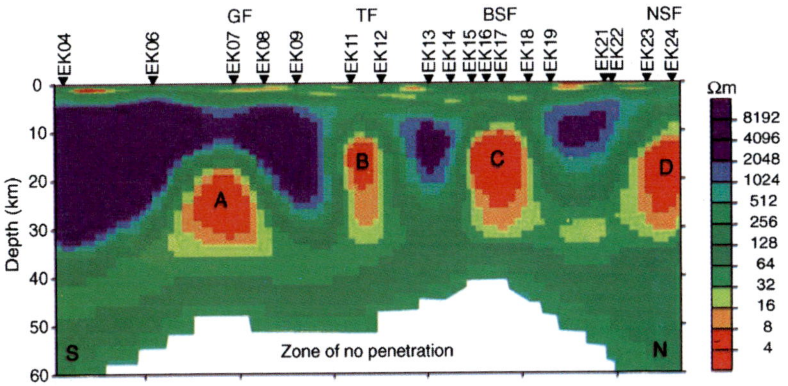

Fig. 7.23 2D inversion results of the MT data along the profile Edlabad-Khandwa (Fig. 7.21a), NSF: Narmada South Fault, BSF: Barwani-Sukta Fault, TF: Tapti Fault, GF: Gavligarh Fault (Patro et al., 2004).

A detailed geoelectrical structure beneath the Dharwar craton is studied using broadband MT data collected at 50 stations over a 400 km long east-west profile across the western Dharwar and Eastern Dharwar terrains (Gokarn et al., 2004). The geoelectric model suggest a suture along the Chitradurga-Gadag schist belt, parallel to the closepet granite belt, with an easterly dip (20°-30°). Further, the MT data revealed low resistivity at a depth of 40 km

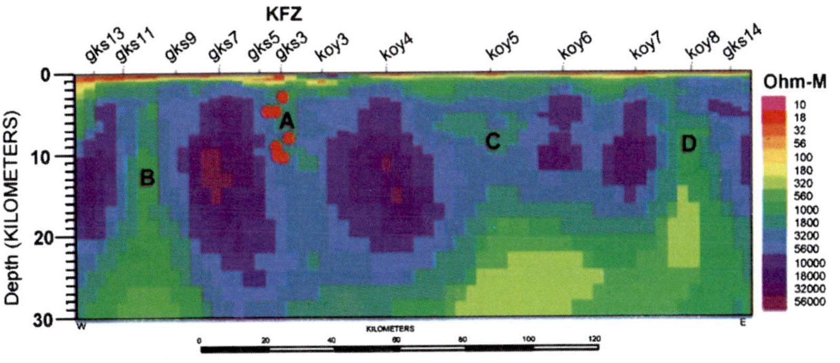

Fig. 7.24 2D inversion results of the MT data across the Koyna seismic zone. Koyna seismicity is indicated in block A (Sarma et al., 2004).

beneath the western Dharwar craton. This seems to be a regional feature and extending to the north beneath the Deccan Plateau. Apart from graphites and sulphides that may cause low resistivity, the fluids generated and trapped in the upper mantle during the passage of the Indian plate over the hotspots in the warming phase of its outbursts could also be a possibility. The asthenosphere is delineated at a depth of 100 km below the eastern Dharwar craton (Gokarn et al., 2004).

Sarma et al. (2004) made wide band MT measurements at 16 sites across the Koyna seismic zone. Besides details of Deccan trap thickness, the results brought out several subsurface geoelctrical features. The Koyna fault is imaged as a moderately conductive zone, and it extends down to 4-5 km; seismic activity is mostly confined in this zone (Fig. 7.24). The conductivity is inferred to the presence of over pressured fluids, that triggers seismic activity or the Reservoir Induced Seismicity (RIS). 2D modelling of the MT data further reveals distinct block structures below the region, that are comparable with the velocity and density observations (Sarma et al., 2004).

7.3.8 Geothermal Maps

A first attempt to prepare heat flow and geothermal gradient maps of India was made by Panda (1985) with a view to evaluating the hydrocarbon and geothermal resource potential. In preparation of these maps, larger emphasis was given on the heat-flow and thermal gradient data in sedimentary basins including about 775 thermal gradient data of ONGC boreholes and various published data from other sources.

Ravishanker (1988) published a heat flow map of India on a larger data base (Fig. 7.25). These data include published data (e.g. Panda, 1985; Gupta and Gaur 1984) as well as GSI borehole data. The map distinguishes five heat flow zones as follows:

Zone I : with over 180 mW/m^2 (>3 times global average)
Zone II : between 100 and 180 mW/m^2 (1.5-3 times global average)
Zone III : between 70 and 100 mW/m^2
Zone IV : between 40 and 79 mW/m^2
Zone V : less than 40 mW/m^2

In the Himalaya, the Zone I covers the parts of Ladak and central Tibet, the Zone II is bounded by the MCT and the Zone III by the MBT to the south. The peninsular shield, the SONATA and its eastern extension are covered by Zone II and Zone I respectively, and the southern part is indicated by the Zone III. The Cambay-Bombay coastal tract including the Cambay rift fall in the Zones III and II. The major Gondwana basins (coal fields) fall in the Zone III and at places in Zone II. Much of the southern peninsular shield, the cratonic region and the Bundelkhund granite massif to the north of SONATA fall in the Zone IV. The granulite belt in the southern part of peninsular India has higher heat flow compared to the adjoining craton.

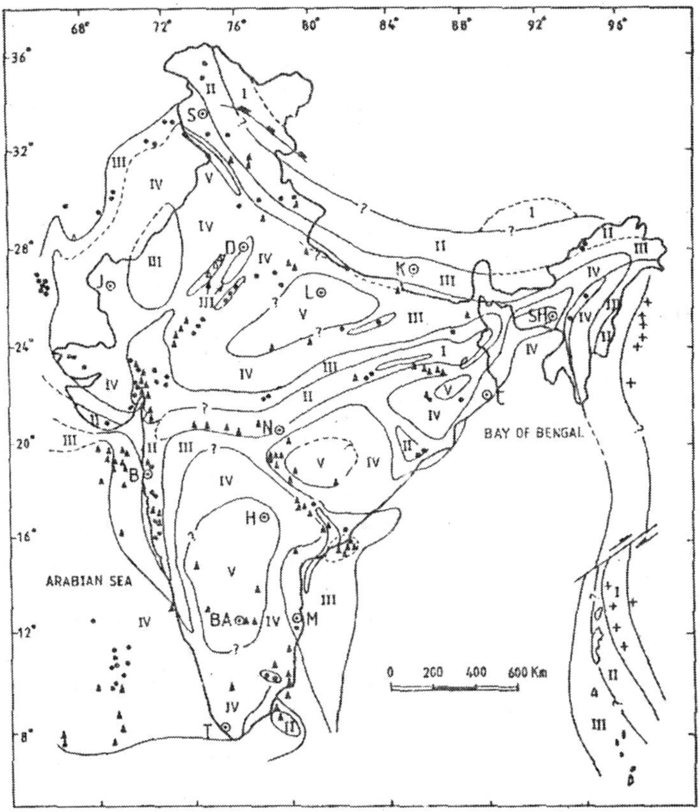

Fig. 7.25 Heat flow map of the Indian subcontinent (Ravishanker, 1988).

Broadly, the heat flow pattern in Indian subcontinent follows the tectonic framework. Heat flow is higher in the mobile belts of Himalaya and post-Palaeozoic rift zones. A major part of the Precambrian craton has lower heat flow, except where some mobility is imposed by recent tectonics. An exceptional deviation is the lowest heat flow Zone V observed along the foothills region extending from Jammu to Nepal. It is suggested that the low heat flow may be due to flushing by waters recharged at higher levels (Ravishanker, 1988). It is, however, also possible that the low heat flow is the result of higher rigidity of the Indian plate below the foothills region.

7.3.9 Tectonic Stress

The stress model of Rajendran et al. (1992) provides a generalised tectonic framework in and around Indian shield (Fig. 7.26). A number of processes may be responsible for tectonic stresses in peninsular India. These include the effect of continental margin, differential crustal movement, hot spots, continental collision etc.

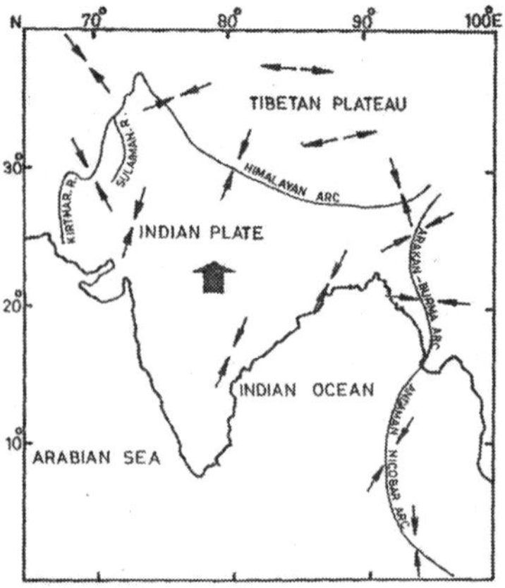

Fig. 7.26 Generalised tectonic stress map of the Indian subcontinent; bold arrow indicates direction of absolute plate motion (Rajendran et al., 1992).

Effect of Continental Margin

Bott and Dean (1972) suggested that differences in thickness and density distribution in the continental and oceanic crust at the continental margin may generate a stress system favourable to the development of normal faults in the continental crust, and thrust faults in the oceanic crust.

Differential Crustal Movement

A number of prominent gravity *'highs'* and *'lows'* are observed in the Bouguer gravity map of peninsular India. Kailasam (1994) suggested that there are zones of marked uplift and subsidence, and these can cause a local stress system.

Hot Spots

The hot spots are surface manifestations of mantle *'Plumes'*, rising columnar currents of hot material. These are regions of volcanic activity, and cause rupturing of continents and opening of new oceans. The hot spots initially cause doming and then rifting in the form of triple junction or some times four-armed junction. Burke and Dewey (1973a) have described seven plume-generated junctions for peninsular India, which in order of increasing age are: Khambhat, Trivandrum, Cuttack, Guntur, Palk, West Bengal and Annupur (see Fig. 7.28).

Khambhat is a four-armed junction, the rifting possibly occurred in Paleocene or Eocene when the lavas of Deccan trap spread over a large area. The four seismic zones, viz., Panvel, Narmada, Cambay and Girnar are structurally related to the four-arms of the Khambhat junction (Chandra, 1977).

The Trivandrum, Cuttack, Guntur and Palk triple junctions are responsible for the formation of continental margin flexures. The continental margin flexures and rifts caused by the plumes provide structures or zones of weakness along which slippage may occur by earthquakes.

Continental Collision

Sykes (1970) considered it reasonable to assume that the high stress generated by the continental collision between the Indian plate and the Eurasian plate may be very extensive spatially. The focal mechanism solutions of the earthquakes in peninsular India show left-lateral motion along a nodal plane which is compatible with the left lateral shear of the Indian plate at the plate margin.

Further, the general N-S trend of the compressional stress in focal mechanism solutions indicates that the stress distribution in the region is greatly influenced by the collision of the Indian continent, by its northward movement, with the Eurasia (Chandra, 1977).

7.4 PAST SEISMIC ACTIVITY

Indian peninsular shield is considered to be SCR, and frequency of occurrence of earthquakes is much less compared to the Himalayan earthquakes. Except the earthquakes in the western margin of the shield, the highest magnitude is also limited to 6.0~6.5 in the peninsular shield area. Further, earthquakes of M 6.0 and above are rare in the southern part of the shield. There are records of about seven earthquakes of magnitude 6.0 and above in the shield area since 1927 (Table 7.4). The occurrence of 1967 Koyna earthquake M 6.7 received much attention of the seismologists to study seismicity of peninsular India. The seismicity map of peninsular India is updated by Rajendran and Rajendran (2003); it is shown in Fig. 7.27. It is observed that the SCR earthquakes in India include both 'rift' and 'non-rift' events; the significant events are listed in Table 7.4.

Table 7.4: Significant recent earthquakes in SCR India

Sl. No.	*Location*	*Date*	*Magnitude*	*Reference*	*Tectonic domain*
01.	Son valley	02.06.1927	6.3 (M)	Chandra (1977)	rift
02.	Satpura	14.03.1938	6.0 (M)	Chandra (1977)	rift
03.	Anjar	21.07.1956	6.0 (Mw)	Chung and Gao (1995)	rift
04.	Midnapur	15.04.1964	5.3 (M)	Chandra (1977)	non-rift
05.	Ongole	27.03.1967	5.4	Chandra (1977)	rift
06.	Koyna	10.12.1967	6.7 (M)	Gupta and Rastogi (1976)	non-rift
07.	Bhadrachalam	13.04.1969	5.7 (Mw)	Chung (1993)	rift
08.	Bankura	03.05.1969	5.7 (M)	Chandra (1977)	non-rift
09.	Mt. Abu	24.10.1969	5.3	Chandra (1977)	rift (?)
10.	Broach	23.03.1970	5.4 (Mw)	Chung (1993)	rift
11.	Killari	30.09.1993	6.1 (Mw)	Gupta et al., 1998; Kayal, 2000	non-rift
12.	Jabalpur	22.05.1997	5.8 (Mw) 6.0 (M)	Bhattacharya et al., 1997, Kayal, 2000	rift
13.	Bhuj	26.01.2001	7.7 (Mw)	Kayal et al., 2002a	rift

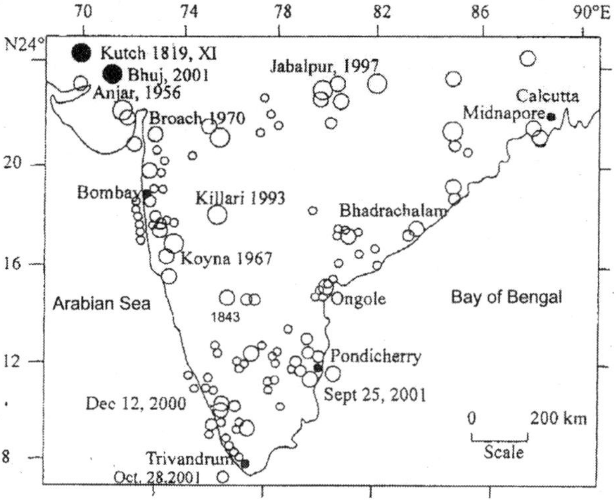

Fig. 7.27 Seismicity map of peninsular India from historical (A.D. 12th century) to recent; circles indicate magnitude are scaled to size, two solid circles indicate earthquakes M≥7.0 (Rajendran and Rajendran, 2003).

Based largely on the macroseismic data and instrumentally recorded earthquake data up to 1973, geological structures and fault-plane solutions, a number of tectonic or seismically active zones were identified by Chandra (1977) (Fig. 7.28). These seismic zones are discussed below. The seismic zones of Indian subcontinent were, however, further classified into some 85 zones by Bhatia et al. (1999).

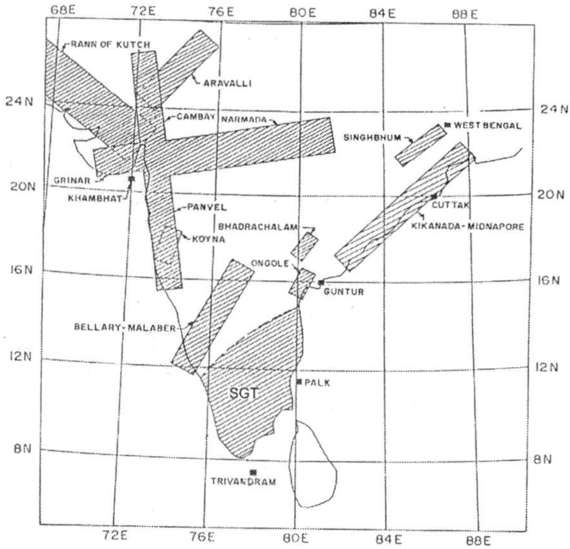

Fig. 7.28 Seismically active zones identified in peninsular India; solid rectangles indicate the hotspots (Chandra, 1977).

7.4.1 Southern Granulite Seismic Zone

This zone includes the Southern Granulite Terrain (SGT) (Fig. 7.28). The irregular distribution of earthquake activity in this zone makes it premature to correlate epicentres with specific structures without monitoring earthquakes with local network. Grady (1971) identified two sets of longer (~200 km) fault systems striking NE; these are the Nilgiri and Annamalai that gave rise to block faulted mountains and graben structures.

Among the significant earthquakes, the February 8, 1900, Coimbatore event, intensity VII, was felt throughout the southern part of India, south of latitude 14°N (Fig. 7.27). Damages to houses were reported.

7.4.2 Bellary Malabar Seismic Zone

This zone extends in northeasterly direction (Fig. 7.28), and cannot be related to any known geological structure. Two significant earthquakes are reported in this zone; the August 22, 1828 Malabar earthquake was well felt along the west coast, and the April 1, 1843 Bellary earthquake was widely felt in the Deccan Plateau. Intensity VII was assigned for both the events (Chandra, 1977).

7.4.3 Western Ghat Seismic Zone

The Western Ghat seismic zone strikes parallel to the west coast. Many hot springs and a system of parallel faults under the Deccan traps are believed to exist in this zone. The major geo-fracture of this terrain is the west coast fault (WCF), which trend NNW, and this fault is considered to be related to the breaking away of the Indian plate from the Gondwanaland. The linear disposition of the west coast and the Western Ghat scarp indicates the geomorphic signature of the WCF. Detailed field survey reveals number of sub parallel faults in close proximity to the WCF. Bouguer gravity anomaly shows steep gradient, and relative positive gravity anomaly near the coast decreases towards east (Fig. 7.15). Chandra (1977), however, divided this zone into two, Panvel and Koyna. Among the historical earthquakes, the August 1764 Mahabaleshwar earthquake, intensity VII, was felt over a large area with maximum intensity at Mahabaleswar. The 1967 Koyna earthquake (M 6.7) is the most significant event in this zone; a short account of this event is given below.

Koyna Earthquake (M 6.7), 1967

The 1967 Koyna earthquake (M 6.7), depth 8 km, epicentre at 17°22′.4N and 73°44′.8E, occurred on December 10, 1967 at about 3 km south of the Koyna dam. The radius of felt area was about 700 km. Maximum intensity reached VIII on the MM scale; the isoseismal map is shown in Fig. 7.29. The meizoseismal area is about 17 km long and 7 km wide, covering an area of about 100 sq km. Most of the houses of stone and brick masonry collapsed in the meizoseismal zone. About 180 people died and 2272 people were injured. Fissures, rock falls and landslides occurred in many places. Guha et al. (1974) have compiled an extensive bibliography of the Koyna earthquake.

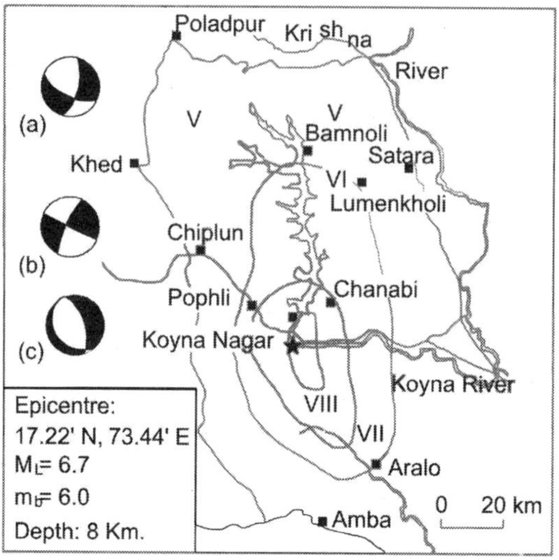

Fig. 7.29 Isoseismal map of the Koyna earthquake, 1967 (GSI, 2000a), and fault plane solutions of the main shock and the largest aftershock (see text).

Large number of foreshocks and aftershocks were recorded at the Koyna observatory. The seismic activity increased from beginning of December 1967. More than six hundred aftershocks were recorded during December 10, 1967 to March 1968. Six aftershocks in the magnitude range 5.5-6.0 were recorded in December, 1967, and four in the following months, the last one on March 5, 1968.

Two fault plane solutions (a and b) of the main shock are shown in Fig. 7.29; left-lateral strike-slip solution on the NE trending fault plane (solution b) is preferred (Langston, 1976; Chandra, 1977). A normal fault solution (solution c) was given for an aftershock M 5.3 that occurred within 48 hours after the main shock (Langston and Franco-Spera, 1985).

7.4.4 SONATA Seismic Zone

This zone follows the ENE trend of the Son-Narmada lineament. Chandra (1977) defined it as Narmada zone. Faults and lineaments are parallel/subparallel to the major tectonic grains. There are two boundary fault systems, Narmada North Fault (NNF) and Narmada South Fault (NSF), (Fig. 7.2), which control the Narmada Rift Basin (NRB). These are mantle reaching faults, and reactivated several times in the past (West, 1962). The Bouguer gravity contours in the NRB show 'highs' and 'lows' (Fig. 7.8c), and the

DSS revealed horst and graben structures for this zone (Verma and Banerjee, 1992; Kaila et al., 1989). Geological and geophysical studies unequivocally established this as a deep fault rift zone cutting across the entire breadth of central India. The entire zone is seismically active (Fig. 7.30). Five significant earthquakes (M>5.0) occurred in this zone. These are briefly given below

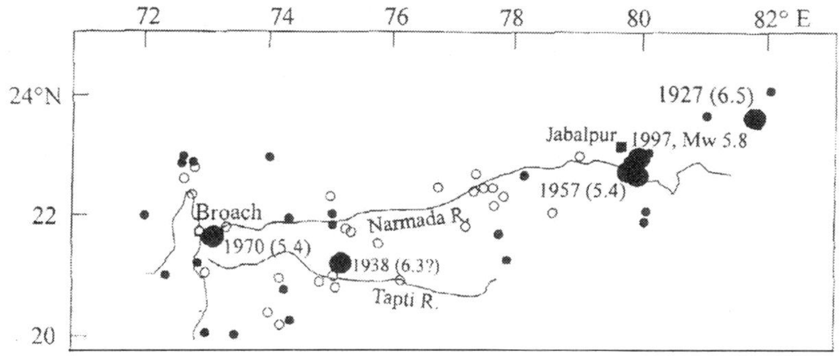

Fig. 7.30 SONATA seismicity map; open circles indicate M 4.0, filled circles: M>4.0, larger circles M>5.4 (Rajendran and Rajendran, 2003).

Son Valley Earthquake (M 6.5), 1927

The earthquake of June 2, 1927, magnitude 6.5, occurred at the northeastern end of the zone in the Son valley (Fig. 7.30). The quake was well felt as far as Ranchi and Allahabad. Radius of perceptibility was about 350 km.

Satpura Earthquake (M 6.3), 1938

The earthquake of March 14, 1938, magnitude 6.3, occurred at the north bank of the Tapti river (Fig. 7.30). It was felt over a large area; radius of perceptibility was about 500 km. It was felt as far as Delhi in the north and Belgaum in the south. A hot spring (near Chopda village) was reported to have disappeared after this shock. Maximum intensity VII (MM scale) was reported. A recent swarm activity is reported in the epicentre zone of this earthquake, in Pandhana area; this is discussed later.

Balaghat Earthquake (M 5.4), 1957

The Balaghat earthquake of August 25, 1957, magnitude 5.4, occurred to the south of Jabalpur (Fig. 7.30). The quake was well felt at Nagpur, Mandla and Jabalpur. Not much damage was reported.

Broach Earthquake (M_W 5.4), 1970

The Broach earthquake of March 23, 1970, moment magnitude M_W 5.4, occurred at the western extremity of the SONATA (Fig. 7.30), near the junction of Panvel, SONATA and Cambay seismic zones. The ISC located the event at 21.6°N and 72.96°E. Shallow hypocentral depth of 8 km was reported for this event. About 26 persons were killed and 200 people were injured in the town of Broach (Chandra, 1977). Radius of perceptibility was about 170 km. Maximum damage was confined to a narrow belt, 10-15 km wide, along the Narmada river. Several ENE and N-S fissures were developed due to this earthquake. A maximum intensity VII (MM scale) was assigned (Fig. 7.31).

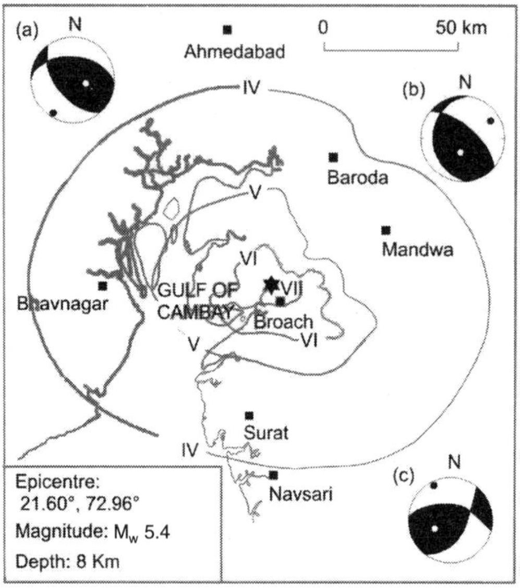

Fig. 7.31 Isoseismal map of the 1970 Broach earthquake (GSI, 2000a); epicentre of the main shock and three fault plane solutions are shown (see text).

Several fault plane solutions were determined for this event (Fig. 7.31). Chandra (1977) used P-wave first motion of the WWSSN and obtained two different solutions (solutions a and b, Fig. 7.31); it was not clear which was the best. Chung (1993) determined a solution with additional constraints from long period modelling and obtained a well constrained reverse faulting solution (solution c, Fig. 7.31). The E-W trending nodal plane is the preferred fault plane, which is comparable with the Narmada North Fault. The focal depth was estimated to be 11 ± 4 km, close to the 8 km reported by ISC. Estimated stress drop $\Delta\sigma$ = 8.75 bar. The aftershock sequence had a b-value 0.4 (Gupta et al., 1972).

Jabalpur Earthquake (M 6.0), 1997

The May 22, 1997 Jabalpur earthquake (M 6.0, M_W 5.8) is the most well studied earthquake in the SONATA seismic zone (Acharya et al., 1998; Rajendran and Rajendran, 1998; Kayal, 2000; Singh et al., 2000). A notable aspect of this earthquake is that the source is well estimated in the lower crust at 35-40 km (Bhattacharya et al., 1997; Kayal, 2000; Singh et al., 2000), similar to the moderate earthquakes reported from the Amazona ancient rift system in SCR of South America (Zoback and Richardson, 1996). This was the first significant earthquake that was recorded by the BB stations established in peninsular India in 1996. Maximum intensity reached VIII in the MSK scale (Fig. 7.32). Fault-plane solution shows a reverse faulting with left-lateral strike-slip motion (Fig. 7.2). Details of the aftershock investigation are given in Section 7.6.

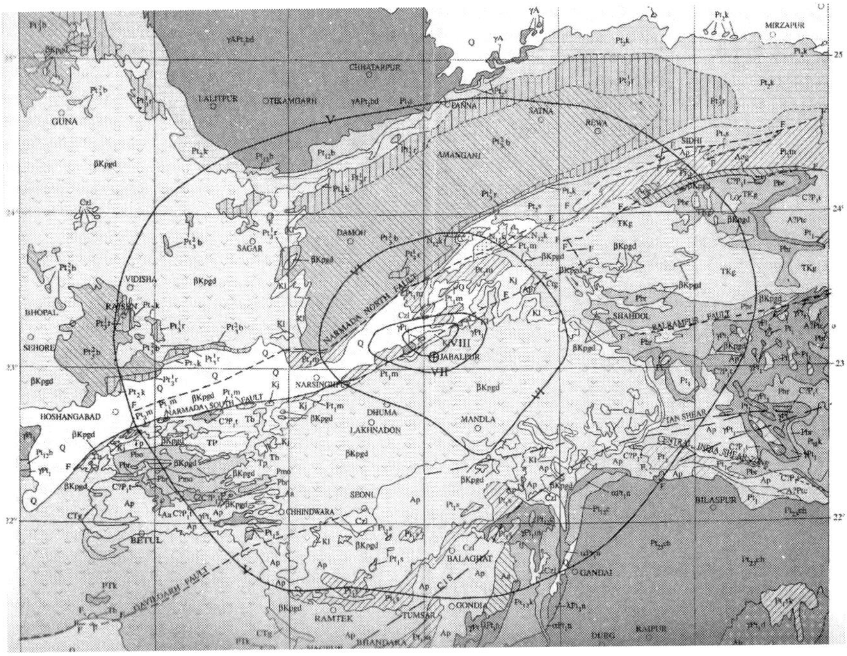

Fig. 7.32 Isoseismal map of the Jabalpur earthquake (GSI, 2000b).

7.4.5 Cambay Seismic Zone

This seismic zone is an extension of the Panvel seismic zone. Structurally this zone is related to the Cambay graben, which is terminated to the south by the Narmada North Fault (Fig.7.2). The graben is bound by the NNW striking deep normal faults. Two significant historical earthquakes occurred in this zone; one occurred on April 26, 1848 in the Mount Abu area, and the other on April 29, 1864 in the Ahmedabad area (Chandra, 1977). Both the earthquakes were well felt, and an intensity VII was assigned for both the events. A significant earthquake, magnitude 5.3, occurred in 1969 in this zone, and is known as the 1969 Mt Abu earthquake.

Mt Abu Earthquake (m_b 5.3), 1969

The October 24, 1969 earthquake, m_b 5.3, depth 15 km, epicentre at 24.8°N and 74.4°E, occurred in the Mt Abu region. Maximum intensity VII was reported. Fault plane solution of this event shows a left lateral strike slip motion (Fig. 7.2).

7.4.6 Aravalli Seismic Zone

This zone is related to the Aravalli range (Fig. 7.28). Seismic activity in this zone is small, and identification of the Aravalli range as a separate seismic zone is questionable (Chandra, 1977). It is possible that this zone extends across the Cambay graben and extension includes the Mount Abu earthquake of April 26, 1848 and the Paliyad earthquake of July 23, 1938, intensity VII. The meizoseismal trend (intensity X) of the recent 2001 Bhuj earthquake in the Kutch is also correlated with the Aravalli trend (Ravishanker and Pande, 2001).

7.4.7 Rann of Kutch Seismic Zone

The Kutch (also called Kuchchh) basin is distinguished by east-west oriented high lands (uplifts) and low lying basins or Ranns (Ranns means uninhabited salt flats that are neither sea nor land, and are flooded periodically).

The seismic zone trends northwest, and it merges with the northeast seismicity trends related to the Owen fracture zone and Kirthar range near Karachi in Pakistan (Fig. 7.28). The Chagos-Laccadive ridge intersects the coastline near this junction. A number of E-W faults control the structural trend of the Kutch Rift Basin (KRB). These are: Nagar Parkar Fault (NPF), the Allah Bund Fault (ABF), Island Belt Fault (IBF), Kutch Mainland Fault (KMF) and North Kathiwar Fault (NKF) (see Section 7.6.4). The NPF is the northern boundary and the NKF is the southern boundary faults of the KRB. The basin is filled with sediments ranging in age from middle Jurassic to Tertiary. The Deccan trap lava divides the Mesozoic and Tertiary stratigraphy of the KRB. After the initial period of extension the KRB has been subjected to N-S compression at least since 20 ma (Talwani and Gangopadhyay, 2001). The Bouguer gravity anomaly in the KRB is high and the contours also show E-W trend (GSI, 2001a). In addition to E-W faults, the basin is transected by N-S to NE-SW and NW-SE tectonic lineaments that include the structural median high (Hinge Zone) to the west of Bhuj, NE-SW Anjar-Rapar lineament, NW-SE Bhachau lineament etc. (Biswas, 1987; Talwani and Gangopadhyay, 2001; Kayal et al., 2002a).

The Kutch falls in the high seismic zone, zone V, in the seismic zoning map of India (Fig. 2.20). Among the past large earthquakes in peninsular India, the largest SCR earthquake was the 1819 Kutch earthquake (M ~8.0). This earthquake occurred before the invention of seismograph. There are several large and significant earthquakes in this zone, these are as follows:

Samaji Earthquake (intensity X), 1668

The May, 1668 earthquake, intensity X (RF scale), occurred in the Samaji area, Delta of Indus (Fig. 2.17). A report described that the Samaji town had sunk into the ground with 30,000 houses by this earthquake (Chandra, 1977).

Kutch Earthquake (M ~8.0), 1819

The Rann of Kutch earthquake of June 16, 1819, M ~8.0, intensity XI (MM scale) was felt over a large portion of Indian subcontinent, as far as Calcutta (Oldham, 1883). This was assigned to be a great earthquake in the Indian shield (Fig. 7.2). Details of this earthquake are given in Section 2.7.1. In recent studies, magnitude of this event is, however, revised to M_W 7.8 (Johnston and Kanter, 1990; Ambraseys and Douglas, 2004). The spectacular feature of this earthquake was the creation of a ~90 km long elevated tract of land called Allah Bund (Dam of God) close to the international border between India and Pakistan. Rajendran et al. (1998) suggested a low angle reverse faulting for this earthquake, and they located the epicentre to the north of Allah Bund Fault (Fig. 7.2).

Evidences for two older events have been obtained from *paleoseismological* studies in the Kutch region (Rajendran et al., 2002). Paleoliquefaction features in Dholka area near Ahmedabad, a site located within the Cambay basin, suggest evidence for occurrence of an earthquake about 3,000 yr B.P.; its source remaining uncertain. The other major event is indicated by Archeological evidence at Dholavira, a major ancient Harappan city (B.C. 2600 to BC 1600) (Joshi and Bisht, 1994).

Anjar Earthquake (M_W 6.0), 1956

The earthquake of July 21, 1956, magnitude M_w 6.0, intensity IX (MM scale), in the Anjar area caused considerable damage to a number of villages in the central mainland of Kutch. About 115 people were killed, hundreds were injured, 1,350 houses destroyed in Anjar town alone. The area of maximum damage was about 2,000 sq km. The radius of perceptibility was about 300 km. The isoseismal map is shown in Fig. 7.33. This event was fairly well studied by Chung and Gao (1995). A reverse fault-plane solution was inferred for this event (Fig. 7.2).

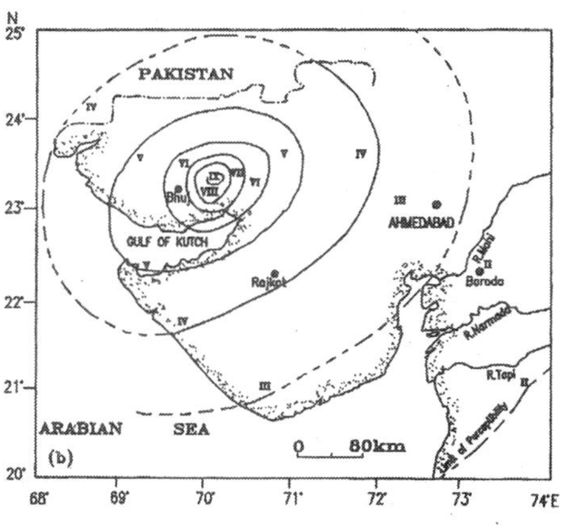

Fig. 7.33 Isoseismal map of the 1956 Anjar earthquake (M_w 6.0) (Tandon, 1956).

Bhuj Earthquake (M_w 7.7), 2001

The January 26, 2001 Bhuj earthquake (M_w 7.7) in western India is one of the most deadly and well reported intraplate earthquakes in the world (e.g. Gupta et al., 2001; Bendick et al., 2001). The moment magnitude (M_w), estimated by different organizations, ranges from 7.5 to 7.7 suggesting a rupture of 15-30 km width, 50-100 km length and average slip 4-6 m at depth. Surface manifestation of such rupture is likely to be a broad zone of distributed uplift and subsidence with secondary surface faulting and cracking (Wesnousky et al., 2001).

Macroseismic surveys were carried out by several groups of scientists immediately after the main shock, and several reports are available (e.g. Ravishanker and Pande, 2001; Rastogi, 2001a; Wesnousky et al., 2001; McCaplin and Thakkar, 2001; Rajendran et al., 2001). A detailed report is published by the Geological Survey of India (GSI, 2003). The earthquake caused a maximum intensity of X (MSK scale) in an area of 780 sq km (Fig. 7.34a). The inhabitants narrated that just before the severe ground shaking there was an explosion or deep rumbling sound, the *earthquake sound*.

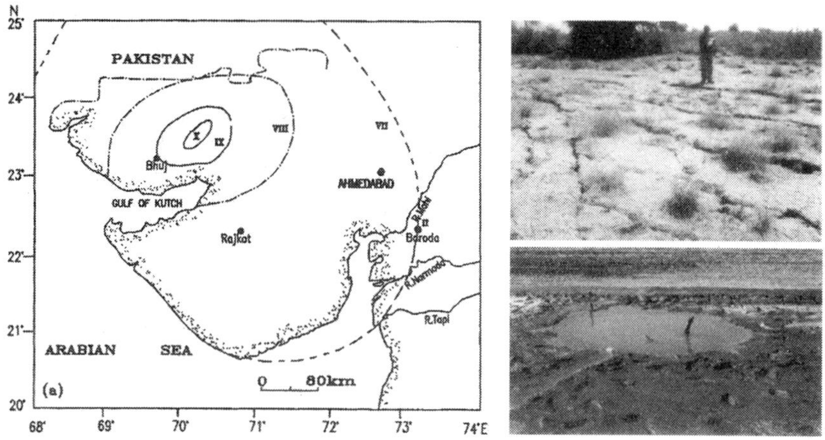

(a) (b)

Fig. 7.34 (a) Isoseismal map of the January 2006 Bhuj earthquake (Pande et al., 2003). (b) Ground cracks and liquefaction in the intensity zone VIII in Bhuj area (Rajendran et al., 2001).

Destruction of civil structures, irrespective of class or type of construction, was near total in the meizoseismal area, isoseist X. The isoseist IX comprised an area of 10,455 sq km (Ravishanker and Pande, 2001). In this zone also, many of the well built structures failed totally. No primary surface rupture was reported. Wesnousky et al. (2001), however, reported about 8 km long secondary rupture along the Bhachau lineament; they called it Manfara rupture which passes through the Manfara town; the rupture showed right lateral motion up to 32 cm of slip. They have also reported a scarp, about 50 cm high, that strikes east-west over several hundred metres. About 100 m south of the scarp, the ground was pervasively fractured by extensional cracks.

Profuse liquefaction in an area of about 50,000 sq km was reported in the form of sand blows/boils, craters, ground fissures, slumps etc. (Ravishanker and Pande, 2001; Rajendran et al., 2001). The sandy plains of Rann of Kutch and the Little Rann, where groundwater level was at shallower depths, have provided the most conducive environment for liquefaction in intensity zone VIII and above (Fig. 7.34b).

7.4.8 Girnar Seismic Zone

This zone extends into the Kathiwar peninsula, almost in line with the SONATA zone (Fig. 7.28). It is one of the four seismic zones, which were probably related to a plume generated four-armed rift junction, described as Khambat junction by Burke and Dewey (1973a). Chandra (1977) suggested that the earthquakes in this zone may be related to the Rann of Kutch zone, but existence of a fault along a direct continuation of the Narmada North fault near the Gir coast should not be ignored to assess the seismic potential of the Girnar zone. The Bhavnagar earthquake of April 21, 1919, intensity VIII, was well felt at Bhavnagar and nearby villages. The historical data indicate that a moderate earthquake (M~6.0) occurred in Surat in 1684 AD (Oldham, 1883).

7.4.9 East Coast Seismic Zone

The area consists part of Dharwar Craton and Eastern Ghat Granulite Belt (EGGB) separated by a prominent shear zone (Fig. 7.28). The principal structural trend is approximately N-S; the region is traversed by fairly dense network of medium and mega lineaments. Gravity trends conform fairly well with the structural trend. E-W gravity trend is observed over the Dharwar craton with values decreasing towards north. The gravity value decreases to –110 mGal near the centre of the Cuddapah basin where the sediment thickness is maximum. The EGGB is marked by N-S gravity trend (Fig. 7.15). A significant earthquake is reported in this zone as given below.

Ongole Earthquake (M 5.4), 1967

The March 27, 1967 Ongole earthquake (m_b 5.4), depth 17 km, epicentre at 15.6°N and 80.1°E, occurred on the NNW-SSE trending Gundlakumma fault traversing the Nellore schist belt. Right lateral strike-slip solution is inferred for this event (Fig. 7.2).

7.4.10 Godavari Valley Zone

The Godavari valley is a Mesozoic rift with a length of ~800 km, that separates the Dharwar craton to the west and the Bhandara craton to the east, and represents a polycyclic structural trend in NNW-SSE direction (Fig. 7.2). The eastern margin of the Godavari rift is bound by NNW-SSE trending basin margin fault. Bouguer gravity anomaly displays conformity with the tectonic domain (Fig. 7.15). The gravity over the graben shows a steep gradient; it drops from 30 mGal at the basin margin to around -100 mGal with low amplitude circular patterns at the centre of the basin where the sediment thickness is maximum. Two DSS profiles across the valley reveals the presence of two sedimentary basins (Kaila et al., 1990); the Godavari graben and the Godavari coastal basin, separated by a basement ridge. It has also indicated the presence of 2.8 km thick lower Gondwana sediments within the graben.

Historical records indicate not many significant earthquakes occurred in the Godavari valley region during the last two hundred years, only a few minor events were reported (Gupta et al., 1970). Seismicity is mostly confined to the Godavari basin, towards southeastern end in the Bhadrachalam area, where a significant event (m_b 5.3) occurred in 1969.

Bhadrachalam Earthquake (m_b 5.3), 1969

On April 13, 1969, the strongest earthquake m_b 5.3, M_S 5.7, occurred in the Godavari valley region, that shook the entire Andhra Pradesh at 8.56 pm (IST). The ISC located the epicentre at 17.9°N and 80.6°E near Bhadrachalam, with an origin time 15h 24m 55.7s (GMT). The focal depth was estimated at 17 km. The maximum intensity VII (MM scale) was reported for this event (Fig. 7.35). The main shock was preceded by a few foreshocks. The aftershocks were confined to a shallow depth 10-15 km within a narrow zone of 10 × 20 sq km (Gupta et al., 1970), which indicate that the main shock was at a shallower depth. The b-value of the sequence was estimated to be 0.51, typical of shield seismicity. Focal mechanism of the main shock was reported by several authors (Fig. 7.35). Chandra (1977), using WWSSN P-wave first motion data, gave two solutions; one strike-slip solution and the other pure reverse faulting, and it was not clear which one is right (a and b, Fig. 7.35). Chung (1993) did waveform modelling of the P and SH waves recorded by the WWSSN, and gave a fault plane solution (c, Fig. 7.35), which is consistent with the strike slip solution (a) of Chandra (1977). The NE-SW nodal plane is conformable with the meizoseismal trend, as well as with the aftershock trend, and it was inferred that the NE-SW trending conjugate fault of the Godavari valley was the seismogenic fault. Estimated stress drop was $\Delta\sigma$ = 16.88 bar (Chung, 1993).

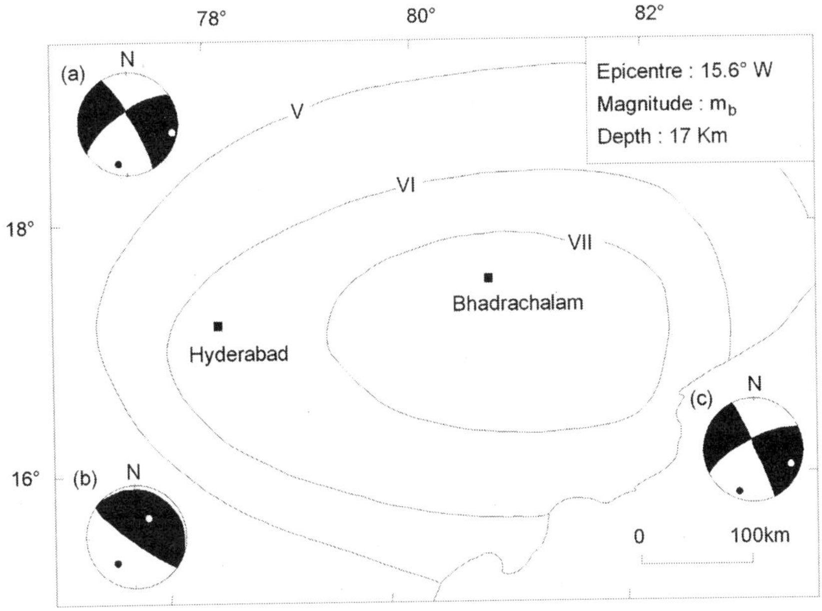

Fig. 7.35 Isoseismal map of the 1969 Bhadrachalam earthquake (GSI, 2000a) and fault plane solutions (a and b: Chandra, 1977; c: Chung, 1993).

7.4.11 Singhbhum Seismic Zone

The earthquake activity is northeast trending in this zone (Fig. 7.28), and it appears to merge with the intense activity of the Himalayan belt to the northeast. Among the significant past earthquakes, the January 23, 1866 Contai earthquake, intensity VI is described to have been accompanied by rumbling sound. Recently two significant earthquakes (M>5.0) are recorded in this zone; one in 1964 in Midnapore area and one in 1969 in Bankura area.

Midnapore Earthquake (M 5.5), 1964

The April 15, 1964 Midnapore earthquake M 5.5, was well felt at Kolkata (Calcutta) and other parts of West Bengal (Fig. 7.36). It was well recorded, and the epicentre was estimated at 21.7°N and 87.0°E with a focal depth at 36 km. A left lateral strike-slip solution was inferred for this event (Fig. 7.2).

Bankura Earthquake (M 5.7), 1969

The May 3, 1969 Bankura earthquake M 5.7, is another significant earthquake in this zone; epicentre was given at 23°N and 86.6°E (for location, Fig. 7.36). It was well felt in Kolkata (Calcutta) and in other parts of West Bengal.

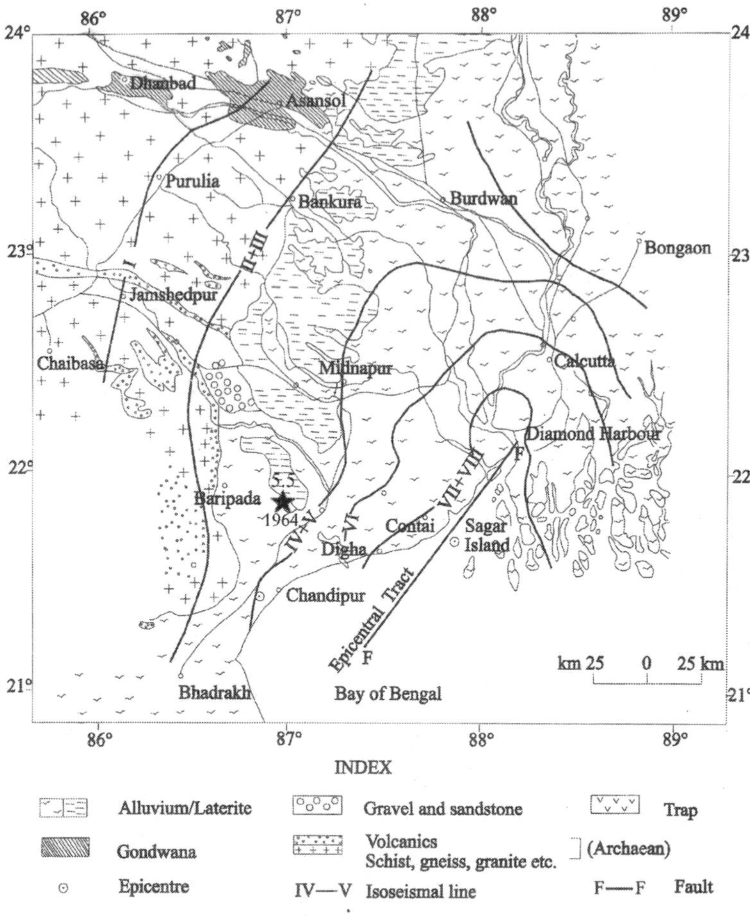

Fig. 7.36 Isoseismal map of the 1964 Midnapore earthquake (GSI, 2000a).

7.4.12 Killari Earthquake Zone

No such zone is classified in the identified seismic zones in peninsular India (Chandra, 1997) (Fig. 7.28). This zone is possibly the most conspicuous after the September 22, 1993 Killari earthquake M_W 6.1, M 6.3 in the eastern margin of the Deccan Plateau (Fig. 7.2). The zone represents a typical mid-cratonic SCR with subdued neotectonic expressions and low background seismicity. The 1993 Killari earthquake nucleated at a shallow depth (6 ± 1 km), and caused a loss about 10,000 human lives and huge property damages. Maximum intensity reached to IX (Fig. 7.37a). The fault plane solution shows a reverse faulting mechanism (Fig. 7.2). Many aftershocks were recorded, and details are given in Section 7.6.

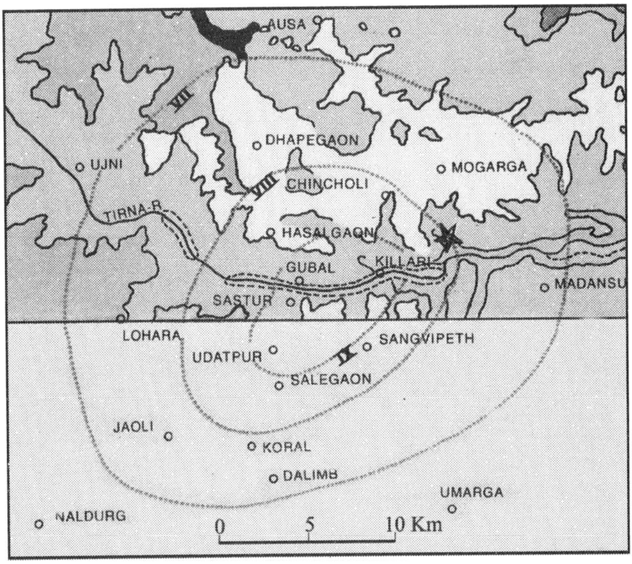

(a)

(b)

Fig. 7.37 (a) Sketch map of the isoseismals of the 1993 Killari earthquake (M_w 6.1), the main shock epicentre is shown by a black star. (b) Photograph shows the ground rupture and deformation in the epicentral area, Killari village (GSI, 1996).

About 1 km long rupture zone was identified in the Killari village area (Fig. 7.37). Trenching excavations across the rupture zone not only exposed fault plane, but also indicated repetition of movements. Preliminary observations, based on TL dating, suggest long recurrence intervals of the order of tens of thousands of years or more for the Killari type earthquake (Rajendran and Rajendran, 1997). Such long recurrence intervals are common for SCR earthquakes (e.g. Crone et al., 1992).

7.5 RECENT SEISMICITY: MICROEARTHQUAKE NETWORKS

Peninsular India although considered to be a stable landmass, various seismic activities are recorded during the last two decades. With the advent of microearthquake recording systems, seismicity in different parts of peninsular India is studied by many investigations. These investigations include monitoring Reservoir Induced Seismicity (RIS), swarm activity, foreshock and aftershock activities, active fault mapping etc. A brief discussion on these studies is made in this section.

7.5.1 Reservoir Induced Seismicity (Koyna Region)

Triggering of earthquakes by filling artificial reservoir water is known for over six decades (Gupta and Rastogi, 1976). Over 90 sites have been globally identified where earthquakes are reported to have triggered by such phenomena. In India, Koyna region has been identified to be the most significant and unique site for Reservoir Induced Seismicity (RIS), where the seismicity that began after the impoundment of a reservoir (Shivsagar lake behind the Koyna dam) in 1961 is persisting till to-date, since last more than 40 years (Gupta, 1992; Talwani, 1997). The first seismological observatory in Koyna region was established in 1963, but following the 1967 destructive earthquake a seismological network of seven stations was established during the years 1967-1972. Then another five-station network was established in 1990 in the Warna reservoir area, about 35 km south of the Koyna Dam. This is one of the regions in peninsular India which is well equipped with microearthquake networks for continuous monitoring of seismicity. In a recent review work, Gupta (2002) has given a detailed status of the Koyna-Warna reservoir seismicity, and he preferred to use the term Reservoir Trigger Seismicity (RTS) rather than RIS.

The December 10, 1967 Koyna earthquake, M 6.7, revised M_W 6.3, was believed to be the RIS (Gupta and Rastogi, 1976). The seismic activity is reported to have enhanced soon after the impoundment of the Koyna reservoir in 1962 and the Warna reservoir in 1992. Since then over 10 earthquakes $M \geq 5.0$, over 150 earthquakes $M \geq 4.0$ and over 100,000 events of $M \geq 0$ have occurred in this region (Gupta, 2002).

Talwani (1997) made a detailed study on the seismicity of the Koyna-Warna dam region using the analog data recorded from 1964 to 1995. The seismicity map (1964-1967) prior to the 1967 Koyna earthquake shows that the activity was confined in the Koyna dam area (Fig. 7.38a); the activity enhanced subsequently, and a NW-SE trend of activity was observed after the December 1967 earthquake during the period 1967-73 (Fig. 7.38b).

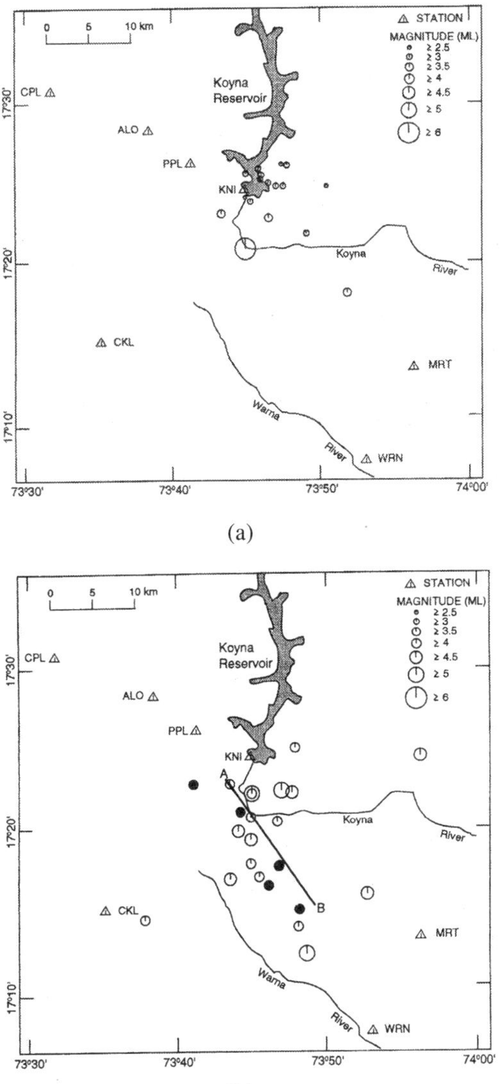

Fig. 7.38 (a) Relocated earthquakes (1964-1967), and epicentre of the December 10, 1967 Koyna earthquake; seismic stations are indicated by triangles, and (b) relocated earthquakes M>4.0 (December 10, 1967-1973) in the Koyna region; the aftershocks (solid circles) of the December 10, 1967 earthquake are aligned along the line AB. (Talwani, 1997).

Till 1992, the seismic activity was mostly confined to a 20 km long seismic zone extending south of Koyna dam (Fig. 7.38b). Two trends of seismicity, NW-SE and NE-SW, were, however, observed during 1993-1995 due to additional filling of the Warna reservoir located SSE of the Koyna dam at a distance of 35 km (Fig. 7.39a). The selected earthquakes (M>3) are shallow, within a depth range 5-15 km, mostly at a depth range 6-8 km. The seismicity is restricted by the Koyna River Fault Zone (KRFZ) to the west, that lies along the N-S portion of the Koyna river and extend to the south by atleast 40 km. The seismicity to the east is bounded by the NE-SW trending Patan fault, and the higher magnitude (M>5.0) earthquakes occurred within this zone (Fig. 7.39b). A similar observation of two seismicity trends was also made by Chadha et al. (1997), who relocated the lower magnitude (M<3.0) earthquakes in the region (Fig. 7.40a).

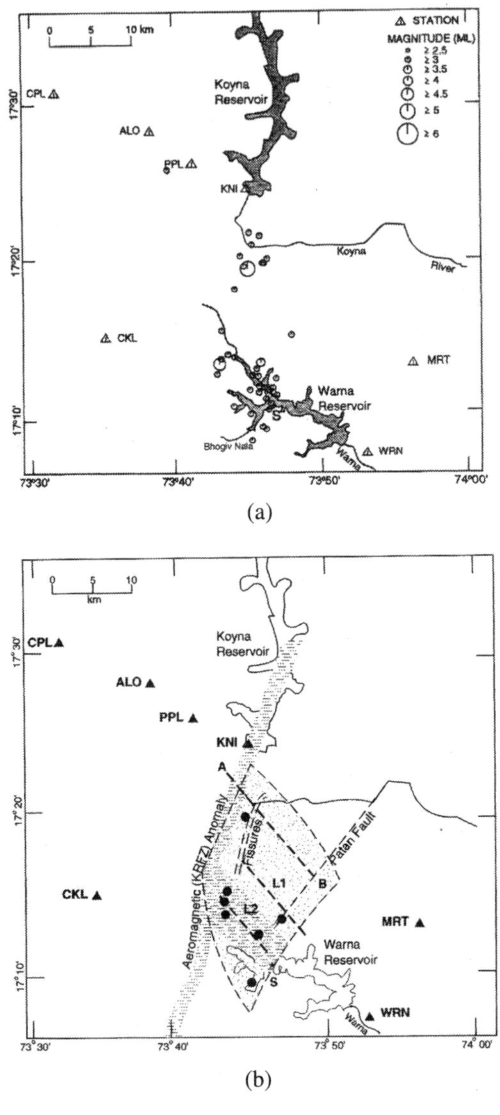

Fig. 7.39 (a) Relocated earthquakes (M>3.0) in Koyna-Warna region for 1993-95; seismic stations are indicated by triangles, and (b) interpreted structural elements of the Koyna-Warna region based on aeromagnetic, geological and seismicity data, the larger events M>5.0 are shown by solid circles (Talwani, 1997).

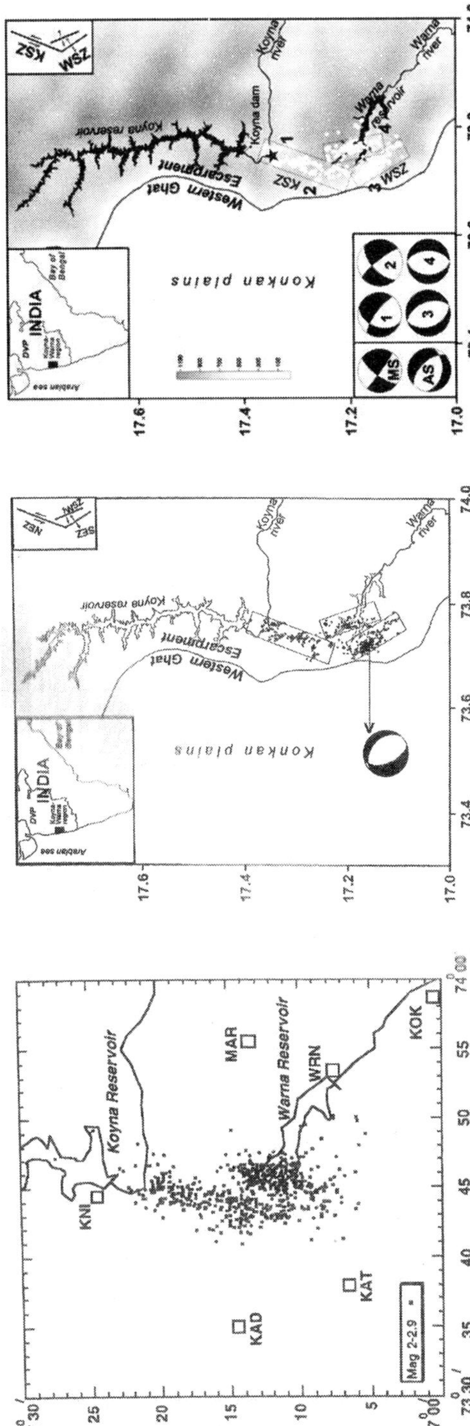

Fig. 7.40 (a) Koyna-Warna microseismicity, October, 1993-December, 1994, relocated by Chadha et al. (1997); seismic stations are shown by small rectangles, (b) microseismicity, April 1996-December 1997, located by a 10-station digital network (Rai et al., 1999b), and (c) the seismicity with fault plane solutions (Gahalaut et al., 2004).

An experiment was mounted deploying a 10-station digital seismic network during April, 1996 through December, 1997 (Rai et al., 1999b). A total of 20 sites over an area of 60 × 80 sq km were occupied by leap frogging with about 50% overlap of the seismic stations. They located 406 events within a zone of 11 × 25 sq km; depth distribution indicated that the base of the seismogenic zone is at ~10 km. They identified three seismic zones: north escarpment zone (NEZ), south escarpment zone (SEZ) and Warna seismic zone (WSZ) (Fig. 7.40 b).

Gahalaut et al. (2004), however, argued that the earthquakes in the Koyna-Warna region may be reservoir triggered, but the high seismicity is also influenced by the geometry of the fault zones and by their interactions through stress transfer. Two distinct fault zones were identified based on the distribution of epicentres and focal mechanisms of the earthquakes. The earthquakes occur by left lateral strike slip on the NNE-SSW trending faults in the Koyna Seismic Zone (KSZ), and by normal faulting on the NNW-SSE trending faults in the Warna Seismic Zone (WSZ) (Fig. 7.40c). They suggested that cumulative slip through occurrence of earthquakes on either of the two fault zones promote failure on the other. They further argued that the 1967 Koyna earthquake occurred by left-lateral strike-slip on the NNE-SSW trending fault (Langston, 1976), and occurrence of the largest aftershock (M 5.3) by normal faulting with slip on the NW-SE oriented plane (Langston and Franco-Sperra, 1985), that occurred 31 hours after the 1967 main Koyna earthquake supports this view (Fig. 7.40c).

7.5.2 Swarm Activity

Table 7.3 gives a list of 11 cases of earthquake swarm activity in peninsular India recorded since 1993. Out of these, three sequences are illustrated in Figs 7.41-7.43, that show burst of seismic activity for a short period, the activity then remains quiet for a longer time. The maximum magnitude of an earthquake in the sequence varied between 3.0 and 3.8 with no particular event as a main shock. The 1994 swarm activity at Medhi (Thane) near Bombay shows three burst of activities; in each burst the maximum magnitude of earthquake was of the order of M 3.0-3.7 (Fig. 7.41), and peak activity level reached about 80-100 events in a day. The swarm activity in Pandhana area during 1998-1999 is one of the longest sequences in central Indian tectonic zone (Fig. 7.42). The peak activity was recorded in October, 1998, with a maximum magnitude M 3.2. The 2001 Haripur swarm activity shows maximum activity in November with a maximum magnitude M 3.2 (Fig. 7.43).

In all the 11 cases listed in Table 7.3, it is observed that the starting month of commencing swarm activity was in the monsoon and post monsoon period between July and December. All the swarm activities were mostly recorded

Table 7.3: Examples of Swarm Activity in peninsular India (Bhattacharya and Dattatrayam, 2003)

Starting date	*Place/area Lat (°N), Lon (°E)*	*Opening period of Local observatory*	*Continued till*	*Total events recorded*	*Max. mag. and data*
1993 Dec. 1st week	Taklikalan, Bagmer (Khandwa Dt.) 21.7, 76.4	1993 Dec., 10-24	Dec. 24 (20 days)	118	2.6
1994 Jul. 27	-do-	1994 Aug. 7-1995 Mar. 31	Oct. 12 (2 mons. 15 days)	75	3.1 on Sep. 16
1994 Jan.	Medhi, (Jawahar Taluka (Thane Dt.) 19.8, 73.1	Dhamni obsy (MERI) existed.	Feb. end (2 months)	487	3.8 on Jan 9 (8 events of mag ≥ 3)
1994 Jul.	-do-	Dhamni obsy (MERI) existed, temp. obsy by IMD & NGRI	Oct. end (3 mons. 15 days)	1805	3.5 on Aug. 19 (12 events of mag ≥ 3)
1994 Aug.	Umarthana (Ratlam Dt.) 23.2, 74.9	1994 Aug. 10 - Oct. 24	Oct. 14 (2 mons. 15 days)	202	2.0 on Aug. 25
1995 Aug.	-do-	1995 Oct. 14-24	Oct. 24 (2 mons. 15 days)	36	2.7 on Oct. 14
1997 Dec. 9	Amarwara (Chhindwara Dt.) 22.3, 79.2	1997 Dec. 19-30	Dec. 30 (21 days)	24	3.2 on Dec. 21 (3 events of mag ≥ 3)
1998 Sept.	Sarangpur (Rajgarh Dt.) 23.6, 76.5	1998 Sep. 16-29	Sep. 29 (20 days)	121	1.2 on Sep. 26
1998 Sept. 11	Pandhana (Khandwa Dt.) 21.7, 76.2	1998 Sep. 17-1999 Feb. 12	1999 Jan. 27 (4 mons. 16 days) 101 events on Oct. 11	1601	3.2 on Oct. 20 (4 events of mag ≥ 3 with event Sep. 16)
1999 Oct. 01	-do-	1999 Oct. 25-Dec. 30	1999 Dec. 25 (2 mons. 25 days)	86	1.9 on Oct. 25
2001 Oct. 23	Haripur (Junagadh Dt.) 21.2, 70.5		2001 Dec. 28 (2 mons. 2 days) 145 events on Nov.10.	1689	3.2 on Nov. 11 (9 events of mag ≥ 3)

at shallower depth (0~3 km) with a maximum depth ~7 km for a few events. The activity is not only confined to a shallower depth but also to a smaller area. Shallow crustal heterogeneity and increase in pore-pressure by fluid flow normally caused the swarm activity. A detailed investigation of the Pandhana swarm activity is discussed below.

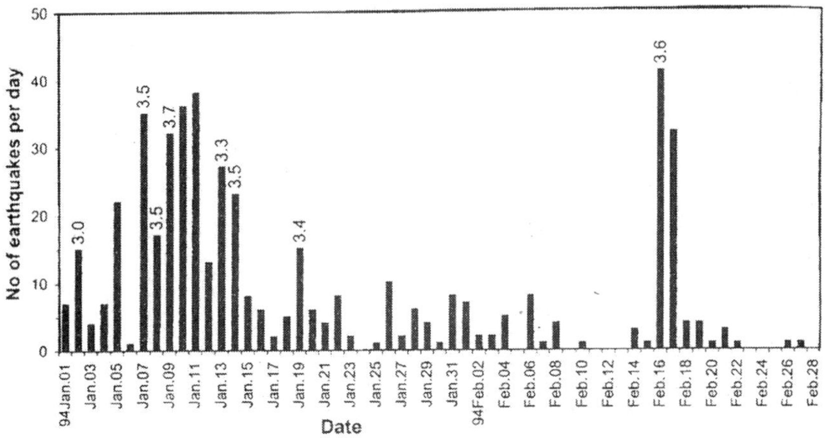

(a)

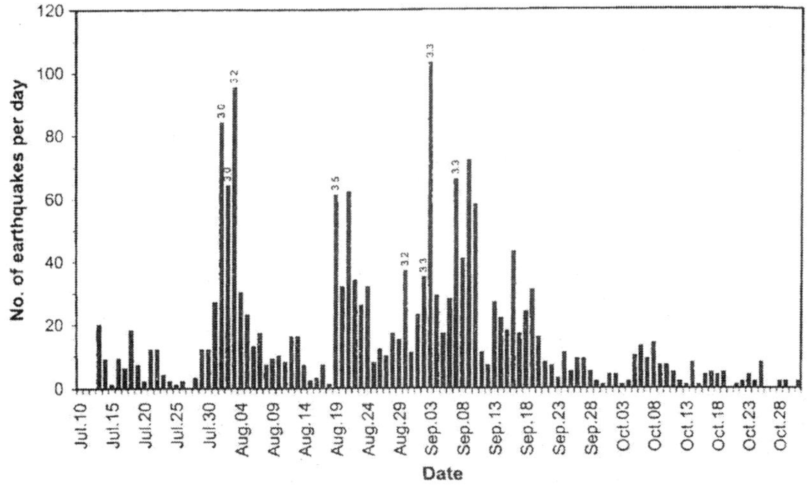

(b)

Fig. 7.41 Swarm at Medhi (Thane), (a) during January-February, 1994, and (b) during July-October, 1994 (Bhattacharya and Dattatrayam, 2003).

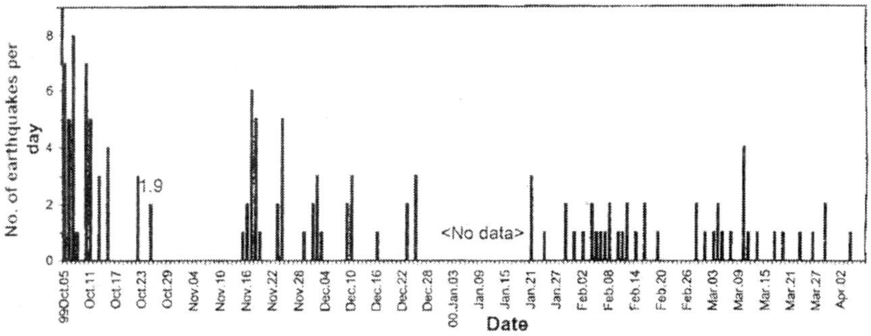

(a)

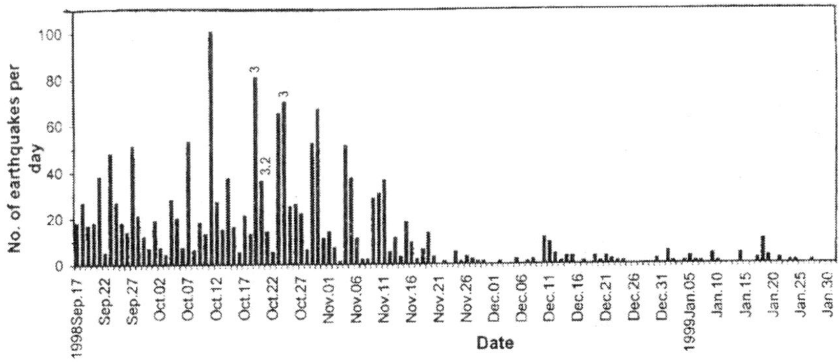

(b)

Fig. 7.42 Swarm at Pandhana, (a) during September 1998-January 1999, and (b) during October-December, 1999 (Bhattacharya and Dattatrayam, 2003).

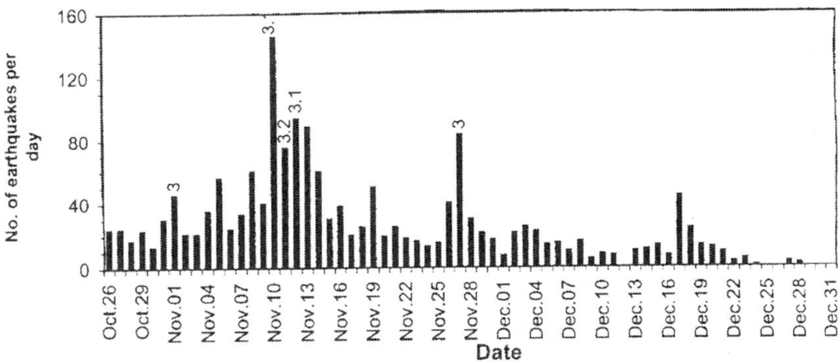

Fig. 7.43 Swarm at Haripur during October-December, 2001 (Bhattacharya and Dattatrayam, 2003).

Pandhana Swarm Activity, 1998-99

A temporary microearthquake network was established to monitor the Pandhana swarm activity, to the south of Khandwa, during 1998-99 (Gaonkar and Srirama, 2003). The location of the swarm activity, Khandwa area, is shown in Fig. 7.21a. The area is bounded by the E-W trending South Narmada fault and Barwani-Sukta fault. The activity was recorded from September, 1998 to April, 1999.

A maximum activity was recorded on 11th October, 1998 with 101 events. Three events of magnitude M~3.0 were recorded with the maximum M 3.2 on October 20, 1998 (Fig. 7.42). About 300 events were located using the temporary network data; the epicentres lie within a small area of 10 sq km (Fig. 7.44 a); a cross section shows maximum events occurred at shallower (0<3 km) depth, a few events were located with fixed depth at 3 km, and a few occurred at a depth range 4-7 km. Two fault plane solutions were obtained; one for the shallower events at a depth range 1-3 km and one for the deeper (4-6 km) cluster (Fig. 7.44c). The events with the fixed depth (3 km) location were not considered for the solutions. The shallower events show a normal fault solution with a south dipping inferred fault plane. The inferred fault plane is comparable with the E-W trend of the major fault(s) in the area. The deeper events also show normal faulting but with N-S trending nodal planes. A transverse conjugate fault at deeper depth is the possible seismogenic fault for the deeper (4-6 km) events. The upper crustal heterogeneity, complex fault interaction and pore pressure changes, particularly during monsoon and post monsoon periods, are the reasons for such type of swarm activity in the shield area.

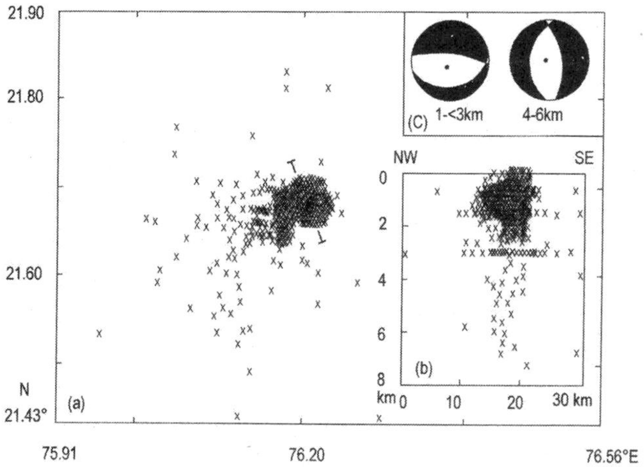

Fig. 7.44 (a) Epicentre map of the Pandhana swarm activity, September 1998-April 1999, (b) seismic cross section of the events, and (c) composite fault plane solutions of the selected events at two different depth ranges (Goankar and Srirama, 2003).

7.5.3 Foreshock-Main shock-Aftershock Sequence

A few earthquake sequences with foreshock-main shock-aftershock pattern are observed in the peninsular shield area; these are listed in Table 7.4. Figure 7.45 shows the sequence of the 1967 Koyna earthquake (M 6.7) and the sequence of 1983 Bhatsa earthquake (M 4.8). The Koyna sequence lasted for about two years, whereas the Bhatsa sequence continued for more than four years with about 11,600 aftershocks for a smaller magnitude of main shock M 4.8 (Chadha, 1995). Rao et al. (1991) reported a still more unusual sequence of the 1986 Valsad earthquake M 4.6; about 20,000 aftershocks were recorded in 11 months (Table 7.4). In all the three cases, 1967 Koyna, 1983 Bhatsa and 1986 Valsad, foreshocks were observed, that occurred about two to three months before the main shocks. Kolvankar (2001) reported the 1991 Valsad earthquake sequence (Fig. 7.46 a), where foreshocks started 35 days before the main shock and about 380 aftershocks were recorded.

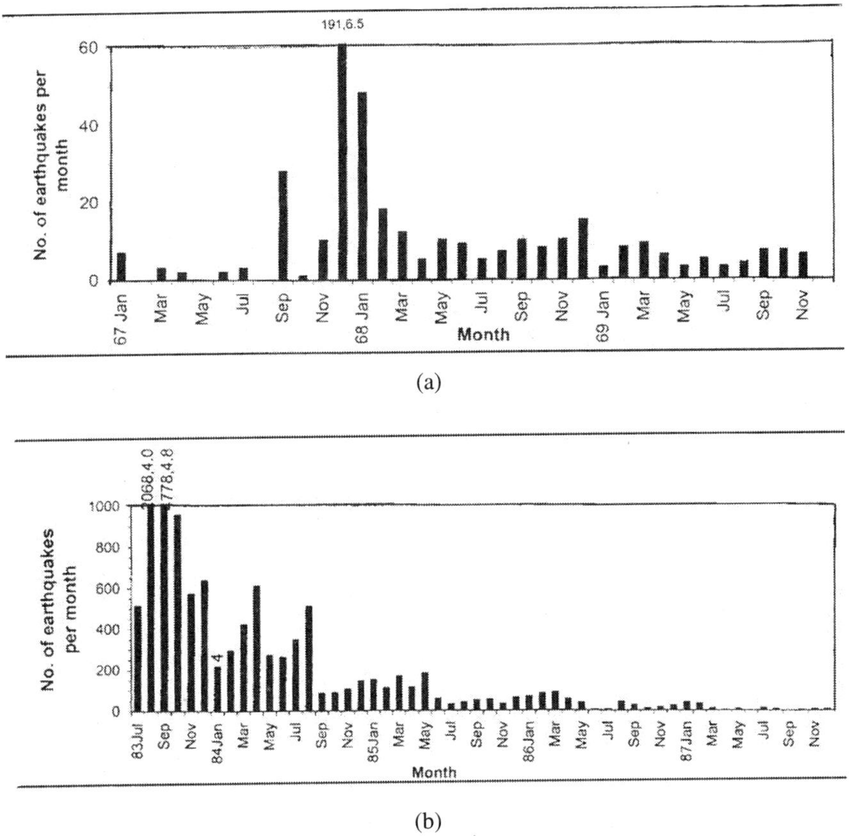

Fig. 7.45 (a) The 1967 Koyna foreshock-main shock-aftershock sequence, and (b) the 1986 Bhatsa foreshock-main shock-aftershock sequence (Bhattacharya and Dattatrayam, 2003).

Table 7.4: Examples of Foreshock-Main shock-Aftershock Sequence in peninsular India (Bhattacharya and Dattatrayam, 2003)

Year, Area, State	*Foreshock*	*Main shock*		*Operation*	*Aftershocks*			
	Started on, Max. mag.	Date, Lat (°N), Long (°E)	Depth Mag.	*period of local network*	Continued till	Total events recorded	Largest, diff. and date	Ref
1967, Koyna Maharashtra	Sept. 12, 1967 (3 months) 5.0 on Sep.13	Dec. 10, 1967 17.37, 73.75	8 km 6.5	MERI obsys, working	Nov.5, 1969 (1 Yr.330 days) Fig.2	375 M≥3	5.0 (-1.5) on Dec. 12, 1967	a
1981-82 Gundipet, Hyderabad	Dec. 3, 1981 (1 mon. 11 days) 2.1 on Jan.6, 82	Jan. 14, 1982 17.43, 78.31	--- 3.5	HYB working + local (Jan. 16, Feb. 21)	Feb.22, 1982 (39 days) Fig. 3	368	3.0 (-0.5) on Jan. 21, 1982	b
1983, Bhatsa, Maharshtra	July 1983 (2 mons 15 days) 4.0 on Aug. 17	Sept. 15, 1983 19.50, 73.29	5 km 4.8	MERI obsys. working	March 1987 (4 yrs. 6 months) Fig. 4	11,606 M≥3	4.0 (-0.8) on Jan. 1, 1984	c
1986, Valsad, Gujarat	Feb. 3, 1986 (22 mons 23 days) 4.3 on Feb.26	Apr. 26, 1986 20.63, 73.47	--- 4.6	GERI obsys started on May 22, 1986	March 1987 (11 months)	20,000 (appx.)	4.1 (-0.5) on Apr. 28, 1986	d
1991, Valsad, Gujarat	Mar. 25, 1991 (1 mon. 5 days) 3.9 on Apr. 14	Apr. 30, 1991	--- 4.6	BARC radio tele-metered network	Jun. 28, 1991 (59 days) Fig. 5	383	3.8(-1.1) on May 4	e,f

(Contd.)

Table 7.4 contd.

1994, Varavoor,	Nov. 28, 1994 (4 days)	Dec. 02, 1994	--- 3.8	Dec. 12, 1994 Jan. 15, 1995	Jan. 05, 1995 (34 days)	20	-----	g
2000 Bhavnagar, Gujarat	Aug. 09, 2000 (one day) 2.5 on Aug. 09	Aug. 10, 2000 21.78, 72.31	5 km 3.6	Aug. 14, 2000 Sept. 08, 2000	Aug. 25, 2000 (15 days) Fig. 6	43	3.1 (-0.5 on Aug. 13, 2000)	h
2000 Bhavnagar, Gujarat	Aug. 09, 2000 (1 mon. 3 days) 3.6 on Aug. 10	Sep. 12, 2000 21.8, 72.2	--- 3.8	Sep. 13, 2000	Oct. 26, 2000 (44 days) Fig. 6	86	3.1 (-0.7 on Sep. 13, 2000)	h

(a) Guha & Basu (1973), (b) Rastogi et al. (1986), (c) Srivastava et al. (1991), Chadha (1995), (d) Rao et al. (1991), Srivastava and Dube (1996), (e) Kolvankar (2001), (f) Bulletin of MERI, (g) & (h) IMD Report

Recently another foreshock-main shock-aftershock sequence was observed during 2000 in Bhavnagar, at the western end of the peninsular shield. The Bhavnagar sequence during 2000 interestingly shows two foreshock-main shock-aftershock sequences (Fig. 7.46 b). There were two main shocks, M 3.6 and M 3.8, those may be considered as *doublets*.

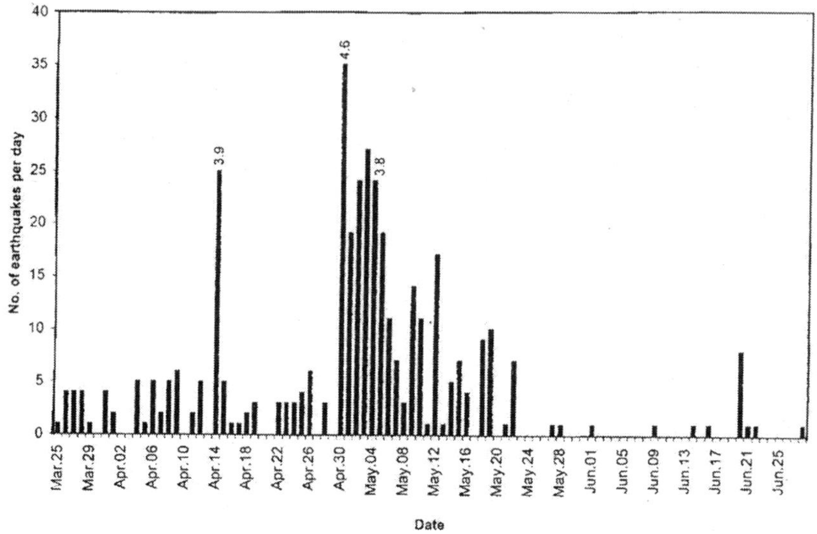

(a)

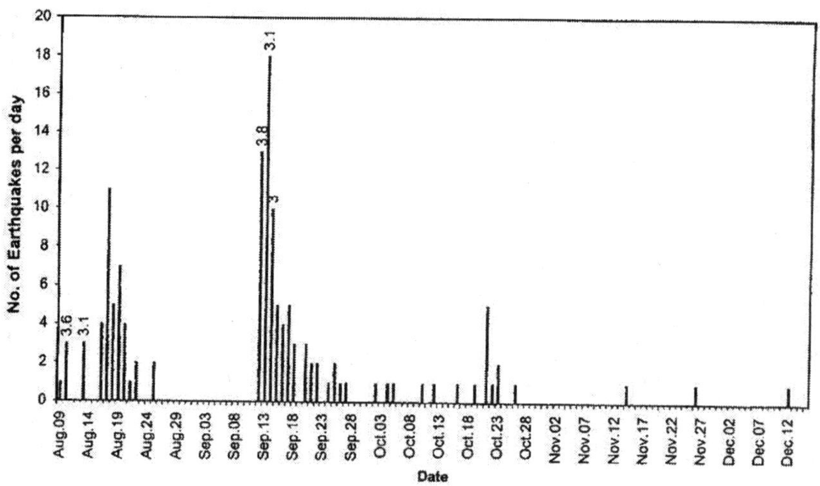

(b)

Fig. 7.46 (a) The 1991 Valsad foreshock-main shock-aftershock sequence (Kolvankar, 2001). (b) The 2000 Bhavnagar foreshock-main shock-aftershock (two) sequences (Bhattacharya and Dattatrayam, 2003).

Magnitude of the largest aftershocks varies with the main shocks; it is ~1.5 unit less for the 1967 Koyna earthquake, but for the other smaller main shocks (M 3.5~4.8) the largest aftershocks are 0.5 to 0.8 unit less than the corresponding main shock magnitudes in the shield area.

7.5.4 Active Fault Investigation

Microearthquake network is a powerful tool to investigate active faults. Such attempt is routinely made in seismically active regions and also in aftershock investigations. The GSI routinely make attempts to investigate active fault(s) in selected intraplate shield areas where neotectonic activity is reported but seismic activity is less; hardly any earthquake of M>4.0 is recorded by the IMD national network from these fault zones. Microearthquake monitoring may identify such active fault(s). An example of microearthquake investigation in shield area for delineation of active fault(s) is illustrated here.

East Margin Fault in Munger, Bihar

The Munger area in the state of Bihar has experienced damage by the 1934 great (M ~8.4) Bihar earthquake as well as by the August, 1988 earthquake (M_W 6.6); both the events occurred in the foothills region, to the south of the Main Boundary Thrust (MBT). It was inferred that the East Patna fault was activated to produce these two events (GSI, 1993) (Fig. 7.47).

A microearthquake investigation was carried out in the Munger area by the GSI (De et al., 2000). The area was reported to be neotectonically active where a chain of hot springs are located along the eastern margin of the Proterozoic massif that is in contact with the Quaternary sediments. A temporary microearthquake network was in operation in the neotectonically active area of Munger for about three months from January to March, 1996. Epicentres of 45 well located earthquakes M 2~3.0 are shown (Fig. 7.47). It was observed that almost no earthquake occurred inside the network in the area of neotectonic activity; only one shallow event (depth ~1 km) was located inside the network and one deeper event (depth ~19 km) to the north of the network. The epicentres are clustered/aligned along the south end of the East Margin fault, that lies far east (~60 km) of the network. The East Margin fault is the east boundary fault of the Munger Saharsa Ridge (MSR), as mapped by gravity survey (Choudhury and Dutta (1975). This fault is the contact zone between the Chotanagpur granite gneiss and the Rajmahal traps to the south, and it demarcates the MSR and the Purnea basin to the north (Fig. 7.47). A 50 km long segment of the fault at the southern end is identified to be active by this microearthquake survey. The earthquakes are all shallow, within a depth range 0-12 km. No reliable fault plane solution was obtained as the earthquakes were all outside the network. The investigation was, however, very useful to delineate the active segment of the fault.

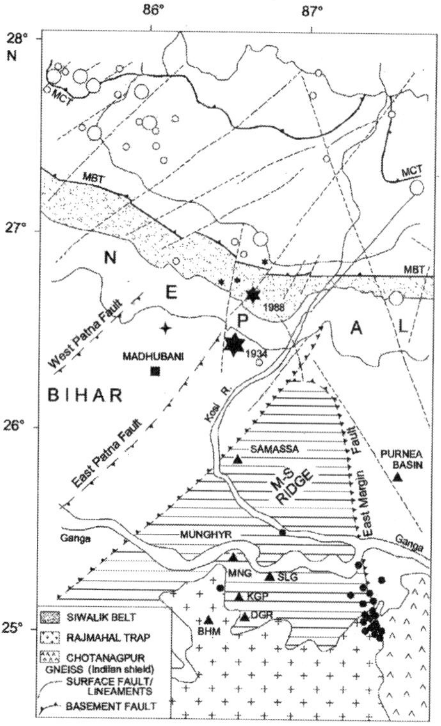

Fig. 7.47 The map showing major tectonic elements, epicentres of the regional earthquakes (open circles), and the two major events (solid stars). Temporary microearthquake stations (solid triangles) and the epicentres of the smaller magnitude (M< 3.0) events (solid circles) recorded by the network along the southern end of the East Margin Fault are shown (De et al., 2000).

7.6 AFTERSHOCK INVESTIGATIONS: RECENT DAMAGING EARTHQUAKES

Since 1993 a list of eight mainshock-aftershock sequences are reported in peninsular India (Table 7.5). Out of these, detailed aftershock investigations were made for four damaging events; two were strong earthquakes M≥6.0, the 1993 Killari earthquake (M 6.3) and the 1997 Jabalpur earthquake (M 6.0), in the south and central part of peninsular shield, one was the large 2001 Bhuj earthquake (M_W 7.7) in the northwestern end of peninsular shield, and the fourth one was the medium magnitude (M_W 5.0) 2000 Idukki earthquake in the state of Kerala, southwest of peninsular shield. Detailed aftershock investigations of these events were carried out by temporary microearthquake networks, that shed new light on understanding of seismotectonics of the SCR events and the active faults in peninsular India. These are discussed in chronological order below.

Table 7.5: Examples of Main shock-Aftershock Sequences in peninsular India (Bhattacharya and Dattatrayam, 2003)

Date of main shock	*Epicentre Lat (^{0}N), Lon (^{0}E) Region*	*Depth, Mag.*	*Operation period of local network*	*Aftershocks*		
				Continued till	*Total events*	*Largest, Diff. and date*
1993 Sep. 29	18.07, 76.02 Killari, Latur	12 km 6.3	Oct. 07, 1993 May. 07, 1994	Mid. Apr. 1994 (165) days)	221 (by 3 obsys.)	5.2 (-1.1) on Oct. 08, 1993
1994 Aug. 30	26.26; 78.38 Gwalior	--- 4.8	Sep. 11, 1994 Sep. 17, 1994	Sep.13, 1994 (14 days)	1	---
1995 Jun. 21	21.7; 85.3 Keonjhargarh	--- 4.8	Aug. 22, 1995 Sep. 09, 1995	Sep. 07, 1995 (65 days)	22	--- (3.0 on Aug. 28)
1996 Jan. 10	22.4; 77.6 Hosangabad	--- 4.5	Feb. 23, 1996 Mar. 04, 1996	Feb. 25, 1996 (46 days)	1	---
1997 May 21	23.08; 8.06 Jabalpur	35 km 6.0	May 23, 1997 Sep. 19, 1997	Aug. 08, 1997 (2 mons. 17 days)	45	4.2 (-1.8) on Jun. 04, 1997
2000 Dec. 12	9.65; 76.74 Idukki	7 km 5.0	KSEB obsys. +Dec. 15, 2000 Jan. 23, 2001	Jan. 04, 2001 (23 days)	47	3.9 (-1.1) on Dec. 16, 2000
2001 Jan. 07	9.63; 76.78 Idukki	8 km 4.8	-do-	Jan. 23, 2001 (16 days)	41	3.6 (-1.2) on Jan. 07, 2001
2001 Jan. 26	23.40; 70.28 Bhuj	25 km 7.7	Jan. 27, 2001 Continued	Continuing beyond April 2002 (approx. 500 days)	1331 till May 2, 2002 (M≥3)	5.7 (-2.0) on Jan. 28, 2001

7.6.1 Killari Earthquake (m_b 6.3), 1993

Main Shock

The Killari earthquake of September 30, 1993 occurred in the Deccan Volcano Province (DVP) in the southern block of the peninsular India shield-area (Fig.7.2). The hypocentral parameters and the CMT solutions of the main shock were reported by the U S Geological Survey (USGS) and Harvard University (HRV) using the global network data. The India Meteorological Department (IMD) and the National Geophysical Research Institute (NGRI), India, subsequently determined the hopocentral parameters of the main shock using the national network as well as the global network data. The body wave magnitude was estimated to be m_b 6.3 by the USGS as well as by the IMD.

The focal depth of the main shock was, however, debated. The USGS reported the focal depth at 6.8 km, the IMD at 6-12 km and the NGRI at 6 km. Chen and Kao (1996) determined the focal depth of the main shock at 7±1 km. Seeber et al. (1996), on the other hand, estimated the centroid depth at 2.6 km, and Ramesh and Estabrook (1998) determined the centroid depth at 6.5 ± 0.1 km. The focal depth 7 km of Chen and Kao (1995) is, however, close to the centroid depth 6.5 km estimated by Ramesh and Estabrook (1998).

Seeber et al. (1996) suggested that the main shock is fundamentally a tectonic event, but might have been triggered by the 14.9 m deep reservoir in Makni (MKN), which is about 15 km away from the main shock epicentre (Fig. 7.48a). Rastogi (1994) argued that the incremental stress due to Makni reservoir was too small to trigger the Killari earthquake. Rajendran et al. (1996) and Rajendran and Rajendran (1997) strongly advocated that the Killari earthquake was generated by reactivation of an existing fault in the area, and not a reservoir triggered event.

The *Centroid Moment Tensor* (CMT) solutions of the main shock were reported by the USGS and HRV, which show a reverse faulting with a small strike-slip component (Fig. 7.48 a). The south dipping near E-W or ESE-WNW striking nodal plane is the inferred fault plane. Gupta et al. (1998) determined a fault plane solution of the main shock based on the P-wave first motions of the national and global network data; the solution was very much compatible with the CMT solutions. Baumbach et al. (1994) estimated a seismic moment $M_0 = 1.8 \times 10^{18}$ Nm and a rupture area 38 sq. km associated with a stress drop of about 7 MPa.

Macroseismic Effects

A detailed macroseismic investigation was carried out by the Geological Survey of India (GSI) (Narula et al., 1996), by the NGRI (Indra Mohan and

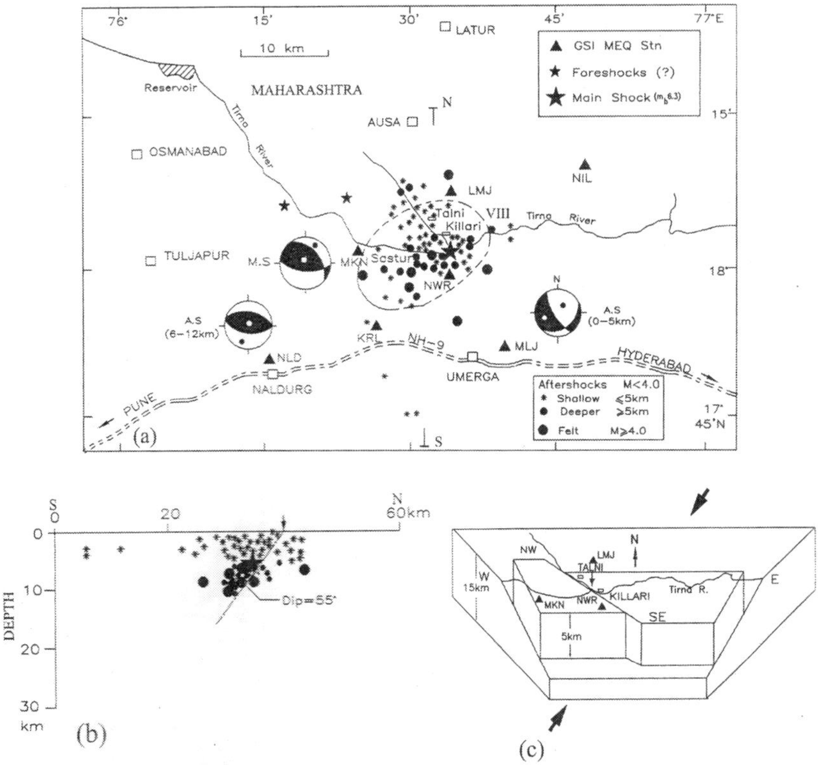

Fig. 7.48 (a) Map showing isoseismal (intensity VIII) of the 1993 Killari earthquake, temporary microearthquake stations, epicentres of the main shock and aftershocks with fault plane solutions, (b) N-S seismic cross section of the main shock and aftershocks, (c) seismotectonic model of the Killari earthquake sequence (Kayal, 2000).

Rao, 1994), and by Seeber et al. (1996). Majority of the houses in about 20 villages in the meizoseismal area totally collapsed and caused maximum casualties. The number of casualties reported by the local authority is well documented to 7,446 in the GSI report (GSI, 1996). Gupta et al. (1998), however, reported an estimated death toll of 11,000 human lives. Most of the houses which collapsed were poorly built with stone-and-mud. In contrast, the well constructed engineering buildings, like water tank or well constructed stone-and-motar houses of the wealthy people had rare collapses. A discontinuous surface rupture of about 1 km was observed between the Killari and Talni villages, along the north bank of the Tirna river. The highest intensity VIII$^+$ (MSK scale) was assigned to this earthquake; the isoseismal VIII shows an elliptical area of 25 × 20 sq. km (Fig. 7.48a). The maximum casualties were caused due to no-experience of any earthquake in the region, poorly constructed stone-and-mud houses, high population density and occurring of the earthquake in the early hours (03h 55m local time) when people were asleep. Ground deformation, gas emanation and subterranean sounds of the aftershocks were reported (e.g. Reddy et al., 1994; Rastogi and Rao, 1994; Gupta et al., 1998). A 50-60 feet deep groundwater well which was used for water supply in the Killari village went dry after the main shock. The most spectacular effect was the ground deformation; scraps of varying height (few cm to 80 cm) appeared at several locations along the 1 km long discontinuous rupture in E-W direction between the Killari and Talni villages. The deformation zone had been the site of many geophysical investigations as well as trenching and drilling (Rajendran et al., 1996; Gupta and Dwivedi, 1996).

Foreshocks

There is no historic record of any moderate or large earthquake of magnitude more than 4.0 in the area during the last 500 years or so till October, 1992. The Koyna earthquake, M 6.7, December 10, 1967, occurred in the west-coast seismic zone, about 300 km away from the Killari area, (Fig. 7.2). The southern part of the peninsular India was demarcated as the seismically least active zone or stable zone in the seismic zoning map of India (Fig. 2.19). In October, 1992, one year before the main shock, two felt earthquakes of magnitude 3.8 and 4.5 were reported by the IMD in the area (Fig. 7.48 a). The local people in the area paniked because these were the first felt earthquakes in their life-time, which caused minor damages to the weak houses in the Killari village. The IMD made a two-month survey with a temporary seismograph station in the area after these two felt earthquakes, but no significant event was reported. The NGRI observatory at Hyderabad, about 250 km away from the Killari area, however, recorded 26 events ($M \geq 2.0 < 4.0$) from October 18 to November 15, 1992 (Gupta et al., 1998). A large earthquake is often preceded by foreshocks of significant magnitude in relation to space, time and magnitude of the main shock. There was no immediate foreshock or felt earthquake before the 1993 Killari main shock. The October, 1992 felt earthquakes were the only two earthquakes of M 3.8 and 4.5 (foreshocks?), which occurred about one year before the main shock of September 30, 1993.

Aftershocks

The main shock was immediately followed by three felt aftershocks, M 4.3-5.0, within five hours. The largest aftershock, M 5.2, occurred on October 9, 1993, about a week after the main shock (Fig. 7.48a). It was about 1.1 magnitude lower than that of the main shock. The immediate felt aftershocks were recorded and located by the IMD national network. Three more felt aftershocks of M 4.0-5.0 occurred in November, 1993, about one and a half months after the main shock. These three felt aftershocks were well recorded by the close-spaced temporary microearthquake network, which was established by the GSI eight days after the main shock. Epicentres of all the felt aftershocks are shown in Fig. 7.48a.

Temporary microearthquake stations, in addition to the GSI network, were established by various organizations including the IMD, Wadia Institute of Himalayan Geology (WIHG), and by the NGRI in collaboration with a German Task Force (GTF) team. The NGRI and the GFT team installed three digital, 14 analog and four strong-motion instruments with an aperture of about 100 km around Killari (Baumbach et al., 1994). Seeing that the aftershocks were very close to the main shock epicentre and were of shallow depth (<10 km), they shifted the three digital instruments near Killari, but these three instruments were run only for five or six days, from October 17 to October 22, 1993. They also closed the analog stations earlier, on December 24, 1993. The IMD and WIHG had run one or two analog stations for a short period only. The GSI made a close-spaced seven-station network with an approximate station spacing of 8-10 km. The station spacing was conformable with the depth of the main shock source. The GSI network was continuously run from October 8, 1993 to February 16, 1994, for 132 days till the aftershocks died down (Kayal et al., 1996).

A detailed analysis of the aftershocks recorded by the GSI temporary network was reported by Kayal et al. (1996) and Kayal (2000). Aftershock characteristics, b-value and p-value, were estimated; the estimated b-value 0.46 is typical for a shield area reported in other parts of the world. The aftershock sequence decayed at $t^{-0.27}$ (p-value 0.27) following the largest aftershock of M 5.2 on October 9, 1993 i.e. nine days after the main shock. The low p-value clearly indicates that the decay was much slow compared to the main shock magnitude. The aftershocks of M 4.0-4.8 occurred even one and a half months after the main shock.

Out of about 450 recorded events, 140 aftershocks which were well recorded by all the close-spaced GSI stations were relocated by the *homogeneous station method* (Kayal, 2000). The well located epicentres of the aftershocks are shown in Fig. 7.48a. Most of the aftershocks (77%) are of shallow origin, depth 0-5 km, and rest of the events occurred at a depth range of 6-15 km (Fig. 7.48b). The epicentres are clustered in a small area, 15 km × 12 km, within the meizoseismal zone of intensity VIII (Fig. 7.48a).

It was noted that almost all the felt aftershocks, magnitude ≥4.0, were deeper (depth >5 km) (IMD report), and occurred to the south of Tirna river; the epicentres of the deeper events are aligned almost in E-W direction (Fig. 7.48a). The shallower (depth ≤5 km) aftershocks, on the other hand, are mostly aligned in the NW-SE, along the Tirna tributary. A N-S cross-section of the hypocentres clearly indicates that the deeper shocks are deepening to the south (Fig. 7.48b). The dipping plane, dipping 55°S, meets the surface rupture. The dipping plane corroborates well with the main shock and the deeper aftershocks at a depth range 6-15 km. Above 6 km to surface, the aftershocks are scattered on both sides of this dipping plane, and are wide spread. Using the aftershocks of a limited time period of six days only (October 17-22, 1993), Baumbach et al. (1994) defined a narrow zone of hypocentres dipping 45°SW.

Two composite fault plane solutions were determined for the aftershocks (Kayal, 2000); one for the deeper aftershocks (depth 6-12 km) and the other for the shallower aftershocks (depth $\leq$ 5 km) (Fig. 7.48a). The composite solution of the deeper aftershocks is in agreement with that of the main shock solution. The south dipping E-W nodal plane is the inferred fault plane. This inference is supported by the south dipping plane in the N-S depth section (Fig. 7.48b). It may be noted that the dip angle (60°S) of the inferred E-W fault-plane in the composite solution is compatible with that of the main shock solution (47°S) as well as with the dipping plane (dip 55°S) in the depth section. The shallower aftershocks, on the other hand, show a right-lateral strike-slip faulting (Fig. 7.48a). The near vertical NW-SE nodal plane, which is compatible with the aftershock trend, is the inferred fault plane. The hypocentres of the shallower aftershocks are not aligned along the south dipping plane; these are scattered and may be representing hypocentres on both sides of a vertical fault plane. Gupta et al. (1998) attempted fault-plane solution of five selected aftershocks, depth range 2-9 km, and reported normal, thrust and strike-slip mechanisms. Their results also indicated that all the aftershocks were not produced by similar mechanism of the main shock.

Tomography of the Source Area

About 2700 precise seismic arrivals, 1500 P and 1200 S, were available for seismic tomography study (Kayal and Mukhopadhyay, 2002). The Local Earthquake Tomography (LET) method was used, which improved the locations of the events and imaged the lateral heterogeneity in the velocity structure at depths. The seismic tomography at a depth of 6 km imaged the main shock source area; the main shock occurred at the contact between an east-west trending low velocity zone (LVZ) and a circular high velocity zone (HVZ) to its southwest (Fig. 7.49a). The east-west trending LVZ indicates the seismogenic fault at depth and the main shock occurred by reverse faulting at the fault end where it meets the HVZ. The seismic image at a depth 3 km shows a northwest-southeast trending LVZ, there are other LVZ to its east and south (Fig. 7.49b). The northwest trending LVZ indicates a shallower fault zone along which shallower aftershocks occurred by strike-slip motion (Fig. 7.48c).

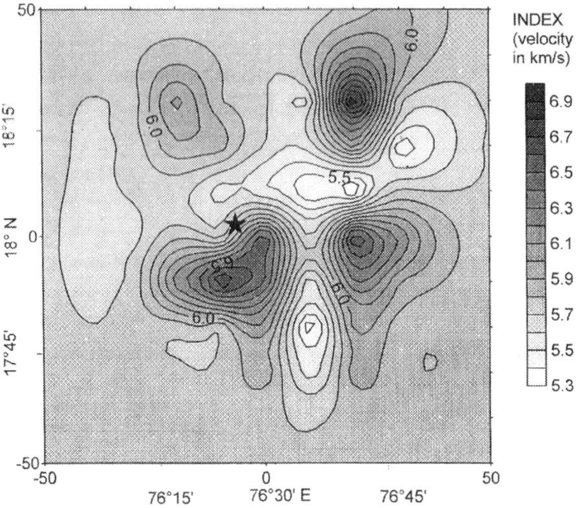

(a) 6 km

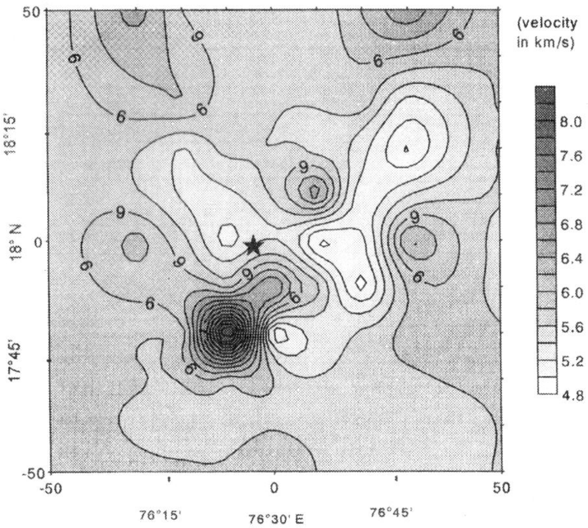

(b) 3 km

Fig. 7.49 Seismic images of the Killari earthquake source area, (a) at 6 km depth, and (b) at 3 km depth (Kayal and Mukhopadhyaya, 2002).

Seismogenic Faults

Based on seismic tomography and fault plane solutions, Kayal and Mukhopadhyay (2002) suggested that the deeper aftershocks (depth 6-15 km) occurred along a south dipping E-W fault plane. This fault plane is correlatable with the Tirna river on the surface (Fig. 7.48a). The fault-plane solution of the shallower aftershocks, on the other hand, shows a near vertical fault plane in the NW-SE direction. This fault plane is comparable with the Tirna tributary as well as with the aftershock trend (Fig. 7.48a).

Detailed gravity surveys were carried out after the main shock in the Killari area (Mishra et al., 1995; Appa Rao et al., 1996); two dominant faults, one in the NW-SE direction and the other in nearly E-W direction, were reported. These observations are very much conformable with the composite fault plane solutions and the observed aftershock trends.

Seeber et al. (1996) described that the Killari earthquake is a unique event on a new generation fault. Based on the deep trenching results, Rajendran et al. (1996), on the other hand, argued that the Killari earthquake occurred as a result of reactivation of a previously existing fault. Post-earthquake magnetotelluric investigation revealed a conductive (~15 Ohm m) zone in the source area (Sarma et al., 1994). Gupta et al. (1998) inferred that a fluid filled rock matrix could be the stress concentrator for the activation of faulting. In a unique experiment, four boreholes, maximum 617 m deep, were drilled by the NGRI and AMD (Atomic Mineral Division) under a DST (Department of Science and Technology) sponsored project to investigate the fault zones in the Killari area (Gupta and Dwivedi, 1996 and Gupta et al., 1998). The boreholes revealed that the basement depth is 347 m in the Killari area. The major outcome is the evidence of slickenside where the boreholes met subsurface rupture. A detailed examination of the lithology indicated a movement of about 6 m. Such large movement, however, cannot be accounted for the Killari earthquake alone. This has further led to the inference that the Killari earthquake occurred due to reactivation of a pre-existing fault (Ramesh and Estabrook, 1998; Gupta et al., 1998, Rajendran and Rajendran, 1999).

A detailed paleoseismological investigation also confirms the occurrence of atleast one earthquake some 2000 ± 200 yrs ago in this fault zone (Sukhija et al., 1998). Rajendran and Rajendran (1997), based on paleoseismological evidences, reported that there was a Killari type earthquake about 1500 years ago at Ter, 40 km northwest of Killari, on a different discrete fault segment measuring 5-6 km on the regional structure. Considering seismic moment, source time duration and the drilling results, Ramesh and Estabrook (1998) advocated pre-existing fault for nucleation, and fracturing of fresh rock right up to the surface. The surface rupture may be caused due to shallow focus with adequate size, thrust mechanism and absence of sediment cover.

Seismotectonic Model

In view of the aftershock trends, fault plane solutions and gravity observations, Kayal (2000) suggested a fault-interaction model (Fig. 7.48c), which was further justified by the seismic tomography study (Kayal and Mukhopadhyay, 2002) (Fig. 7.49). The model explains the reverse faulting for the main shock as well as for the deeper aftershocks at a depth 6-15 km along the major E-W trending fault, and the right-lateral strike-slip faulting for the shallower aftershocks at 0-5 km along the NW-SE trending fault, which intersects the E-W fault at Killari. The main shock was originated by reverse faulting at the intersection of the two faults at a depth 7±1 km, and the deeper aftershocks at 6-15 km depth were also generated by reverse faulting.

The shear adjustment at shallower depth (0-5 km) generated 77% aftershocks of lower magnitude (1.0 to 3.9) by strike-slip mechanism along the NW-SE trending fault, which were triggered by the main shock. No surface traces of these faults are mapped due to thick basaltic cover, but the post-earthquake gravity observations indicate these faults (Mishra et al., 1995). The E-W Tirna river and NW-SE Tirna tributary are possibly the surface geomorphological evidences of these two faults (Kayal, 2000). Further, it may be noted that the NW striking Tirna tributary joins approximately at 120° with the E-W trending Tirna river in Killari; this indicates a strike-slip regime along the tributary. The fault plane solutions of the main shock and the aftershocks consistently reveal a NNE-SSW compressional stress, which is compatible with the regional tectonic stress due to NNE ward movement of the Indian plate.

7.6.2 Jabalpur Earthquake (M 6.0), 1997

Main Shock

The Jabalpur earthquake of May 22, 1997 (M 6.0, M_W 5.8) occurred in the SONATA zone. It was well recorded by the 50-station national network including the 10 GSN (Global Standard Network) stations equipped with broadband instruments; the 10 GSN stations were installed in peninsular India in 1996. This is the first significant event in the Indian peninsular shield area to be well recorded by the newly installed 10-station broadband network. Immediately after the main shock, the IMD reported the hypocentral parameters based on the broadband seismograms. The IMD determined the location of the main shock about 20 km southeast of Jabalpur town (Fig. 7.50). The USGS and the HRV located the main shock using moment tensor analysis based on the global network data. All these locations are compatible. A detailed study of the main shock using the broadband seismograms was reported by Bhattacharya et al. (1997), and the focal depth was well determined at 35±1 km using P_n and sP_n phases. Later, Ramesh and Estabrook (1998) estimated the centroid depth at 37.8±0.1 km using the global broadband network data. It may be mentioned• that there was no *foreshock* for the Jabalpur earthquake either.

Fault plane solution of the main shock was determined by the IMD using the P-wave first motion data (Bhattacharya et al., 1997). It shows a reverse faulting with strike-slip motion (Fig. 7.50). The USGS as well as the HRV determined the fault plane solution by moment tensor analysis. These solutions are compatible with the IMD solution. The southeast dipping ENE trending nodal plane strike: 63°– 83°, dip: 60°–70°, is correlatable with the Narmada South Fault (NSF) and is inferred as the fault plane (Bhattacharya et al., 1997). This inference is further supported by the aftershock investigation (Acharyya et al., 1998). A fault area of about 30 sq km with an average dislocation of about 0.35 m and a stress drop of 7 Mpa were estimated for this event (Bhattacharya et al., 1997).

Macroseismic Effects

The macroseismic investigation immediately after the main shock was carried out by the GSI (Acharyya et al., 1998), and by the NGRI (Gupta et al., 1997). The earthquake was felt over an area having radius of about 350 km. The meizoseismal area, highest intensity VIII, forms an elliptical zone with longer axis measuring 35 km and shorter axis 15 km (Fig. 7.50). All A-type construction (mud houses) suffered total collapse in Kosamghat and Ghana villages in the meizoseismal area. The Jabalpur town with about 25,000 buildings of class A to D comes under the isoseist VIII. Depending upon the foundation as well as soil condition the degree of damage varied. In some parts of the town there was even total collapse of C-type construction. On the other hand, in many areas A-type construction suffered partial collapse or shear cracks only. This observation indicates variations in *site amplifications*. It was, however, very shocking to note that no building was left without any partial damage or crack. Thus in terms of property loss it was more extensive. There were reports of *earthquake sound.* In several villages people have described it as passing train or bulldozer. Although electric supply was disrupted by the earthquake, miraculously the *earthquake sound* was recorded on a cassette which was running on a battery at a marriage party. There were also reports of fissures, ground cracks, groundwater percolation in a dry tributary of the Narmada river, change in groundwater level in wells etc. (Acharyya et al., 1998).

Aftershocks

Only one felt aftershock M 3.5 occurred within five hours of the main shock. This event was recorded by the IMD permanent observatory at Bhopal; the P-arrival time was 08h 55m 26s (local time) on May 22, 1997. This aftershock, however, could not be located as it was not well recorded by the other distant permanent observatories of the IMD (IMD, 1997).

A five-station temporary microearthquake network, with station spacing about 30-35 km, was established by the GSI in the epicentral area, three days after the main shock (Acharyya et al., 1998). The network continued till July 15, as long as the aftershock activity persisted. The station spacing was judged with the knowledge of the known main shock depth at 35 km. The NGRI and WIHG established one or two seismograph stations each in Jabalpur area for a few days, and the IMD established two stations for about a month.

Two severely felt aftershocks (M≥4.0) were recorded by the GSI temporary network as well as by the IMD permanent observatories including the GSN stations of the national network on June 4 and 5 respectively, two weeks after the main shock. The earthquake of June 5 is the largest magnitude (M 4.4) aftershock, which is about 1.6 magnitude unit lower than the main shock. Three mildly felt aftershocks (M 3.0-4.0) were also recorded by the temporary network; one within a week of the main shock, on May 29, and the other two on June 15 and July 2 respectively. After July 2, i.e. after 12 days of the main shock, no felt aftershock (M>3.0) was reported or recorded. In addition to the five felt aftershocks, 23 aftershocks in the magnitude range 1.5-3.0 were recorded by the temporary network during the 52-day survey from May 25 to July 15, 1997. All the 28 aftershocks were relocated by the *homogeneous station method* using at least four precise P-phases and two reliable S-phases (Fig. 7.50) (Kayal, 2000).

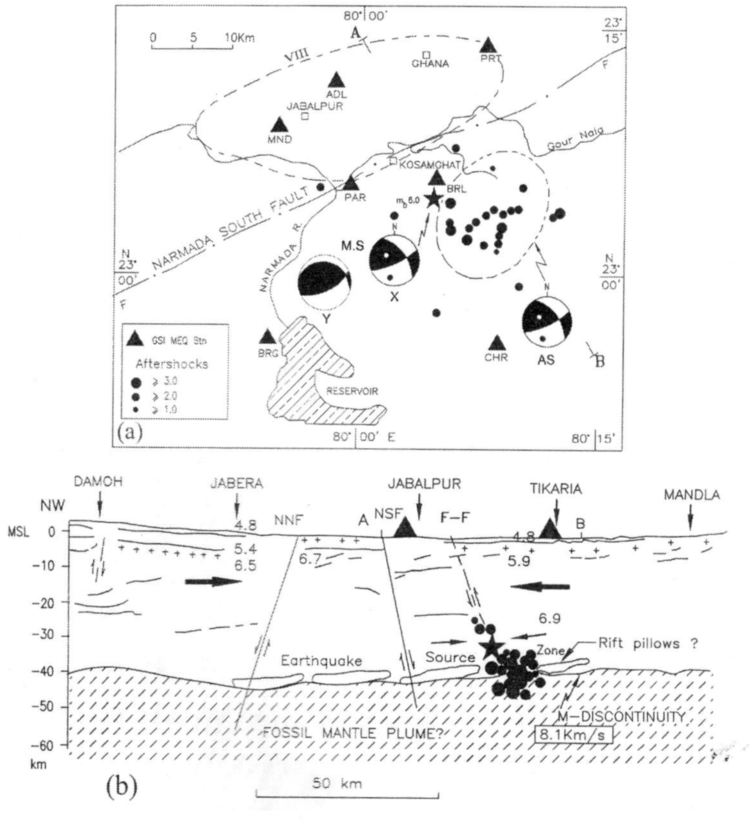

Fig. 7.50 (a) Map showing isoseismal VIII of the 1997 Jabalpur earhquake, temporary seismic stations and epicentres of the main shock and aftershocks; fault plane solutions of the main shock (MS) X: IMD, Y: USGS and aftershock (AS) are shown (Kayal, 2000) and (b) seismic cross section of the main shock and aftershocks (Kayal, 2000), with crustal section of the velocity structure (DSS model by Kaila et al., 1985).

The aftershock characteristics, b-value and p-value, were estimated (Kayal, 2000). The estimated b-value 0.49 is slightly higher than that of the Killari aftershock sequence (0.46). The b-values for both the sequences are, however, very much in the range of SCR events. The attenuating rate of the aftershocks, p-value, is found to be 1.30 which is much higher than that of the Killari aftershock sequence. The fast attenuation of the aftershocks implies that the maximum stress was released by the main shock, and the aftershocks died down quickly in this tectonically active SONATA zone. The observation suggests that the aftershock decay-dynamics is constrained, at least to some extent, by an active fault system; the decay is quicker in an active fault zone (Kayal, 2000). The shorter aftershock activity may also be related to the deeper crustal source. Depth dependency of aftershock production has been reported by Page (1968); shallower events produce larger number of aftershocks compared to deeper events.

A composite focal-mechanism of the aftershocks, which are clustered in the 10 × 15 km elliptical zone near the main shock, reveals the same mechanism as shown by the main shock; it shows a well constrained reverse-fault solution with left-lateral strike-slip motion (Fig. 7.50a). The southeast dipping (dip 67°) ENE-WSW nodal plane (strike 74°) is very much compatible with the inferred fault plane of the main shock solution. The seismic section of the main shock and the aftershocks also indicates a steeply dipping fault plane (Fig. 7.50b). This plane meets the Narmada South Fault (NSF) at the surface. Surface trace of the fault was confirmed by the macroseismic investigation (Acharyya et al., 1998). The focal mechanism solution further reveals a NNE-SSW compressional stress, which is similar to that for the Killari earthquakes.

Seismogenic Fault

Seismic tomography of the source area could not be done due to meagre aftershock data. The Bouguer gravity map, however, reflects the crustal structure in the SONATA zone. Verma and Banerjee (1992) reported chain of gravity highs, 25-30 km wide, running parallel to the SONATA, almost 1000 km long. The residual Bouguer anomaly map shows a major gravity high in Jabalpur region (Fig. 7.8c). The NW-SE DSS profile and the observed gravity anomaly across Jabalpur clearly revealed the two boundary faults, the NNF and NSF (Figs 7.8b and c, and 7.50).

The results of the Jabalpur earthquake investigation indicate that the main shock and the aftershocks were generated by an ENE-WSW trending fault (F-F′) at the NSF system. The macroseismic study confirmed the surface trace of the fault to the south of Jabalpur (Acharyya et al., 1998). The aftershock investigation clearly reveals the presence of this near vertical seismogenic fault, which has been activated at the crust-mantle boundary by a reverse-fault mechanism. It may be noted that the DSS did not indicate the presence of this seismogenic fault (F-F′) that lies about 15 km south of Jabalpur (Fig. 7.50). Nayak (1990), however, identified this fault based on gravity and deep resistivity data. The deep rooted fault F-F′ in the NSF system is very clearly delineated by the aftershock investigation.

In a reactivation hypothesis we may assume that the NSF, which developed initially as a ENE-WSE trending normal fault and had a history of repeated reactivation, appears to have failed under the present compressive regime as a reverse fault with left-lateral strike-slip movement (Fig. 7.50). A 'stationary model', i.e. the concept of old faults controlling centres of rock failure, specially in the paleorifts, may well explain the Jabalpur earthquake. Rajendran and Rajendran (1998) gave a schematic model showing that the 'rift pillows', i.e. excess mass at the crust-mantle boundary in the SONATA zone, may act as a 'stress concentrator', leading to the Jabalpur earthquake. Mahadevan and Subbarao (1999), on the other hand, reported that 'rift cushions' or 'rift pillows' are absent below this zone. They suggested that the Precambrian faults and the numerous feeder dykes may cause lower crustal underplating, which provide a tectonic milieu for the stress concentration. The aftershock investigation indicates that there is no controversy about reactivation of the NSF for generating the Jabalpur earthquake and its aftershocks at the lower crust. The geological evidence of fault gouges (Nair et al., 1985) at the NSF is suggestive of lower shear strength for brittle fracture and reactivation. Therefore, a 'stationary concept', i.e. reactivation of existing fault, the NSF, is preferred over any other model for the Jabalpur earthquake (Kayal, 2000).

Seismically the Narmada rift or the SONATA zone is relatively active. More than 30 earthquakes in the magnitude range 3.0-6.5 are reported from this region during the last 70 years (Rajendran and Rajendran, 2003) (Fig. 7.30). This includes four significant/damaging earthquakes of magnitude ≥ 5.4 as mentioned earlier. The focal depth (35 km) of the Jabalpur earthquake is, however, unusually deeper compared to the common SCR earthquakes. Most SCR earthquakes originate at a focal depth range of 10-15 km, with a few at a deeper (> 25 km) focus (Johnston, 1996). Chen and Molnar (1983) suggested a maximum focal depth of about 25 km for continents older than 800 m.y. However, a good number of deeper events are reported to be associated with ancient rift systems (Stein et al., 1989). The deep focus Jabalpur earthquake may be attributed to the ancient Narmada rift system, and to the reactivation of the NSF in the rifted passive margin. The depth of the 1938 Satpura earthquake (M 6.3) at the southern margin of the rift system (Fig. 7.30) was estimated to be about 40 km (Mukherjee, 1942). These depths are comparable with the deeper SCR earthquakes in the Palaeozoic rift in Central Brazil (Assumpcao and Suarez, 1988), and also with the 2001 Bhuj earthquake in the Kutch rift basin as discussed below.

7.6.3 Idukki (Kerala) Earthquake (M 5.0), 2000

The Kerala state falls in Zone III in the seismic zoning map of India (Fig. 2.20). The 1900 Coimbatore earthquake M 6.0, intensity VII, is the largest known earthquake in this zone (Fig. 7.28). The December 12, 2000 Idukki earthquake (M 5.0) is the recent significant earthquake in this zone, which has been fairly well studied by macroseismic and aftershock investigations.

Main Shock

The December 12, 2000 earthquake M 5.0 (IMD), epicentre at 9°42′.5N and 76°44′.6N, depth 5.8 km (Rastogi, 2001b), occurred ~27 km southwest of the Idukki dam (Fig. 7.51). It caused partial collapse of few houses and minor to small cracks in about 10,000 houses in Idukki and adjoining districts. A macroseismic investigation indicates that the maximum intensity reached VI (MM scale); the isoseismal map is shown in Fig. 7.51. Subterranean sound was heard, and severe shaking was felt in the meizoseismal zone. No foreshock was reported prior to the main shock.

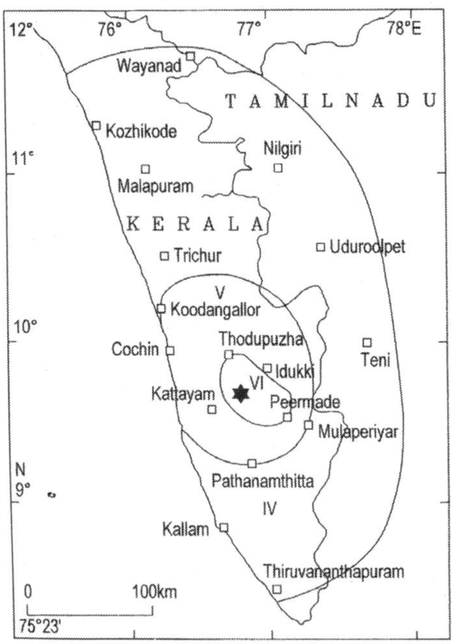

Fig. 7.51 Isoseismal map of the 2000 Idukki earthquake (M 5.0), Kerala; the solid star shows the epicentre of the main shock (Rastogi, 2001b).

Aftershocks

The main shock was followed by an aftershock M 4.8 on January 7, 2001. It was reported that two persons were killed in a wall collapse by this aftershock. The wall was already damaged by the main shock, soaked overnight by heavy rains and collapsed by the strong aftershock (M 4.8). This was the largest aftershock, which was almost of equal magnitude of the main shock.

The GSI made a temporary network to monitor aftershocks. About 50 aftershocks M 1-4.0 were recorded by the GSI network during December 22, 2000 to March 15, 2001 (GSI, 2001; unpub report). Aftershocks prior to December 22 were recorded by the Kerala State Electricity Board (KSEB) permanent network in the Idukki dam area. Aftershocks of two sequences were observed; one sequence is from December 12, 2000 to January 4, 2001 after the main shock, and the other from January 7 to March 15, 2005 after the largest aftershock. The aftershocks of the first sequence decayed at $t^{-0.4}$ (p=0.40) and the second sequence decayed at $t^{-0.54}$ (p=0.54). The b-value was estimated to be 0.40 (GSI, 2001; unpub report). The observed p and b values are in the typical range that is observed in the peninsular India shield area (Kayal, 2000). The aftershocks were reported to be clustered in a small area, 15 km × 15 km, and most of the events occurred within 4-8 km depth. A reverse faulting was inferred from a composite fault plane solution of the aftershocks (GSI, 2001; unpub. report).

Although this event occurred within 30 km of the 169 m high Idukki dam, which started filling in 1974 and impounded in 1982, is not considered as a Reservior Trigger Seismicity (RTS). Rastogi (2001b) argued that this event neither had foreshocks nor long sequence of aftershocks; the sequence had a typical b-value (0.40) of shield seismicity. Normally reservoir triggered main shock is preceded by foreshocks, and followed by longer aftershock sequence of higher b-value.

7.6.4 Bhuj Earthquake (M_w 7.7), 2001

Tectonic Setting

The January 26, 2001 Bhuj earthquake (M_W 7.7) in the Kutch region falls in the highest seismic zone (Zone V) in the seismic zoning map of India (Fig. 2.20). The Kutch, Cambay and Narmada rifts are the three major rift-basins in the western margin of peninsular India craton (Fig. 7.52). These rift basins are bounded by three important Precambrian tectonic trends: the NNW-SSE Dharwar trend, the NE-SW Delhi-Aravalli trend and the ENE-WSW SONATA, which dominate the structural fabric of the region (e.g. Oldham, 1926; Biswas, 1987; Talwani and Gangopadhyay, 2001).

In the Kutch region, the tectonic trend is E-W, along which rifting resulted in the formation of the Kutch Rift Basin (KRB). The NE-SW Aravalli trend to its north, across the Cambay basin, continues into Saurashtra horst to form the southwesterly plunging Saurashtra arch. The Cambay Rift Basin (CRB) is considered to have formed along the northern extension of the Dharwar trend. The Narmada Rift Basin (NRB) extends into the Cambay graben, and continues into the southern part of Saurashtra. All the three rift basins along the Precambrian structural trends originated in different periods during the Mesozoic. The Cambay basin, however, though originated during the Mesozoic, subsided at a greater rate during the Tertiary (GSI, 2000a).

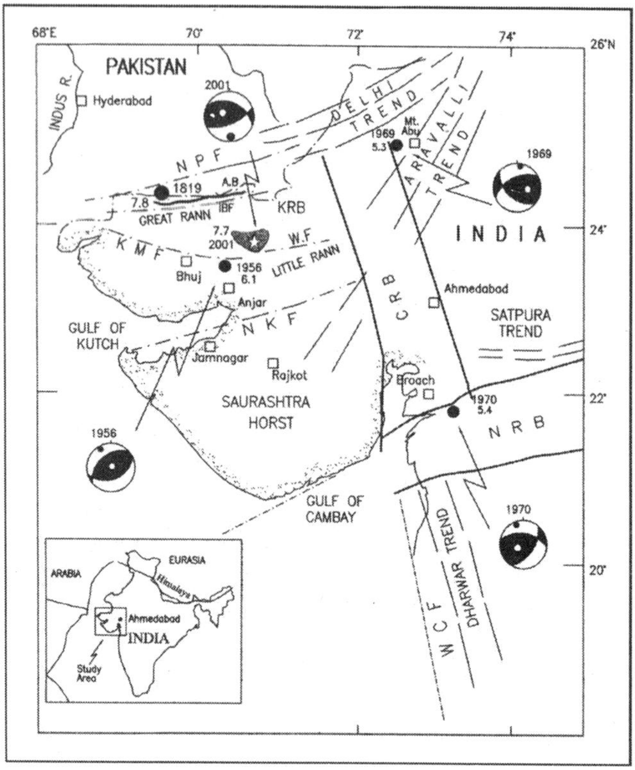

Fig. 7.52 Tectonic map of the western peninsular India showing major tectonic features; KRB: Kutch Rift Basin, CRB: Cambay Rift Basin, NRB: Narmada Rift Basin, NPF: Nagar Parkar Fault, IBF: Island Belt Fault, KMF: Kutch Mainland Fault, NKF: North Kathiwar Fault, WCF: West Coast Fault, AB: Allah Bund (thicker line). Epicentres and fault plane solutions of the past significant earthquakes M> 5.0 are shown, the oval shaped shaded area shows the aftershock area of the 2001 large earthquake (Kayal et al., 2002a). Inset: study area is shown.

Main Shock

The epicentre of the main shock is estimated at latitude 23.4°N and longitude 70.28°E, and at depth 25 km (IMD, 2002). Teleseismic body waves were inverted to understand source mechanism and to map slip on the fault (e.g. Yagi and Kikuchi, 2001; Mori, 2001; Antolik and Dreger, 2003). Source mechanism of the main shock was studied by different groups; the solutions are shown in Fig. 7.53. The inversion of the waveforms studied by the Earthquake Research Institute (ERI, Japan) group (Yagi and Kikuchi, 2001) suggests that the slip occurred on a steeply dipping thrust fault, but it did not break the surface (Bendick et al., 2001). The moment tensor solution reported by the USGS and the P-wave first-motion fault-plane solution reported by the IMD (2002) are comparable with the ERI solution (Fig. 7.53).

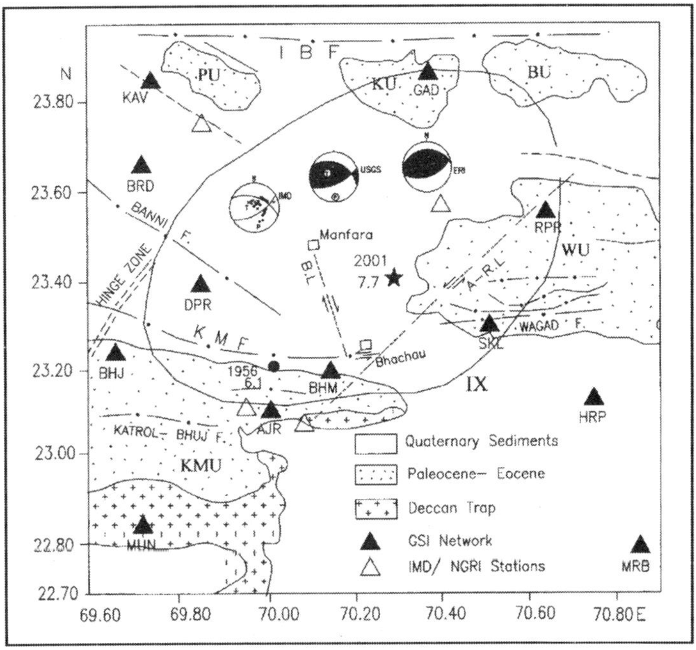

Fig. 7.53 Map showing detailed geology and temporary network in the 2001 Bhuj earthquake epicentre area; fault plane solutions of the main shock obtained by various institutes are shown. ARL: Anjar-Rapar Lineament, BL: Bhachau Lineament, KMU: Kutch Mainland Uplift, WU: Wagad Uplift, BU: Bela Uplift, KU: Khadir Uplift, PU: Pachham Uplift. The high intensity zone, isoseismal IX, is shown (Kayal et al., 2002a).

Details of the macroseismic surveys are given above (Section 7.4.7); no primary rupture was reported. Coseismic secondary fractures were, however, observed; these ruptures cut across the ground surface near the epicentre area with left-lateral and right lateral displacements. Based on the isoseismal geometry, Ravishanker and Pande (2001) suggested that the main shock rupture propagated to the northeast.

Gahalaut and Bürgmann (2004) analysed pre- and post-earthquake satellite images of the epicentral region, and they suggested that maximum uplift is located about 15 km north of the main shock epicentre. Comparison of GPS-1997 and GPS-2001 coordinates at Jamnagar, that lies 150 km south of the main shock epicentre, gives the coseismic displacement vector of 16 ± 8 mm at N35°E; this is the only GPS-GPS based estimate of coseismic slip of the Bhuj earthquake (Jade et al., 2002).

Chandrasekhar and Mishra (2002) and Chandrasekhar et al. (2004), based on repeat microgravity and geodetic measurements, reported maximum uplift of 1.57 ± 0.5 m and a corresponding gravity change of -393 ± 18 μGal in the epicentre region. In a recent work on gravity and magnetic data, Mishra et al. (2005) delineated major E-W and NW-SE oriented faults and deep crustal inhomogeneity in the epicentre region.

Singh et al. (2004) studied source time function (STF) and radiated seismic energy (E_R) using the empirical Green's function (EGF) technique. They reported STF as a function of azimuth and nearly constant duration of 18 sec. The estimated E_R is 2.1×10^{23} erg and the corresponding estimate of slip velocity at the centre of the fault is 156 cm/sec. They also reported a high static stress drop of 200 bar which is a common feature for large intraplate earthquakes.

The main shock was well recorded by 13 strong-motion seismograph stations at a distance from 40 to 290 km from the epicentre (IMD, 2002). The spatial variation of the PGA (Peak Ground Acceleration) was estimated; it was 0.55 g at the nearest (43 km) station at Anjar (rock), and 0.04 g at Anand (alluvium) at a distance of 290 km. The Ahmedabad station (alluvium) recorded 0.13 g at a distance of 240 km. It may be noted that many reinforced concrete frame buildings in the Ahmedabad city collapsed. The earthquake had left behind a large stock of buildings in the Bhuj and adjoining Ahmedabad area, which needed retrofitted and safety evaluation.

Intraplate earthquakes are though much complex, unlike the interplate earthquakes, some parallels have been reported between the 2001 Bhuj earthquake and the 1811-1812 New Madrid earthquakes (M_w 7.5-8.0) that struck the central United States two centuries ago (e.g. Hough et al., 2002; Schweig et al., 2003). Though it is not clear whether the strain rates and the tectonic settings of the two regions are analogous, but both the intraplate Bhuj earthquake of January 26, 2001 and the largest New Madrid event (M 8.0) of February 7, 1812 occurred on paleorift thrust fault(s) in the lower crust, that did not manifest surface rupture (e.g. Jhonston, 1996; Mueller and Pujol, 2001). In a recent reinterpretation, Hough et al. (2002) argued that the magnitudes of these two events might have been similar.

Aftershocks

Aftershock investigations were carried out by several organizations including foreign institutes like USGS and ERI (e.g. Bodin and Horton, 2004; Negishi et al., 2002) and national organisations like GSI, NGRI, IMD etc. A 12-station microearthquake network was systematically operated by the GSI for about two and a half months immediately after the main shock to monitor the aftershocks (Kayal et al., 2002a). The GSI network azimuthally covers the epicentre zone of the main shock region quite well (Fig. 7.53). About 3000 aftershocks (M 1.0~5.0) were recorded by the close-spaced temporary network. The aftershocks were thoroughly scanned and the phase data were fairly sufficient for inversion for precise locations, seismic imaging and for fault-plane solutions (Kayal et al., 2000a and b, Kayal and Mukhopadhyay, 2006).

Preliminary locations of the aftershocks were made using the Hypo-71 program of Lee and Lahr (1975) and the 1-D velocity model estimated by Kayal et al. (2002a). About 800 events ($M \geq 2.0$) were selected, that were recorded by at least four stations, on the basis of preliminary analysis (Kayal et al., 2002a). Out of about 800 events, 560 were relocated by simultaneous inversion (Fig. 7.54a). In the preliminary locations, using the Hypo-71 with the 1D velocity model, the average rms error was 0.5 s for the 800 events, and a large number of events were located with depth fixed at the initial value. In the simultaneous inversion method, the average rms error for the 560 relocated events was reduced to 0.19 s. The average error estimates in latitude, longitude and depth for these events were 1.2 km, 1.1 km and 2.3 km respectively. The relocated epicentres, two selected seismic cross sections and frequency of aftershocks versus depth are shown in Fig. 7.54.

The epicentres of the 560 relocated events show that the main shock as well as most of the aftershocks occurred to the north of the Kutch Mainland Fault (KMF) and within a V-shaped zone delimited by the Bhachau Lineament (BL) to the west and by the Anjar Rapar Lineament (ARL) to the east (Fig. 7.54a). The aftershocks occurred in an area with a maximum length of 70 km in the east-west and 35 km in the north-south. Two intense clusters of aftershocks, one to the northeast and one to the northwest, on both sides of the main shock, are observed (Fig. 7.54a).

Two cross sections of the aftershocks, one along east-west and the other along north-south, are illustrated in Fig. 7.54b and c. In the east-west section, two major clusters of aftershocks flanked both sides of the main shock. In the north-south section a steeply south dipping plane is envisaged (Fig. 7.54c). The frequency of aftershocks versus depth is shown in Fig. 7.54d. It is observed that the base of the seismogenic zone lies at about 37 km, while the main shock occurred at 25 km depth. Most of the aftershocks occurred within 12 to 37 km depth range. A bimodal distribution of the aftershocks, above and below the main shock, is also observed (Fig. 7.54d).

Aftershock characteristics, b-value and p-value, are determined by Kayal et al. (2002a). The estimated b-value for the aftershocks in the magnitude range 3.0~5.0 is found to be 1.21, which is higher than the normal value of 1.0. The activity decayed at $t^{-0.91}$; the p = 0.91 is near normal value 1.0. These two estimated parameters clearly depict that the KRB earthquake characteristics are different from that of the Killari or Jabalpur earthquake sequences in the shield area. The KRB that falls in the zone V in the zoning map of India, is seismically very active, unlike the Jabalpur or Killari area.

Seismogenic Faults

Moment-tensor solution of the main shock (USGS) and 10 composite fault-plane solutions of the best located aftershocks are shown in Fig. 7.55. The

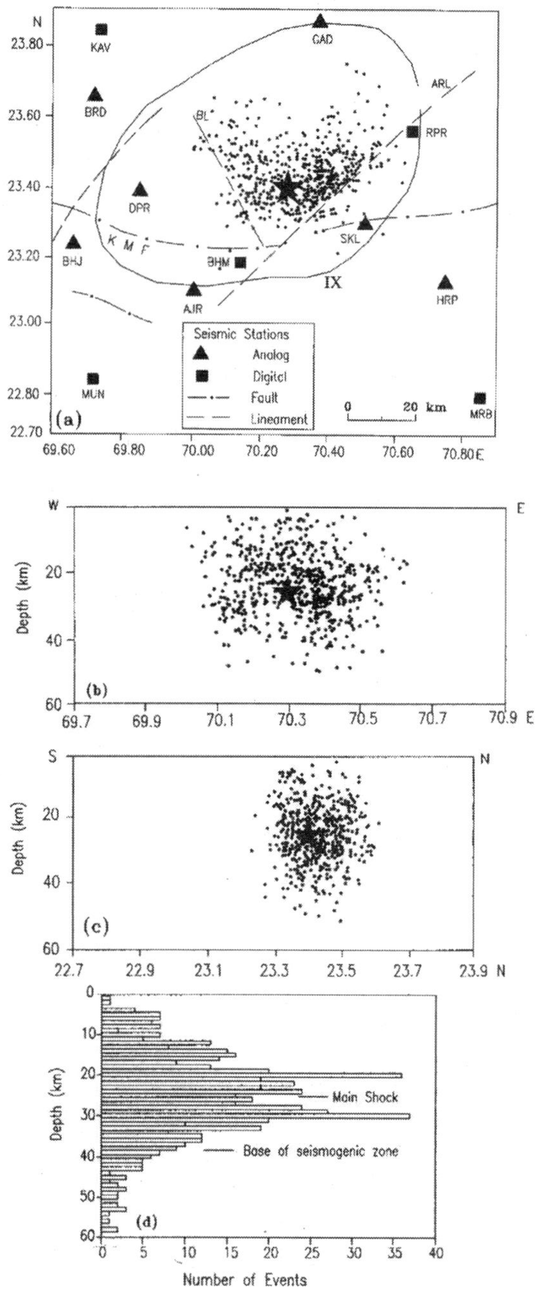

Fig. 7.54 (a) Relocated Bhuj aftershocks (small dots) by simultaneous inversion; main shock is shown by the solid star, major tectonic features and the high intensity zone are shown, (b) E-W depth-section and (c) N-S depth-section of the main shock and aftershocks, and (d) frequency of aftershocks at each 1 km depth interval (Kayal and Mukhopadhyay, 2006).

best located epicentres clearly show two trends of the aftershocks, one northeast and one northwest. The composite fault-plane solutions were obtained for the clusters of aftershocks along the two trends at different depth levels (Kayal et al., 2002a). The fault-plane solutions vary for the two trends, and also vary with depth (Fig. 7.55). The source processes of the aftershocks at two time intervals were also examined; five solutions were studied for the sequence in February, 2001, and five solutions for the sequence in March-April, 2001.

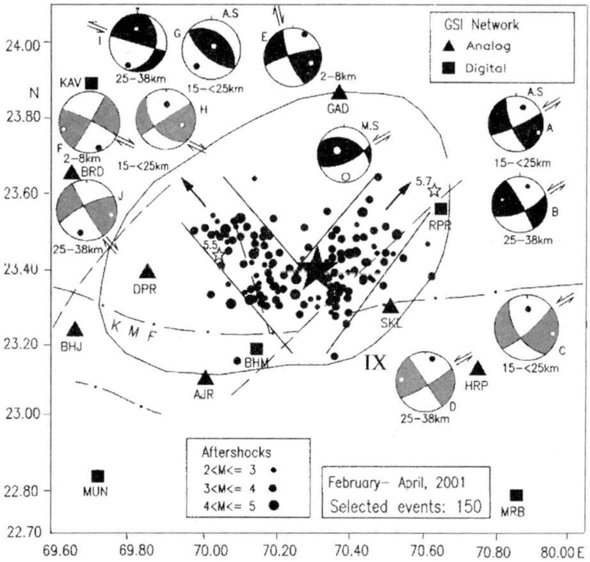

Fig. 7.55 Fault-plane solutions of the main shock (MS) and selected aftershocks (AS) are shown with usual notations; the darker solutions are for the February 2001 sequence and the lighter shades for the March-April 2001 sequence, two largest aftershocks (M 5.7 and 5.5) are shown by open stars, two trends of seismicity are marked, other explanations are same as in Fig. 7.53 (Kayal and Mukhopadhyay, 2006).

The main shock shows a reverse faulting with a left-lateral strike-slip motion along the east-northeast trending south dipping inferred fault plane (Fig. 7.55); the inferred fault plane is supported by the south dipping plane envisaged in the north-south cross section of the aftershocks (Fig. 7.54c).

Four composite solutions (A-D) for the northeast trending aftershocks, that occurred in the mid crust (depth 15-25 km) and in the lower crust (depth 25-38 km), also show reverse faulting with left-lateral strike-slip motion along the northeast trending inferred fault (Fig. 7.55). There is not much change in the fault-plane solutions with time or with depth; the solutions are fairly consistent. It may be noted that there was almost no event at a shallower depth (<10 km) in the northeast for a fault-plane solution.

Six composite solutions (E-J) were obtained for the northwest trending aftershocks (Fig. 7.55). A few upper crustal (depth 2-8 km) aftershocks were recorded in the northwest, two composite fault-plane solutions (E and F) indicate reverse faulting with right-lateral strike-slip motion along the northwest trending inferred fault. There is not much change in the solutions (E-F) with time. In the mid crustal depth (15-25 km), the northwest trending aftershocks recorded during February, 2001, show dominantly reverse faulting (solution G), but the later aftershocks recorded during March-April, 2001, on the other hand, show reverse faulting with right-lateral strike-slip motion (solution H). In the lower crust, at 25-38 km depth, the solution of the aftershocks of February 2001 shows left-lateral strike slip (solution I), and the aftershocks of March-April, 2001 show right-lateral strike-slip (solution J). It may be noted that all the fault-plane solutions show NNE-SSW compressional stress.

Seismic Tomogaphy

A good data set was available for seismic tomography of the Bhuj earthquake source area. A total of 3813 high precision digital data, 1948 P and 1865 S phases were used for the tomography study (Kayal et al., 2002b). The tomographic images at 10, 15 and 20 km depth reveal two LVZs, and a HVZ in between (Fig. 7.56). The two LVZs may be representing two fault/rupture zones. The images at 25 km depth show no LVZ; the main shock source area is identified as a HVZ (Fig. 7.56). It implies that the tectonic stress accumulated at the HVZ, and the main shock nucleated at the fault ends,

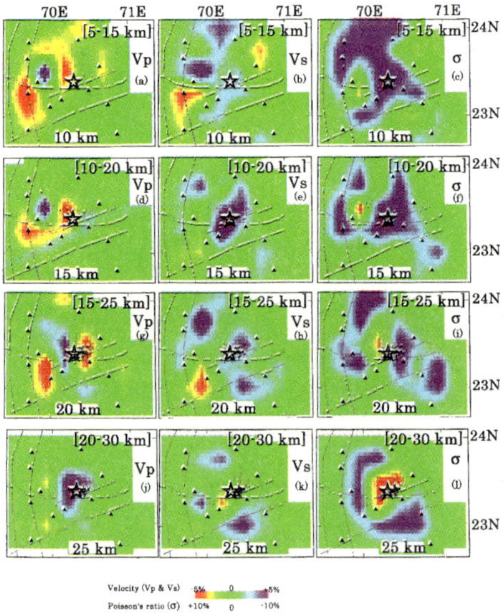

Fig. 7.56 Seismic images (Vp, Vs and Vp/Vs) of the Bhuj earthquake source area (Kayal et al., 2002b).

within the HVZ. The seismic images of the S-waves are more or less conformable with the P-wave images, except that S-wave velocity is lower at the main shock source zone. Further, the source area at 25 km depth is well reflected as a higher Vp/Vs, i.e. higher Poisson's ratio zone. This observation implies that the source zone is partly saturated with fluid (Kayal et al., 2002b).

Seismotectonic Model

The Bhuj earthquake is a paleorift zone earthquake, which occurred at a deeper depth (25 km) compared to common shallower (10-15 km) intraplate events. The deeper Bhuj earthquake source zone in the Kutch rift basin is comparable with that of the deep focus (35 km) 1997 Jabalpur earthquake (M_w 6.0) in the Narmada rift basin in central India (Kayal, 2000). The well recorded large data set of the Bhuj earthquake sequence, however, provided an opportunity for simultaneous inversion for precise locations of the events and for seismic imaging. The large sequence of the aftershocks also provided well constrained fault plane solutions; these were examined with space, time and depth (Fig. 7.55).

The high resolution seismic tomography revealed that the Bhuj earthquake source area falls within the high Vp and high Vp/Vs zone (Fig. 7.56) (Kayal et al., 2002b), and it is an uplifted block (Mishra et al., 2005; Mandal et al., 2004). By waveform inversion, Mori et al. (2001) identified this zone as an asperity, and they argued that the asperity extends both upward and downward from the main shock hypocentre. This suggestion supports the observation of the bimodal distribution of aftershocks (Fig. 7.54d), which indicates that the rupture propagated both upward and downward from the main shock hypocentre.

Negishi et al. (2002), based on aftershock data, reported that the depth range of faulting is 10-35 km; the main shock and the aftershocks occurred on buried fault(s). The deeper source zone of the Bhuj earthquake is very much evident from the intense aftershock activity (81%) at a depth range 12-37 km (Fig. 7.54c). The base of the seismogenic zone is envisaged at 37 km (Fig. 7.54d). Sibson (1986) and Lay and Wallace (1995) suggested that rupture for large crustal earthquake initiates at the base of a seismogenic zone. Such an observation was made for the 1997 Jabalpur earthquake (M_w 6.0) in the Narmada rift basin in central India; it occurred at the base of the seismogenic zone at a depth of 35 km and a downward rupture was reported (Kayal, 2000). The Bhuj earthquake was, however, initiated at a shallower (25 km) depth compared to the base of the seismogenic zone at 37 km. The bimodal distribution of the aftershocks suggests that the Bhuj earthquake rupture initiated in the mid-crust, and it propagated both down-dip near the Moho, and up-dip, but not to the surface.

Further, the aftershock epicentres show a 'V' shaped area (Figs 7.54a and 7.55); a similar observation was reported by Wesnousky et al. (2001). Within the V-shaped aftershock area, two intense clusters of aftershock epicentres were observed, one in the northeast and the other in the northwest of the main shock (Fig. 7.54a); this is more evident with the best located larger events (Fig. 7.55). It may be noted that the IMD (2002) located two largest aftershocks (M 5.7 and 5.5) immediately after the main shock, one to the northeast and the other to the northwest (Fig. 7.55); this observation further supports the suggestion that the main shock rupture propagated to the northeast as well as to the northwest. Although no primary surface evidence of rupture was reported, surface manifestations were observed in the form of liquefaction and secondary ruptures, both in the northeast and in the northwest of the main shock epicentre (Ravishanker and Pande, 2001; Karanth et al., 2001; Rajendran et al., 2002; Wesnousky et al., 2001). Multifractures and fault interactions for large intraplate earthquakes are common (Talwani and Gangopadhyay, 2001), and it is well exemplified by the Bhuj earthquake sequence (Kayal et al., 2002a).

Kayal et al. (2002b) reported a number of crustal blocks based on the seismic velocity images. Mandal et al. (2004) reported uplift of a high velocity block in the epicentre region. A gravity study in the area supports a block tectonics in the region (Mishra et al., 2005) (Fig. 7.57). The regional isostatic anomaly suggests larger thickness of sediments towards south and general uplift of the basement towards north of the KMF, descending stepwise towards the south indicating post-rift vertical tectonics (Chandrasekhar and Mishra, 2002). Based on the analysis of gravity data, Chandrasekhar et al. (2004) and Mishra et al. (2005) further suggested that there are basement uplifts controlled by reverse faults with about 2-5 km up throw in this region.

The fault-plane solutions of the main shock and aftershocks support the upward reverse movement in the source area (Fig. 7.55). The main shock shows reverse faulting with a left-lateral strike-slip motion along the south dipping ENE-WSW trending inferred fault. The northeast trending aftershocks also show reverse faulting with left-lateral strike-slip motion in the mid crust (15-25 km) as well as in the lower crust (25-38 km) along the inferred fault. There is not much change in mechanisms (A-D) of the aftershocks with depth or with time for the northeast trending aftershocks (Fig. 7.55). Kayal et al. (2002a) argued that the northeast trending aftershocks occurred along the main rupture at the main shock depth level. The northwest trending aftershocks, on the other hand, show different solutions; these solutions vary with depth and also with time. At the lower crust (25-38 km), reverse fault solutions with left-lateral strike-slip as well as with right-lateral strike-slip were obtained (solutions I and J respectively). At the mid-crust (15-25 km), a pure reverse faulting and a thrust faulting with right-lateral slip were obtained (solutions G and H respectively). At the upper crust (2-8 km), the two solutions (E and F) show almost consistent right-lateral strike-slip (Fig. 7.55). It was suggested that the aftershocks along the northwest were generated by conjugate fault(s) which show dominantly reverse faulting with right-lateral slip, except one (solution I) at the lower crust (25-38 km) that shows left-lateral motion, and one at the mid-crust that shows pure reverse faulting (solution G) for the sequence in February, 2001. The NNE-SSW compressional axis of all the solutions is consistent with the regional stress due to north-northeast movement of the Indian plate.

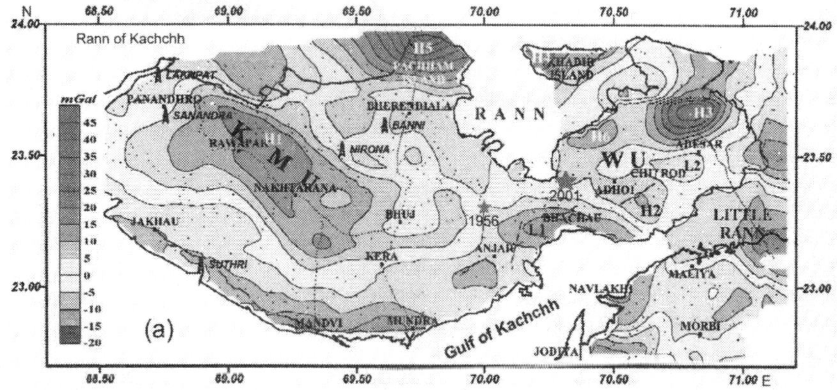

(a)

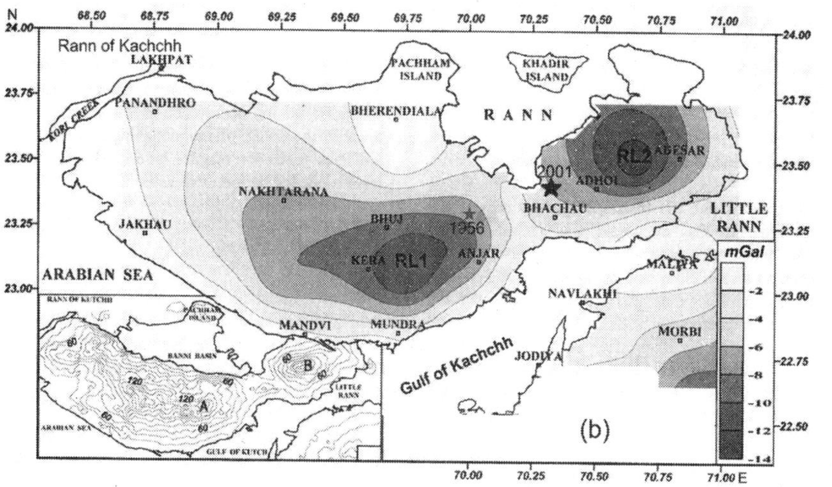

(b)

Fig. 7.57 (a) Residual Bouguer anomaly map of Kutch area; H1 represents Kutch Mainland uplift (KMU), H2 represents Wagad uplift (WU), L1 and L2 local basins, H3, H4 and H5 semicircular/circular gravity highs representing northern part of Wagad, Khadir and Pachham uplifts, respectively. (b) Regional isostatic anomaly corresponding to regional elevation. RL1 and RL2 represent regional gravity lows corresponding to elevated zones of the eastern part of KMU and WU indicating deep seated mass deficiency. Inset shows the elevation in metres; A and B represent KMU and WU, respectively (Mishra et al., 2005).

Ground observations along the two known surface lineaments are compatible with the above fault-plane solutions. Wesnousky et al. (2001) reported a northwest striking 8 km long secondary rupture with a right-lateral strike-slip showing a maximum offset of 32 cm (Fig. 7.58). Ravishanker and Pande (2001), based on the isoseismal geometry (Fig. 7.53), suggested that the main rupture propagated towards northeast, and reported left-lateral ground motion along the ARL (Fig. 7.58). They also reported right-lateral ground motion along the Bhachau lineament (Fig. 7.58). The differential ground movement in the epicentre area is also observed by campaign mode GPS survey (Likhar et al., 2006). Pande et al. (2003) further reported that the isoseismal trend of the intensities IX and VIII corresponds to the Delhi-Aravalli structural grain (Fig. 7.52), and it was suggested that the causative fault for the main shock could be parallel to this Precambrian fold belt. The isoseismals of the 1956 Anjar earthquake also showed similar trend (Fig. 7.33). The isoseismal VII of the 2001 Bhuj earthquake, after maintaining a fair degree of parallelism with the isoseismals in the Saurashtra region, abruptly adopts a N-S trend in the Mainland Gujarat (Fig. 7.34). This conspicuous swing was attributed to the trend of CRB (Fig. 7.52), (Pande et al., 2003). Mishra and Zhao (2003) reported that the fracture along northeast had a higher degree of saturation at the main shock source depth. These observations further support the suggestion that the main rupture propagated along the northeast and a conjugate rupture to the northwest; this observation was illustrated by a fault interaction model (Fig. 7.58) by Kayal et al. (2002a) and Kayal and Mukhopadhyay (2006). It was suggested that these two ruptures caused a V-shaped aftershock area (Figs 7.54a and 7.55).

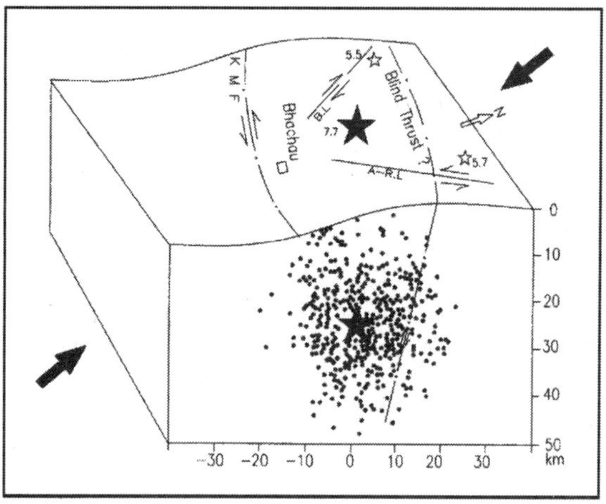

Fig. 7.58 Seismotectonic model of the 2001 Bhuj earthquake sequence; depth section of the events are taken from the Fig. 7.54c; solid arrows indicate the regional tectonic stress (modified by Kayal and Mukhopadhyaya, 2006 from Kayal et al., 2002a).

There is no mapped fault along the surface projection of the inferred seismogenic fault (Negishi et al., 2002; Kayal et al., 2002a). The aftershocks are mostly confined to deeper depths (12-37 km). A similar observation of the 1811-12 New Madrid large earthquake was reported (Hough et al., 2002); these findings suggest that such large intraplate earthquake can occur by hidden fault(s) without producing surface rupture. The main shock and the aftershocks occurred in a high velocity block filled with fluid (?), this is similar to that observed for other large earthquakes in the world (Mendoza and Hartzell, 1988; Zhao et al., 2002). The precise locations and fault-plane solutions provide a new insight into the Bhuj earthquake source processes. The rupture propagation and different fault-plane solutions with depth, space and time are well reflected by the aftershocks.

Lateral heterogeneity as well as complex velocity structure, a high velocity and low velocity contact zone at a shallow depth (7 km ± 1 km), (Fig. 7.49), was reported to be the source area for the 1993 Killari earthquake (m_b 6.3), a damaging intraplate earthquake in central India (Kayal and Mukhopadhyay, 2002). Hauksson and Haase (1994) reported four earthquakes (M>5.9) in the Los Angles basin, California, at the boundary between low and high velocity zones; they interpreted that the high velocity zone forms the upper block of a thrust fault. A similar observation is reported for the 2001 Bhuj earthquake source area (Kayal et al., 2002b; Mandal et al., 2004), and it is supported by gravity and magnetic observations (Chandrasekhar and Mishra, 2002; Chandrasekhar et al., 2004 and Mishra et al., 2005).

Kayal et al. (2002b) inferred that the rocks of lower rigidity or the fluid-filled rock matrix might have contributed to the initiation of the Bhuj main shock. Zhao et al. (2002) reported the role of fluid/magma filled rocks in nucleation of earthquakes in Japan. Mishra and Zhao (2003) reported high crack density, high porosity and high saturation in the Bhuj earthquake source area. Ray (2004), on the other hand, suggested that activation of a fault that causes a large earthquake generates enough flush heat to melt the fragmented rocks at the frictional interfaces. The block structures, velocity heterogeneity and melt rock (?)/fluid filled rock (?) in the source area are, however, imaged after the main shock. The fluid filled (?) or melt rock (?) in source area may be a pre-earthquake or a post-earthquake phenomena. This phenomena cannot be resolved unless both the pre-earthquake (main shock) and post earthquake tomography studies are made.

The Kutch paleorift basin comprises a series of uplifts, master faults and up-thrusts (Fig. 7.53). Biswas (1987) postulated that the uplifts are the result of the differential movements of discrete basement blocks due to compression along these faults. Ravishanker (1995) suggested that differential thermal and crustal structures provide the motive force for relative movements between various crustal blocks under the NNE-SSW compressive stress due to northward movement of the Indian plate. Pre-earthquake and post-earthquake gravity and magnetic study confirmed a block uplift in the source area (Chandrasekhar and Mishra, 2002; Mishra et al., 2005). A coseismic displacement of 16±8 mm at N35°E during the Bhuj earthquake is reported by Jade et al. (2002) from GPS (1997) - GPS (2001) measurement at Jamnagar. The Jamnagar area is about 150 km south of the main shock epicentre, and lies on the hanging wall of the reverse fault that caused the earthquake. This indicates that the post seismic deformation controls this shortening. They further reported shortening of the baseline lengths of five GPS points within the KRB irrespective of their orientation relative to the NNE-SSW compressional direction.

7.7 SRI LANKA: TECTONIC CONFIGURATION

Rifting between India and Sri Lanka

Since the breakup of the Indian subcontinent from Gondwanaland in early Cretaceous, its kinship with Madagascar, Sri Lanka, Antarctica and Australia has been a fascinating geo-scientific problem. It was suggested that unlike the present geographical positions, southeast India and northwest Sri Lanka were once a single landmass (Katz, 1974, 1978a, b). He suggested two proto-boundary faults with a separation of about 200 km between the two landmasses, which earlier closed up the Gulf of Mannar and Palk Strait (Fig. 7.59).

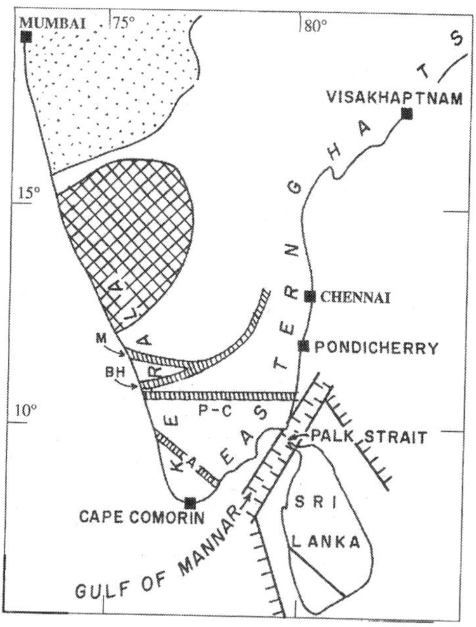

Fig. 7.59 Present position of Sri Lanka with respect to India. M: Moyar, BH: Bhavani, P-C: Palaghat-Cauvery and A: Achankovil shear zones respectively (Agarwal and Pandey, 1999).

Yoshida et al. (1992), however, pointed out the juxtaposition of Sri Lanka with Lutzow-Holm Bay of Antarctica (Fig. 7.60), that required 80 km northward movement and 30° clockwise rotation of Sri Lanka from its present position. This assembly shows a good correlation of geological units among peninsular India, Sri Lanka and east Antarctica. It was hypothesized that the Palk Strait is a triple junction that joined the east coast of India at Cape Comorin in the west and Pondicherry in the east.

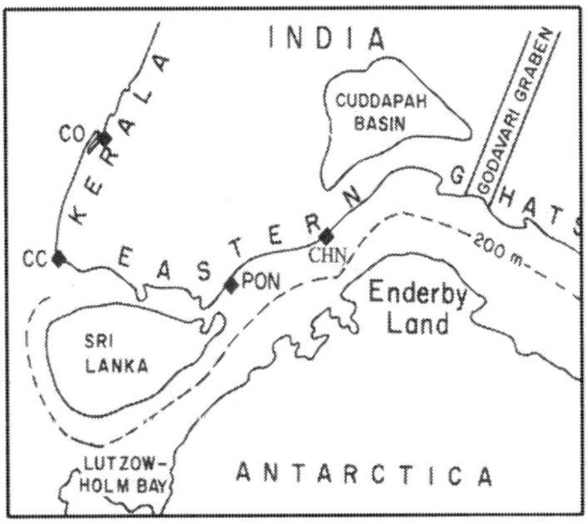

Fig. 7.60 Pre-break up position of Sri Lanka with respect to Antarctica and India (Yoshida et al., 1992). CO: Cochin, CC: Cape Comorin, PON: Pondicherry and CHN: Chennai (Madras).

Volcanism

Continental break up are normally associated with surface magmatism (Courtillot et al., 1999); for example Rajmahal traps reflect the break up of India from Antarctica at ~115 Ma. Cretaceous flood basalts of Madagascar are related to the break up of Madagascar from Dharwar craton, south India at ~88 Ma (Storey et al., 1995), Deccan traps to the break up of Seychelles from west coast of India at ~65 Ma. No such volcanism is, however, reported with the break up (?) of Sri Lanka from India. The 200 m isobath around the east coast of India and Sri Lanka, on the other hand, indicates non-oceanic crust between the two.

Gravity Anomaly

Based on Bouguer gravity anomaly (Fig. 7.61), Agarwal and Pandey (1999) suggested that the gentle nature of anomalies with a few closures near the east coast of India does not support any rifting or a fault below the east coast area. The continuation and one to one correspondence of gravity contours between Sri Lanka and India, on the other hand, suggest an identical nature of crust or lithosphere below the two areas.

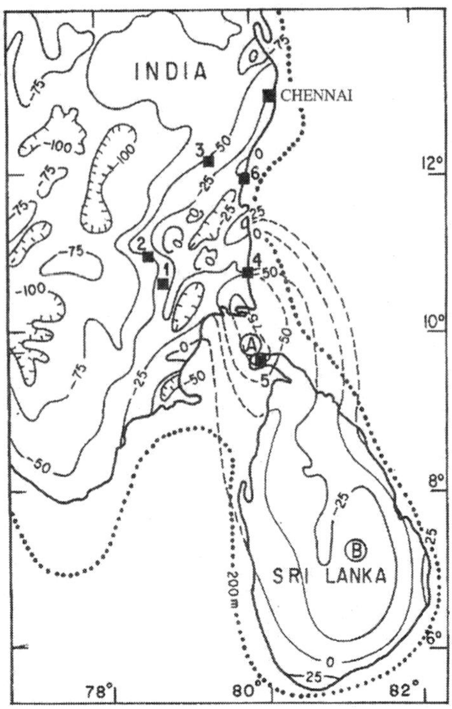

Fig. 7.61 Bouguer anomaly (mGal) map of southeast Indian shield and Sri Lanka (Agarwal and Pandey, 1999). 1. Nartamali, 2. Tiruchirapalli, 3. Gingee, 4. Nagapattinam, 5. Jaffna peninsula, 6. Pondicherry. A: gravity anomaly over Palk Srait, B: gravity anomaly over eastern Vijayan series, Sri Lanka.

The gravity anomaly -75 mGal marked as A over the Palk Srait, between Nagapattinam in the east of India and Jaffna peninsula of Sri Lanka strikes almost right angle to the east coast of India. Agarwal and Pandey (1999) suggested that this could be the result of asthenosheric upwelling during the break up between Antarctica and India at ~130 Ma and Marion hot spot activity at ~90Ma. Another anomaly of -25 mGal, marked as B (Fig. 7.61), over eastern Vijayan group of rocks may be a local anomaly, and does not seem to have any apparent relationship with the break up or rifting.

Magnetic Anomaly

Reddi et al. (1988) prepared a total intensity map of peninsular India region, south of 12° latitude (Fig. 7.62). They adopted 2D low pass filtering that resulted in a predominantly longer wavelength essentially reflecting the deeper structures. The gentle nature of the magnetic contours does not suggest any rift structure at the east coast.

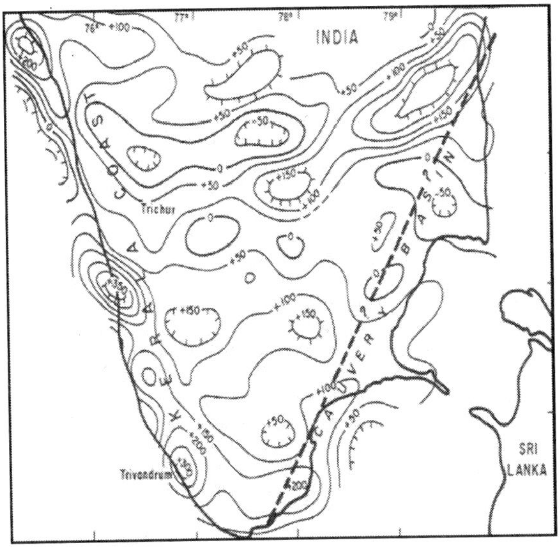

Fig. 7.62 Long wave length total intensity magnetic map of south India (Reddi et al., 1988); thick dashed line indicates protoboundary fault proposed by Katz (1978a).

In the west coast, on the other hand, the trends are clearly suggestive of faulting parallel to the Kerala coast. It may be worth mentioning that some of the NW-SE mega-lineaments in south India seem to be extending to Sri Lanka (Fig. 7.63).

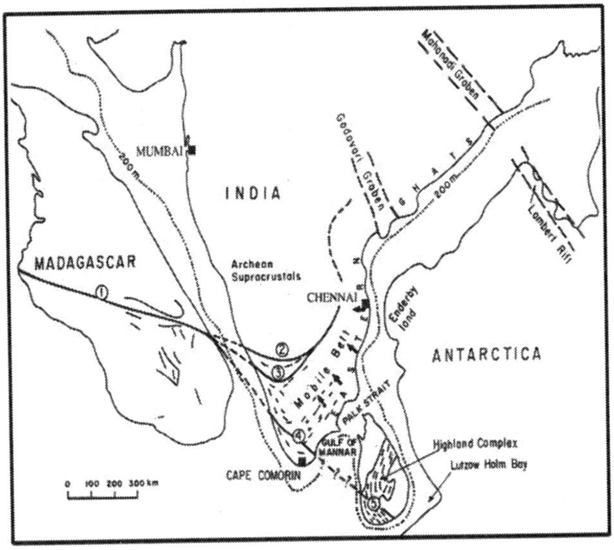

Fig. 7.63 Correlations between Madagascar, India, Sri Lanka and Antarctica (Agarwal et al. 1992). 1. Bongolava-Ranotsara lineament, 2. Moyar shear zone, 3. Bhavani shear zone, 4. Achankovil shear zone, 5. major fault/lineament in Sri Lanka (Katz, 1978b, Vitanage, 1986).

Seismicity

Seismicity of south India (Rao, 1992) and microearthquake epicentres in Sri Lanka (Fernando and Kulasinghe, 1986) are shown in Fig. 7.64. The proposed protoboundary faults over the two landmasses are also shown. No seismicity is observed along the proposed protoboundary faults. A higher degree of seismicity is, observed at the latitude 12-14°N in south India, which may be due to structural heterogeneities in the area. In Sri Lanka, there is not much seismic activity; microseismicity is observed in the geologically complex southern part, that is bounded by two lineaments, and practically no seismicity is observed in the north. Only one permanent BB station (part of the GSN) is now in operation in the southern part of Sri Lanka since last one decade. The ISC catalog recorded some five earthquakes M 4.5-5.0 in the last 100 years, one in the main land of Sri Lanka (Fig. 7.64) and rest in the surrounding ocean floor of Sri Lanka, which could be attributed to intraplate earthquakes in this region.

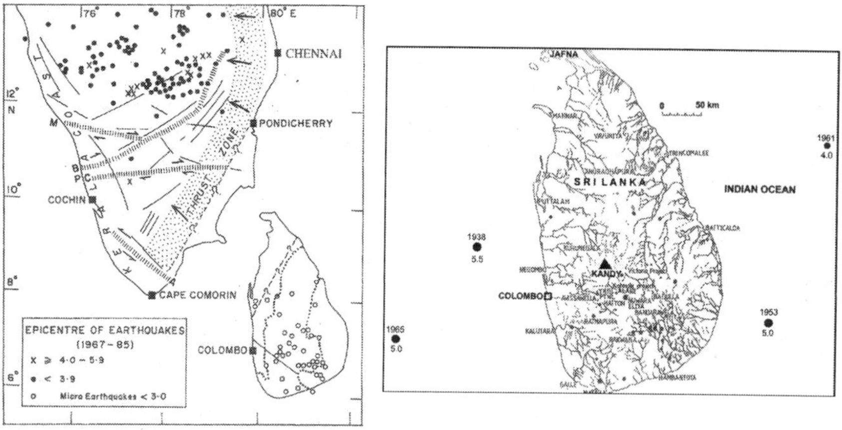

Fig. 7.64 (a) Epicentral distribution of earthquakes in southern India (Rao, 1992) and in Sri Lanka (Fermando and Kulasinghe, 1986). (b) Geomorphological map of Sri Lanka with epicentres of earthquakes (M > 4.0) in and around Sri Lanka (Agarwal and Pandy, 1999).

Tectonic Interpretaion

The above evidences suggest that Sri Lanka remained a part of super peninsular India, occupying the similar position as it was having in the erstwhile Gondwanaland about 180 Ma ago (Fig. 7.60). The two fragments accompanied each other throughout their geodynamic mobility after they broke away from Antarctica at ~130 Ma ago. The paleo-assembly seems reasonable as evidenced by the present geological, tectonic and geophysical information (Fig. 7.63). The magsat and gravity data suggest that Madagascar was a part of Dharwar craton of south India and the Bhavani lineament matches well with the Bongolava-Ranotsara lineament in Madagascar (Fig. 7.63). Similarly, there is a possibility that the Lambert rift (Antarctica) joins the Mahanadi graben in India (Fig. 7.63). Burke and Dewey (1973b) anticipated the presence of triple junctions at Gulf of Mannar and Cambay, and it matches well with the concept of Marion and Reunion plume outburst at about 89 Ma and 65 Ma respectively which resulted in the separation of Madagascar and Seychelles from greater India.

It appears that before the break up of the Gondwanaland, the island of Sri Lanka rested in the Lutzow-Holm Bay, and its paleo-position with respect to India may have been same as it is now, although a small scale rotational movement may not be ruled out as suggested by Yoshida et al. (1992). Sri Lanka and India travelled together to its present position, the available geological and geophysical data do not support rifting between the two landmasses or existence of protoboundary faults. Seismically the island of Sri Lanka typically falls in the category of SCR. Although there is no historical record of large earthquake in Sri Lanka, which is limited only for 100-150 years, occurrence of strong intraplate earthquake M ~6.0 in Sri Lanka cannot be ruled out. Recurrence period of such SCR earthquake is normally of the order of 1000 years, and we have no historical data to be sure that such event never happened in Sri Lanka unless verified by some paleoseismological studies.

7.8 CONCLUSION AND RECOMMENDATION

The Indian peninsular shield is no longer considered to be a 'stable land' or seismically inactive as was shown in the seismic zonation map earlier (Fig. 2.19). The devastating earthquakes in the recent past, like the 1967 Koyna earthquake (M 6.7), 1993 Killari earthquake (m_b 6.3), 1997 Jabalpur earthquake (M 6.0) and the most devastating 2001 Bhuj earthquake (M_w 7.7) have radically changed the seismic zonation map of India (Fig. 2.20).

Significantly enhanced activity of moderate magnitude earthquakes (M $\simeq$ 6.0) in recent years compared to that in the past century (Gaur, 2003) called for modernisation of the national network (IMD) and for enough fund and instruments for digital microearthquake networks, stand alone seismic stations/temporary networks around the country. In this chapter, it has been attempted to highlight the results of temporary stations/network data to shed light in understanding of the intraplate seismicity in peninsular India.

Among the four damaging earthquakes mentioned above, the 1967 Koyna and the 1993 Killari earthquakes are the typical shield earthquakes, originated at shallower depth (<10 km) by reactivation of thrust/reverse fault(s). Seismic tomography study revealed crustal heterogeneities, and it is observed that the existing faults had been the source areas for these earthquakes.

The 1997 Jabalpur and the 2001 Bhuj earthquakes, on the other hand, are found to be generated at deeper depths (25-35 km), at the base of the paleo-rift zones, NRB and KRB respectively, by reactivation of the existing surface faults and/or hidden faults. The normal faults of these rifts developed in the geological past are now reactivating as reverse faults due to accumulation of compressional stress at the fault end by the plate movements.

In addition to the detailed aftershock study and review of a few strong earthquakes M~6.0 in peninsular India, several studies on RTS, active faults, swarm activity are made by microearthquake networks, that have shed more light on the earthquake processes in different tectonic blocks of the shield area.

The results of these investigations revealed that the SCR seismicity occur in the vicinity of stress concentrators in the pre-existing hidden fault zones and failed rift zones. The stress concentrators include intersecting faults, heterogeneity in velocity structures at the fault end, rift pillows etc. The SCR seismicity have longer return periods compared to their plate boundary counterparts. Talwani and Gangopadhyay (2003) studied about 40 SCR earthquakes around the world and suggested that the earthquakes occur in failed rifts of all ages, whereas those not associated with failed rifts occur only in the Precambrian crust. The failed rift earthquakes are deeper (~20-30 km) and have shorter return periods (~500 years), the Precambrian crust events, on the other hand, are shallower (~10 km) and have longer return periods (~1000 years).

Our knowledge in seismotectonics is, however, limited to a few earthquakes in peninsular India; much remains to be done. Our present national network (IMD) is capable to locate earthquakes M≥4.0. The capability on the national network needs to be increased to locate earthquakes down to M 3.0. This would require more close-spaced (<100 km) seismic stations in the country compared to the present 150-200 km spacing of the stations in the national network. In the earthquake prone/high risk zones, telemetry systems need to be implemented and in some areas the existing networks need to be upgraded. Permanent digital telemetry and stand alone digital seismic network data need to be analysed by R&D (Research and Development) effort more rigorously not only for tectonic study but also for precursor study.

The seismic tomography study is only limited to the Killari and Bhuj earthquake source areas. We did very little to understand the deep crustal structure of the continent except a few studies on receiver function analysis. We need to have high precision large quantity digital data set of the whole country to bring out detailed seismic images at least down to the mantle depth for identification of the earthquake source areas for hazard mitigation. This is an important task for the future seismologists. There are now about 50 permanent BB stations in peninsular India that are run by various organizations for about five to six years, these data are although used to some extent for receiver function analysis, no effort is made yet for detailed seismic tomography study to identify the earthquake source areas beneath the shield area. Seismic tomography may shed light on the hidden seismogenic structures, fault(s), velocity heterogeneities, stress concentrators etc. beneath the crust. This should be a challenge for the future seismologists.

Several international projects are now taken up for seismological studies in the country. Our future research should focus not only on the seismotectonics of the earthquakes but also on the crustal deformation studies by the GSP measurements, microzonation, seismic hazard studies etc. Real time monitoring of earthquakes is very much essential; our national network needs to be upgraded for this purpose immediately. Along with high precision seismic data/networks, earthquake precursor studies should be taken up as a challenging task in future. Multi-parameter observatory should be established in seismically active zones to monitor seismological, geophysical, geochemical and geodetic precursors. Finally, it is an indisputable fact that although we develop modernisation or scientific knowledge to understand earthquake genesis or its source areas, it is the adequate engineering knowledge and capabilities to construct or design human habitats that would provide less human loss. Earthquakes do not kill people, it is the poorly constructed houses that kill people. We have had enough experience of this ignorance in 1993 Killari earthquake or in 1997 Jabalpur and 2001 Bhuj earthquakes. Even if the scientists are able to answer when and where the next earthquake might strike, it may save human lives, but not the man-made properties, houses and buildings, which are not properly constructed. Knowledge of seismology and seismotectonics may shed light on our understanding of earthquakes, but cannot prevent the earthquakes to occur.

References

Abercrombie, R.E., Antolik, M. and Ekstrom, G., 2003. The June 2000 Mw 7.9 earthquake south of Sumatra: Deformation in the India-Australia plate. *J. Geophys. Res.*, **108(B),** 2018, doi: 10.1029/2001JB 000674.

Acharyya, S.K., 1986. Tectono stratigraphic history of Naga hill ophiolites. *Geol. Surv. India Mem.*, **119,** 94-103.

Acharyya, S.K., Ghosh, S.C. and Ghosh, R.N., 1983. Geological framework of the eastern Himalaya in parts of Kameng, Subansiri, Siang districts, Arunachal Pradesh. *Geol. Surv. India Misc. Pub.*, **43,** 145-152.

Acharyya, S.K., Mitra, N.D. and Nandy, D.R., 1986. Regional geology and tectonic setting of northeast India and adjoining region. *Geol. Surv. India Mem.*, **119,** 61-72.

Acharyya, S.K., Kayal, J.R., Roy, A. and Chaturvedi, R.K., 1998. Jabalpur earthquake of May 22, 1997: Constraint from aftershock study. *J. Geol. Soc. India*, **51,** 295-304.

Adams, R.D., 1979. T-phase recordings at Rarotonga from underground nuclear explosion. *Geophys. J .R. Astro. Soc.*, **58,** 361-369.

Agarwal, B.N.P., Das, L.K., Chakraborty, K. and Sivaji, CH., 1995. Analysis of the Bouguer anomaly over Central India: A regional perspective. *Geol. Soc. India Mem.*, **31,** 469-493.

Agarwal, P.K. and Pandey, O.P., 1999. Was there an intercontinental rift between India and Sri Lanka? *J. Geol. Soc. India*, **54,** 237-249.

Agarwal, P.K., Pandey, O.P. and Negi, J.G., 1992. Madagascar: A continental fragment of the paleosuper Dharwar craton of India. *Geology*, **20,** 543-546.

Agarwal, R.K., 1977. Structure and tectonics of Indo Gangetic Plains – Geophysical case histories of India. *J. Asoc. Expl. Geophys.*, **1,** 29-46.

Airy, G.B, 1855. On the computations of the effect of the attraction of the mountain Manes as disturbing the apparent astronomical latitude of stations in geodetic surveys. *Phil. Trans. Roy. Soc.* London, **14,** 101.

Aki, K., 1965. Maximum-likelihood estimate of b in the formula log N = a – bM and its confidence limits. *Bull. Earthquake Res. Inst.*, Tokyo Univ., **43,** 237-239.

Aki, K., 1966. Earthquake generating stress in Japan for the years 1961 to 1963 obtained by smoothing the first motion radiation patterns. *Bull. Earthquake Res. Inst.*, Tokyo Univ., **44,** 447-471.

Aki, K., 1988. Local site effects on strong ground motion. *Proc. Earthquake Engg. Soil Dyn.*, 103-155.

Aki, K., and Lee, W.H.K., 1976. Determination of three dimensional velocity anomalies under a seismic array using first P arrival times from local earthquakes: A homogeneous initial model. *J. Geophys. Res.*, **81,** 4381-4399.

Aki, K. and Richards, P.G., 1980. *Methods of Quantitative Seismology*, Freeman, California.

Aki, K., Christoffersson, A. and Husebye, E.S., 1977. Determination of three dimensional seismic structure of the lithosphere. *J. Geophys. Res.*, **82,** 277-296.

Allegre, C.J. et al., 1984. Structure and evolution of the Himalaya-Tibet Organic Belt. *Nature*, **307,** 17-22.

Ambraseys, N.N., 2000. Reappraisal of north Indian earthquakes at the turn of the 20th century. *Curr. Sci.*, **79,** 101-114.

Ambraseys, N.N. and Bilham, Roger, 2003. Reevaluated intensities for the Great Assam earthquake of 12th June 1897, Shillong, India. *Bull. Seism. Soc. Am.*, **93(2),** 655-673.

Ambraseys, N.N. and Douglas, J., 2004. Magnitude calibration of north Indian earthquakes. *Geophys. J. Int.*, **159,** 165-206.

Ammon, C.J., 1991. The isolation of receiver effects from teleseismic P waveforms. *Bull. Seism. Soc. Am.*, **81,** 2504-2510.

Ammon, C.J., Ji, C., Thio, H., Robinson, D., Ni, S., Hjorteifsdottir., Kanamori, H., Lay, T., Das, S., Helmberger, D., Ichinose, G., Polet, J. and Wald, D., 2005. Rupture process of the 2004 Sumatra-Andaman earthquake. *Science*, **308,** 1133-1139.

Anderson, E.M., 1951. *The dynamics of faulting and dyke formation with applications to Britain*. 2nd ed., Oliver and Boyd, London.

Anderson, J.A. and Wood, H.O. 1925. Description and theory of the torsion seismometer. *Bull. Seism. Soc. Am.*, **15,** 1-72.

Angulo-Brown, F., Ramirez-Guzman, A.H., Yepez, E., Rudiof-Navarroa, A. and Pavia-Miller, C.G., 1998. Fractal geometry and seismicity in the Mexican subduction zone. *Geofisica International*, **37,** 29-33.

Antolik, M. and Dreger, D.S., 2003. Rupture process of the 26 January 2001 Mw 7.6 Bhuj, India earthquake from teleseismic broadband data. *Bull. Seism. Soc. Am.*, **93,** 1235-1248.

Appa Rao, M., Sastri, C.B.K. and Reddy, R.A., 1996. Gravity, magnetic and resistivity investigations around Killari. *Geol. Surv. India Sp. Pub.*, **37,** 193-214.

Argand, E., 1924. La Tectonic de I'Asie. *Int. Geol. Cong. Rep.*, **13(1),** 170-732.

Armijo, R., Tapponnier, P., and Han, T., 1989. Late Cenozoic right-lateral strike-slip faulting in southern Tibet. *J. Geophys. Res.*, **94,** 2787-2838.

Armijo, R., Tapponnier, P., Mercirr, J.L. and Han, T., 1986. Quaternary extension in southern Tibet: Field observations and tectonic implications. *J. Geophys. Res.*, **91,** 13803-13872.

Armstrong, B.H., 1969. Acoustic emission prior to rock bursts and earthquakes. *Bull. Seism. Soc. Am.*, **59,** 1259-1279.

Arora, B.R. and Singh, B.P., 1992. Geomagnetic and geoelectric investigations for seismicity and seismotectonics of the Himalayan region. *In:* Himalayan Seismicity, G.D Gupta (ed.), *Geol. Soc. India Mem.*, **23,** 1-21.

Arya, A.S., 2000. Recent development towards earthquake risk development. *Curr. Sci.*, **79,** 1270-1277.

Asada, T., 1957. Observations of nearby microearthquakes with ultrasensitive seismometers. *J. Phys. Earth*, **5,** 83-113.

Assumpcao, M. and Suarez, G., 1988. Source mechanisms of moderate size earthquakes and stress orientation in mid-plate South America. *Geophys. J. Int.*, **92,** 253-267.

Atchuta Rao, D., Sanker Narayan, P.V. and Hari Narain, 1970. A study of the aeromagnetic profiles across the Cuddapah basin, *Geophys. Res. Bull.*, **8,** 11-25.

Avouac, J.P. and Tapponier, P., 1993. Kinemetic model of active deformation in central Asia. *Geophys. Res. Lett.*, **20,** 895-898.

Avouac, J.P., Tapponier, P., Bai, M., You, H. and Wang, G., 1993. Active thrusting and folding along the northern Tien Shan and Cenozoic rotation of the Tarim relative to Dzingaria and Kazakhstan. *J. Geophys. Res.*, **98,** 6755-6804.

Avouac, J.P., Ayoub, F., Leprince, S., Konca, O. and Helmberger, D.V., 2006. The 2005 Mw 7.6 Kashmir earthquake: Sub-pixel correlation of ASTER images and seismic waveform analysis. *Earth Planet Sci. Lett.*, **249,** 514-528.

Bacon, C.F., 1975. Acoustic emission along San Andreas fault in southern central California. *Calif. Geol.*, **28,** 147-154.

Baksi, S.K., 1965. Stratiography of Barail series in southern part of Shillong Plateau, Assam, India. *Am. Assoc. Pet. Geol. Bull.*, **49,** 2282-2294.

Bakun, W.H. and Lindh, A.G., 1977. Local magnitudes, seismic moments and coda durations for earthquakes near Oroville, California. *Bull. Seism. Soc. Am.*, **67,** 615-629.

Balakrishnan, S., Christopher, G. and Rama Rao, A.V., 1967. Regional magnetic and gravity studies over Cuddapah basin. Proc. UMP Symp., Hyderabad, GRB & NGRI Publ. No. 8, pp. 303-319.

Banerjee, B. and Shaw, R.K., 2001. Marine magnetic anomalies near the Barren Islands volcano, Andaman sea. *Curr. Sci.*, **81,** 816-818.

Banerjee, P. and Bürgmann, R., 2002. Convergence across the northwest Himalaya from GPS measurements, *Geophys. Res. Lett.*, **29(13),** 30.1-30.4.

Banerjee, P., Pollitz, F.F. and Burgmann, R., 2005. The size and duration of the Sumatra-Andaman earthquake from far-field static offsets. *Science*, **308,** 1769-1772.

Banghar, A.R., 1990. Mechanism solutions of two recent Indian earthquakes. Symposium on Recent Advances in Seismology and their Applications, Abs. vol., Bangalore, India, July 16-19.

Banghar, A.R., 1991. Mechanism solution of Nepal-Bihar earthquake of August 20, 1988. *J. Geol. Soc. India*, **37,** 25-30.

Bapat, A., Kulkarni, R.C. and Guha, S.K., 1983. Catalogue of earthquakes in India and neighbourhood from historical record up to 1979. Indian Soc. Earthq. Tech., Roorkee, 211 p.

Baranowski, J., Armbruster, J., Seeber, L. and Molnar, P., 1984. Focal depths and fault plane solutions of earthquakes and active tectonics of the Himalaya. *J. Geophy. Res.*, **89,** 6918-6928.

Barazangi, M. and Ni, J., 1982. Velocities and propagation characteristics of Pn and Sn beneath the Himalayan arc and Tibetan Plateau: Possible evidence for understanding of the Indian continental lithosphere beneath Tibet. *Geology*, **10,** 179-185.

Båth, M., 1973. *Introduction to Seismology*, Birkhauser Verlag. Basel, 355 p.

Båth, M., 1979. *Introduction to Seismology*, Birkhauser Verlag. Basel, 428 p.

Båth, M. and Shahidi, M., 1974. T phase from Atlantic earthquakes. *Pure and Appl. Geophys.*, **94,** 74-114.

Baumbach, M., Grosser, H., Schmidt, H.G., Paulat, A., Rietbrock, A., Rao, C.V.R., Raju, P.S., Sarkar, D. and Mohan, I. 1994. Study of the foreshocks and aftershocks of the intraplate Latur earthquake of September 30, 1993, India. *Geol. Soc. India Mem.*, **35,** 33-63.

Behera, L., Sain, K. and Reddy, P.R., 2004. Evidence of underplating from seismic and gravity studies in the Mahanadi delta of eastern India and its tectonic significance. *J. Geoph. Res.*, **109,** B12311, doi: 10.1029/2003 JB 002764.

Beloussov, V.V., Belyaevsky, N.A., Borsov, A.A., Volvovsky, B.S., Volvovsky, I.S., Resvoy, D.P., Talvirsky, B.B., Khamrabaev, I.Kh., Kaila, K.L., Hari Narain, Marussi, A. and Finetti, J., 1980. Structure of the lithosphere along the deep seismic sounding profile. Tien-Shan Pamirs-Karakorum-Himalayas. *Tectonophysics*, **70,** 193-221.

Ben Menahem, A., Aboodi, E. and Schild, R., 1974. The source of the great Assam earthquake – An interplate wedge motion. *Phys. Earth Planet. Inter.*, **9,** 265-289.

Bender, F., 1983. Geology of Burma. Borntrager, Berlin, 293 p.

Bendick, R., Bilham, R., Fielding, E., Gaur, V.K., Hough, S.E., Keir, G., Kulkarni, M.N., Martin, S., Mueller, K. and Mukul, M., 2001. The 26 January "Republic Day" Earthquake, India. *Seism. Res. Lett.*, **72(3),** 328-335.

Bevis, M. and Isacks, B.L., 1984. Hypocentral trend surface analysis: Probing the geometry of Wadati-Benioff zones. *J. Geophys. Res.*, **89,** 6153-6170.

Bhatia, S.C., Kumar, R. and Gupta, H.K., 1999. A probabilistic seismic hazard map of India and adjoining regions (GSHAP). *Curr. Sci.*, **77(3),** 447-450.

Bhattacharya, Pankaj M. and Kayal, J.R., 2003. Mapping the b-value and its correlation with the fractal dimension in the northeast region of India. *J. Geol. Soc. India*, **62,** 680-695.

Bhattacharya, Pankaj M., Majumdar, R.K. and Kayal, J.R., 2002. Fractal dimension and b-value mapping in northeast India. *Curr. Sci.*, **82(12),** 1486-1491.

Bhattacharya, P.M., Pujol, J., Majumdar, R.K. and Kayal, J.R., 2005a. Relocation of earthquakes in the northeastern region using joint hypocentre determination method. *Curr. Sci.*, **89,** 1404-1413.

Bhattacharya, P.M., Mukhopadhyay, S., Majumdar, R.K. and Kayal, J.R., 2005b. Crustal velocity structure studies in the northeast India by joint hypocentre determination and simultaneous inversion. Proc. Seminar *East Crust*, Indian School of Mines, Dhanbad.

Bhattacharya, S.N., 1974. The crust-mantle structure of the Indian peninsula from surface wave dispersion. *Geophys. J.R. Astr. Soc.*, **36,** 273-283.

Bhattacharya, S.N., 1981. Observation and inversion of surface wave group velocities across central India. *Bull. Seismol. Soc. Am.*, **71,** 1489-1501.

Bhattacharya, S.N., 1992. Crustal and upper mantle velocity structure of India from surface wave dispersion. *Curr. Sci.*, **62,** 94-100.

Bhattacharya, S.N. and Dattatrayam, R.S., 2003. Some characteristics of recent earthquake sequences in peninsular India. *Gond. Geol. Magz.*, **5,** 67-85.

Bhattacharya, S.N. and Kayal, J.R., 2005. Seismicity of the Himachal Himalaya: Constraint from local seismic network. *Geol. Surv. India Sp. Pub.*, **85,** 71-79.

Bhattacharya, S.N., Prakash, C. and Srivastava, H.N., 1986. Microearthquake observations around Thein dam in northwest Himalayas. *Phys. Earth Planet. Inter.*, **44,** 169-178.

Bhattacharya, S.N., Ghosh, A.K., Suresh, G., Baidya, P.R. and Saxena, R.C., 1997. Source parameters of Jabalpur earthquake of May 22, 1997, *Curr. Sci.*, **73,** 855-863.

Bilek, S.L., 2007. Using earthquake rupture variations along the Sumatra-Andaman subduction system to examine fault zone variations. *Bull. Seism. Soc. Am.*, **97,** 502-570.

Bilham, R., 2001. Slow tilt reversal of the Lesser Himalaya between 1862 and 1992 at 78° E and bounds to the southeast rupture of the 1905 Kangra earthquake. *Geophys. J. Int.*, **144,** 1-23.

Bilham, R., 2005. A flying start, then a slow slip. *Science*, **308,** 1126-1127.

Bilham, R. and England, Phillip, 2001. Plateau 'pop up' in the great 1897 Assam earthquake. *Nature*, **410,** 806-809.

Bilham, R. and Hough, S.E., 2006. Future earthquakes on the Indian continent inevitable hazard, preventive risk. *South Asian J.*, **12,** 1-9.

Bilham, R., Blume Frederick, Bendick Rebecca and Gaur, V.K., 1998. Geodetic constraints on the Translation and deformation of India: Implications for future great Himalayan earthquakes. *Curr. Sci.*, **74,** 213-229.

Bilham, R., Gaur, V.K. and Molnar, P., 2001. Earthquakes: Himalayan seismic hazard. *Science*, **293,** 1442-1444.

Bilham, R., Engdahl, E.R., Feldl, N. and Satyabala, S.P., 2005. Partial and complete rupture of the Indo-Andaman plate boundary 1847-2004. *Seism. Res. Letts.*, **76,** 299-311.

Billings, M.P., 1954. *Structural Geology*. 2nd ed., Prentice Hall, New York, 514 p.

Billington, S., Isacks, B.L. and Barazangi, M., 1977. Spatial distribution and focal mechanisms of mantle earthquakes in the Hindu Kush-Pamir region: A contorted Benioff zone. *Geology*, **5,** 699-704.

Bird, P., 1978. Interaction of intra-continental subduction in the Himalaya. *J. Geophys. Res.*, **83,** 4975-4987.

Bird, P. and Toksöz, M.N., 1977. Strong attenuation of Rayleigh waves in Tibet. *Nature*, **266,** 161-163.

BIS, 2004. Seismic Zoning Map of India. Bureau Indian Standard publication, New Delhi, India.

Biswas, S.K., 1987. Regional tectonic framework, structure and evolution of the western marginal basins of India. *Tectonophysics*, **135,** 307-327.

Bisztricsany, E., 1958. A new method for determination of the magnitude of earthquakes. *Geofiz Kozlemen*, **7,** 69-96.

Bodin, P. and Horton, S., 2004. Source parameters and tectonic implications of aftershocks of the Mw 7.6 Bhuj earthquake of 26 January 2001. *Bull. Seism. Soc. Am.*, **94,** 818-827.

Bolt, B.A., 1999. *Earthquakes, A Primer*, 4th edition. W.H. Freeman NY, 366 p.

Borman, P., 1971. Location of teleseismic events by means of body-wave records at station Moxa (in German: Ortung teleseismischer Ereignisse aus Raumwellenregistrierungen der Station Moxa), *Monatberichte der ADW*, **13,** 10-12, 847-852.

Bott, M.H.P. and Dean, D.S., 1972. Stress systems at young continental margins. *Nature*, **235,** 23-25.

Bowin, C., 1973. Origin of the Ninety East Ridge from the studies near the equator. *J. Geophys. Res.*, **78,** 6029-6043.

Brown, L., Zhao, W., Nelson, K., Hauck, M., Alsdorf, D., Ross, A., Cogan, M., Clark, M., Liu, X. and Che, J., 1996. Bright spots, structure and magmatism in southern Tibet from INDEPTH seismic reflection profiling. *Science*, **274,** 1688-1690.

Brune, J.N., 1970. Tectonic stress and the spectra of seismic shear waves from earthquakes. *J. Geophys. Res.*, **75,** 4997-5009.

Brune, J.N., 1976. The physics of earthquake strong motion. *In:* Seismic Risk and Engineering Decisions, C. Lomnitz and E. Rosenblueth (eds.). Elsevier, Amsterdam, pp. 141-177.

Brune, J.N. and Allen, C., 1967. A microearthquake survey of the San Andreas fault system in southern California. *Bull. Seism. Soc. Am.*, **57,** 277-296.

Brune, J.N. and Singh, D.D., 1986. Continent-like crustal thickness beneath the Bay of Bengal sediments. *Bull. Seism. Soc. Am.*, **76,** 191-203.

Bullen, K.E., 1939. The crustal structure of New Zealand region as inferred from the study of seismic waves. Proc. 6th Pacific Sci. Cong., 103-110.

Bullen, K.E., 1965. *An introduction to the theory of seismology* (3rd ed.), Cambridge Univ. Press, London and New York.

Bullen, K.E. and Bolt, B.A., 1985. *An introduction to the theory of seismology* (4th ed.). Cambridge Univ. Press, U.K.

Burdick, L.J. and Langston, C.A., 1977. Modeling crustal structure through the use of the converted phases in teleseismic body waveforms. *Bull. Seismol. Soc. Am.*, **67,** 677-691.

Burke, K. and Dewey, J.F., 1973a. Plume-generated triple junctions: Key indicators in applying plate tectonics to old rocks. *J. Geophys. Res.*, **81,** 406-433.

Burke, K. and Dewey, J.F., 1973b. An outline of Precambrian plate movement. *In:* D.H. Tarling and S.K. Runcorn (eds.), Implications of Continental Drift to the Earth Sciences, v. 2, Academic Press, London, pp. 1035-1045.

Burke, K., Dewey, L.F., Edelstein, A., Kidd, W.S.F., Nelson, K.D., Sengor, A.M.C. and Stroup, J., 1978. Rift and sutures of the world (NASA-CR-175201) compiled for Goddard Space Flight Center, Greenbelt, Maryland. Albany Global Tectonics Group, State Univ. New York, Albany, NY, 238 p.

Burrard, S.G., 1915. Origin of the Indo-Gangetic trough, commonly called Himalayan foredeep. *Proc. Roy. Soc. Lond.*, **91A,** 220-238.

Burtman, V.S. and Molnar, P. 1993. Geological and geophysical evidence for deep subduction of continental crust beneath the Pamir. Tech. Rep Sp. Paper 281, Geol. Soc. Am.

Calhaem, I.M., Haines, A.J. and Lowry, M.A., 1977. An intermediate-depth earthquake in the central region of the South Island used to determine a local crustal thickness. *N. Z. J. Geol. Geophys.*, **20,** 353-361.

Catchings, R.D. and Kohler, W.M., 1996. Reflected seismic waves and their strong shaking during the 1989 Loma Prieta, California earthquake. *Bull. Seism. Soc. Am.*, **86,** 1401-1416.

Chadha, R.K., 1994. Estimation of T-waves in the Indian Ocean between Srilanka and Southern India. *Pure and Appl. Geophys.*, **142(2),** 319-328.

Chadha, R.K., 1995. Role of dykes in induced seismicity at Bhatsa reservoir, Maharashtra, India. *Pure and App. Geophys.*, **145,** 155-165.

Chadha, R.K., Gupta, H.K., Kumpel, H.J., Mandal, P., Nageswara Rao, A., Narendra Kumar, Radhakrishna, I., Rastogi, B.K., Raju, I.P., Sarma, C.S.P., Satyamurthy, C. and Satyanarayana, H.V.S., 1997. Delineation of active faults, nucleation process and pore pressure measurements at Koyna (India). *Pure and Appl Geophys.*, **150,** 551-562.

Chander, R., 1988. Interpretation of observed ground level changes due to the 1905 Kangla earthquake, northwest Himalaya. *Tectonophysics*, **149,** 289-298.

Chander, R., 1989. Southern limits of major earthquake ruptures along the Himalaya between 75° and 90° E. *Tectonophysics*, **170,** 115-123.

Chandra, U., 1975. Seismicity, earthquake mechanisms and tectonics of Burma, 20°N-28°N. *Geophys. J. Roy Astro. Soc.*, **40,** 367-381.

Chandra, U., 1977. Earthquakes of peninsular India – A seismotectonic study. *Bull. Seism. Soc. Am.*, **67,** 1387-1413.

Chandra, U., 1978. Seismicity, earthquake mechanisms and tectonics along the Himalayan mountain range and vicinity. *Phys. Earth Planet. Inter.*, **16,** 109-131.

Chandra, U., 1992. Seismotectonics of Himalaya. *Curr. Sc.*, **62,** 40-71.

Chandrasekaran, A.R., 1994. Evaluation of design earthquake parameters for a site and utilization of strong motion data. *Curr. Sci.*, **67,** 353-357.

Chandrasekaran, A.R. and Das, J.D., 1990. Strong motion arrays in India and characteristics of recent recorder events. *Bull. Ind. Soc. EQ. Tech.*, **27,** 1-66.

Chandrasekaran, A.R. and Das, J.D., 1992. Strong motion arrays in India and analysis of data from Shillong array. *Cur. Sci.*, **62,** 233-250.

Chandrasekaran, A.R. and Das, J.D., 1994. Analysis of strong motion accelerograms of northeast India earthquake of August 6, 1988. *Inst. Engg. Rep.*, **75,** 1-11.

Chandrasekhar, D.V. and Mishra, D.C., 2002. Some geodynamic aspects of Kutch basin and seismicity: An insight from gravity studies. *Curr. Sci.*, **83(4),** 492-498.

Chandrasekhar, D.V., Mishra, D.C., Singh, B., Vijayakumar, V. and Burgmann, R., 2004. Source parameters of the Bhuj earthquake, India of January 26, 2001 from height and gravity changes. *Geophys. Res. Lett.*, **31,** L 19608, doi: 10.1029/2004 GLO 20768.

Chang Ta, 1963. The geology of China. U.S. Deptt. of Commerce, Joint Publication Research Service, Washington DC., 548 p.

Chatelain, J.L., Roecker, S.W., Hatzfeld, D. and Molnar, P., 1980. Microearthquake seismicity and fault plane solutions in the Hindukush region and their tectonic implications. *J. Geophys. Res.*, **85,** 1365-1387.

Chatterjee, S.N. and Bhattacharya, S.N., 1992. Microearthquake surveys and seismicity in western Himalaya. *Geol. Soc. India Mem.*, **23,** 23-44.

Chaudhury, H.M. and Chatterjee, S.N., 1981. Microearthquake survey in the hot spring area in Parbati valley, Himachal Pradesh, India. *J. Volc. Geothermal Res.*, **9,** 29-30.

Chen, W.P. and Kao, H., 1996. Seismotectonics of Asia: Some recent progress. *In:* The Tectonic Evolution of Asia, A. Yin and M. Harrison (ed). Cambridge Univ. Press, New York, pp 37-62.

Chen, W.P. and Molnar, P., 1977. Seismic moments of major earthquakes and the average rate of slip in central Asia. *J. Geophys. Res.*, **82,** 2945-2969.

Chen, W.P. and Molnar, P., 1981. Constraints on the seismic wave velocity structure beneath the Tibetan plateau and their implications. *J. Geophys. Res.*, **86,** 5937-5962.

Chen, W.P. and Molnar, P., 1983. Focal depths of intracontinental and intraplate earthquakes and their implications for the thermal and mechanical properties of the lithosphere. *J. Geophys. Res.*, **88,** 4183-4214.

Chen, W.P. and Molnar, P., 1990. Source parameters of earthquakes and intraplate deformation beneath the Shillong Plateau and northern Indo-Burma ranges. *J. Geophys. Res.*, **95,** 12,527-12,552.

Choubey, V.D., 1971. Narmada-Son Lineament, India. *Nature*, **232,** 38-40.

Choudhury, S.K., 1975. Gravity and crustal thickness in Indo-Gangetic plains and Himalaya region, India. *Geophys. J. R. Astro Soc.*, **40,** 441-452.

Choudhury, S.K. and Dutta, A.N., 1973. Bouguer gravity and its geological evaluation in the western part of the Bengal basin and adjoining area, India. *Geophysics*, **38,** 691-700.

Choudhury, S.K. and Datta, A.N., 1975. Crustal thickness in north India and Himalayan region and its geological significance. *Geophys. Res. Bull.*, **13,** 29-37.

Chugh, R.S., 1974. Study of recent crustal movements in India and future programs, *Int. Symp. Recent. Crustal Movements*, Abs. vol., pp. 20, Zürich.

Chun, K.Y., 1986. Crustal block of the western Ganga basin: A fragment of oceanic affinity. *Bull. Seism. Soc. Am.*, **76,** 1687-1698.

Chun, K.Y. and Yoshi, T., 1977. Crustal structure of the Tibet plateau: A surface wave study by a moving window analysis. *Bull. Seism. Soc. Am.*, **67,** 735-750.

Chung, Wai-Ying, 1993. Source parameters of two rift-associated intraplate earthquakes in peninsular India: Bhadrachalam earthquake of April 13, 1969 and Broach earthquake of March 23, 1970. *Tectonophysics*, **225,** 219-230.

Chung, Wai-Ying and Gao, H., 1995. Source parameters of the Anjar earthquake of July 21, 1956, India, and its seismotectonic implications for the Kutch rift basin. *Tectonophysics*, **242,** 281-292.

Claerbout, J.F., 1985. *Imaging the Earth's Interior*. Blackwell Scientific Publications, Oxford, 398 p.

Condie, K.C., 1982. *Plate Tectonics and Crustal Evolution.* 2nd ed. Pergamon Press, New York, 288 p.

Cook, K.L., Algermissen, S.T. and Costain, J.L., 1962. The status of Ps converted waves in crustal studies. *J. Geophys. Res.*, **67,** 4769-4778.

Cormack, A.M., 1963. Representation of a function by its line integral with some radiological applications. *J. Appl. Geophys.*, **34,** 2722-2727.

Cotton, F., Campillo, M., Deschamps, A. and Rastogi, B.K., 1996. Rupture history and seismotectonics of the 1991 Uttarkashi, Himalayan earthquake. *Tectonophysics*, **258,** 35-51.

Courtillot, V., Jaupart, C., Manighetti, I., Tapponnier, P. and Besse, J., 1999. On causal links between flood basalts and continental breakup. *Earth. Planet. Sci. Lett.*, **166,** 177-195.

Crawford, A.R., 1974. The Indus Suture line, the Himalaya, Tibet and Gondwanaland. *Geol. Mag.*, **11,** 369-383.

Crone, A.J., Machette, M.N. and Bowman, J.R., 1992. Geologic investigations of the 1988 Tennant Creek, Australia, earthquakes – Implications for paleoseismicity in stable continental regions. *U.S. Geol. Surv. Bull.* 2032-A, U.S. Govt. Printing Office, Washington, D.C., 51 p.

Crosson, R.S., 1972. Small earthquakes, structure and tectonics of the Puget Sound Region. *Bull. Seism. Soc. Am.*, **62,** 1171-1182.

Crosson, R.S., 1976a. Crustal structure modeling of earthquake data Part 1: Simultaneous least squares estimation of hypocenter and velocity parameters. *J. Geophys. Res.*, **81,** 3036-3046.

Crosson, R.S., 1976b. Crustal structure modeling of earthquake data. Part 2: Velocity structures of the Puget Sound Region, Washington. *J. Geophys. Res.*, **81,** 3036-3046.

Curray, J.R., 1989. The Sunda Arc: A model for oblique plate convergence. *J. Sea Res.*, **24,** 131-140.

Curray, J.R., 2005. Tectonics and history of the Andaman Sea region. *J. Asian Earth Sci.*, **25,** 187-228.

Curray, J.R. and Moore, D.G., 1971. Growth of the Bengal deep sea fan and denudation in the Himalayas. *Bull. Geol. Soc. Am.*, **82,** 563-572.

Curray, J.R. and Moore, D.G., 1974. Sedimentary and tectonic process in the Bengal deep-sea fan and geosyncline. *In:* The Geology of Continental Margins, C.A. Burke and C.L. Drake (eds.), Springer-Verlag, New York, pp. 617-628.

Curray, J.R., Moore, D.G., Lawver, L.A., Emmel, F.J., Raitt, R.W., Henry, M. and Kieckhefer R., 1979. Tectonics of the Andaman Sea and Burma. *In:* Geological and Geophysical Investigations of Continental Margins, J.S. Watkins, L. Montadert and P. Dickenson (eds), *Am. Asoc. Pet. Geol. Mem.*, pp. 189-198.

Curray, J.R., Emmel, F.J., Moore, D.G. and Raitt, R.W., 1982. Structure tectonics and geological history of the northeastern Indian Ocean. *In:* The Ocean Baisns and Margins, vol. VI. The Indian Ocean, A.E.M. Nairn and F.G. Stehli (eds), pp. 399-450, Plenum, New York.

Das, D., Mehra, G., Rao, K.G.C., Roy, A.L. and Narayana, M.S., 1979. Bouguer, Free Air and Magnetic anomalies over northwestern Himalayas. *Geol. Surv. India Misc. Pub.*, **41,** 141-149.

Das, K.K. and Viswanathan, S., 1964. On the possibility of the existence of a Cuddapah rift valley in the Penganga–Godavari basin. *Bull. Geol. Soc. India,* **1(2),** 1-6.

Das, L.K. Agarwal, B.N.P. and Bose, R.N., 1996. Moho relief in central India from spectral analysis of gravity data. *Geol. Soc. India Mem.*, **36,** 225-234.

Dasgupta, A.B., 1977. Geology of Assam–Arakan region. *Quart. J. Geol. Min. Met. Soc. India*, **49,** 1-53.

Dasgupta, S. and Mukhopadhyay, M., 1993. Seismicity and plate deformation below the Andaman arc, northeastern Indian Ocean. *Tectonophysics*, **225,** 529-542.

Dasgupta, S., Mukhopadhyay, M. and Nandy, D.R., 1987. Active transverse features in the central portion of the Himalaya. *Tectonophysics*, **136,** 255-264.

Dasgupta, S., Mukhopadhyaya, M., Bhattacharya, A. and Jana, T.K., 2003. The geometry of the Burmese-Andaman subducting lithosphere. *J. Seismol.*, **7,** 155-174.

De Paolo, D.J., Stopler, E.M. and Thomas, D.M., 1991. Physics and Chemistry of mantle plumes. *EOS.* Vol. 72, No. 21, May 21, 1991, Am. Geophys. Union. pp. 236-237.

De, Reena and Kayal, J.R., 1990. Crustal P-wave velocity and velocity-ratio study in northeast India by a microearthquake survey. *Pure and Appl. Geophys.*, **134,** 93-108.

De, R. and Kayal, J.R., 2003. Seismotectonic model of the Sikkim Himalaya: Constraint from microearthquake surveys. *Bull. Seism. Soc. Am.*, **93(3),** 1395-1400.

De, Reena and Kayal, J.R., 2004. Seismic activity at the MCT in Sikkim Himalaya. *Tectonophysics*, **386,** 243-248.

De, Reena, Chowdhury, S.N. and Banerjee, D., 2000. Delineation of an active fault by a MEQ survey in Indian peninsular shield area, Munger, Bihar. Abs Vol., Second International Seminar and Exhibition, Geophysics Beyond 2000, AEG, Hyderabad, November 15-20, pp 47-48.

DeMets, C., Gordan, R.G., Argus, D.F. and Stein, S., 1990. Current plate motions. *Geophys. J. Int.*, 101, 425-478.

DeMets, C., Gordan, R.G., Argus, D.F. and Stein, S., 1994. Effect of recent revisions to the geomagnetic reversal time scale on estimates of current plate motions. *Geophys. Res. Lett.*, **21,** 2191-2194.

Desikachar, S.V., 1974. A review of the tectonic and geological history of eastern India in terms of plate tectonics theory. *J. Geol. Soc. India*, **15,** 137-149.

Dewey, J.F. and Bird, J.M., 1970. Mountain belts and the new global tectonics. *J. Geophys. Res.*, **75,** 2625-2647.

Dewey, J.F., Shackleton, R.M., Chang, C.F. and Sun, Y.Y., 1988. The tectonic evolution of the Tibetan plateau. *Phil. Trans. R. Soc. London*, **A327,** 379-413.

Dickinson, W.R. and Seely, R., 1979. Structure and stratigraphy of forearc regions. *Am. Assoc. Pet. Geol. Bull.*, **63(1),** 2-31.

Dietz, R.S. and Holden, J.C., 1970. Rcconstruction of Pangea break-up and dispersion of continents, Permian to Present. *J. Geophys. Res.*, **94,** 4939-4956.

Doornbs, D.J., 1989. Seismic Diffraction. *In:* Encyclopedia of Solid Earth Geophysics, D.E. James (ed.), Van Nostrand-Reinhold, N.Y., pp 1018-1024.

Douglas, A., 1967. Joint Hypocentre Determination. *Nature*, **215,** 47-48.

Draper, N.R. and Smith, H., 1966. *Applied Regression Analysis*. John Wiley & Sons, N.Y., 407 p.

Dresen, L. and Freystatter, S., 1976. Rayleigh channel waves for the in-seam seismic detection of discontinuities. *J. Geophys. Res.*, **42,** 111-129.

Dube, R.K., Bayana, J.C. and Chaudhury, H.M., 1973. Crustal structure of the peninsular India. *Pure Appl. Geophys.*, **109,** 1717-1727.

Dube, R.K., Dattatrayam, R.S., Singh, M. and Srivastava, H.N., 1986. Seismicity of northeast India with reference to the Cachar earthquake of December 1984. Proceedings Int. Symp. Neotectonics in South Asia, Dehradun, India. Feb. 18-21, pp 378-395.

Dufumier, H., 1996. On the limits of linear moment tensor inversion of teleseismic body wave spectra. *Pure Appl. Geophys.*, **147,** 467-482.

Dunn, J.A., Auden, J.B., Ghosh, A.M. and Wadia, D.N., 1939. The Bihar-Nepal earthquake of 1934. *Geol. Surv. India Mem.*, **73,** 1-139.

Durrheim, R.J. and Mooney, W.D., 1994. Evolution of the Precambrian lithosphere: Geological and geochemical constraints. *J. Geophys. Res.*, **99,** 15359-15374.

Dutta, T.K., 1964. Seismicity of Assam belts of tectonic activities. *Bull. National Geophys. Res. Instt.*, **2,** 152-163.

Dutton, C.E., 1888. The Charleston earthquake of August 31, 1886. *US Geol. Surv., 9th Annual Report*, 203-258.

Dziewonski, A.M. and Anderson, D.L., 1981. Preliminary reference Earth model. *Phys. Earth Planet. Inter.*, **25,** 297-356.

Dziewonski, A.M. and Anderson, D.L., 1983. Travel times and station corrections for P waves at teleseismic distances. *J. Geophys. Res.*, **88,** 3295-3314.

Dziewonski, A.M., Friedman, A. and Woodhouse, J.H. 1985. Centroid-moment tensor solutions for April-June, 1984. *Phys. Earth Planet. Inter.*, **37,** 87-96.

Dziewonski, A.M., Ekstrom, G., Woodhouse, J.H. and Zwart, G., 1988. Centroid-moment tensor solutions for January-March 1988. *Phys. Earth Planet. Inter.*, **54,** 22-32.

Dziewonski, A.M., Ekstorm, G., Woodhouse, J.H. and Zwart, G., 1989. Centroid moment tensor solution for July-September 1988. *Phys. Earth. Planet Int.*, **56,** 165-180.

Eaton, J.P., 1962. Crustal structure and volcanism in Hawaii. *Geophys. Monogr., Am. Geophys. Union*, **6,** 13-29.

Eaton, J. P., 1992. Determination of amplitude, duration magnitudes and site residuals from short period seismographs in northern California. *Bull. Seism. Soc. Am.*, **82(2),** 533-579.

Eaton, J.P., O'Neill, M.E. and Murdock, J.N., 1970a. Aftershocks of the 1966 Parkfield-Cholame, California earthquake: A detailed study. *Bull. Seismol. Soc. Am.*, **60,** 1151-1197.

Eaton, J.P., Lee, W.H.K. and Pakiser, L.C., 1970b. Use of microearthquakes in the study of the mechanics of earthquake generation along the San Andreas fault in central California. *Tectonophysics*, **9,** 259-282.

Eberhart-Phillips, D., 1993. Local earthquake tomography: Earthquake source region. *In:* Seismic Tomography, Theory and Practice, H.M. Iyer and K. Hirahara (eds), Chapman and Hall, London, pp. 613-643.

Elsasser, W.M., 1967. Convection and stress propagation in the upper mantle. *In:* The Application of Modern Physics to the Earth and Planetory Interior, S.K. Runcorn (ed.), Wiley, N.Y. 1969, pp. 223-246.

Engdahl, E.R. and Lee, W.H.K., 1976. Relocation of local earthquakes by seismic ray-tracing. *J. Geophys. Res.*, **81,** 4400-4406.

England, P.C. and Houseman, G.A., 1989. Extension during continental convergence with applications to the Tibetan Plateau. *J. Geophys. Res.*, **94,** 17561-17579.

Eremenko, N.A. and Negi, B.S. et al. (eds), 1968. Tectonic map of India, 1 : 2 million. Oil and Natural Gas Commission, Dehra Dun, India.

Evans, P., 1964. The tectonic framework of Assam. *J. Geol. Soc. India*, **5,** 80-96.

Evans, P. and Crompton, W., 1946. Geological factors in gravity interpretation illustrated by evidence from India and Burma. *Quatr. J. Geol. Soc. London*, **102,** 211-249.

Evison, F.F., 1955. A coal seam as a guide for seismic energy. *Nature*, **176,** 1224-1225.

Evison, F.F., 1982. Generalised precursory swarm hypothesis. *J. Phys. Earth*, **30,** 155-170.

Ewing, M., Press, F. and Worzel, J.L., 1952. Further study of T-phase. *Bull. Seism. Soc. Am.*, **42,** 37-51.

Ewing, M., Jardetzky, W.S. and Press, F., 1957. *Elastic Waves in Layered Media*, McGraw Hill, N.Y., 380 p.

Fan, G., Ni, J.F. and Wallace, T.C., 1994. Active tectonics of the Pamirs and Karakoram. *J. Geophys. Res.*, **99,** 7131-7160.

Farah, A., 1973. Example of application of geophysical techniques in determining geological environment of a coal deposit. *Geol. Soc. Am. Bull.*, **84,** 2433-2444.

Fernando, M.J. and Kulasinghe, A.N.S., 1986. Seismicity of Srilanka. *Phys. Earth Planet. Inter.*, **44,** 99-106.

Field, E.H. and Jacob, K.H., 1995. A comparison and test of various site response estimation techniques, including three that are not reference site dependent. *Bull. Seism. Soc. Am.*, **85,** 1127-1143.

Field, E.H., Jacob, K.H. and Hough, S.H., 1992. Earthquake site response estimation: A week motion case study, *Bull. Seism. Soc. Am.*, **82,** 2283-2307.

Fitch, T.J., 1970. Earthquake mechanisms in the Himalaya, Burmese and Andaman regions and continental tectonics in central Asia. *J. Geophys. Res.*, **75,** 2699-2709.

Fitch, T.J., 1972. Plate convergence, transcurrent faults and internal deformation adjacent to southeast Asia and the western Pacific. *J. Geophys. Res.*, **77(23),** 4432-4460.

Forsythe, G.E., Malcolm, M.A. and Moler, C.B., 1977. *Computer Methods for Mathematical Computations*. Prentice Hall, USA.

Francheteau, J. et al., 1984. High heat flow in southern Tibet. *Nature*, **307,** 32-36.

Frankel, A., Fletcher, J., Vernon, F., Haar, L., Berger, J., Hanks, T. and Burne, J., 1986. Rupture characteristics and tomographic source imaging of earthquakes near Anza, Southern California. *J. Geophys. Res.*, **91,** 12633-12650.

Freymueller, J., Bilham, R., Bürgmann, R., Larson, K.M., Paul, J., Jade, S. and Gaur V.K., 1996. Global Positioning System measurements of Indian plate motion and convergence across the Lesser Himalaya. *Geophys. Res. Lett.*, **23,** 3107-3110.

Frolich, C. and Davis Scott, D., 1993. Teleseismic b values. *J. Geophys. Res.*, **98,** 631-644.

Fuchs, G.R., 1975. Contribution to the geology of the northwestern Himalaya. *Abh. Geol. Bull.*, **32,** 1-59.

Gahalaut, V.K. and Bürgmann, R., 2004. Constraints on the source parameters of the 26th January 2001 Bhuj India earthquake from satellite images. *Bull. Seism. Soc. Am.*, **96(6),** 2407-2413.

Gahalaut, V.K. and Chander, R., 1997. On interseismic elevation changes and strain accumulation for great thrust earthquakes in the Nepal Himalaya. *Geophys. Res. Letts.*, **24,** 1011-1014.

Gahalaut, V.K., Kalpana and Singh, S.K., 2004. Fault interaction and earthquake triggering in the Koyna-Warna region, India. *Geophys. Res. Lett.*, **31,** L11614 doi: 10.1029/2004 GLO19818.

Gahalaut, V.K., Nagrajan, B., Catherine, J.K. and Kumar, S., 2006. Constraints on 2004 Sumatra-Andaman earthquake rupture from GPS measurements in Andaman-Nicobar Islands. *Earth Planet. Sci. Lett.*, **242,** 365-374.

Galanopolous, A.G., 1961. On magnitude determination by using macroseismic data. *Ann. Geofis*, **14,** 225-253.

Gansser, A., 1964. *The Geology of the Himalayas*. Wiley Interscience, London, 289 p.

Gansser, A., 1966. The Indian Ocean and the Himalayas: A geological interpretation. *Eclogne Geol. Helv.*, **59,** 831-848.

Gansser, A., 1982. The morphogenic phase of mountain building. *In:* Mountain Building Process, K.J. Hsu (ed.), London Acad. Press, pp. 221-228.

Gaonkar, S.G. and Srirama, B.V., 2003. A perspective of microearthquake studies for aftershock sequences of recent earthquakes and swarm activities in central India. *Gond Geol. Mag.*, Sp. V. **5,** 213-226.

Garland, G.D., 1979. *Introduction to Geophysics*. W.B. Saunders Co. USA, 494 p.

Gaur, V.K., 2003. Seismic hazard in central India shield. *Gond. Geol. Mag.*, Sp. V. **5,** 1-6.

Gaur, V.K. and Bhattacharji, J.C., 1983. Gravimetric determination of the shape of Mohorovičić discontinuity in peninsular and northeastern India. Presented at the I.U.G.G. General Assembly, Aug. 15-27, Hamburg, Germany.

Gaur, V.K., Khattri, K.N., Kumar, A. and Verma, J.K., 1982. Recent microearthquake activity in northeast India. Proc. VIIth Symp. Earthquake Engg. Roorkee, **2,** 1-5.

Gee, E.R., 1934. Dhubri earthquake of 3rd July 1930. Geol. Surv. India Mem., **65,** 106.

Geiger, L., 1912. Probability method for the determination of earthquake epicenters from the arrival time only. *Bull. St. Louis Univ.*, **8,** 60-71.

Geller, R.J., 1991. Supercomputers in seismology: Determining 3-D Earth structure. *SIAM News*, **24,** no. 5.

Gibowicz, S.J. and Kijko, A., 1994. *An Introduction to Mining Seismology*. Academic Press, USA.

Gibowicz, S.J., Latter, J.H. and Sutton, G.K., 1974. Earthquake swarm associated with volcanic eruption, Curacoa Ruf area, Northern Tonga, July 1973. *Ann. Geophys.*, **27,** 443-475.

Gill, P.E., Murray, W. and Wright, M.H., 1981. *Practical Optimization.* Academic Press, London, 401 p.

Gokarn, S.G., Rao, C.K., Singh, B.P. and Nayak, P.N., 1992. Magnetotelluric studies across the Kuduwardi gravity features. *Phys. Earth Planet. Inter.*, **72,** 58-67.

Gokarn, S.G., Rao, C.K., Gautam Gupta, Singh, B.P. and Yamashita, M., 2001. Deep crustal structure in the Damoh-Jabalpur-Mandla region using magnetotelluric studies. *Geophys. J. Int.*, **144,** 685-694.

Gokarn, S.G., Gupta, G. and Rao, C.K., 2004. Geoelectric structure of the Dharwar Craton from magnetotelluric studies: Archaean suture identified along the Chitradurga-Gadag schist belt. *Geophys. J. Int.*, **158,** 712-728.

Got, J. L., Fréchet, J. and Klein, F.W., 1994. Deep fault plane geometry inferred from multiplet relative location beneath the south flank of Kilauea. *J. Geophys. Res.*, **99,** 15375-15386.

Grady, J.C., 1971. Deep main faults in India. *J. Geol. Soc. India*, **12,** 55-62.

Grant, W., 1819. Earthquake in Kutch and the adjoining region on 16th June, 1819. *Asiatic J.*, **8,** 610-611.

Grassberger, P. and Procaccia, I., 1983. Characterisation of strange attractors. *Phys. Rev. Lett.*, **50,** 346-349.

Greenhalgh, S.A., Denham, D., McDougall, R. and Rynn, J.M., 1989. Intensity relations for Australian earthquakes. *Tectonophysics*, **166,** 255-267.

Griscom, M. and Arabasz, W.J., 1979. Local magnitude (M_L) in the Wasatch Fronts and Utah region: Wood-Anderson calibration, coda duration estimates of M_L and M_L versus m_b. *In:* Earthquake Studies in Utah 1850-1978, W.J. Arabasz, R.B. Smith and W.D. Richins (eds.), pp. 433-443.

GSI, 1939. The Bihar-Nepal earthquake of 1934. *Geol. Surv. India Mem.*, **73,** 280 p.

GSI, 1992. Uttarkashi Earthquake, October 20, 1991. Geol. Surv. India Sp. Pub., 30, C.P. Vohra and D. Gupta Sarma (eds.), 217 p.

GSI, 1993. Bihar-Nepal Earthquake, August 20, 1988. Geol. Surv. India, Sp. Pub., 31, D.R. Nandy, A.K. Choudhury, C. Chakraborty and P.L. Narula (eds.), 104 p.

GSI, 1996. Killari Earthquake, September 30, 1993. Geol. Surv. India, Sp. Pub., 37, P.L. Narula, S.K. Shome and B.S.R. Murty (eds.), 282 p.

GSI, 2000a. Seismotectonic Atlas of India and its Environs. Geol. Surv. India, Sp. Pub., P.L. Narula, S.K. Acharya and J. Banerjee (eds.), 86 p.

GSI, 2000b. Jabalpur Earthquake 22 May 1997: A geoscientific study. Geol. Surv. India Sp. Pub., 51, 239 p.

GSI, 2001. Chamoli Earthquake of March 29, 1999. Geol. Surv. India, Sp. Pub., 33.

GSI, 2003. Kutch (Bhuj) Earthquake 26 January, 2001. Geol. Surv. India, Sp. Pub., 76, Prabhas Pande and J.R. Kayal (eds.), 282 p.

GSI, 2006. A Manual of the Geology of India. Geol. Surv. India Sp. Pub., 77, v. 1, Fourth Edition, 572 p.

GSI, 2007. Sumatra-Andaman earthquake and tsunami, 26 December 2004. Sujit Dasgupta (ed.), Geol. Surv. India Sp. Pub. 89, Kolkata, 239 p.

Gubin, I.E., 1968. Seismic zoning of Indian Peninsula, *Bull. Int. Inst. Seism. Earthquake Engg.*, **5,** 109-139.

Guha, S.K., 2000. *Induced Earthquakes*. Kluwer Acad. Pub., 314 p.

Guha, S.K. and Basu, P.C., 1973. Catalogue of Earthquakes in Peninsular India. Tech. Docu. No. TD/CSE-1, Atomic Energy Regulatory Board, Bombay.

Guha, S.K. and Bhattacharya, U., 1984. Studies on prediction of seismicity in northeast India. Proc. World Conf. on Earthquake Engineering, July, 21-27, San Francisco, USA.

Guha, S.K., Gosabi, P.D., Nand, K., Pande, J.G. and Marwadi, S.C., 1974. Koyna earthquakes (October 1963-December, 1973). Govt of India, Central Water and Power Research Station, Poona.

Gulatee, B.L., 1956. Gravity data in India. Survey of India Tech. Pub., 10(a), 95 p.

Guo, Z. and Ogata, Y., 1995. Correlation between characteristic parameters of aftershock distribution in time, space and magnitude. *Geophys. Res. Lett.*, **22,** 993-996.

Guo, Z. and Ogata, Y., 1997. Statistical relations between the parameters of aftershocks in time, space and magnitude. *J. Geophys. Res.*, **102,** 2857-2873.

Gupta, H.K., 1992. *Reservoir Induced Earthquakes*. Elsevier, Amsterdam, 364 p.

Gupta, H.K., 2002. A review of recent studies of triggered earthquakes by artificial water reservoirs with special emphasis on earthquakes in Koyna, India. *Earth Science Reviews*, **58,** 279-310.

Gupta, H.K. and Dwivedi, K.K., 1996. Drilling at Latur earthquake region exposes a peninsular gneiss basement. *J. Geol. Soc. India*, **47,** 129-131.

Gupta, H.K. and Hari Narain, 1967. Crustal structure in the Himalayan and Tibet Plateau region from surface-wave dispersion. *Bull. Seisml. Soc. Am.*, **57,** 235-248.

Gupta, H.K. and Rajendran, K., 1986. Large artificial water reservoir in the vicinity of Himalayan foothills and reservoir induced seismicity. *Bull. Seism. Soc. Am.*, **76,** 205-215.

Gupta, H.K. and Rastogi, B.K., 1976. *Dams and Earthquakes*. Elsevier, Amsterdam, 229 p.

Gupta, H.K. and Singh, H.N., 1986. Seismicity of northeast India region. Part II: Earthquake swarm precursory to moderate magnitude to great earthquakes. *J. Geol. Soc. India*, **28,** 367-406.

Gupta, H.K. and Singh, H.N., 1989. Precursory to moderate magnitude to great earthquakes in northeast India region. *Tectonophysics*, **167,** 255-298.

Gupta, H.K., Mohan, I. and Narain, H., 1970. The Godavari earthquake sequence of April 1960. *Bull. Seism. Soc. Am.*, **69,** 601-615.

Gupta, H.K., Mohan, I. and Hari Narain, 1972. The Broach earthquake magnitude, intensity, energy and acceleration. *Bull. Seism. Soc. Am.*, **62,** 47-61.

Gupta, H.K., Rastogi, B.K. and Singh, D.D., 1982. Seismological investigations in the Himalaya and nearby regions. *Geophys. Res. Bull.*, **20(3),** 141-165.

Gupta, H.K., Rajendran, K. and Singh, H.N., 1986. Seismicity of northeast India region. Part I: The Data Base. *J. Geol. Soc. India*, **28,** 345-365.

Gupta, H.K., Fleitout, L. and Froidevaux, C., 1990. Lithospheric subduction beneath the Arakan-Yoma fold belt: Quantitative estimates using gravimetric and seismic data. *J. Geol. Soc. India*, **35,** 235-250.

Gupta, H.K., Chadha, R.K., Rao, M.N., Narayana, B.L., Mandal, P., Ravi Kumar, M. and Kumar, N., 1997. The Jabalpur Earthquake of May 22, 1997. *J. Geol. Soc. India*, **50,** 85-91.

Gupta, H.K., Rastogi, B.K., Indra Mohan, Rao, C.V.R.K., Sarma, S.V.S. and Rao, R.H.M., 1998. An investigation into Latur earthquake of September 29, 1993 in southern India. *Tectonophyicss*, **28,** 299-318.

Gupta, H.K., Rao, P.N., Rastogi, B.K. and Sarkar, D., 2001. The deadliest intraplate earthquake. *Science*, **291,** 2101-2102.

Gupta, M.L., 1995. Thermal regime of the Indian shield. *In:* Terrestrial heat flow and geothermal energy in Asia, M.L. Gupta and M. Yamano (eds.). Oxford & IBH, pp. 63-81.

Gupta, M.L. and Gaur, V.K., 1984. Surface heat flow and probable evolution of Deccan volcanism. *Tectonophysics*, **105,** 309-318.

Gupta, S., Rai, S.S., Prakásam, K.S., Srinagesh, D., Bansal, B.K., Chadha, R.K., Priestley, K. and Gaur, V.K., 2003. The nature of the crust in southern India: Implications for Precambrian crustal evolution. *Geophys. Res. Lett.*, **30,** 8,1419, doi: 10.1029/2002GL016770.

Gutenberg, B., 1914. Uber Erdbebenwellen VIIIA Boebachtungen an Registrierungen Von Frenbebeh in Gottingen und Folgerungen uber die Konstitution des Erdkorpers. Nachr. Der. Konig. Gesell. Der Wiss Zu Gottingen, Math Phys. Kl - 1.

Gutenberg, B., 1945a. Amplitudes of surface waves and magnitudes of shallow earthquakes. *Bull. Seism. Soc. Am.*, **35,** 3-12.

Gutenberg, B., 1945b. Amplitudes of P, PP and S and magnitudes of shallow earthquakes. *Bull. Seism. Soc. Am.*, **35,** 57-69.

Gutenberg, B. and Richter, C.F., 1941. Seismicity of the Earth. *Geol. Soc. Am. Sp Pap.*, **34,** 1-133.

Gutenberg, B. and Richter, C.F., 1944. Frequency of earthquakes in California. *Bull. Seism. Soc. Am.*, **34,** 185-188.

Gutenberg, B. and Richter, C.F., 1954. *Seismicity of the Earth and Associated Phenomena.* Princeton University Press, New Jersey, 310 p.

Gutenberg, B. and Richter, C.F., 1956. Earthquake magnitude, intensity, energy and acceleration. *Bull. Seism. Soc. Am.*, **46,** 105-145.

Haines, S.S., Klemperer, S.L., Brown, L., Jingru, G., Meehie, J., Meissner, R., Ross, A. and Wenjin, Z., 2003. INDEPTH III seismic data: From surface observations to deep crustal process. *Tectonics*, **22,** doi: 10.1029/2001 TC001305.

Haldar, D., Laskar, T., Bandopadhyay, P.C., Sarkar, N.K. and Biswas, J.K., 1992. Volcanic eruption of the Barren Island Volcano, Andaman sea. *J. Geol. Soc. India*, **39,** 407-419.

Hagiwara, T., 1964. Brief description of the project proposed by the earthquake prediction group in Japan. Proc. US-Japan Conference on research related to earthquake prediction problems (Tokyo), 10-12.

Hamburger, M.W., Sarewitz, D.R., Pavlis, T.L. and Pooandopulo, G.A., 1992. Structural and seismic evidence for intracontinental subduction in the Peter the First Range, central Asia. *Geol. Soc. Am. Bull.*, **104,** 397-408.

Hanks, T.C. and Kanamori, H., 1979. A moment magnitude scale. *J. Geophys. Res.*, **84,** 2348-2350.

Hari Narain and Kaila, K.L., 1982. Inferences about the Vindhyan Basin from Geophysical Data. *In:* Geology of Vindhyanchal, Hindustan Publishing Corporation, Delhi. pp. 179-192.

Harikumar, P., Rajaram, Mita and Balakrishnan, T.S., 2000. Aeromagnetic study over peninsular India. *Proc. Indian Acad. Sci (E & P)*, **109,** 381-391.

Hartzell, S.H., 1992. Site response estimation from earthquake data. *Bull. Seism. Soc. Am.*, **82,** 2308-2327.

Hauksson, E. and Haase, J., 1994. Three dimensional Vp and Vp/Vs velocity models of the Los Angles basin and central Transverse ranges, California. *J. Geophys. Res.*, **102,** 5423-5453.

Havaskov, J. and Ottemoller, L., 2000. SEISAN: The earthquake analysis software. Institute of Solid Earth Physics, Norway.

Hayden, H.M., 1918. Relationship of the Himalaya to the Indo-Gangetic plain and the Indian Peninsula. *Geol. Surv. India Rec.*, **43,** 138-157.

Helmberger, D.V., 1983. Theory and application of synthetic seismograms. *In:* Earthquakes: Observation, Theory and Interpretation. pp 174-222. *Soc. Ital. Fis.*, Bologna.

Herman, G.T., 1980. *Image Reconstruction from Projections.* Academic Press, New York, 316 p.

Herrin, E., Arnold, E.P., Bolt, B.A., Clawson, G.E., Engdahl, E.R., Freedman, H.W., Gordon, D.W., Hales, A.L., Lobdell, J.L., Nuttli, O., Romney, C., Taggart, J. and Tucker, W., 1968. Seismological Tables for P phases. *Bull. Seism. Soc. Am.*, **58,** 1193-1241.

Herrmann, R.B., 1975. The use of duration as a measure of seismic moment and magnitude. *Bull. Seism. Soc. Am.*, **65,** 899-913.

Herrmann, R.B., Park, S. and Wang, C., 1981. The Denver earthquakes of 1967-1968. *Bull. Seism. Soc. Am.*, **71,** 731-745.

Hiller, K. and Elahi, M., 1984. Structural development and hydrocarbon entrapment in the Surma basin/Bangladesh (northwest Indo Burman fold belt). Proc. Offshore South East Asia (SEAPEX) Conf. 5th, 6.50-6.63.

Hirata, T., 1987. Omori's power law aftershock sequences of microfracturing in rock fracture experiment. *J. Geophys. Res.*, **92,** 6215-6221.

Hirata, T., 1989. A correlation between the b-value and the fractal dimension of earthquakes. *J. Geophys. Res.*, **94,** 7507-7514.

Hirn, A. et al., 1995. Seismic anisotropy as an indicator of mantle flow beneath the Himalayas and Tibet. *Nature*, **375,** 571-574.

Hirn, A., Nercessian, A., Sapin, M., Jobert, G., Xu, Z., Yuan, G.E., Yuan, L.D. and Wen, T.J., 1984. Lasha block and bordering structures – A continuation of a 500 km Moho traverse through Tibet. *Nature*, **307,** 25-27.

Ho-Liu, P., Kanamori, H. and Clayton, R.W., 1998. Applications of attenuation tomography to Imperial Valley and Cosco-Indian Wells region, Southern California. *J. Geophys. Res.* **93,** 10501-10520.

Holmes, A., 1965. *Principles of Physical Geology,* 2nd Edn. ELBS and Nelson.

Holt, W.E. and Wallace, T.C., 1990. Crustal thickness and upper mantle velocities in the Tibetan Plateau region from the inversion of regional Pnl waveforms: Evidence for a thick upper mantle lid beneath southern Tibet. *J. Geophys. Res.*, **95,** 12,499-12,525.

Holt, W.E., Ni, J.F., Wallace, T.C. and Haines, A.J., 1991. The active tectonics of the eastern Himalayan syntaxis and surrounding regions. *J. Geophys. Res.*, **96,** 14,595-14,632.

Honda, H., 1962. Earthquake mechanism and seismic waves. *J. Phys. Earth,* **10,** 1-97.

Honda, H. and Emura, K., 1958. Some charts for studying the mechanism of earthquakes. SCI Rep. Tohoku Univ., 5th Ser. *Geophys.*, **9,** 113-115.

Hooper, P.R., 1990. The timing of crustal extension and the eruption of continental basalts. *Nature*, **345,** 246-249.

Hori, M., 1973. Determination of earthquake magnitude of the local and near earthquake by the Dodaria Microearthquake Observatory. *Bull. Res. Inst. Univ. Tokyo*, **10,** 1-4 (in Japanese with English Abstract).

Houck, S.T., Guerrero, J.D., Miller, A.E. and Lee, W.H.K., 1976. Handbook for USGS central California microearthquake network, 1969-1975. *U.S. Geol. Surv. Open File Report*, 76-282, 1-241.

Hough, S.E., Bilham, R., Ambraseys, N. and Feldi, N., 2005. The 1905 Kangra and Dehradun earthquakes. Geol. Surv. India, Sp. Pub., **85,** 15-23.

Hough, S.E., Stacey, M., Bilham, R., Atkinson, G.M., 2002. The 26th January 2001 M 7.6 Bhuj India earthquake: Observed and predicted ground motions. *Bull. Seism. Soc. Am.*, **92,** 2061-2079.

Houndsfield, G.N., 1973. Computerised traverse axial scanning tomography. Part 1: Description of the system. *Br. J. Radiol*, **46,** 1016-1022.

Huang, H.J. and Mitchell, B.J., 1987. Shear velocities, Q_b, and the frequency dependence of Q_b in stable and tectonically active regions from surface wave observations. *Geophys. J.R. Astr. Soc.*, **90,** 575-613.

Huang, W.C., Ni, J.F. and Project INDEPTH team, 2000. Seismic polarization anisotrophy beneath the central Tibet Plateau. *J. Geophys. Res.*, **105,** 27979-27989.

IASPEI, 2002. *New Manual of Seismological Observatory Practice*, Peter Borman (ed.). Geo Forschungs Zentrum Potsdam.

Idziak, A. and Teper, L., 1996. Fractal dimension of faults network in the upper Silesian coal basin, Poland: Preliminary studies. *Pure and Appl. Geophys.*, **147,** 239-247.

IMD, 1980. Microearthquake survey around the Puga geothermal project area in June-August 1976. Tech. Report, IMD, New Delhi.

IMD, 1988. Microearthquake survey around Salal dam. Tech. Report, IMD, New Delhi.

IMD, 1992. Instrumental record of Uttarkashi earthquake of October 20, 1991. *In:* Uttarkashi Earthquake, October 20, 1991, Geol. Surv. India Sp. Pub. 30, C.P. Vohra and D. Gupta Sarma (eds.), 217 p.

IMD, 1997. Jabalpur earthquake of May 22, 1997 and its aftershocks: A consolidated document. India Met. Dept. Unpub Rep., New Delhi.

IMD, 2000. Chamoli earthquake of March 29, 1999 and its aftershocks: A consolidated document. Meteorological Monograph Seism. 2/2000, 70 p.

IMD, 2002. Bhuj Earthquake of January 26, 2001: A consolidated document. India Met. Dept. Monograph, New Delhi, 107 p.

Indra Mohan and Rao, M.N., 1994. A field Survey of Latur earthquake of 30th September, 1993. *Geol. Soc. India. Mem.*, **35,** 7-32.

Isacks, B.L. and Molnar, P., 1971. Distribution of stresses in the descending lithosphere from global survey of focal mechanism solutions of mantle earthquakes. *Rev. Geophys. Space Phys.*, **9,** 103-174.

Iyer, H.M. and Hirahara, K. (Editors), 1994. *Seismic Tomography: Theory and Practice*. Prentice Hall, USA.

Iyer, H.M., Gaur, V.K., Rai, S.S., Ramesh, D.S., Rao, C.V.R., Srinagesh, D. and Suryaprakasham, K., 1989. High velocity anomaly beneath the Deccan Volcanic Province: Evidence from seismic tomography. *Proc. Indian Acad. Sci. (Earth Planet Sci)*, **98(1),** 31-60.

Jackson, J.A., 1980. Reactivation of basement faults and crust shortening in orogenic belts. *Nature*, **283,** 343-346.

Jade Sridevi, Mukul, M., Parvez I.A., Ananda, M.B., Kumar, P.D. and Gaur, V.K., 2002. Estimates of coseismic displacement and post-seismic deformation using global positioning system geodesy for the Bhuj earthquake of 26 January 2001. *Curr. Sci.*, **82(6),** 748-752.

Jeanloz, R., 1990. The nature of the earth's core. *Annu. Rev. Earth Planet. Sci.*, **18,** 357-386.

Jeffreys, H., 1935. Some deep focus earthquakes. *MNRAS, Geophys. Suppl.*, **3,** 310-343.

Jin, Y., McNutt, M.K. and Zhu, Y-S, 1996. Mapping the descent of Indian and Eurasian plates beneath the Tibetan plateau from gravity anomalies. *J. Geophys. Res.*, **101(B5),** 11,275-11,290.

Johnson, B.D., Powell, C.McA and Veevers, J.J., 1976. Spreading history of the Eastern Indian Ocean and Greater India's northward flight from Antarctica and Australia. *Geol. Soc. Am. Bull.*, **87,** 1560-1566.

Johnson, C.E. 1979. I. CEDAR: An approach to the computer automation of short period local seismic networks. II. Seismotectonics of the Imperial Valley of Southern California. Ph.D. Thesis, California Instt. of Technology, Pasadena, USA.

Johnson, C.E. and Hadley, D.M., 1976. Tectonic implications of the Brawley earthquake swarm, Imperial Valley California, January 1975. *Bull. Seism. Soc. Am.*, **66,** 1133-1144.

Johnson, R.H., Northrop, J. and Eppley, R., 1963. Source of Pacific T-phase. *J. Geophys. Res.*, **68,** 4251-4260.

Johnston, A.C., 1989. The effect of large ice sheets on earthquake genesis. *In:* Earthquakes in North Atlantic passive margins: Neotectonics and post-glacial rebound, S. Gregersen and P.W. Basham (eds.). Kluver Academic Pub. pp 581-599.

Johnston, A.C., 1996. Seismic moment assessment of earthquakes in stable continental regions. *Geophys. J. Int.*, **124,** 381-414.

Johnston, A.C. and Kanter, L.R., 1990. Stable continental earthquakes. *Science*, **262,** 68-75.

Jones, L.M. and Molnar, P., 1979. Some characteristics of foreshocks and their possible relationship to earthquake prediction and premonitory slips on faults. *J. Geophys. Res.*, **84,** 3596-3608.

Joshi, J.P. and Bisht, R.S., 1994. *India and the Indus Civilization.* National Museum Institute, New Delhi, pp. 23-31.

Kagan, Y.Y. and Knopoff, L. 1980. Spatial distribution of earthquakes: The two-point correlation function. *Geophys. J. R. Astro. Soc.*, **62,** 303-320.

Kaila, K.L., 1988. Mapping the thickness of Deccan trap flows in India from DSS studies and inferences about a hidden Mesozoic basin in the Narmada-Tapti region. *Geol. Soc. India Mem.*, **10,** 91-116.

Kaila, K.L. and Krishna, V.G, 1983. Upper Mantle Velocity Structure in the Hindukush Region. Boll. Dl. Geofisics Teorica Ed Applicata. V.XXV., N. 99-100 pp. 277-289.

Kaila, K.L. and Krishna, V.G., 1992. Deep seismic sounding studies in India and major discoveries. Seismology in India – An overview. *Current Science*. Sp. Vol. **62,** 117-154.

Kaila, K.L. and Sain, K., 1997. Variation of crustal velocity structure in India as determined from DSS studies and their implications on regional tectonics. *J. Geol. Soc. Ind.*, **49,** 395-407.

Kaila, K.L., Reddy, P.R. and Hari Narain, 1968. P-Wave travel times from shallow earthquakes recorded in India and inferred upper mantle structure. *Bull. Seism. Soc. Am.,* **58,** 1879-1897.

Kaila, K.L., Krishna, V.G. and Hari Narain, 1969. Upper mantle velocity structure in the Hindukush region from travel time studies of deep earthquakes using a new analytical method. *Bull. Seism. Soc. Am.*, **59,** 1949-1967.

Kaila, K.L., Roy Chowdhury, K., Reddy, P.R., Krishna, V.G., Hari Narain, Subbotin, S.I., Sollogub, V.B., Chekunov, A.V., Kharetchko, G.E., Lazarenko, M.A. and Ilchenko, T.V., 1979. Crustal structure along Kavali Udipi profile in the Indian peninsular shield from deep seismic sounding. *J. Geol. Soc. India*, **20,** 307-333.

Kaila, K.L., Tripathi, K.M. and Dixit, M.M., 1984. Crustal structure along Wular lake-Gulmarg-Naoshera profile across Pir Panjal range of the Himalayas from deep seismic soundings. *J. Geol. Soc. India.*, **25(11),** 706-718.

Kaila, K.L., Reddy, P.R., Dixit, M.M. and Koteswara Rao, P., 1985. Crustal structure across the Narmada-Son lineament, central India from deep seismic soundings. *J. Geol. Soc. India*, **26,** 465-480.

Kaila, K.L., Murthy, P.R.K., Mall, D.M., Dixit, M.M. and Sarkar, D., 1987. Deep Seismic soundings along Hirapur-Mandla profile, central India. *Geophys. J. R. Astro. Soc.*, **89,** 399-404.

Kaila, K.L., Murthy, P.R.K. and Mall, D.M., 1989. The evolution of Vindhyan basin vis-à-vis the Narmada-Son lineament, central India from deep seismic soundings. *Tectonophysics*, **162,** 277-289.

Kaila, K.L., Tewari, H.C., Krishna, V.G., Dixit, M.M., Sarkar, D. and Reddy, M.S., 1990. Deep seismic sounding studies in the north Cambay and Sanchor basin, India. *Geophys. J. Int.*, **103,** 621-637.

Kaila, K.L., Reddy, P.R., Mau, D.M., Venkateswar, N., Krishna, V.G. and Prasad, A.S.S.S.R.S., 1992. Crustal structure of the West Bengal basin, India from deep seismic sounding investigations. *Geophys. J. Int.*, **3,** 45-66.

Kailasam, L.N., 1975. Epeirogenic studies in India with reference to recent vertical movement. *Tectonophysics*, **29,** 505-521.

Kailasam, L.N., 1976. Geophysical studies of the major sedimentary basins of the Indian craton, their deep structural features and evolution. *Tectonophysics*, **76,** 225-245.

Kailasam, L.N., 1979. Plateau uplift in Peninsular India. *Tectonophysics*, **61,** 243-269.

Kailasam, L.N., 1994. Geophysical and geodynamical aspects of the Maharashtra earthquake of September 30, 1993. *Curr. Sci.*, **65,** 736-739.

Kanamori, H., 1980. The size of earthquakes. *Earthquake Information Bulletin*, **12(1),** 10-15.

Kanamori, H., Mori, J. and Heaten, T.H., 1990. The 3 December 1988 Pasadena earthquake (M_L 4.9) recorded with the very broadband system in Pasadena. *Bull. Seism. Soc. Am.*, **80,** 483-487.

Kanamori, H. and Cippar, J.J., 1974. Focal process of the great Chilean earthquake, May 22, 1960. *Phys. Earth Planet. Inter.*, **9,** 128-136.

Karanth, R.V., Sohini, P.S., Mathew, G. and Khadikar, A.S., 2001. Geological observations of the 26 January 2001 Bhuj earthquake. *J. Geol. Soc. India*, **58,** 193-202.

Karig, D.E., Suparka, S., Moore, G.F. and Hchanussa, P.E., 1979. Structure and Cenozoic evolution of the Sunda arc in the central Sumatra region. *In:* Geological and Geophysical Investigations of Continental Margins, J.S. Watkins, L. Montadert and P. Dickerson (eds.), *AAPG Memoir*, **29,** 223-237.

Karnik, V., 1969. *Seismicity of the European Area. Part I.* Reidel Publishing Company, Dordrecht, 364 p.

Karnik, V., Kondorskaya, N.V., Riznichenko, Yu.V., Savarensky, Ye.F., Soloviev, S.L., Shebalin, N.V., Vanek, J. and Zatopek, A., 1962. Standardisation of the earthquake magnitude scales. *Studia Geophysica et Geodaetica*, **6,** 41-48.

Katz, M.B., 1974. Paired metamorphic belts in Precambrian granulite rocks in Gondwanaland. *Geology*, **2,** 237-241.

Katz, M.B., 1978a. Sri Lanka in Gondwanaland and the evoluion of the Indian Ocean. *Geol. Mag.*, **115,** 237-244.

Katz, M.B., 1978b. Tectonic evolution of the Archean granulite facies belt of Sri Lanka-south India. *J. Geol. Soc. India*, **19,** 185-205.

Kawasumi, H., 1951. Measures of earthquake danger and expectancy of maximum intensity throughout Japan as inferred from the seismic activity in historical times. *Bull. Earthq. Res. Inst.*, **29,** 469-482.

Kayal, J.R., 1979. Electrical and gamma ray logging in Gondwana and Tertiary coalfields of India. *Geoexploration*, **17,** 243-258.

Kayal, J.R., 1983. A microearthquake study in a subduction zone: Southeast Wellington Province, New Zealand. Ph.D. Thesis, Victoria University of Wellington, New Zealand.

Kayal, J.R., 1984. Microseismicity and tectonics at the Indian Pacific Plate Boundary: Southeast Wellington Province, New Zealand. *Geophys. J. R. Astr. Soc.*, **77,** 567-592.

Kayal, J.R., 1986. Analysis of strong phases other than P and S from a microearthquake survey in the Wellington region, New Zealand. *Bull. Seism. Soc. Am.*, **76,** 1347-1354.

Kayal, J.R., 1987. Microseismicity and source mechanism study: Shillong Plateau, northeast India. *Bull. Seism. Soc. Am.*, **77,** 184-194.

Kayal, J.R. 1989. Subduction structure at the India/Burma plate boundary: Seismic and gravity evidences. 28th Int. Geol. Cong., USA, Abs vol. 2, 164-165.

Kayal, J.R., 1991. Earthquake prediction in northeast India – A review. *Pure and Appl. Geophys.*, **136(2/3)**, 297-313.

Kayal, J.R., 1996a. Precursor seismicity, foreshocks and aftershocks of the Uttarkashi earthquake of October 20, 1991 at Garhwal Himalaya. *Tectonophysics*, **263,** 339-345.

Kayal, J.R., 1996b. Earthquake source process in northeast India: A review. *J. Himalayan Geol.*, **17,** 53-69.

Kayal, J.R., 1998. Seismicity of northeast India and surroundings – Development over the past 100 years. *J. Geophysics*, **XIX**, 9-34.

Kayal, J.R., 2000. Seismotectonic study of the two recent SCR earthquakes in central India. *J. Geol. Soc. India*, **55,** 123-138.

Kayal, J.R., 2001. Microearthquake activity in some parts of the Himalaya and the tectonic model. *Tectonophysics*, **339,** 331-351.

Kayal, J.R., 2002. The nature of seismic sources and prediction of earthquakes: Notes. *J. Geol. Soc. India*, **60,** 467-470.

Kayal, J.R. and Banerjee, B., 1988. Anomalous behaviour of precursor resistivity in Shillong area, northeast India. *Geophys. J. Int.*, **94,** 97-103.

Kayal, J.R. and Bhattacharya, S.N., 2006. Seismicity of the Himachal Himalaya: Constraint from local Seismic Network. *Geol. Surv. India Sp. Pub.*, **85,** 71-79.

Kayal, J.R. and De, Reena, 1987. Pn velocity study using a temporary seismograph network in the Shillong Plateau, northeast India. *Bull. Seism. Soc. Am.*, **77,** 1718-1727.

Kayal, J.R. and De, Reena, 1991. Microseismicity and tectonics in northeast India. *Bull. Seism. Soc. Am.*, **81,** 131-138.

Kayal, J.R. and Mukhopadhyay, S., 2002. Seismic tomography structure of the 1993 Killari earthquake source area. *Bull. Seism. Soc. Am.*, **92(5),** 2036-2039.

Kayal, J.R. and Mukhopadhyay, S., 2006. Seismotectonics of the 2001 Bhuj Earthquake (M_w 7.7) in western India: Constraint from aftershocks. *J. Indian Geophys. Union*, **10(1),** 45-57.

Kayal, J.R. and Zhao, Dapeng, 1998. Three dimensional seismic structure beneath Shillong Plateau and Assam Valley, northeast India. *Bull. Seism. Soc. Am.*, **88,** 667-676.

Kayal, J.R., De, Reena and Chakraborty, P., 1993. Microearthquakes at the main boundary thrust in eastern Himalaya and the present-day tectonic model. *Tectonophysics*, **218,** 375-381.

Kayal, J.R., Ghosh, Biman, Chakraborty, P. and De, Reena, 1995. Aftershock study of the Uttarkashi earthquake of October 20, 1991 by a temporary microearthquake network. *Geol. Soc. India Mem.*, **30,** 25-41.

Kayal, J.R., De, Reena, Das, B. and Chowdhury, S.N., 1996. Aftershock monitoring and focal mechanism studies. *In:* Killari Earthquake 30 September 1993. *Geol. Surv. India Sp. Pub.*, **37,** 165-185.

Kayal, J.R., Nath, S.K., Goswami, T.R., Roy, S., Sagina Ram, Srirama, V., 2002. Site response study by shear wave spectral analysis using the 1999 Chamoli earthquake sequence in Garhwal Himalaya. *J. Himalayan Geol.*, **23,** 45-50.

Kayal, J.R., De, Reena, Sagina Ram, Srirama, B.V. and Gaonkar, S.G., 2002a. Aftershocks of the 26 January, 2001 Bhuj earthquake in western India and its seismotectonic implications. *J. Geol. Soc. India*, **59,** 395-417.

Kayal, J.R., Zhao, D., Mishra, O.P., De, Reena and Singh, O.P., 2002b. The 2001 Bhuj earthquake: Tomography evidence for fluids at hypocenter and its implications for rupture nucleation. *Geophys. Res. Lett.*, **29(24)**, 51-54.

Kayal., J.R., Sagina Ram, Singh, O.P., Chakraborty, P.K. and Karunakar, G., 2003a. Aftershocks of the March 1999 Chamoli earthquake and seismotectonic structure of the Garhwal Himalaya. *Bull. Seism. Soc. Am.*, **93(1)**, 109-117.

Kayal, J.R., Sagina Ram, Singh, O.P., Chakraborty, P.K. and Karunakar, G., 2003b. The March 1999 Chamoli earthquake in the Garhwal Himalaya: Aftershock characteristics and tectonic structure. *J. Geol. Soc. India*, **62,** 558-580.

Kayal, J.R., Gaonkar, S.G., Chakraborty, G.K. and Singh, O.P., 2004. Aftershocks and seismotectonic implications of the 13th September 2002 earthquake (M_W 6.5) in the Andaman Sea basin. *Bull. Seism. Soc. Am.*, **94(1),** 326-333.

Kayal, J.R., Arefiev, S.S., Baruah, S., Hazarika, D., Gogoi, N., Kumar, A., Chowdhury, S.N. and Kalita, S., 2006. Shillong Plateau earthquakes in northeast India region: Complex tectonic model. *Curr. Sci.*, **91(1),** 109-114.

Kelkar, Y.N., 1968. Earthquake in Maharashtra in last three hundred years. *Kesari Daily*, Poona, 07.01.68 (In Marathi).

Kennett, B.L.N., Engdahl, E.R. and Buland, R., 1995. Constraints on seismic velocities in the Earth from traveltimes. *Geophys. J. Int.*, **122,** 108-124.

Kennett, B.L.N. and Engdahl, E.R., 1991. Traveltimes for global earthquake location and phase identification. *Geophys. J. Int.*, **105,** 429-465.

Kennett, B.L.N. and Widiyantoro, S., 1999. A low seismic wave speed anomaly beneath northwestern India: A seismic signature of the Deccan Plume Earth Planet. *Sci. Lett.*, **165,** 145-155.

Khan, A.H. and Azad, J., 1963. The geology of Pakistan gas fields. Proc. 2nd Symp. Development of Petroleum Resources. *ECAFE Res. Dev. Ser.*, **18,** 275-282.

Khan, M.A.M., Ismail, M. and Ahmad, M., 1988. Geology and hydrocarbon prospects of the Surma Basin, Bangladesh. Proc. Offshore South East Asia (SEAPEX) Conf. 7th, 364-384.

Kharshiing, A.D., Khattri, K.N., Moharir, P.S. and Chander, R., 1986. Neotectonics in Shillong massif and neighbouring regions. Proc. Int. Symp. Neotectonics in South Asia, India, pp 407-416.

Khattri, K.N., 1992. Local seismic investigations in the Garwal-Kumaon Himalaya. *Geol. Soc. India Mem.*, **23,** 45-66.

Khattri, K.N., 1999. Probabilities of occurrence of great earthquakes in the Himalaya. *Proc. Indian Acad. Sci. (Earth Planet. Sci.)*, **108,** 1442-1444.

Khattri, K.N., 2006. A need to review the current official Seismic Zoning map of India. *Curr. Sci.*, **90(5)**, 1-3.

Khattri, K. and Wyss, M., 1978. Precursory variation of seismicity rate in Assam area, India. *Geology*, **6,** 685-688.

Khattri, K.N. and Tyagi, A.K., 1983. Seismic patterns in the Himalayan plate boundary and identification of areas of high seiemic potential. *Tectonophysics*, **96,** 281-297.

Khattri, K., Wyss, M., Gaur, V.K., Saha, S.N. and Bansal, B.K., 1983. Local seismic activity in the region of the Assam gap, northeast India. *Bull. Seism. Soc. Am.*, **73,** 459-469.

Khattri, K.N., Rogers, A.M., Perkins, D.M. and Algermissen, S.T., 1984. A seismic hazard map of India and adjacent areas. *Tectonophysics*, **108,** 93-134.

Khattri, K.N., Mukhopadhyay, S., Subrahmaniam, S. and Appa Rao, M., 1986. Seismotectonic studies in the eastern part of Shillong massif region. Proc. VIIIth Symp. Earthquake Engg., Roorkee, pp 103-110.

Kind, R., Yuan, X., Saul, J., Nelson, D., Sobolev, S.V., Mechie, J., Zhao, W., Kosarev, G., Ni, J., Achuer, U. and Jiang, M., 2002. Seismic images of crust and upper mantle beneath Tibet: Evidence for Eurasian plate seduction. *Science*, **298,** 1219-1221.

Kisslinger, C. and Jones, L.M., 1991. Properties of aftershocks in southern California. *J. Geophys. Res.*, **96,** 11947-11958.

Kolvankar, V.G., 1988. Multichannel waveform display system. Presented at the Seminar on Seismic Instrumentation, Chidambaram, BARC.

Kolvankar, V.G., 2001. Earthquake sequence of 1991 from Valsad region, Gujarat. Technical Report. BARC/2001/E/006, BARC, Mumbai.

Kono, M., 1974. Gravity anomalies in east Nepal and their implications to the crustal structure of the Himalayas. *Geophys. J. R. Astro Soc.*, **39,** 283-299.

Korvin, G., 1992. *Fractal Methods in Earth Sciences*. Elsevier, Amsterdam.

Kosarev, G., Kind, R., Sobolev, S.V., Yuan, X., Hanka, W. and Oreshin, S., 1999. Seismic evidence for a detached Indian lithospheric mantle beneath Tibet. *Science*, **283,** 1306-1309.

Kravanja, S., Panza, G.F. and Sileny, J., 1999. Robust retrievel of a seismic point-source function. *Geophys. J. Int.*, **136,** 385-394.

Krey, T., 1963. Channel waves as a tool of applied geophysics in Coal mining. *Geophysics*, **28,** 701-714.

Krishnan, M.S., 1953. The structure and tectonics of India. *Geol. Surv. India Mem.*, **81,** 1-109.

Krishnan, M.S., 1960. *Geology of India and Burma*. Higginbothams, Madras, India, 553 p.

Krishnaswamy, V.S., 1981. Status report of the work carried out by Geological Survey of India in the framework of the International Geodynamics Project. *In:* Zagros Hindukush, Himalaya Geodynamic Evolution, H.K. Gupta and Delany (eds.), *AGU Geodynamic Sr.* **3,** 169-188.

Kumar, A., Pandey, A.D., Sharma, M.L., Gupta, S.C., Jindal, A.K. and Jain, S.K., 1997. Contemporary local seismicity of the Garhwal Himalaya. Proc. Workshop on Earthquake Disaster preparedness, October 13-14, Roorkee, pp. 39-47.

Kumar, A., Pandey, A.D., Sharma, M.L., Gupta, S.C., Jindal, A.K. and Jain, S.K., 1998. Pattern of two earthquake swarms in the Garhwal Himalaya. Proc. Eleventh Symp. on Earthquake Engineering, University of Roorkee, Roorkee, pp. 67-74.

Kumar, B. and Gautam, U.P., 1998. Focal mechanism of central Nepal earthquake of 31/01/1997. ASC 98 Symposium, December 1-3, 1998, Hyderabad, India, Abstract vol., pp. 57.

Kumar, M.R. and Mohan, G., 2005. Mantle discontinuities beneath Deccan volcanic province. *Earth Planet. Sci. Lett*, **217,** 252-263.

Kumar, M.R. and Rao, N.P., 1995. Significant trends related to slab seismicity and tectonics in the Burmese arc region from Harvard CMT solutions. *Phys. Earth Planet. Int.*, **90,** 75-80.

Kumar, M.R., Rao, N.P. and Chalam, S.V., 1996. A seismotectonic study of the Burma and Andaman Arc regions using centroid moment tensor data. *Tectonophysics*, **253,** 155-165.

Kumar, M.R., Saul, J., Sarkar, D., Kind, R. and Shukla, K.A. 2001. Crustal structure of the Indian shield: New constraints from teleseismic receiver functions. *Geophys. Res. Lett.*, **28,** 1339-1342.

Kumar, M.R., Solomon Raju, P., Uma Devi, E., Saul, J. and Ramesh, D.S., 2004. Crustal structure variations in northeast India from converted phases. *Geophys. Res. Lett.*, **31,** L17605, doi: 10.10.29 2004 GL020576.

Kumar, P., Tewari, H.C. and Khandekar, G., 2000. An anomalous high velocity layer at shallow crustal depths in the Narmada zone, India. *Geophys. J. Int.*, **142,** 95-107.

Kumar, P., Yuan, X., Kind, R. and Kosarev, G., 2005. The lithosphere-asthenosphere boundary in the Tien Shan-Karakoram region from S receiver functions: Evidence for continental subduction. *Geophys. Res. Lett.*, **32,** L07305, doi: 10.1029/ 2004GL022291.

Kumar, P., Yuan, X., Kind, R. and Ni, James, 2006. Imaging the colliding Indian and Asian lithospheric plates beneath Tibet. *J. Geophys. Res.*, **111**, B06308, doi: 10.1029/2005JB003930.

Kumar, S., Wesnousky, S.G., Rockwell, T.K., Ragona, D., Thakur, V.C. and Seitz, G.G., 2001. Earthquake recurrence and rupture dynamics of Himalayan frontal thrust. *Science*, **294**, 2328-2331.

Kuthánek, O., 1990. *Anatomy of Seismograms*. Elsevier Science, Amsterdam.

Lahiri, A.K. and Chakraborty, P., 2003. Repeat gravity surveys in Sikkim-Darjeeling Himalayas: Preliminary results. *J. Him. Geol.*, **24(2)**, 113-115.

Lahiri, A.K, Chakraborty, P.K., Singh, N.P. and Kayal, J.R., 2007. Seismic risk in Sikkim Himalaya: An assessment of repeat microgravity observations during last five years. *Curr. Sci.* (in press).

Lahr, J.C., 1992. Local earthquake location programs. *In:* A course on PC based seismic networks, W.H.K. Lee and D.A. Dodge (eds.). Open file report 92-441, USGS, 226-250.

Lanczos, C., 1961. *Linear Different Operators*. Van Nostrand, London, 564 p.

Langston, C.A., 1976. A body wave inversion of the Koyna, India earthquake of December 10, 1967 and some implications for body wave focal mechanism. *J. Geophys. Res.*, **81(41),** 2517-2529.

Langston, C.A. and Franco-Sperra, M., 1985. Modeling of the Koyna, India, aftershocks of 12th December 1967. *Bull. Seism. Soc. Am.*, **75,** 651-660.

Larson, K., Bürgmann, R., Bilham, R. and Freymueller, J.T., 1999. Kinematics of the India-Eurasia collision zone from GPS measurements. *J. Geophys. Res.*, **104,** 1077-1093.

Lave, J. and Avouac, J.P., 2000. Active folding of fluvial terraces across the Siwalik Hills, Himalayas of central Nepal. *J. Geophys. Res.*, **105,** 5735-5770.

Lay, T., 1989. Mantle, lower structure. *In:* The Encyclopedia of Solid Earth Geophysics, D. James (ed.). Van Nostrand-Reinhold, New York, 770-775.

Lay, T., Kanamori, H., Ammon, C.J., Nettles, M., Ward, S.N., Aster, R.C., Beck, S.N., Bilek, S.L., Brudzinski, M.R., Buttler, R., Deshon, H.R., Ekstrom, G, Satake, K. and Sipkin, S., 2005. The great Sumatra-Andaman earthquake of 26 December 2004. *Science*, **308,** 1127-1133.

Lay, T. and Wallace, Terry C., 1995. *Modern Global Seismology*. Academic Press, New York, USA, 521 p.

Le Dain, A.Y., Tapponnier, P. and Molnar, P., 1984. Active faulting and tectonics of Burma and surrounding regions. *J. Geophys. Res.*, **89,** 453-472.

Le Fort, P., 1975. Himalayas: The collided range, present knowledge of the continental arc. *Am. J. Sci.*, **275-A,** 1-44.

Le Pichon, X., 1968. Sea-floor spreading and continental drift. *J. Geophys. Res.*, **73,** 3661-3697.

Le Pichon, X., Francheteau, J. and Binnin, J., 1976. *Plate Tectonics*. Elsevier, Amsterdam, 311 p.

Lee, W.H.K. and Lahr, J.C., 1975. Hypo-71: A computer program for determining hypocenter, magnitude and first motion pattern of local earthquakes. Open file report, U.S. Geol. Surv. Rev. Ed.

Lee, W.H.K. and Peareyra, V., 1993. *Mathematical Introduction to Seismic Tomography*. Academic Press, New York.

Lee, W.H.K. and Stewart, S.W., 1981. *Principles and Applications of Microearthquake Networks.* Academic Press, New York, 293 p.

Lee, W.H.K., Bennett, R.E. and Meagher, K.L., 1972. A method of estimating magnitude of local earthquakes from signal duration. Open File Report, U.S. Geol. Surv.

Lee, W.H.K., Yerkes, R.F. and Simirenko, M., 1979. Recent earthquake activity and focal mechanisms in western Transverse Ranges. California. U.S. Geol. Surv. Circ., **799-A,** 1-26.

Lee, W.H.K., Kanamori, H., Jennings, P.C. and Kisslinger, C. (Eds.), 2002. *Earthquake and Engineering Seismology*. Academic Press, N.Y.

Lehmann, I., 1935. P^1. Pub. Bur. Cent. Seism. Internat, A, 14, 3.

Lemonnier, C., Marquis, G., Perrier, F., Avouac, J.P., Chitrakar, G., Kafle, B., Sapkota, S., Gautam, U., Tiwari, D. and Bano, M., 1999. Electrical structure of the Himalaya of Central Nepal: high conductivity around the mid-crustal ramp along the MHT. *Geophys. Res. Lett.*, **26,** 3261-3264.

Lepine, J.C., Hirn., A., Pandey, M.R. and Tater, J.M., 1984. Features of the P-waves propagated in the crust of the Himalayas. *Ann. Geophysicae*, **2,** 119-121.

Likhar, S., Kulkarni, M.N. and Kayal, J.R., 2006. Interpretation of post geodetic and seismic data of the 2001 Bhuj earthquake, Mw 7.7. *Curr. Sci.*, **91(2),** 225-229.

Linehan, 1940. Earthquakes in the West Indian region. *Trans Am. Geophys. Union*, pp 229-252.

Lines, L.R. (Editor), 1988. Investigation of Geophysical Data, Geophysics Reprint Series No. 9 SEG, USA 543 p.

Louvari, E.K. and Kiratzi, A.A., 1997. Rake: A windows program to plot earthquake focal mechanisms and the orientation of principal stresses. *Comput. Geosci.*, **23,** 851-857.

Love, A.E.H., 1911. *Some Problems of Geodynamics*. Cambridge Univ. Press, London.

Lyon-Caen and Molnar, P., 1983. Constraints on the structure of the Himalaya from an analysis of gravity anomalies and a flexture model of the lithosphere. *J. Geophys. Res.*, **88,** 8171-8191.

Mahadevan, T.M., 1994. Deep Continental Structure of India: A Review. *Geol. Soc. India Mem.*, **28,** 569 p.

Mahadevan, T.M., 1998. Tentative chronology of development of the deep continental structure of the Indian peninsular shield from the Precambrian through the

Phanerozoic. *In:* The Indian Precambrian, B.S. Paliwal (ed.). Scientific Publishers (India) Jodhpur, pp. 86-103.

Mahadevan, T.M. and Subbarao, K.V., 1999. Seismicity of the Deccan Volcanic Province – An evaluation of some endogenous factors. *Geol. Soc. India Mem.*, **43,** 453-484.

Mailing, D.H., 1973. *Coordinate Systems and Map Projection.* George Philip & Son, London.

Main, I.G., 1991. Damage mechanism with long range interactions: Correlation between the seismic b-value and the fractal two-point correlation dimension. *Geophys. J. Int.*, **107,** 531-541.

Mandal, P. and Rastogi, B.K., Satyanarayana, H.V.S. and Kousalya, M., 2004. Results from local earthquake velocity tomography: Implications toward the source process involved in generating the 2001 Bhuj earthquake in the lower crust beneath Kachchh (India). *Bull. Seism. Soc. Am.*, **94,** 633-649.

Mandelbort, B.B., 1982. *The Fractal Geometry of Nature.* W.H. Freeman, New York.

Marquering, H., Dhalen, F.A. and Nolet, G, 1999. Three-dimensional sensitivity kernels for finite-frequency traveltimes: The banana-doughnut paradox. *Geophys. J. Int.*, **137,** 805-815.

Marrusi, A., 1976. Geotectonics Della zone orogenish de Kashmir Himalaya, Karakoram-Hindukush-Pamir. Accademia Waxionals dei Lincel, pp 17-25.

Martin, S.S., 2005. Intensity distribution from the 2004 M 9.0 Sumatra-Andaman earthquake. *Seism. Res. Lett.*, **76,** 321-330.

Mathur, L.P. and Kohli, G., 1963. Exploration and development for oil in India. World Pet. Cong., **6,** 633 p.

Mathur, L.P. and Evans, P., 1964. Oil in India (Sp.Brochure). 22nd Int. Geol. Congr., New Delhi, 85 p.

McNamara, D.E., Owens, T.J., Silver, P.G. and Wu, F.T., 1994. Shear wave anistropy beneath the Tibetan Plateau. *J. Geophys. Res.*, **99,** 13,655-13,665.

McCaplin, J.P. and Thakkar, M.G., 2001. The 2001 Bhuj-Kachchh earthquake: Surface faulting and its relation with neotectonics and regional structures, Gujarat, western India. *Annals of Geophysics*, **46,** 937-956.

McCloskey, J., Nalbant, S.S. and Steacy, S., 2005. Indonesian earthquake: Earthquake risk from co-seismic stress. *Nature*, **434,** doi: 10.1038/434291a.

McKenzie, D.P., 1967. The viscosity of the mantle. *Geophys. J. Roy. Astro. Soc.*, **14,** 297-305.

McKenzie, D.P., 1969. Speculations on the consequences and causes of plate motions. *Geophys. J. R. Astro. Soc.*, **18,** 1-32.

McKenzie, D.P. and Sclater, J.G., 1971. The evolution of the Indian ocean since the late Cretaceous. *Geophys. J. R. Astro. Soc.*, **24,** 437-528.

Mellors, R.J., Pavlis, G.L., Hamburger, M.W., Al-Shukri, H.J. and Lukk, A.A., 1995. Evidence for high velocity slab associated with the Hindukush seismic zone. *J. Geophys. Res.*, **100,** 4067-4078.

Mendoza, C. and Hartzell, S.H., 1988. Aftershock patterns and main shock faulting. *Bull. Seism. Am.*, **78,** 1438-1449.

Menke, W., 1984. *Geophysical Data Analysis: Discrete Inverse Theory.* Academic Press, Orlando, 200 p.

Metivier, F., Gaudemer, Y., Tapponier, P. and Klein, M., 1999. Mass accumulation rates in Asia during the Cenozoic. *Geophys. J. Int.*, **137,** 280-318.

Middlemiss, C.S., 1910. The Kangra earthquake of 4th April 1905. *Geol. Surv. India Mem.*, **38,** 409 p.

Milne, J., 1911. Catalogue of destructive earthquakes – A.D. 7 to A.D. 1809. British Assoc. Adv. Sci. Rep., Portsmouth meeting, Appendix 1, 649-740.

Milne, A.R., 1959. Comparison of spectra of an earthquake T-phase with similar signals from nuclear explosions. *Bull. Seism. Soc. Am.*, **49,** 317-329.

Mishra, D.C., 1992. Midcontinental gravity high in central India and the Gondwana tectonics. *Tectonophysics*, **212,** 153-161.

Mishra, D.C. and Tiwari, R.K., 1981. Spectral study of the Bouguer anomaly map of a rift valley and adjacent areas in Central India. *Pure and Appl. Geophys.*, **119,** 1051-1062.

Mishra, D.C., Laxman, G., Rao, M.B.S.V. and Gupta, S.B., 1995. Analysis of gravity-magnetic data around Nagaur-Jhalawar geotransect. *In:* Continental crust of NW and central India, S. Sinha-Roy and K.R. Gupta (eds.). *Geol. Soc. India Mem.*, **31,** 345-352.

Mishra, D.C., Chandrasekhar, D.V. and Singh, B., 2005. Tectonics and crustal structures related to Bhuj earthquake of January 26, 2001: Based on gravity and magnetic surveys, constrained from seismic and seismological studies. *Tectonophysics*, **396,** 195-207.

Mishra, O.P. and Zhao, D., 2003. Crack density saturation rate and porosity at the 2001 Bhuj, India, earthquake hypocenter: A fluid driven earthquake. *Earth Planet. Sci. Lett.*, **212,** 393-405.

Mishra, O.P., Kayal, J.R., Chakraborty, G.K., Singh, O.P. and Ghosh, D., 2007. Aftershock investigation in the Andaman Nicobar islands of India and its seismotectonic implications. *Bull. Seism. Soc. Am.*, **97,** 1A, S71-S85.

Mitchell, A.H.G., 1981. Phanerozoic plate boundaries in mainland SE Asia, the Himalayas and Tibet. *Geol. Soc. London J.*, **138,** 109-122.

Mitchell, A.H.G. and McKerrow, W.S., 1975. Analogous evolution of the Burma orogen and the Scottish caledonides. *Geol. Soc. Am. Bull.*, **86,** 305-315.

Mitra, S., Pristley, K., Bhattacharya, A. and Gaur, V.K., 2005. Crustal structure and earthquake focal depths beneath northeastern India and southern Tibet. *Geophys. J. Int.*, **160,** 227-248.

Mitra, S., Priestley, K., Gaur, V.K., Rai, S.S. and Haines, J., 2006. Variation of Rayleigh wave group velocity dispersion and seismic heterogeneity of the Indian crust and uppermost mantle. *Geophys. J. Int.*, **164,** 88-98.

Mogi, K., 1962. On the time distribution of aftershocks accompanying the recent major earthquakes in and near Japan. *Bull. Earthquake Res. Inst.*, **40,** 107-124.

Mohan, G., Rai, S.S. and Panza, G., 1992. Seismic heterogeneities in the Indian lithosphere. *Phys. Earth Planet. Interior*, **73,** 189-198.

Molnar, P., 1984. Structure and tectonics of the Himalaya: Constraints and implication of geophysical data. *Ann. Rev. Earth Planet. Sci.*, **12,** 489-518.

Molnar, P., 1987. The distribution of intensity associated with the great 1897 Assam earthquake and bounds on the extent of the rupture zone. *J. Geol. Soc. India*, **30,** 13-27.

Molnar, P., 1988. Continental tectonics in the aftermath of plate tectonics. *Nature*, **335,** 131-137.

Molnar, P., 1990. A review of the seismicity and the rates of active underthrusting and deformation at the Himalaya. *J. Him. Geol.*, **1,** 131-154.

Molnar, P. and Chen, W.P., 1982. Seismicity and Mountain building. *In:* Mountain building process. Academic Press, London.

Molnar, P. and Deng, Q., 1989. Faulting associated with large earthquakes and the average rate of deformation in central and eastern Asia. *J. Geophys. Res.*, **89,** 6203-6227.

Molnar, P. and Lyon-Caen, H., 1989. Fault plane solutions of earthquakes and active tectonics of the Tibetan Plateau and its margins. *Gephys. J. R. Astro. Soc.*, **99,** 123-153.

Molnar, P. and Pandey, M.R., 1989. Rupture zones of great earthquakes in the Himalaya region. *Proc. Indian Acad. Sci. (Earth Planet. Sci.)*, **98,** 61-70.

Molnar, P. and Tapponnier, P., 1977. Relation of the tectonics of eastern China to the India-Eurasia collision: Application of slip-line field theory to large-scale control tectonics. *Geology*, **5,** 212-216.

Molnar, P., Fitch, T.J. and Wu, F.T., 1973. Fault plane solutions of shallow earthquakes and contemporary tectonics in Asia. *Earth Planet. Sc. Lett.*, **19,** 101-112.

Molnar, P., Chen, W.P., Fitch, T.J., Tapponnier, P., Warsi, W.E.K. and Wu, F.T., 1977. Structure and tectonics of the Himalaya: A brief summary of relevant geophysical observations. *In:* Himalaya. Science de la Terre, Paris, pp. 269-294.

Monsalve, G., Sheehan, A., Schulte-Pelkum, V., Rajaure, S., Pandey, M.R. and Wu, F., 2006. Seismicity and 1-D velocity structure of the Himalayan collision zone. Earthquakes in the crust and upper mantle. *J. Geophys. Res.*, **111,** B10301, doi: 10.1029/2005 JB004062.

Morelli, A. and Dziewonski, A.M., 1993. Body wave traveltimes and a spherically symmetric P- and S-wave velocity model. *Geophys. J. Int.*, **112,** 178-194.

Morgan, J.P. and McIntyre, W.G., 1959. Quaternary geology of the Bengal basin. *Geol. Soc. Am Bulletin*, **70,** 319-342.

Mori, J., Negishi, H. and Sato, T., 2001. Slip distribution of the 2001 west India earthquake from inversion of teleseismic data. *In:* Int. Conf. Seismic Hazard with particular reference to Bhuj earthquake of January 26, 2001. IMD-DST, New Delhi, Oct. 3-5, 2001.

Mueller, K. and Pujol, J., 2001. Three dimensional geometry of the Reeltoot blind thrust: Implications for moment release and earthquake magnitude in the New Madrid Seismic Zone. *Bull. Seism. Soc. Am.*, **91,** 1563-1573.

Mukherjee, S.N., 1942. Seismological features of the Satpura earthquake of the 14th March, 1938. *Proc. Ind. Acad. Sci.*, **16,** 167-175.

Mukhopadhyay, M., 1974. Relationship of gravity and seismicity to tectonics in northeastern India. Ph.D. Thesis, Indian School of Mines, Dhanbad, India.

Mukhopadhyay, M., 1984. Seismotectonics of transverse lineaments in the eastern Himalaya and its foredeep. *Tectonophysics*, **109,** 227-240.

Mukhopadhyay, M. and Dasgupta, S., 1988. Deep structure and tectonics of the Burmese arc: Constraints from earthquake and gravity data. *Tectonophysics*, **149,** 299-322.

Mukhopadhyay, M., Verma, R.K. and Ashraf, M.H., 1986. Gravity field and structures of the Rajmahal hills: Example of the Paleo-Mesozoic continental margin in eastern India. *Tectonophysics*, **131,** 353-367.

Mukhopadhyay, S. and Kayal, J.R., 2003. Seismic tomography structure of the 1999 Chamoli earthquake in the Garhwal Himalaya. *Bull. Seism. Soc. Am.*, **93(4),** 1854-1861.

Mukhopadhyay, S., Khattri, K.N. and Chander, R., 1995. Seismic velocity and related elastic parameters of the crust in the Shillong Massif. *J. Him. Geol.*, **6(1),** 1-8.

Mukhopadhyay, S., Mishra, O.P., Zhao, D. and Kayal, J.R., 2006. 3-D seismic structure of the source area of the 1993 Latur, India, earthquake and its implications for rupture nucleations. *Tectonophysics*, **415,** 1-16.

Murthy, M.V.N., 1970. Tectonic and mafic igneous activity in northeast India in relation to the upper mantle. Proc. 2nd Symp. Upper Mantle Project, NGRI, Hyderabad, pp 287-304.

Murthy, M.V.N., Talukdar, S.C., Bhattacharya, A.C. and Chakrabarty, C., 1969. The Dauki fault of Assam. *Bull. Oil & Natural Gas Comm.*, **6(2),** 57-64.

Murti, K.V.S., 1969. Regional analysis of seismic data from upper Assam valley. 2nd Indian Pet. Conf., Baroda, India.

Nair, K.K.K., Jain, S.C. and Yedekar, D.B., 1985. Geology, structure and tectonics of Son-Narmada-Tapti lineament zone. *Geol. Surv. India Rec.*, **117,** 138-147.

Nakata, T., 1972. Geomorphic history and crustal movements of foothills of the Himalaya. Sendai Institute of Geography, Tohuku University, 77 p.

Nakata, T., 1989. Active fault of the Himalayas of India and Nepal. *In:* Tectonics of Western Himalayas, L.L. Malinconico Jr. and R. Lillie (eds.). *Geol. Soc. Am. Sp. Pap.*, **232,** 243-264.

Nakata, T., Otsuki, K. and Khan, S.H., 1990. Active faults, stress field and plate motion along Indo-Eurasian plate boundary. *Tectonophysics*, **181,** 83-95.

Nandy, D.R., 1976. The Assam syntaxis of the Himalayas – A re-evaluation. *Geol. Surv. India Misc. Pub.*, **24(2),** 363-367.

Nandy, D.R., 1983. The eastern Himalaya and the Indo-Burman orogen in relation to the Indian plate movement. *Geol. Surv. India Misc. Pub.*, **43,** 153-159.

Nandy, D.R., 2001. *Geodynamics of Northeastern India and the Adjoining Region*, ACB Publication, Kolkata, 209 p.

Nanjo, K., Nagahama, H. and Satmura, M., 1998. Roles of aftershock decay and the fractal structure of active fault systems. *Tectonophysics*, **287,** 173-186.

Narula, P.L., Shome, S.K., Kumar, S. and Pandey, P., 1995. Damage patterns and delineation of isoseismals of Uttarkashi earthquake of 20th October, 1991. *In:* Uttarkashi Earthquake, H.K. Gupta and G.D. Gupta (eds.). *Geol. Soc. India Mem.*, **30,** 1-18.

Natawidjaja, D.H., Sieh, K., Ward, S.N., Cheng, H., Edwards, R.L., Galetzka, J. and Suwargadi, B.W., 2004. Paleogeodetic records of seismic and aseismic subduction from central Sumatran microattolls, Indonesia. *J. Geophys. Res.*, 109, doi: 1029/2003JB002398.

Nath, S.K., Sengupta, P. and Kayal, J.R., 2002. Determination of site response at Garhwal Himalaya, from aftershock sequence of 1999 Chamoli earthquake. *Bull. Seism. Soc. Am.*, **92,** 1072-1081.

Nayak, P.N., 1990. Deep crustal configuration of central India. Geol. Surv. India Sp. Pub., **28,** 67-98.

Negishi, H., Mori, J:, Sato, T., Singh, R., Kumar, S. and Hirata, N., 2002. Size and orientation of the fault plane for the 2001 Gujarat, India earthquake Mw 7.7 from aftershock observations: A high stress drop event. *Geophys. Res. Lett.*, **29(20),** 10-1-10-4.

Nelson, K.D., Zhao, W. and Project INDEPTH Team 1996. Partially molten middle crust beneath southern Tibet: Synthesis of project INDEPTH Results. *Science*, **274,** 1684-1688.

Nersesov, I.O., Semenov, A.N. and Simbireva, I.G., 1971. Space-time distribution of ratios of travel times of P and S waves in the Garm Region. *In:* Experimental Seismology. Science Publishers, Moscow, pp 334-345.

NGRI, 1974. Bouguer gravity anomaly map of India, 1:5 million scale. National Geophys. Res. Inst., Hyderabad, India.

Ni, J.F., 1989. Active tectonics of the Himalaya. *Proc. Indian Acad. Sci. (Earth Planet Sci)*, **98,** 71-89.

Ni, J.F. and Barazangi, M., 1984. Seismotectonics of the Himalayan collision zone: Geometry of the underthrusting Indian Plate beneath the Himalaya. *J. Geophys. Res.*, **89,** 1147-1163.

Ni, J.F., Guzman-Speziale, M., Bevis, M., Holt, W.E., Wallace, T.C. and Seager, W., 1989. Accretionary tectonics of Burma and the three dimensional geometry of the Burma subduction zone. *Geology*, **17,** 68-71.

Nolet, G. (Editor), 1987. *Seismic Tomography*. D. Reidal Pub. Co., Dordrecht, 386 p.

Norin, E., 1946. Geological exploration in western Tibet. Rep. Sion-Swedish Expedition, Stockholm, **16,** 1-229.

Ogata, Y., 1988. Statistical model of aftershocks temporal behaviour. *J. Am. Satist. Assoc.*, **83,** 9-27.

Oldham, R.D., 1899. Report on the great earthquake of the 12th June 1897. *Geol. Surv. India Mem.*, **29,** Reprinted, 1981, Geological Survey of India, Calcutta, 379 p.

Oldham, R.D., 1906. The constitution of the interior of the Earth, as revealed by earthquakes. *Q. J. Geol. Soc.*, **62,** 456-475.

Oldham, R.D., 1928. The Cutch earthquake of 16th June 1819 with a revision of the great earthquake of the 12th June 1897. *Geol. Surv. India Mem.*, **46,** 80-147.

Oldham, T., 1883. Catalogue of Indian earthquakes. *Geol. Surv. India Mem.*, **19,** 163-215.

Oldham, T. and Oldham, R.D., 1882. The Cachar earthquake of 10th January 1869, R.D. Oldham (ed.). *Geol. Surv. India Mem.*, **19,** 1-98.

Oliver, J. and Isacks, B., 1967. Deep earthquake zones, anomalous structure in the upper mantle and the lithosphere. *J. Geophys. Res.*, **72,** 4259-4275.

Oliver, J., Ryall, A., Brune, J.N. and Slemmons, D.B., 1966. Microearthquake activity recorded by portable seismographs of high sensitivity. *Bull. Seism. Soc. Am.*, **56,** 899-924.

Omori, F., 1894. On aftershocks (in Japanese). *Rep. Imp. Earthquake Invest Comm.*, **2,** 103-139.

Ortiz, M. and Bilham, R., 2003. Source area and rupture parameters of 31 December, 1881, Mw 7.9 Car Nicobar earthquake estimated from Tsunamis recorded in the Bay of Bengal. *J. Geophys. Res.*, **108,** 2215-2230.

Ouchi, T. and Uekawa, T., 1986. Statistical analysis of the spatial distribution of earthquakes – Variation of the spatial distribution of earthquakes before and after large earthquakes. *Phys. Earth Planet. Inter.*, **44,** 211-225.

Owens, T.J., Tayolr, S.R. and Zandt, G., 1987. Crustal structure at regional seismic test network stations determined from inversion of broadband teleseismic P waveforms. *Bull. Seism. Soc. Am.*, **77,** 631-662.

Pacheco, J.F., Scholz, C.H. and Sykes, L.R., 1992. Changes in frequency-size relationship from small to large earthquakes. *Nature*, **355,** 71-73.

Page, R., 1968. Focal depths of aftershocks. *J. Geophys. Res.*, **73,** 3897-3903.

Paige, C.C. and Saunders, M.A., 1982. LSQR: Sparse linear equations and least squares problems. *ACM Transactions on Mathematical Software*, **8(2),** 195-209.

Panda, P.K., 1985. Geothermal maps of India and their significance in resources assessment. *Pet. Asia J.*, **VIII,** 202-210.

Pande, P., Kayal, J.R., Joshi, Y.C. and Ghevariya, Z.G., 2003. Lithotectonic framework of Gujarat and adjoining regions. *In:* Kutch (Bhuj) earthquake 26 January 2001. P. Pande and J.R. Kayal (eds.), *Geol. Surv. India Sp. Pub.*, **76,** 5-9.

Pande, P., Joshi, Y.C., Sharda, Y.P., Joshi, K.C., 2003. Isoseismal map and isoseist characteristics. *In:* Kutch (Bhuj) earthquake 26 January 2001. P. Pande and J.R. Kayal (eds.), *Geol. Surv. India Sp. Pub.*, **76,** 147-156.

Pandey, M.R. and Molnar, P., 1988. The distribution of intensity of the Bihar Nepal earthquake of 15th January, 1934 and bounds on the extent of the rupture zone. *J. Geol. Soc., Nepal*, **5,** 22-44.

Pandey, M.R., Tandukar, R.P., Avouac, J.P., Lave, J. and Massot, J.P., 1995. Interseismic strain accumulation on the Himalayan crustal ramp (Nepal). *Geophys. Res. Lett.*, **22,** 751-754.

Pandey, M.R., Tandukar, R.P., Avouac, J.P., Vergne, J. and Heritier, T.H., 1998. Characteristics of seismicity of Nepal and their seismotectonic implications, Extd Abstract. ASC 98 Symposium, December 1-3, 1998, Hyderabad, India.

Pandey, M.R., Tandukar, R.P., Avouac, J.P., Vergne, J. and Heritier, T.H., 1999. Seismotectonics of the Nepal Himalaya from local seismic network. *J. Asian Earth Sc.*, **17,** 703-712.

Park, J., Anderson, K., Aster, R., Butler, R., Lay, T. and Simpson, D., 2005. Global seismographic network records the great Sumatra Andaman earthquake. *EOS Transaction, Am. Geophys. Union*, **86(6)**, 57, 60-61.

Patro, B.P.K., Harinarayana, T., Sastry, R.S., Rao, M., Manoj, C., Naganjaneyulu, K. and Sarma, S.V.S., 2004. Electrical imaging of Narmada-Son lineament zone, central India from magnetotellirics. *Phys. Earth. Plan. Int.*, **148,** 215-232 .

Patton, H., 1980. Crust and upper mantle structure of the Eurasian continent from the phase velocity measurements and Q of surface waves. *Rev. Geophys.*, **18,** 605-625.

Paul, A. and Pant, C.C., 2005. On the seismicity and seismic hazard estimation of the Kumaun Himalaya. *Palaeontol. Soc. Ind. Sp Pub.*, **2,** 41-51.

Paul, D.D. and Lian, H.M., 1975. Tertiary basins of southeast Asia, Bay of Bengal to South China Sea. Proc. World Pet. Congr. 9th, 107-121.

Pavlis, G. and Booker, J., 1983. Progressive multiple event location (PMEL). *Bull. Seism. Soc. Am.*, **73,** 1753-1777.

Pegler, G. and Das, S., 1998. An enhanced image of the Pamir-Hindukush seismic zone from relocated hypocenters. *Geophys. J. Int.*, **134,** 573-595.

Person, W.J. (Ed.), 1989. Seismological notes – July-August 1988. *Bull. Seism. Soc. Am.*, **79,** 925-932.

Pham, V.N., Boyer, D., Therme, P., Yuan, X.C., Li, L. and Jin, G.Y., 1986. Partial melting zones in the crust in southern Tibet from magnetotelluric results. *Nature*, **319,** 310-314.

Pines, I., Teng, T.L., Rosenthal, R. and Alexander, S., 1980. A surface wave dispersion study of the crustal and upper mantle structure of China. *J. Geophys. Res.*, **85,** 3829-3844.

Plafker, G. and Galloway, J.P., 1989. Lessons learned from the Loma Prieta California earthquake of October 17, 1989. USGS open file Rep.

Poddar, M.C., 1950. Preliminary report of the Assam earthquake of 15th August 1950. *Geol. Surv. India Bull. Ser.*, **B(2),** 1-40.

Powell, C.M., 1979. A speculative tectonic history of Pakistan and surroundings – Some constraints from Indian Ocean. *In:* Geodynamics of Pakistan, A. Farah and A.K. Dee Jong (eds.). Geological Survey of Pakistan, Quetta, pp 5-24.

Powell, C.M., Johnson, B.D. and Veevers, J.J., 1980. A revisited fit of east and west Gondwanaland. *Tectonophysics*, **63,** 13-29.

Powers, P.M., Lillie, R.J. and Yeats, R.S., 1998. Structure and shortening of the Kangra and Dehradun reentrants. *Geol. Soc. Am. Bull.*, **110,** 1010-1027.

Pratt, J.H., 1855. On the attraction of the Himalayan mountains and of elevated regions beyond them, upon the plumb line in India. *Phil. Trans. R. Soc. London*, **145,** 53-100.

Pratt, J.H., 1859. On the computations of the effect of the attraction of the mountatin masses of disturbing the apparent astronomical latitude stations in geodetic surveys. *Phil. Trans. R. Soc. Lond.*, **147,** 747-763.

Pujol, J., 1988. Comments on the joint determination of hypocenters and station corrections. *Bull. Seism. Soc. Am.*, **78,** 1179-1189.

Pulli, J.J. and Upton, Z.M., 2002. Hydroacoustic observations of Indian earthquakes provide new data on T-waves. *EOS Transactions, Am Geophys. Union*, **83(13)**.

Purucker, M.E., Langel, R.A., Rajaram, Mita and Raymond, C., 1998. Global magnetization models with apriori information. *J. Geophys. Res.*, **103,** 2563-2584.

Qureshy, M.N., 1969. Thickening of the basalt layer as a possible cause for the uplift of the Himalaya – A suggestion based on gravity data. *Tectonophysics*, **7,** 137-157.

Qureshy, M.N., 1970. Relation of gravity to elevation, geology and tectonics in India. Proc. Second Symposium on Upper Mantle Project, NGRI, Hyderabad, pp. 1-20.

Qureshy, M.N., 1971. Relation of gravity to elevation and rejuvenation of blocks in India. *J. Geophys. Res.*, **76(12),** 545-557.

Qureshi, M.N., 1981. Gravity anomalies, isostasy and crust mantle relations in the Deccan trap and contignous regions (India). *Geol. Soc. India Mem.*, **3,** 184-197.

Qureshy, M.N. and Midha, R.K., 1986. Deep crustal signature in India and contignous regions from satellite and ground geophysical data. *In:* Reflection seismology, the continental crust, M. Barazangi and L. Brown (eds.). *Geodynamic series 14,* AGU Pub., pp. 77-94.

Qureshy, M.N. and Warsi, W.E.K., 1975. Role of regional gravity surveys in a concept oriented exploration programme. *J. Geol. Soc. India*, **16(1),** 44-54.

Qureshy, M.N. and Warsi, W.E.K., 1980. A Bouguer anomaly map of India and its relation to broad tectonic elements of the sub-continent. *Geophys. J.R. Astron. Soc.*, **61,** 235-242.

Radhakrishna, B.P. and Naqvi, S.M., 1986. Precambrian continental crust of India and its evolution. *J. Geol. Soc. India*, **94,** 145-166.

Ragan, D.M., 1973. *Structural Geology : An introduction to geometrical techniques*, 2nd ed. John Wiley.

Raha, P.K. and Sastry, M.V.A., 1982. Stromatolites and Precambrian stratigraphy in India. *Precambrian Research*, **18,** 293-318.

Rai, S.S., Ramesh, D.S., Srinagesh, D., Suryaprakasham, K., Mohan, G., Rajagopala Sarma, P.V.S.S. and Satyanarayana, Y., 1992. Seismic tomography of the south Indian shield. *In:* Seismology in India – An overview. *Current Science, Special Issue*, **62,** 213-226.

Rai, S.S., Singh, S.K., Rajagopal Sarma, P.V.S., Srinagesh, D., Reddy, K.N.S., Suryaprakasham, K. and Satynarayana, Y., 1999a. What triggers Koyna region earthquakes? Preliminary results from seismic tomography digital array. *Proc Earth Planet. Sci.*, **108,** 1-14.

Rai, S.S., Suryaprakasham, K. and Agarwal, N., 1999b. Pn wave velocity and Moho geometry in northeastern India. *Proc. Indian Acad. Sci. (Earth Planet. Sci.)*, **108(4),** 297-304.

Rai, S.S., Priestley, K., Suryaprakasham, K., Srinagesh, D., Gaur, V.K. and Du, Z., 2003. Crustal shear velocity structure of the south Indian shield. *J. Geophys. Res.*, **108**(B2), 2088, doi: 10.1029/2002JB001776.

Rai, S.S., Vijay Kumar, T. and Jagadeesh, S., 2005. Seismic evidence for significant crustal thickening beneath Jabalpur earthquake, 21 May 1997, source region in Narmada-Son Lineament, central India. *Geophys. Res. Lett.*, **32,** L22306, doi: 10.1029/2005GL023580.

Raiverman, V., Kunte, S.V., Mukherjee, A., 1983. Basin geometry, Cenozoic sedimentation and hydrocarbon prospects in north western Himalaya and Indo-Gangetic Plains. *Pet. Asia J.*, **2,** 67-92.

Rajal, B.S., Virdi, N.S. and Hasija, N.L., 1986. Recent crustal uplift in the Doon Valley, *Proc. Int. Symp. on Neotectonics in S. Asia*, Dehra Dun, India, pp. 146-159.

Rajaram, Mita, 2000. Research highlights of Magsat Studies. *In:* Research highlights in Earth System Sciences with focus on Deep Continental Studies, O.P. Verma and T.M. Mahadevan (eds.), Dept. Sci. Tech., Vol. I, pp. 107-112.

Rajaram, Mita, Anand, S.P. and Erram, V.C., 2000. Magnetic studies over Krishna-Godavari basin in Eastern Continental margin of India. *Gondwana Research*, **3,** 385-393.

Rajaram, Mita and Anand, S.P., 2003. Crustal structure of south India from aeromagnetic data. *In:* Geophysics: Window to Indian Geology, Mita Rajaram (ed.). *Virtual Explorer*, **12,** 72-82.

Rajendran, C.P. and Rajendran, K., 1997. Geological investigations at Killari and Ter, central India and their implications for paleoseismicity in the shield region. *Tectonophysics*, **308,** 67-81.

Rajendran, C.P. and Rajendran, K., 2001. Characteristic deformation and past seismicity associated with 1819 Kutch seismic zone. *Bull. Seism. Soc. Am.*, **91,** 407-426.

Rajendran, C.P. and Rajendran, K., 2002. Historical constraints on previous seismic activity and morphologic changes near the source zone of the 1819 Rann of Kutch earthquake: Further light on the penultimate event. *Seism. Res. Lett.*, **73(4),** 470-479.

Rajendran, C.P. and Rajendran, K., 2003. Earthquake recurrence in peninsular India: Status and prospects. *Gond. Geol. Magz.*, Spl. V. 5, 107-124.

Rajendran, C.P. and Rajendran, K., 2004. Studying earthquake recurrence in the Kachchh region, India. *EOS Transactions*, *Am Geophys Union*, **84,** 529-536.

Rajendran, C.P. and Rajendran, K., 2005. The status of central seismic gap: A perspective based on spatial and temporal aspects of large Himalayan earthquakes. *Tectonophysics*, **395,** 19-39.

Rajendran, C.P., Rajendran, K. and John, B., 1996. The 1993 Killari (Latur), central India, earthquake: An example of fault reactivation in the Precambrian crust. *Geology*, **24,** 651-654.

Rajendran, C.P., Rajendran, K. and John, B., 1998. Surface deformation related to the 1819 Kachchh earthquake: Evidence for recurrent activity. *Curr. Sci.*, **75,** 623-626.

Rajendran, C.P., Earnest, A., Rajendran, K., Dev Das, R. and Sreekumari, K., 2003. The 13 September 2002 North Andaman (Diglipur) earthquake: An analysis in the context of regional seismicity. *Curr. Sci.*, **84(7),** 919-924.

Rajendran, C.P., Rajendran, K., Durah, B.P., Baruah, S. and Earnest, A., 2004. Interpreting the style of faulting and paleoseismicity associated with the 1897 Shillong, northeast India earthquake: Implications for regional tectonism. *Tectonics*, **23,** TC4009, doi 1029/2003 TC001605.

Rajendran, K. and Rajendran, C.P., 1998. Characteristics of the 1997 Jabalpur earthquake and their bearing on its mechanism. *Curr. Sci.*, **74,** 168-174.

Rajendran, K., Talwani, P. and Gupta, H.K., 1992. State of stress in the Indian subcontinent – A review. *Curr. Sci.*, **62,** 86-93.

Rajendran, K., Rajendran, C.P., Thakkar, M. and Tuttle, M.P., 2001. The 2001 Kutch (Bhuj) earthquake: Coseismic surface features and their significance. *Curr. Sci.*, **80(11),** 1397-1405.

Rajendran, K., Rajendran, C.P., Thakkar, M. and Gartia, R.K., 2002. Sand blows from the 2001 Bhuj earthquake reveal clues on past seismicity. *Curr. Sci.*, **83,** 603-610.

Rajendran, K., Rajendran, C.P. and Earnest, A., 2005. The great Sumatra-Andaman earthquake of 26, December 2004. *Curr. Sci.*, **88(1),** 11-12.

Rajput, S., Gahalaut, V.K., Raju, P.S. and Kayal, J.R., 2005. Rupture parameters of the 1999 Chamoli earthquake in Garhwal Himalaya: Constraints from aftershocks and change in failure stress. *Tectonophysics*, **404,** 23-32.

Ram Babu, H.V., 1999. Gravity image of India. *Curr. Sci.*, **76(12),** 1533-1535.

Ramakrishna, T.S. and Chayanulu, A.Y.S.R., 1988. A geophysical appraisal of the Purana basins of India. *J. Geol. Soc. India.*, **32,** 48-60.

Ramesh, D.S. and Estabrook, C.H., 1998. Rupture histories of two stable continental region earthquakes of India. *Proc. Indian Acad. Sci (Earth Planet Sci.)*, **107,** L225-L233.

Ramesh, D.S., Kumar, M.R., Devi, E.U., Raju, P.S. and Yuan, X., 2005. Moho geometry and upper mantle images of northeast India. *Geophys. Res. Lett.*, **32,** 14301-14304, doi: 10.1029/2005GL022789.

Rana, B.S.J.B., 1935. Nepalko Maha Bhukampa (The great earthquake of Nepal), Kathmandu, Nepal (in Nepali).

Rao, A., 1983. Geology and hydrocarbon potential of a part of Assam-Arakan basin and its adjacent region. *In:* Petroliferous basins of India, B.P. Radhakrishna (ed.). *Pet. Asia J.*, pp. 127-158.

Rao, D.T., Jambusaria, B.B., Srivastava, S., Srivastava, N.P., Hamid, A., Desai, B.N. and Srivastava, H.N., 1991. Earthquake swarm activity in south Gujarat. *Mausam*, **42,** 89-98.

Rao, M.B.R., 1973. The sub-surface geology of the Indo-Gangetic plains. *J. Geol. Soc. India*, **14,** 217-242.

Rao, N.P. and Chary, H., 2005. What caused the great Sumatran earthquakes of 26 December 2004 and 28 March 2005? *Curr. Sci.*, **89,** 449-452.

Rao, N.P. and Kalpana, 2005. Deformation of the subducted Indian lithospheric slab in the Burmese arc. *Geoph. Res. Lett.*, **32,** L05301, doi: 10.1029/2004GL022034.

Rao, N.P., Kumar, P., Tsukuda, T., Kosuga, M., Bhatia, S.C. and Suresh, G., 2002. Deep crustal earthquakes in central India: Inferences from analysis of regional broadband data of the 1997 May 21, Jabalpur earthquake. *Geophys. J. Int.*, **148,** 132-138.

Rao, N.P., Kumar, P., Kalpana, Tsukuda, T. and Ramesh, D.S., 2006. The devastating Muzaffarabad earthquake of 8 October 2005: New insights into Himalayan seismicity and tectonics. *Gond. Res.*, **9,** 365-378.

Rao, Ramkrishna B., 1992. Seismicity and geodynamics of the low-to-high grade transition zone of peninsular India. *Tectonophysics*, **20,** 175-185.

Raper, F.V., 1810. Narratives of a survey for the purpose of discovering the resources of the Ganges. *Asiat. Res.*, **11,** 446-563.

Rastogi, B.K., 1986. Risk of reservoir induced seismicity and necessary investigations. Proc. Sem. Engg. Geophys. Perspectives and Prospects. Indian Geophys. Union, Hydearbad, 7-15.

Rastogi, B.K., 1994. Latur earthquake: Not triggered. *Geol. Soc. India Mem.*, **35,** 131-138.

Rastogi, B.K., 1995. Seismological studies of Uttarkashi earthquake of October 20, 1995. *Geol. Soc. India Mem.*, **30,** 43-50.

Rastogi, B.K., 2000. Chamoli earthquake of magnitude 6.6 on March 29, 1999. *J. Geol. Soc. India*, **55,** 505-514.

Rastogi, B.K., 2001a. Ground deformation study of Mw 7.7 Bhuj earthquake of 2001. *Episodes*, **24(3),** 160-165.

Rastogi, B.K., 2001b. Erattupetta earthquake of 12 December 2000 and seismicity of Kerala. *J. Geol. Soc. India*, **57,** 273-278.

Rastogi, B.K. and Chadha, R.K., 1995. Intensity and isoseismals of Uttarkashi earthquake of October 20, 1991. *Geol. Soc. India Mem.*, **30,** 19-24.

Rastogi, B.K. and Rao, M.N. 1994. After effects of Latur earthquake smoke-gas emanations and subterranean sounds/microearthquakes. *Geol. Soc. India Mem.*, **35,** 139-149.

Rastogi, B.K., Singh, J. and Verma, R.K., 1973. Earthquake mechanisms and tectonics in the Assam-Burma region. *Tectonophysics*, **18,** 355-366.

Rastogi, B.K., Chadha, R.K. and Rajagopalan, G., 1993. Palaeoseismicity studies in Meghalaya. *Curr. Sci.*, **64,** 933-935.

Raval, U., 1989. On hotspots, Meso-Cenozoic tectonics, and possible thermal networking beneath the Indian continent. Proc. Seminar on Advances in Geophysical Research in India, 8-10 Feb., Indian Geophys. Union, Hyderabad. pp. 314-330.

Ravishanker, 1988. Heat-flow map of India and discussions on its geological and economic significance. *Indian Minerals*, **42(2)**, 89-110.

Ravishanker, 1995. Fragmented Indian shield and recent earthquakes. *Geol. Surv. India Sp. Pub.*, **27,** 41-48.

Ravishanker and Narula, P.L., 1999. Chamoli earthquake of 29th March, 1999 – A preliminary appraisal. *Indian Minerals*, **52,** 141-150.

Ravishanker and Pande, P., 2001. Geoseismological studies of Kutch (Bhuj) earthquake of 26, January 2001. *J. Geol. Soc. India*, **58,** 203-208.

Ravishanker, 1995. Fragmented Indian shield and recent earthquakes. *Geol. Surv. India Sp. Pub.*, **27,** 41-48.

Ray, S., Lyngdoh, S., Gheevarghese, K.K. and Sengupta, S.N., 1964. Seismic prospecting for oil in the southeastern part of the Brahmaputra valley of upper Assam: Geological results of applied geophysics. Proc. 22nd Intl. Geol. Congr., New Delhi, pp. 277-291.

Ray, S.K., 2000. Culmination zones in Eastern Himalaya. *Geol. Surv. India Sp. Pub.*, **55,** 85-94.

Ray, S.K., 2004. Melt-clast interaction and power law size distribution of clasts in Pseudotachylytes. *J. Str. Geol.*, **26(10),** 1831-1843.

Rayleigh, L., 1885. On waves propagated along the plane surface of an elastic solid. *Proc. Lond. Math. Soc.*, **17,** 4-11.

Real, C.R. and Teng, T.L., 1973. Local Richter magnitude and total signal duration in southern California. *Bull. Seism. Soc. Am.*, **63,** 1809-1827.

Reasenberger, P.A. and Oppenheimer, D., 1985. FPFIT, FPPLOT and FPPAGE: Fortran computer programs for calculating and displaying earthquake fault-plane solutions. U.S. Geological Survey Open-File Report, 85-739, 109 p.

Reddi, A.G.B., Mathew, P.M. and Kailasam, L.N. 1967. Geophysical studies in the Cuddapah Basin. Proc. Symp. Upper Mantle Project, 4-8 January, Geophys. Res. Board, NGRI, Hyderabad, pp. 286-302.

Reddi, A.G.B., Mathew, M.P., Singh, B. and Naidu, P.S., 1988. Aeromagnetic evidence of crustal structure in the granulite terrain of Tamil Nadu-Kerala. *J. Geol. Soc. India*, **32(5),** 368-381.

Reddy, G.K., Rao, G.V., Rao, R.U.M. and Gopalan, K., 1994. Surface rupture of Latur earthquake: The soil gas Helium signature. *Geol. Soc. India Mem.*, **35,** 83-99.

Reddy, P.R., Koteswara Rao, P. and Sain, K., 1999. Crustal configuration of the Narmada-Son lineament in central India from Deep Seismic Sounding studies. *Geol. Soc. India Mem.*, **43,** 353-365.

Reddy, P.R., Murthy, P.R.K., Rao, I.B.P., Mall, D.M. and Koteswara Rao, P., 2000. Coincident Deep Seismic Reflection and Refraction profiling central India. *In:* Research Highlights in Earth System Sciences, O.P. Varma and T.M. Mahadevan (eds.). Dept. Sci. Tech., Sp. Pub., V. 1, Indian Geol. Cong., pp. 49-53.

Reddy, P.R., Rajendra Prasad, B., Rao, V.V., Sain, K., Prasada Rao, P., Khar, P. and Reddy, M.S., 2003. Deep seismic reflection and refraction/wide angle reflection studies along Kuppam-Palani transect in southern granulite terrain in India. *Geol. Soc. India Mem.*, **50,** 79-106.

Reid, H.F., 1910. The California earthquake of April 18, 1906, vol. 2: The mechanics of the earthquakes. Carnegie Inst., Washington, D.C.

Rex, J., Searle, M.P., Tirrul, R., Crawford, M.B., Prior, D.J., Rex, D.C. and Barnicoat, A., 1988. The geochemical and tectonic evolution of the central Karakoram, north Pakistan. *Phil. Trans. R. Soc., London*, **A326,** 229-255.

Richter, C.F., 1935. An instrumental earthquake magnitude scale. *Bull. Seism. Soc. Am.*, **25,** 1-32.

Richter, C.F., 1958. *Elementary Seismology.* W.H. Freeman and Co., San Francisco, USA, 768 p.

Roberts, R.G., Christoffersson, A. and Cassidy, F. 1989. Real time event detection, phase identification and source location estimation using single station three components seismic data. *Geophys. J. Roy. Astro. Soc.*, **97,** 471-480.

Roecker, S.W., 1982. Velocity structure of the Pamir-Hindukush region: Possible evidence of subducted crust. *J. Geophys. Res.*, **87,** 945-959.

Roecker, S.W., Soboleva, V., Nersesov, I.L., Lukk, A.A., Hatzfeld, D., Chatlain, J.L. and Molnar, P., 1980. Seismicity and fault plane solutions of intermediate depth earthquakes in the Pamir-Hindukush region. *J. Geophys. Res.*, **85,** 1358-1364.

Rowley, D.B., 1996. Age of initiation of collision between India and Asia: A review of the stratigraphic data. *Earth Planet. Sci. Lett.*, **145,** 1-13.

Roy, B.C. and Hasija, N.L., 1995. Coseismic and postseismic crustal deformation in the Doon valley as determined from repeat leveling. *Geol. Soc. India Mem.*, **30,** 173-178.

Roy, A. and Bandopadhyay, B.K., 1988. Tectonic significance of ultramafic and associated rocks near Tal in the Bijawar belt, Sidhi district, M.P. *J. Geol. Soc. India*, **32,** 397-410.

Ruud, B.O., Husebye, E.S., Ingate, S.F. and Christoffersson, A., 1988. Event location at any distance using seismic data from a single, three-component station. *Bull. Seism. Soc. Am.*, **78,** 308-325.

Saha, S.N., Gaur, V.K., Bansal, B.K., Wyss, M. and Khattri, K.N., 1981. Microearthquakes in northeast India. Proc. Symp on Earthquake Disaster mitigation, Roorkee, V. 2, 65-72.

Sahu, O.P. and Saikia, M.M., 1994. The b-value before the 6th August 1988 India-Myanmar border region earthquake – A case study. *Tectonophysics*, **234,** 349-354.

Sain, K., Zelt, C.A. and Reddy, P.R., 2002. Imaging of subvolcanic Mesozoics using traveltime inversion of wide-angle seismic data in the Saurashtra peninsula of India. *Geophys. J. Int.*, **150,** 820-826.

Sanford, A.R. and Holmes, C.R., 1962. Microearthquakes near Socorro, New Mexico. *J. Geophys. Res.*, **67,** 4449-4459.

Santo, T., 1969. On the characteristic seismicity in South Asia from Hindukush to Burma. *Bull. Int. Instt. Seism. EQ Engg.*, **6,** 81-93.

Sarkar, D., Chandrakala, K., Padmavati Devi, P., Sridhar, A.R., Sain, K. and Reddy, P.R., 2001. Crustal velocity structure of western Dharwar craton, South India. *J. Geodyn.*, **31,** 227-241.

Sarkar, D., Sain, K., Reddy, P.R., Catchings, R.D. and Mooney, W.D., 2007. Seismic reflection images of the crust beneath the 2001 M = 7.7 Kutch (Bhuj) epicentral region, western India. *J. Geol. Soc. Am.* (in press).

Sarma, S.V.S., Virupakshi, G., Harinarayana, T., Murthy, D.N., Rao, S.P.E., Veeraswamy, K., Rao, M., Sarma, M.V.C. and Gupta, K.R.B., 1994. A wide band magnetotelluric study of the Latur earthquake region, Maharashtra, India. *Geol. Soc. India Mem.*, **35,** 101-118.

Sastri, V.V., Bhandari, L.L., Raju, A.T.R. and Dutta, A.N., 1971. Tectonic framework and subsurface stratigraphy of the Ganga basin. *J. Geol. Soc. India*, **12,** 222-233.

Sato, Y. and Skoko, D., 1965. Optimum distribution of seismic observation points. *Bull. Earthquake Res. Inst.,* Univ. Tokyo, **43,** 451-457.

Satyabala, S.P., 2003. Oblique plate convergence in the Indo-Burma Myanmar subduction region. *Pure and Appl. Geophys.*, **160,** 1611-1650.

Saul, J., Kumar, M.R. and Sarkar, D., 2000. Lithospheric and upper mantle structure of the Indian shield from teleseismic receiver functions. *Geophys. Res. Lett.*, **27,** 2357-2360.

Savage, M.K., 1999. Seismic anistrophy and mantle deformation. What have we learnt from shear wave splitting? *Rev. Geophys.*, **37,** 63-106.

Schelling, D., 1992. The tectono-stratigraphy and structure of the eastern Nepal Himalaya. *Tectonics*, **11,** 925-943.

Scholtz, C.H., 1968. The frequency magnitude relation of microfacturing in rock and its relation to earthquakes. *Bull. Seism. Soc. Am.*, **58,** 399-415.

Scholtz, C.H., 1990. *The Mechanics of Earthquakes and Faulting*, Cambridge Univ. Press, New York, 439 p.

Schulte-Pelkum, V., Sheehan, A., Wu, F. and Billham, R., 2005. Imaging the Indian subcontinent beneath the Himalaya. *Nature*, **435,** 1222-1225.

Schwartz, S.Y., Dewey, J.W. and Lay, T., 1989. Influence of fault plane heterogeneity on the seismic behaviour in southern Kurile Island arc. *J. Geophys. Res.*, **94,** 5637-5649.

Schweig, E., Gomberg, J., Petersen, M., Ellism, M., Bodin, P., Mayrose, L. and Rastogi, B.K., 2003. The Mw 7.7 Bhuj Earthquake: Global lessons for earthquake Hazard in Intra-plate regions. *J. Geol. Soc. India*, **61,** 277-282.

Sclater, J.G. and Fisher, R.L., 1974. Evolution of the east central Indian ocean with emphasis on the tectonic setting of the Ninetyeast Ridge. *Bull. Geol. Soc. Am.*, **85,** 683-702.

Searle, M.P., 1991. *Geology and Tectonics of Karakoram.* John Wiley, N.Y.

Searle, M.P., 1996. Geological evidence against large scale Pre-Holocene offsets along the Karakoram fault: Implications for the limited extrusion of the Tibetan Plateau. *Tectonics*, **15,** 171-186.

Searle, M.P. and Tirrul, R., 1991. Structure and thermal evolution of the Karakoram crust. *J. Geol. Soc. London*, **148,** 65-82.

Seeber, L. and Armbruster, J.G., 1981. Great detachment earthquakes along the Himalayan arc and long-term forecasting. *In:* Earthquake Prediction – An International Review. *Manrice Ewing Ser.*, Am. Geophys. Union, **4,** 259-277.

Seeber, L., Armbruster, J.G. and Quittmeyer, R., 1981. Seismicity and continental subduction in the Himalayan Arc. *In:* Zagros, Hindu Kush, Himalaya, Geodynamic Evolution, *Geodyn. Ser.*, Vol. 3, H.K. Gupta and F.M. Delany (eds.). pp. 215-242.

Seeber, L., Ekstrom, G., Jain, S.K., Murty, C.V.R., Chandak, N. and Armbruster, J.G., 1996. The Killari earthquake in central India: A new fault in Mesozoic basalt flows? *J. Geophys. Res.*, **101,** 8543-8560.

Semenov, A.N., 1969. Variations in travel time of transverse and longitudinal waves before violent earthquakes. *Izv. Acad. Sci. USSR, Phys. Solid Earth* (English Trans.), **4,** 245-248.

Sengupta, S., 1966. Geological and geophysical studies in western part of Bengal basin, India. *Am. Assoc. Petrol. Geol. Bull.*, **50(5),** 1001-1017.

Shebalin, N.V., 1955. O Suyazi mezhdu energy, ballnost'yui glubinoy ochaga zemletyaseniya. *Izv. Akd. Sci. USSR, Ser. Geofiz.*, 377-380.

Shi, Y. and Bolt, B., 1982. The standard error of magnitude-frequency b value. *Bull. Seism. Soc. Am.*, **72,** 1677-1687.

Shimazaki, T. and Nagahama, H., 1995. Do earthquakes occur at random, collectivities and individual of earthquakes. *Kagakusu*, Tokyo, **65,** 241-256.

Sibson, R.H., 1986. Earthquakes and rock deformation in crustal fault zones. *Ann. Rev. Earth Planet. Sci.*, **14,** 149-175.

Sieh, K. and Natawidjaja, D., 2000. Neotectonics of the Sumatra fault, Indonesia. *J. Geophys. Res.*, **105,** 28,295-28,326.

Sikka, S.K., Roy, F. and Nair, G.J., 1998. Indian Explosion of 11 May, 1998: An analysis of global seismic body wave magnitude estimates. *Curr. Sci.*, **75,** 486-491.

Silver, P.G., 1996. Seismic anisotropy beneath the continents: Probing the depths of geology. *Ann. Rev. Earth Planet. Sci.*, **24,** 385-432.

Simpson, D.W., 1976. Seismicity changes associated with reservoir loading. *Engg. Geol.*, **10,** 123-150.

Singh, A., Kumar, M.R., Raju, P.S. and Ramesh, D.S., 2006. Shear wave anisotrophy of the northeast Indian lithosphere. *Geophys. Res. Lett.*, **33,** L16302, doi: 10.1029/2006GL026106.

Singh, B.P. and Gokarn, S.G., 1989. *Deep continental studies through geoelectromagnetism.* Dept Sci. Tech. Report, pp. 1-10.

Singh, B.P. and Rajaram, M., 1990. Magsat studies over the Indian region. *Proc. Indian Acad. Sci. (Earth Planet Sci.)*, **99,** 619-637.

Singh, D.D., 1988. Quasi-continental oceanic structure beneath the Arabian fan sediments from observed surface wave dispersion studies. *Bull. Seism. Soc. Am.*, **78,** 1510-1521.

Singh, D.D., 1991. Anelasticity of the crust and mantle beneath north and central India from the inversion of the Love and Rayleigh wave attenuation data. *Pure and Appl. Geophys.*, **135(4),** 545-558.

Singh, D.D. and Gupta, H.K., 1980. Source dynamics of two great earthquakes of Indian subcontinent: The Bihar-Nepal earthquake of January 15, 1934 and the Queta earthquake of May 30, 1935. *Bull. Seism. Soc. Am.*, **70(3)**, 757-773.

Singh, S.K., Dattatrayam, R.S., Shapiro, N.M., Mandal, P., Pacheco, J.F. and Midha, R.K., 2000. Crustal and Upper Mantle Structure of Peninsula India and Source Parameters of the 21 May 1997, Jabalpur earthquake (Mw = 5.8): Results from a New Regional Broadband Network. *Bull. Seism. Soc. Amer.*, **89,** 1631-1641.

Singh, S.K., Pacheco, J.F., Bansal, B.K., Perez-Campos, X., Dattatrayam, R.S. and Suresh, G., 2004. A source study of the Bhuj, India Earthquake of 26 January 2001 (Mw 7.6). *Bull. Seism. Soc. Am.*, **94(4),** 1195-1206.

Sinha, A.K. and Upadhyay, R., 1995. Himalaya: Geological aspects. *Paleobotanist*, **44,** 9-28.

Sinha Roy, S., 1982. Himalaya main central thrust and its implications for Himalayan inverted metamorphism. *Tectonophysics*, **84,** 197-224.

Sipkin, S.A., 1982. Estimation of earthquake source parameters by the inversion of waveform data: Synthetic waveforms. *Phys. Earth Planet. Inter.*, **30,** 242-259.

Sitaram, M.V.D., John, G., Rao, P.G. and Saikia, M.M., 1986. Pn wave velocities in the uppermost mantle beneath northeast India. *In:* Earthquake prediction – Present status, S.K. Guha (ed.), University of Poona, India, pp. 267-275.

Smith, E.G.C., 1983. Joint determination of seismic velocity ratios: Theory and application to an aftershock sequence. *Bull. Seism. Soc. Am.*, **73,** 405-417.

Snoke, J.A., Sacks, I.S. and Okada, H., 1977. Determination of the subducting lithosphere boundary by use of converted phases. *Bull. Seism. Soc. Am.*, **67,** 1051-1060.

Solovev, S.L., 1965. Seimicity of Sakhalin. *Bull. Earthquake Res. Inst. Univ. Tokyo*, **43,** 95-102.

Srivastava, H.N. and Ramchandran, K., 1985. New catalogue of earthquakes for Peninsular India. *Mausam*, **36(3),** 351-358.

Srivastava, H.N., Rao, D.T. and Singh, M., 1991. Seismicity pattern for earthquakes near Bhatsa reservoir, India. *Tectonophysics*, **196,** 141-156.

Stauder, W., 1962. The focal mechanism of earthquakes. *Adv. Geophys.*, **9,** 1-76.

Steeples, D.W., 1979. Microearthquake network in Kansas. *In:* Terrestial and space techniques in Earthquake Prediction Research, A. Vogel (ed.), Vieweg Braunschweig, pp. 115-130.

Steim, J.M., 1986. The very broadband seismograph. Ph.D thesis. Harvard Univ., Massachusetts, 184 p.

Stein, S., Cloeting, S., Sleep, N.H. and Wortel, R., 1989. Passive margin earthquakes, stress and rheology. *In:* Earthquakes of North Atlantic Passive margins: Neotectonic and post-glacial rebound, S. Gregerson and P.W. Basham (eds.). Kluwer Acad. Pub., pp. 563-579.

Stocklin, J., 1980. Geology of Nepal and its regional frame. *J. Geol. Soc. London*, **137,** 1-34.

Storey, M., Mahoney, J.J., Saunders, A.D., Duncan, R.A., Kumar, S.P. and Coffin, M.F., 1995. Timing of hot spot related volcanism and the breakup of Madagascar and India. *Science*, **267,** 852-855.

Stuart, M., 1926. The Srimangal earthquake of 8th July, 1918. *Geol. Surv. India Mem.*, **46** (pt. 1), 1-70.

Stump, B.W. and Johnson, L.R., 1977. The determination of source properties by linear inversion of seismograms. *Bull. Seism. Soc. Am.*, **67,** 1489-1502.

Subrahmanyam, C. and Verma, R.K., 1982. A gravity interpretation of the Dharwar greenstone gneiss granite terrain in the southern Indian shield and its geological implication. *Tectonophysics*, **84,** 225-245.

Sukhija, B.S., Rao, M.N., Reddy, D.V., Nagabhusanam, P., Laxmi, B.V., Hussain, S. and Gupta, H.K., 1998. Evidence of prehistoric seismic event near Latur (Killari) India. Chapman Conference on SCR earthquakes, NGRI, Hyderabad, Abs, pp. 35.

Sukhija, B.S., Rao, M.N., Reddy, D.V., Nagabhusanam, P., Hussain, S., Chadha, R.K. and Gupta, H.K., 1999a. Timing and return of major paleoseismic events in the Shillong Plateau, India. *Tectonophysics*, **308,** 53-65.

Sukhija, B.S., Rao, M.N., Reddy, D.V., Nagabhushanam, P., Hussain, S., Chadha, R.K. and Gupta, H.K., 1999b. Paleoliquefaction evidence and periodicity of large prehistoric earthquakes in Shillong Plateau, India. *Earth Planet. Sci. Lett.*, **167,** 269-282.

Sukhija, B.S., Rao, M.N., Reddy, D.V., Nagabhushanam, P. and Devender Kumar, 2005. Paleoseismological studies in the Himalayan belt. NGRI Annual Report.

Sunmonu, A. and Dimri, V.P., 1999. Fractal analysis and seismicity of Bengal basin and Tripura fold belt, northeast India. *J. Geol. Soc. India.*, **53,** 587-592.

Suteau, A.M. and Whitcomb, J.H., 1979. A local earthquake coda magnitude and its relation to duration, moment Mo and local Richter magnitude M_L. *Bull. Seism. Soc. Am.*, **69,** 353-368.

Suzuki, Z., Tsumura, K., Oike, K. and Matsumura, K. (Eds), 1979. *Microearthquake Observatories in Japan.* 2nd ed.

Swindell, W. and Barrett, H.H., 1977. Computerised tomography taking sectional X-rays. *Phys. Today*, **30(12),** 32-41.

Sykes, L.R., 1970. Seismicity of the Indian ocean and a possible nascent island arc between Ceylon and Australia. *J. Geophys. Res.*, **75,** 5041-5055.

Takahasi, H. and Hamada, K., 1975. Deep borehole observation of the earth's crust activities around Tokyo – Introduction of the Iwatsuki observatory. *Pure and Appl. Geophys.*, **113,** 311-320.

Talebi, S. and Cornet, F.H., 1987. Analysis of the microseismicity induced by a fluid injection in a Granitic rock mass. *Geophys. Res. Lett.*, **14,** 227-230.

Talebi, S., Nechtschein, S. and Boone, T.J., 1998. Seismicity and casing failures due to steam stimulation in oil sands. *Pure and Appl. Geophys.*, **153,** 219-233.

Talukdar, S.N., 1982. Geology and hydrocarbon prospects of east coast basins of India and their relationship to evolution of the Bay of Bengal. *Offshore SE Asia, Conf.*, Singapore, Expl. I, Abs vol., pp. 1-8.

Talwani, P., 1997. On the nature of reservoir induced seismicity. *Pure and Appl. Geophys.*, **150,** 473-492.

Talwani, P. and Gangopadhyay, A., 2001. Tectonic framework of the Kachchh earthquake of January 26, 2001. *Seism. Res. Lett.*, **72,** 336-345.

Talwani, P. and Gangopadhyay, A., 2003. Symptomatic features of intraplate earthquakes. *Seism. Res. Lett.*, **74,** 863-883.

Tan, T.K., 1987. Geodynamics and tectonic evolution of the Panxi rift. *Tectonophysics*, **133,** 287-304.

Tandon, A.N., 1954. A study of Assam earthquake of August 1950 and its aftershocks. *Indian J. Meteorol. Geophys.*, **5,** 95-137.

Tandon, A.N., 1955. Direction of faulting in the great Assam earthquake of 15 August 1950. *Indian J. Meteorol. Geophys.*, **6,** 61-64.

Tandon, A.N., 1956. Zone in India liable to earthquake damage. *Indian J. Meteorol. Geophys.*, **7(1),** 93.

Tandon, A.N. and Choudhury, H.M., 1967. Phase velocity of Rayleigh waves over the Punjab. *Indian. J. Meteorol. Geophys.*, **18,** 431-434.

Tandon, A.N. and Srivastava, H.N., 1974. Earthquake occurrence in India. *In:* Earthquake Engineering, Jai Krishna volume, A.N. Tandon and H.N. Srivastava (eds.). Sarita Prakashan, Meerut.

Tandon, A.N. and Srivastava, H.N., 1975. Focal mechanisms of some recent Himalayan earthquakes and regional plate tectonics. *Bull. Seism. Soc. Am.*, **65,** 963-969.

Tapponnier, P. and Molnar, P., 1976. Slip-line field theory and large scale tectonics. *Nature*, London, **264,** 319-324.

Tapponnier, P., Peltzer, G., Le Dain, A.Y., Armijo, R. and Cobbold, P., 1982. Propagating extrusion tectonics in Asia: New insights from simple experiments with plasticine. *Geology*, **10,** 611-616.

Tapponnier, P., Peltzer, G. and Armijo, R., 1986. On the Mechanics of the collision between India and Asia. *In:* Collisional Tectonics, M.P. Coward and A.C. Ries (eds.). *Geol. Soc. London Sp. Pub.*, **19,** 115-157.

Tarantola, A., 1987. *Inverse Problem Theory*. Elsevier Pub., Amsterdam, 613 p.

Teng, W. and Zhang, M., 1981. Characteristics of geothermal activities in Xizang Plateau and their controlling influence on Plateau's tectonic model. *In:* Geological and Ecological studies of the Oinghai – Xizang Plateaus, vol. 1. Science Press, Beijing, pp. 841-846.

Thakur, V.C., 1981. Regional framework and geodynamic evolution of the Indus-

Tsangpo suture zone in the Ladakh Himalayas. *Trans. Royal Soc. Edinburgh: Earth Sciences*, **72,** 89-97.

Thomson, A., 1929. Earthquake sounds heard at great distances. *Nature*, **124,** 686-688.

Thurber, C.H., 1981. Earth structure and earthquake locations in the Coyote Lake area, Central California. Ph.D. thesis, Massachusetts Institute of Technology.

Thurber, C.H., 1983. Earthquake location and three dimensional crustal structure in the Coyote Lake area, central California. *J. Geophys. Res.*, **88**(B10), 8226-8336.

Thurber, C.H. and Aki, K., 1987. Three dimensional seismic imaging. *Ann. Rev. Earth Planet. Sci.*, **15,** 115-139.

Tinti, S., Vittori, T. and Mulargia, F., 1987. On the macroseismic magnitudes of the largest Italian earthquakes. *Tectonophysics*, **138,** 159-178.

Tipper, G.H., 1911. The geology of Andaman islands with reference to Nicobars. *Geol. Surv. India Mem.*, **35(4),** 195-212.

Tiwari, H.C. and Geotransect team (NGRI), 1998. Nagaur-Jhalawar geotransect across the Delhi-Aravalli Fold Belt in northwest India. *J. Geol. Soc. India*, **52,** 153-162.

Toksöz, M.N., Minear, J.W. and Julian, B.R., 1971. Temperature field and geophysical effects of a downgoing slab. *J. Geophys. Res.*, **76,** 1113-1138.

Tappozada, T.R., 1975. Earthquake magnitude as a function of intensity data in California and Western Nevada. *Bull. Seism. Soc. Am.*, **65(5),** 1223-1238.

Tosi, P., 1998. Seismogenic structure behaviour revealed by spatial clustering of seismicity in the Umbria-Marche region (Central Italy). *Ann. De Geofisica*, **41(2),** 215-224.

Tsumura, K., 1967. Determination of earthquake magnitude from total duration of oscillation. *Bull. Earthquake Res. Inst.*, **15,** 7-18.

Uhrhammer, R.A., 1980. Analysis of small seismographic station networks. *Bull. Seism. Soc. Am.*, **70,** 1369-1380.

Um, J. and Thurber, C.H., 1987. A fast algorithm for two-point seismic ray tracing. *Bull. Seism. Soc. Am.*, **77,** 972-986.

Utsu, T., 1957. Magnitude of earthquakes and occurrence of their aftershocks, Zishin. *J. Seism. Soc. Japan*, **10,** 35-45.

Utsu, T., 1961. A statistical study on the occurrence of aftershocks. *Geophys. Mag.*, **30,** 521-605.

Utsu, T., 1965. A method for determining the b-value in a formula log N = a – bm showing magnitude frequency relation for earthquakes. *Hokkaido Univ. Geophys. Bull.*, **13,** 99-103 (in Japanese with English abstract).

Utsu, T., 1969. Aftershocks and earthquake statistics. 1: Some parameters which characterises an aftershock sequence and their interrelations. *Hokkaido Univ. Geophys. Bull.*, **3,** 129-195.

Utsu, T., 2002. A list of deadly earthquakes in the world: 1500-2000, *In:* International Handbook of Earthquake and Engg. Seismology, W.H.K. Lee, H., Kanamori, Paul C., Jennings, Carl Kisslinger (eds.). Acad. Press, pp. 691-717.

Utsu, T. and Seki, A., 1954. A relation between the area of aftershock region and the energy of main shock. *J. Seismol. Soc. Japan.*, **7,** 233-240.

Uyeda, S. and Kanamori, H., 1979. Back-arc opening and mode of subduction. *J. Geophys. Res.*, **83,** 1049-1061.

Valdiya, K.S., 1976. Himalayan transverse faults and their parallelism with subsurface structures of north Indian plains. *Tectonophysics*, **32,** 352-386.

Valdiya, K.S., 1980. The two intracrustal boundary thrusts of the Himalaya. *Tectonophysics*, **66,** 323-348.

Valdiya, K.S., 1984. Aspects of tectonics: Focus on south central Asia, Tata McGraw-Hill, New Delhi, 319 p.

Valdiya, K.S., 1987. Trans-Himadri Thrust and domal upwards immediately south of collision zone and tectonic implications. *Curr. Sci.*, **56,** 200-209.

Valdiya, K.S., 1988. Tectonics and evolution of the central sector of the Himalaya. *Phil. Trans. Roy. Soc., London*, **A-326**, 151-175.

Verma, R.K. and Mukhopadhyay, M., 1977. An analysis of the gravity field in northeastern India. *Tectonophysics*, **42,** 283-317.

Verma, R.K. and Banerjee, P., 1992. Nature of continental crust along the Narmada-Son lineament inferred from gravity and deep seismic sounding data. *Tectonophysics*, **202,** 375-397.

Verma, R.K. and Prasad, K.A.V.L., 1987. Analysis of gravity fields in the northeastern Himalayas and Kohistan region using deep seismic sounding data. *Geophys. J.R. Astro. Soc.*, **91,** 869-889.

Verma, R.K. and Subrahmanyam, C., 1984. Gravity anomalies and the Indian lithosphere: Review and analysis of existing gravity data. *Tectonophysics*, **105,** 141-161.

Verma, R.K., Mukhopadhyay, M. and Ahluwalia, M.S., 1976a. Seismotectonics of northeast India and northern Burma. *Bull. Seism. Soc. Am.*, **66,** 1683-1694.

Verma, R.K., Mukhopadhyaya, M. and Ahluwalia, M.S., 1976b. Earthquake mechanisms and tectonic features of northern Burma. *Tectonophysics*, **32,** 387-399.

Verma, R.K., Mukhopadhyay, M. and Roy, B.N., 1977. Seismotectonics of the Himalaya and the continental plate convergence. *Tectonophysics*, **42,** 319-335.

Verma, R.K., Mukhopadhyay, M. and Bhuin, N.C., 1978. Seismicity, gravity and tectonics in the Andaman Sea. *J. Phys. Earth*, **26,** Suppl. S233-S248.

Verma, R.K., Sarma, A.U.S. and Mukhopadhyay, M., 1984. Gravity field over Singhbhum, its relationship to geology and tectonic history. *Tectonophysics*, **106,** 87-107.

Verma, R.K., Mukhopadhaya, M., Subba Rao, P.B.V. and Satyanarayan, Y., 1988. Analysis of the gravity field over parts of south Singhbhum, Sukinda ultramafic complex and nature of Sukinda thrust. *J. Geol. Soc. India*, **32,** 334-342.

Verma, R.K., Roonwal, G.S. and Gupta, Y., 1993. Statistical analysis of seismicity of NE India and northern Burma during the period 1979-1990. *J. Him. Geol.*, **4(1),** 71-79.

Vigny, C., Simons, W.J.F., Abu, S., Bamphenyu, R., Satirapod, C., Choosakul, N., Subarya, C., Socquet, A., Omar, K., Abidin, H.Z. and Ambrosius, B.A., 2005. Insight into the 2004 Sumatra-Andaman earthquake from GPS measurements in southeast Asia. *Nature*, **436,** doi: 10.1038/nature03937.

Vijay Rao, V., Rajendra Prasad, B., Reddy, P.R., and Tiwari, H.C., 2000. Evolution of Proterozoic Aravalli Delhi fold belt in northwestern Indian shield from seismic studies. *Tectonophysics*, **327,** 109-130.

Vinnik, L.P., Lukk, A.A. and Nersesov, L.L., 1977. Nature of the intermediate seismic zone in the mantle of the Pamir-Hindukush. *Tectonophysics*, **38,** T9-T14.

Vitanage, P.W.F., 1986. Morphological, geological and geophysical study of lineaments (linears) in Srilanka and their correlation with lineaments in south India. *In:*

Deep Seismic Soundings and Crustal Tectonics, K.L. Kalia and H.C. Tiwari (eds.), Assoc. Expl. Geophys. Hyderabad, pp. 176-190.

Wadati, K., 1933. On the travel time of earthquake waves, Part II. *Geophys. Mag.*, **7,** 101-111.

Wadia, D.N., 1961. *The Geology of India*, 3rd ed. Macmillan, London.

Waldhauser, F., 2001. HypoDD: A computer program to compute double difference hypocenter locations. USGS open file report, 01-113, 25 p.

Waldhauser, F. and Ellsworth, W.L., 2000. A double-difference earthquake algorithm: Method and application to the Northern Hayward Fault, California. *Bull. Seism. Soc. Am.*, **90(6),** 1353-1368.

Wallace, K., Bilham, R., Blume, F., Gaur, V.K. and Gahalaut V.K., 2005. Surface deformation in the region of 1905 Kangra Mw 7.8 earthquake in the period 1846-2001. *Geophys. Res. Lett.*, **32,** L15307, doi: 10.1029GL022906.

Wallace, T.C. and Helmberger, D.V., 1982. Determining source parameters of moderate size earthquakes from regional waveform. *Phys. Earth Planet Inter.*, **30,** 185-196.

Wallace, T.C., Velasco, A., Zhang, J. and Lay, T., 1991. Loma Prieta, California earthquake: Evidence for deep slow slip. *Bull. Seism. Soc. Am.*, **81,** 1622-1646.

Wang , C.Y., Shi, Y. and Zhon, W.H., 1982. On the tectonics of the Himalaya and the Tibet Plateau. *J. Geophys. Res.*, **87,** 2949-2957.

Ward, S.N., 1980. A technique for the recovery of seismic moment tensor applied to the Oaxaca, Mexico earthquake of November, 1978. *Bull. Seism. Soc. Am.*, **70,** 717-734.

Warsi, W.E.K. and Molnar, P., 1977. Gravity anomalies and plate tectonics in the Himalaya, Colloques International de CNRS No. 268. *In:* Himalaya: Science de la Terre (ed. du Centre National de la Recherche Scientifique, Paris), 463-478.

Wellman, H., 1996. Active wrench faults of Iran, Afghanistan and Pakistan. *Geol. Rundsch.*, **55,** 716-735.

Wesnousky, S.G., Seeber, L., Rockwell, T.K., Thakur, V.C., Briggs, R., Kumar, S. and Ragona, D., 2001. Eight days in Bhuj: Field report bearing on surface rupture and genesis of the 26 January 2001 earthquake in India. *Seism. Res. Lett.*, **72(5),** 514-524.

Wesnousky, S., Kumar, S., Mohindra, R. and Thakur, V.C., 1999. Holocene slip rate of the Himalayan frontal thrust of India: Observations near Dehra Dun. *Tectonics*, **18,** 967-976.

West, W.D., 1958. The petrography and petrogenesis of fortyeight flows of Deccan Traps penetrated by borings in western India. *Trans. Nat. Inst. Sci. India*, **4,** 1-56.

West, W.D. 1962. The line of the Narmada and Son valleys. *Curr. Sci.*, **31,** 143-144.

White, R.S. and McKenzie, D.P., 1981. Magnetism at rift zone: The generation of volcanic continental margins and flood basalts. *J. Geophys. Res.*, **94,** 7685-7729.

Wiecher, E. and Zöppritz, K., 1907. Uber Erddebenwellen II. Nachder Köning. Gesellsch. Der Wissensch. Zu Göttingen, Math-Phys., K1, 529 p.

Wiemer, S. and Wyss, M., 1997. Mapping the frequency-magnitude distribution in asperities: An improved technique to calculate recurrence times. *J. Geophys. Res.*, **102,** 15115-15128.

Wiemer, S., Mcnutt, S.R. and Wyss, M., 1998. Temporal and three-dimensional spatial analysis of the frequency-magnitude distribution near Long Valley Caldera, California. *Geophys. J. Int.*, **134,** 409-421.

Willmore, P., 1979. Manual of Seismological Observatory Practice, Tech. report SE-20, World Data Centre for Earth Geophysics, Boulder, Colorado.

Wilson, J.T., 1965. A new class of faults and their bearing on continental drift. *Nature*, **207,** 343-347.

Wu, F.T., 1996. Estimation of rupture velocity of the 1989 Loma Prieta main shock from Love wave group velocity. USGS Sp Paper, A, 279-288.

Wu, F.T., Levshin, A.L. and Kozhevnikov, V.M., 1997. Rayleigh wave group velocity tomography of Siberia, China and the vicinity. *Pure and Appl. Geophys.*, **149,** 447-473.

Wynne, A.B., 1872. Memoir on the Geology of Kutch. *Geol. Surv. India Mem.*, **9,** 29-47.

Wyss, M., 2005. Human losses expected in Himalayan earthquakes. *Natural Hazards*, **34,** 305-314.

Xu, Y., 1992. A study on characteristics of the information dimension of the temporal and spatial distribution of earthquakes in an active fault zone. *Acta Seism. Sinica*, **5(2),** 389-398.

Yagi, Y. and Kikuchi, M., 2001. Results of rupture process for January 26, 2001 western India earthquake (Ms 7.9), (revised). *httpll wwweic.eri-u.tokyo.ac.jp.*

Yeats, R.S. and Lillie, R.J., 1991. Contemporary tectonics of the Himalayan frontal fault system: folds, blind thrusts and the 1905 Kangra earthquake. *J. Struct. Geol.*, **13,** 215-225.

Yeats, R.S. and Thakur, V.C., 1998. Reassessment of earthquake hazard based on a fault-bend fold model of the Himalayan plate-boundary fault. *Curr. Sci.*, **74,** 230-233.

Yoshida, M., Funaki, M. and Vitanage, P.W., 1992. Proterozoic to Mesozoic east Gondwana : The Juxtaposition of India, Sri Lanka and Antarctica. *Tectonics*, **11,** 381-391.

Yuan, X., Ni, J., Kind, R., Mechie, J. and Sandvol, E., 1997. Lithospheric and upper mantle structure of southern Tibet from a seismological passive source experiment. *J. Geophys. Res.*, **102,** 27,491-27,500.

Zeitler, P.K., 1985. Cooling history of the NW Himalaya, Pakistan. *Tectonics*, **4,** 127-151.

Zeitler, P.K., Johnson, N.M., Naeser, C.W. and Tahirkheli, A.K., 1982. Fission track evidence for Quaternary uplift of the Nanga Parbat region, Pakistan. *Nature*, **298,** 255-257.

Zhao, D., 2001. Seismic structure and origin of hotspots and mantle plumes. *Earth. Planet. Sci. Lett.*, **192,** 251-265.

Zhao, D. and Kanamori, H., 1993. The 1992 Landers earthquake sequence: Earthquake occurrence and structural heterogeneities. *Geophys. Res. Lett.*, **20,** 1083-1086.

Zhao, D. and Kanamori, H., 1995. The 1994 Northridge earthquake: 3-D crustal structure in the rupture zone and its relation to the aftershock locations and mechanisms. *Geophys. Res. Lett.*, **22,** 763-766.

Zhao, D. and Kayal, J.R., 2000. Impact of seismic tomography on Earth Sciences. *Curr. Sci., Sp. Vol.: Seismology 2000*, **79,** 1208-1214.

Zhao, D., Hasegawa, A. and Horiuchi, S., 1992. Tomographic Imaging of P and S wave velocity structure beneath northeast Japan. *J. Geophys. Res.*, **97,** 19,909-19,928.

Zhao, D., Kanamori, H. and Humphreys, E., 1996. Simultaneous inversion of local and teleseismic data for the crust and mantle structure of southern California. *Phys. Earth Planet. Inter.*, **93,** 191-214.

Zhao, D., Tani, H. and Mishra, O.P., 2004. Crustal heterogeneity in the 2000 western Tattori earthquake region: Effect of fluids for slab dehydration. *Phys. Earth Planet. Inte*r., **145,** 161-177.

Zhao, D., Mishra, O.P. and Sande, R., 2002. Influence of fluids and magma on earthquakes: Seismological evidence. *Phys. Earth Planet. Int.*, **132,** 249-267.

Zhao, W., Nelson, K.D. and Project INDEPTH Team, 1993. Deep seismic reflection evidence for continental under thrusting beneath south Tibet. *Nature*, **366,** 557-559.

Zoback, M.L., 1992. First and second order patterns of tectonic stress: The World Stress Map Project. *J. Geophys. Res.*, **97,** 11703-11728.

Zoback, M.L. and Richardson, R.M., 1996. Stress perturbation associated with the Amazons and other continental rifts. *J. Geophys. Res.*, **101,** 5459-5475.

Index